Molecular Mechanisms of Steroid Hormone Biosynthesis and Action

Molecular Mechanisms of Steroid Hormone Biosynthesis and Action

Editor

Jacques J. Tremblay

MDPI • Basel • Beijing • Wuhan • Barcelona • Belgrade • Manchester • Tokyo • Cluj • Tianjin

Editor
Jacques J. Tremblay
Obstetrics, Gynecology and
Reproduction
Université Laval
Québec City
Canada

Editorial Office
MDPI
St. Alban-Anlage 66
4052 Basel, Switzerland

This is a reprint of articles from the Special Issue published online in the open access journal *International Journal of Molecular Sciences* (ISSN 1422-0067) (available at: www.mdpi.com/journal/ijms/special_issues/Steroid_Biosynthesis).

For citation purposes, cite each article independently as indicated on the article page online and as indicated below:

LastName, A.A.; LastName, B.B.; LastName, C.C. Article Title. *Journal Name* **Year**, *Volume Number*, Page Range.

ISBN 978-3-0365-7003-7 (Hbk)
ISBN 978-3-0365-7002-0 (PDF)

Contents

About the Editor

Jacques J. Tremblay

Jacques J. Tremblay is currently Full Professor in the Department of Obstetrics, Gynecology and Reproduction at the Faculty of Medicine of Université Laval, and Scientist at the Centre de recherche du CHU de Québec—Université Laval in Quebec City. He is also a member of the Centre for Research in Reproduction, Development and Intergenerational Health (CRDSI) and the Québec Reproduction Network (RQR). At Université Laval, Professor Tremblay leads an active research program studying the molecular mechanisms and signaling pathways regulating Leydig cell differentiation and function in the mammalian testis. He is also interested in the molecular mechanisms of action of endocrine disruptors known to cause Leydig cell dysfunction. His research is currently funded by the Canadian Institutes of Health Research (CIHR) and the Natural Sciences and Engineering Research Council of Canada (NSERC). Professor Tremblay is involved at various levels in scientific activities such as organization of international scientific conferences, member of various committees for different scientific organizations, member of peer review committees for various organizations from Canada and abroad, and Editorial Board member and Reviewer for numerous scientific journals. In recognition of his contributions to the field, Professor Tremblay was awarded the 2013 American Society of Andrology (ASA) Young Andrologist Award.

Preface to "Molecular Mechanisms of Steroid Hormone Biosynthesis and Action"

Dear Colleagues,

Steroids are essential hormones that regulate biological processes across the lifespan. Steroid hormone production and action must be tightly controlled otherwise there can be detrimental effects on physiological function leading to disease.

The biosynthesis of steroids and their actions in target tissues and cells are complex but coordinated processes that rely on a series of diverse cellular and molecular events that include proper steroidogenic cell differentiation and gene expression, hormone transport, hormone processing, and hormone action via interaction with specific receptors. The regulation of these events requires the concerted action of various hormones, growth factors, transcription factors, and other signaling and regulatory molecules to ensure adequate genomic, cellular and/or physiological responses. Many of these processes are also targets for endocrine disruption.

This reprint contains original (15) and review (4) articles by experts at the cutting edge of their fields published in a Special Issue on the "Molecular Mechanisms of Steroid Hormone Biosynthesis and Action" of the *International Journal of Molecular Sciences*. The topics are broad and include: 1) Steroid hormone biosynthesis, including substrate availability, mechanism of steroidogenic enzyme action and impact of mutations; 2) Molecular and cellular regulation of steroidogenesis, including steroidogenic cell response to hormone stimulation, signaling pathways, kinases, transcription factors, and gene expression; 3) Molecular mechanisms of endocrine disruptor action on steroidogenic cells; and 4) New tools and approaches to detect and study steroid hormones.

I sincerely hope that the readers will appreciate the breath of information this reprint contains in addition to encouraging them to further explore this constantly evolving and fascinating field.

Finally, I would like to express my sincere gratitude to all the authors and referees for their dedicated contribution to this Special Issue.

Jacques J. Tremblay
Editor

International Journal of
Molecular Sciences

Review

Role of STAR and SCP2/SCPx in the Transport of Cholesterol and Other Lipids

Melanie Galano [1], Sathvika Venugopal [2] and Vassilios Papadopoulos [1,2,*] 

[1] Department of Pharmacology and Pharmaceutical Sciences, School of Pharmacy,
University of Southern California, Los Angeles, CA 90089, USA

[2] Department of Medicine, The Research Institute of the McGill University Health Centre, McGill University,
Montreal, QC H4A 3J1, Canada

* Correspondence: vpapadop@usc.edu; Fax: +1-323-442-1681

Abstract: Cholesterol is a lipid molecule essential for several key cellular processes including steroidogenesis. As such, the trafficking and distribution of cholesterol is tightly regulated by various pathways that include vesicular and non-vesicular mechanisms. One non-vesicular mechanism is the binding of cholesterol to cholesterol transport proteins, which facilitate the movement of cholesterol between cellular membranes. Classic examples of cholesterol transport proteins are the steroidogenic acute regulatory protein (STAR; STARD1), which facilitates cholesterol transport for acute steroidogenesis in mitochondria, and sterol carrier protein 2/sterol carrier protein-x (SCP2/SCPx), which are non-specific lipid transfer proteins involved in the transport and metabolism of many lipids including cholesterol between several cellular compartments. This review discusses the roles of STAR and SCP2/SCPx in cholesterol transport as model cholesterol transport proteins, as well as more recent findings that support the role of these proteins in the transport and/or metabolism of other lipids.

Keywords: cholesterol; cholesterol transport; steroidogenic acute regulatory protein; sterol carrier protein 2; sterol carrier protein-x

Citation: Galano, M.; Venugopal, S.; Papadopoulos, V. Role of STAR and SCP2/SCPx in the Transport of Cholesterol and Other Lipids. *Int. J. Mol. Sci.* **2022**, *23*, 12115. https://doi.org/10.3390/ijms232012115

Academic Editor: Jacques J. Tremblay

Received: 7 September 2022
Accepted: 8 October 2022
Published: 11 October 2022

Publisher's Note: MDPI stays neutral with regard to jurisdictional claims in published maps and institutional affiliations.

1. Introduction

Cholesterol is a lipid molecule that has been associated with many diseases such as cardiovascular disease, atherosclerosis, Alzheimer's disease, and different types of cancers [1–3]. While much of the cholesterol research has focused on its dysregulation and role in the pathology of these diseases, it is also an essential lipid molecule critical to the maintenance of a variety of indispensable functions and pathways. Cholesterol is a 27-carbon molecule containing four fused rings and an 8-carbon tail, making it a hydrophobic molecule [4]. As such, cholesterol is a major component of cellular membranes, where cholesterol interacts with other lipids to regulate the membrane permeability of certain ions and solutes, the rigidity of the membrane to act as a scaffold for membrane proteins, and the fluidity of the membrane for rapid diffusion, budding to form vesicles, or fusion with other membranes for trafficking [5]. Furthermore, cholesterol is the precursor of all steroid hormones, which play vital roles in reproduction, salt and water balance, and stress response [4]. It is also metabolized into various oxysterols that then give rise to bile acids, which are critical for the absorption and digestion of lipids [4,6]. Because of its role in several key cellular processes, cholesterol must be properly trafficked throughout the cell, which occurs via both vesicular and non-vesicular pathways, of which the latter involves cholesterol transport proteins [7]. Cholesterol interacts with a variety of cholesterol transport proteins for its intracellular trafficking, such as the steroidogenic acute regulatory protein (STAR; STARD1), a key protein in acute steroidogenesis, and sterol carrier protein 2/sterol carrier protein-x (SCP2/SCPx), which are non-specific lipid transport proteins that may play roles in cholesterol transport and metabolism [8,9]. While STAR is classically known to transport cholesterol for mitochondrial steroidogenesis, recent data suggest that STAR may also function in the transport and/or metabolism of other lipids [10]. Additionally, since the discovery that STAR and not SCP2 is the key cholesterol transporter for steroid biosynthesis, the general role of SCP2/SCPx in the transport of cholesterol and other lipids has been ignored [11–13]. In this review, we discuss intracellular cholesterol distribution and current evidence for the multifunctional roles of STAR and SCP2/SCPx as model cholesterol transport proteins.

1.1. Synthesis

Cholesterol is derived from two sources: de novo synthesis or dietary intake. While all mammalian cells are able to synthesize cholesterol, steroidogenic cells of the adrenals, testis, ovaries, and brain synthesize cholesterol at the highest rates [14,15]. De novo cholesterol biosynthesis occurs in the endoplasmic reticulum (ER) by a complex process involving several, tightly regulated enzymes of the mevalonate pathway [4,16,17].

In addition to de novo cholesterol synthesis, cholesterol is also supplied to the cell via dietary intake. Cholesterol derived from food is absorbed by the small intestine, packed into chylomicrons, and transported to the liver, where cholesterol is processed into very low-density lipoproteins (VLDLs) [18]. Through circulation, VLDLs become low-density lipoproteins (LDLs), which bind to LDL receptors (LDLR) on the cell surface. These complexes are then endocytosed through clathrin-coated pits into the cytoplasm, which then fuse to endosomal compartments for processing the LDLs, while LDLR is recycled back to the cell surface. The cholesteryl esters derived from LDLs are subsequently de-esterified to free cholesterol by lysosomal acid lipase [4,19]. In contrast, another mechanism of cholesterol uptake into the cell occurs via a non-endocytic pathway involving the uptake of high-density lipoproteins (HDLs) by scavenger receptor class B, type I (SR-BI), which is mostly expressed in steroidogenic cells of the testes, ovaries, and adrenals [20]. HDL particles bind SR-BI, which transfers HDL cholesterol to the plasma membrane [21]. The cholesteryl esters derived from HDLs are then hydrolyzed by hormone-sensitive lipase, forming free cholesterol [22].

1.2. Distribution

Regulation of cellular cholesterol levels occurs at many levels. Cholesterol biosynthesis can be regulated by the sterol regulatory element-binding protein 2, which regulates transcription of genes important for cholesterol biosynthesis, and by HMG-CoaA reductase enzyme (HMGCR) [1]. Additionally, cholesterol homeostasis may be regulated by controlling LDLR-mediated uptake or by controlling cholesterol efflux by cholesterol efflux transporters such as ATP-binding cassette subfamily A member 1 [1]. Further, levels of cholesterol can be controlled by cholesterol esterification through the regulation of acyl-coenzyme A: cholesterol acyltransferases (ACAT1 and ACAT2) [23]. While active, free cholesterol is most commonly found in membranes or trafficked to other organelles for other functions, cholesterol esterification prevents the accumulation of free cholesterol and primes esterified cholesterol for storage in lipid droplets [24].

Lipids in biological mammalian membranes include sphingolipids, glycerol-based lipids, and cholesterol, and the lipid composition of a specific membrane is determined in part by its subcellular location [25]. In the cell, about 65%–80% of cellular cholesterol is present at the plasma membrane, while the Golgi apparatus and endosomal recycling compartment (ERC), which are closely associated with the plasma membrane, have intermediate cholesterol levels [26]. However, the ER has only 0.1%–2% of total cellular cholesterol despite this being the site of cholesterol biosynthesis [27]. This has led to the notion that newly synthesized cholesterol in the ER is rapidly transported to other organelles for other functions or stored in lipid droplets as cholesteryl esters. Additionally, mitochondria are also cholesterol-poor, despite mitochondria being the initiation site of steroidogenesis [28,29]. Therefore, these low intrinsic concentrations necessitate the rapid transport of cholesterol from intracellular stores into mitochondria for steroid production [30].

1.3. Intracellular Trafficking: Vesicular and Non-Vesicular Cholesterol Transport

The transport of cholesterol between organelles or from intracellular stores is vital for many cellular processes; therefore, proper cholesterol trafficking is finely regulated. Because cholesterol itself is insoluble, it is trafficked in the cell in two general ways: vesicular pathways and non-vesicular pathways. Vesicular cholesterol transport involves the delivery of membrane cholesterol via a vesicle formed by budding off of the donor membrane, followed by vesicle fusion with another membrane [31]. Vesicular cholesterol transport, such as the trafficking of cholesterol via internalization of LDL-derived cholesterol, vesicle formation, and movement of cholesterol through the endosomal and lysosomal pathways, requires an intact cytoskeleton and metabolic energy [27,32].

Although much work has been done elucidating mechanisms of vesicular cholesterol trafficking, evidence shows that inhibiting vesicular transport by genetic or pharmacological means does not inhibit intracellular cholesterol transport [33–35]. Furthermore, work has shown that cholesterol trafficking between the ER and plasma membrane occurs much faster than vesicular trafficking via membrane proteins [36], implying robust non-vesicular trafficking. Non-vesicular cholesterol trafficking involves the transport of cholesterol from a donor membrane to an acceptor membrane that does not require metabolic energy. This process can occur spontaneously via passive diffusion, the rate of which is dependent on the donor membrane's lipid composition; by cholesterol-transport proteins, either soluble or membrane-bound; or by membrane contact sites that involve brief interactions between membranes [37]. Although passive diffusion of cholesterol is a slow process, cholesterol-transport proteins or membrane-contact sites can accelerate cholesterol transport [32,37]. This review focuses on the role of cholesterol-transport proteins in intracellular cholesterol trafficking for various cellular processes.

1.4. Cholesterol Transport in Steroid Biosynthesis

One important transport route of cholesterol is the delivery of cholesterol from intracellular stores into mitochondria for steroid biosynthesis. The rate-limiting step of acute steroid biosynthesis is the transport of cholesterol derived from intracellular stores at the outer mitochondrial membrane (OMM) to the matrix side of the IMM where cytochrome P450 side chain cleavage enzyme CYP11A1 resides [38–40]. This process is initiated by stimulation of steroidogenic cells by the pituitary trophic hormones, luteinizing hormone (LH), follicle stimulating hormone (FSH), and adrenocorticotropic hormone (ACTH), which induce cyclic AMP (cAMP) production [41]. Previous work in our laboratory showed that hormonal stimulation and cAMP induction initiates the formation of a multi-protein complex called the transduceosome, consisting of cytosolic and OMM proteins that together transport cholesterol across the OMM [42]. The cytosolic components of the transduceosome include acyl-CoA binding domain-containing 3, protein kinase A regulatory subunit 1, and hormone-induced STAR, which are anchored to TSPO and voltage dependent anion channel 1 (VDAC1), the OMM components of the transduceosome. While the exact mechanism by which cholesterol moves through the transduceosome is still unclear, disruption of the interactions between these proteins and knockdown or deletion studies have demonstrated the importance of each of these proteins in cholesterol transport for steroid biosynthesis [10,42–45]. Once at the OMM, proteins of the steroidogenic metabolon, which include TSPO and VDAC along with the IMM proteins ATPase family AAA domain-containing protein 3A and CYP11A1, facilitate the transport of cholesterol to the matrix side of the IMM, where cholesterol is converted to pregnenolone by CYP11A1 [46].

2. Model Cholesterol-Transport Proteins

There are several proteins and protein families that facilitate non-vesicular cholesterol transport between membranes. Here, we discuss STAR, which is critical for cholesterol transport into the mitochondria for hormone-induced steroidogenesis, and SCP2/SCPx, which are non-specific lipid transfer proteins that play a role in the transport and/or metabolism of many lipids including cholesterol, as model cholesterol-transport proteins.

2.1. STAR

Hormonal stimulation of steroidogenic cells results in the rapid delivery of cholesterol, the precursor of all steroids, to the IMM where cholesterol is converted to pregnenolone, the first step in steroidogenesis [38]. One of the proteins indispensable for steroid biosynthesis is STAR, which is known to facilitate the transfer of cholesterol to mitochondria upon hormonal stimulation [47]. In addition to STAR, there are many other proteins and enzymes involved in steroid biosynthesis, spanning several cellular compartments. The direct interaction between STAR and several of these proteins was shown to be critical for steroidogenesis [30,48].

2.1.1. Role of STAR in Steroidogenesis

Because steroidogenic cells store only very low amounts, steroids must be synthesized rapidly in response to hormonal stimulation. While basal steroidogenesis involves the slow process of transcribing steroidogenic enzymes, hormone-stimulated steroidogenesis involves the rapid transfer of cholesterol to CYP11A1 [39,40,49]. STAR was first discovered when it was shown that acute steroidogenic responses paralleled the synthesis of a 37 kDa phosphoprotein, steroidogenic acute regulatory protein or STAR [50,51]. Overexpression of STAR in MA-10 mouse tumor Leydig cells was found to induce steroidogenesis to a similar extent as cAMP, and introduction of STAR into non-steroidogenic COS-1 cells transfected with the CYP11A1 system increased steroidogenesis 6-fold [47,52–54]. Furthermore, the key role of STAR in cholesterol transport for acute steroidogenesis was shown when STAR mutations in humans caused congenital lipoid adrenal hyperplasia (lipoid CAH), a disease characterized by severe deficiency in steroid production and accumulation of cholesterol in steroidogenic cells [55,56]. *Star* knockout (KO) in mice has a similar phenotype as humans with *STAR* mutations; however, gonadal function was less affected in mice [57].

2.1.2. STAR Protein Activity

STAR is synthesized as a 37 kDa cytosolic preprotein composed of an N-terminal mitochondrial targeting sequence and a C-terminal cholesterol-binding STAR-related lipid transfer (START) domain [52,58,59]. While STAR is constitutively expressed under basal conditions, hormonal stimulation parallels a rapid increase in STAR levels and leads to translocation of STAR to the OMM [50–52]. STAR is synthesized from pre-existing mRNA, as inhibition of gene transcription did not affect induction of steroidogenesis by cAMP and only newly synthesized STAR protein is active [60,61]. Active 37 kDa STAR is rapidly processed at the OMM to an inactive 30 kDa mature protein, which is then imported to the mitochondrial matrix where it is degraded [48,62]. Active STAR has a half-life of 3–5 min in the cytoplasm, while inactive STAR has an average half-life of 2–4 h in the matrix [63,64]. Deletion of the N-terminal mitochondrial targeting sequence of STAR (N-62 STAR) resulted in no change in activity, although N-62 STAR haphazardly inserted cholesterol into other membranes, suggesting that STAR's targeting sequence is vital for confining its activity to mitochondria [13,65]. STAR functions solely at the OMM and does not need to enter the mitochondria for its activity [66,67]. Further work has shown that the residence time of STAR at the OMM is proportional to its steroidogenic activity [68]. This activity has been shown to be tightly regulated by various mechanisms, particularly phosphorylation at Ser-194, which induces its cholesterol transfer activity by 50%, and interaction with 14-3-3γ, which negatively regulates steroid production by blocking phosphorylation of STAR at Ser-194 [60,69–71].

2.1.3. STAR Function in Cholesterol Transport and the START Domain

While the exact mechanism by which STAR facilitates cholesterol transport into the mitochondria is unknown, several studies utilizing cell-free systems have shown that STAR binds cholesterol and transfers it between membranes [12,13,72]. As mentioned above, STAR contains a C-terminal cholesterol-binding START domain, a hydrophobic sterol-binding pocket composed of four α-helices and nine antiparallel β-sheets, which is common among other closely related lipid transfer proteins [59,72–74]. Of the START proteins, the closest homolog to STAR is STARD3, also known as metastatic axillary lymph node protein 64 (MLN64). The START domains of STAR and STARD3 have 35% sequence identity. However, while both these proteins have been shown only to bind cholesterol, STAR contains a mitochondrial targeting sequence, whereas STARD3 contains an N-terminal domain targeting the protein to late endosomes suggesting distinct functions in cholesterol transport due to differences in subcellular localization [59,75].

One current model of cholesterol transport by STAR is called the molten globule model, which has the C-helix of STAR's START domain interacting with protonated phospholipid head groups at the OMM, inducing the C-helix to swing open [76,77]. According to this model, this conformational change allows STAR to bind and then release cholesterol [76,77]. In addition to its proposed role in binding and releasing cholesterol itself, another model suggests that STAR can trigger IMM importation of cholesterol bound to the cholesterol-binding domain of TSPO [43,76].

2.1.4. Other Roles of STAR

Whereas most studies investigating the function of STAR have focused on the role of hormone-induced STAR in cholesterol transport for steroid biosynthesis, we developed a STAR KO MA-10 mouse tumor cell line (STARKO1) to investigate the role of constitutive STAR, i.e., STAR protein present under basal conditions independent of hormonal stimulation [10]. We showed that the absence of constitutive STAR altered lipid droplet content, specifically leading to dramatic increases in the amounts of cholesteryl ester, diacylglycerol, and phosphatidylcholine in STARKO1 cell lipid droplets [10]. Alterations in lipid droplet content paralleled alterations in the levels of many lipid-related genes. These data suggested that STAR functioned in the transport and/or metabolism of various other lipids, independent of its role in cholesterol transport for steroidogenesis. Furthermore, our recent data suggested that absence of constitutive STAR led to alterations in mitochondrial structure and function, which were exacerbated by reintroduction of STAR into STARKO1 cells [78]. Taken together these results show that STAR may have other distinct functions in addition to its known classical role in cholesterol transport for hormone-induced steroidogenesis (Figure 1).

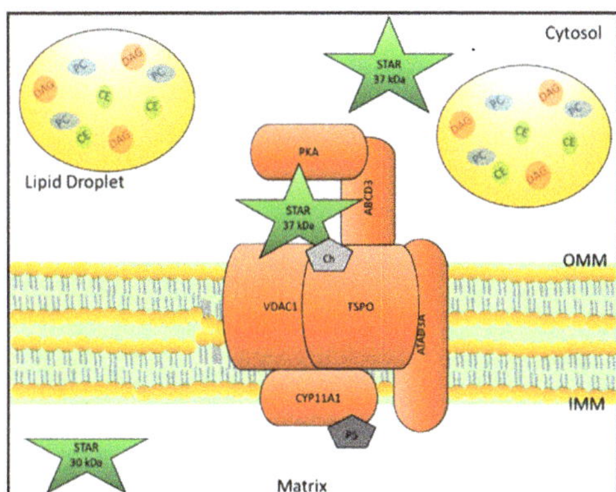

Figure 1. Function of hormone-induced STAR and constitutive STAR. Hormonal stimulation induces STAR to localize to the OMM where it interacts and works with other transduceosome proteins to transport cholesterol to CYP11A1 at the IMM for acute steroidogenesis. CYP11A1 converts cholesterol to the first steroid, pregnenolone. Upon reaching the OMM, the mitochondrial targeting sequence of STAR is cleaved, inactivating the protein, and inducing its import into the matrix. Constitutive STAR plays a role in the transport and/or metabolism of cholesteryl esters, diacylglycerol, and phosphatidylcholine as knockout of STAR results in the accumulation of these lipids. Abbreviations: Ch: cholesterol; P5: pregnenolone; DAG: diacylglycerol; CE: cholesteryl ester; PC: phosphatidylcholine.

2.2. SCP2/SCPx

Unlike STAR, SCP2 and SCPx have broad specificity for various lipids, and for this reason, they are referred to as non-specific lipid transfer proteins [9]. Since it is sometimes hard to tease out separate functions between the two, we use SCP2/SCPx to indicate nonspecificity. Both proteins have been shown to function in intracellular cholesterol transport between several sites such as mitochondria, ER, and plasma membrane [11,79,80]. In addition to cholesterol, previous studies have shown that SCP2/SCPx are involved in the transport and metabolism of other lipids including cholesteryl esters, fatty acids, fatty acyl-CoAs, and phospholipids [81].

2.2.1. SCP2/SCPx Gene and Protein Products

The *SCP2* gene contains two distinct transcription initiation sites and encodes a 15 kDa pro-SCP2 protein and the 58 kDa SCPx, which have identical C-termini [82]. Upon further processing, the 15 kDa pro-SCP2 protein is cleaved to form the mature 13 kDa SCP2. The 58 kDa SCPx is also post-transcriptionally processed into 46 kDa SCPx, as well as the same mature 13 kDa SCP2 [83,84]. SCP2 and SCPx are both highly expressed in adrenals, testis, ovaries, liver, and intestine, all tissues with high rates of cholesterol metabolism [85]. While SCP2 and SCPx have lipid transfer activity, SCPx has been shown to be a critical enzyme in peroxisomal β-oxidation [86]. Owing to the peroxisomal targeting AKL sequence at the common C-terminus, SCP2/SCPx are both found in peroxisomes, but SCPx is exclusively localized to peroxisomes, in line with its critical function in β-oxidation, while SCP2 is also found in the cytosol [87–89]. Additionally, there is evidence suggesting that these proteins contain a predicted mitochondrial targeting sequence at the N-terminus, suggesting dual targeting [81,90].

2.2.2. Role of SCP2/SCPx in Cholesterol Transport

Because SCP2/SCPx are both synthesized via expression of a single gene, most genetic manipulation studies described here are non-specific as to whether the findings pertain to SCP2 alone, SCPx alone, or both proteins. Several lines of evidence have supported the role of SCP2/SCPx in intracellular cholesterol trafficking. Studies have shown that recombinant human SCP2 binds cholesterol at a single binding site, however, the reported Kd values between these studies are drastically different, with one reporting a Kd of 4.2 nM and the other reporting a Kd of 0.3 μM [91,92]. Furthermore, work suggesting that SCP2/SCPx function in cholesterol transport includes in vitro studies showing that SCP2/SCPx are effective in enhancing sterol trafficking ~27-fold from plasma membranes to microsomal membranes and ~12-fold from plasma membranes to mitochondria [93]. Additionally, since cholesterol transport between lysosomes to plasma membrane occurs within two minutes in intact cells, which is inconsistent with vesicular transport, a cholesterol transport protein-mediated mechanism involving SCP2/SCPx was postulated [94]. It was shown that cholesterol transport from lysosomal membranes to plasma membranes isolated from mouse L-fibroblasts was enhanced 364-fold by SCP2/SCPx [79]. This study also showed that in L-cells plasma membranes with *Scp2* overexpressed, cholesterol levels/mg protein decreased by 38%, consistent with other data showing that SCP2 is involved in distributing cholesterol away from the plasma membrane to other cellular sites such as lipid droplets [79,95]. Additionally, cholesterol/mg protein decreased by 17% in lysosomal membranes isolated from these over-expressing L-cells, while there was a 2.2-fold increase in cholesterol/mg protein in ER membranes from the same cells [79,94]. In intact cells, transfection with human *SCP2* increased exogenous cholesterol uptake by 1.9-fold and total cholesterol mass by 1.4-fold [83]. In addition, SCP2/SCPx enhanced cholesterol transport from the plasma membrane to ER for esterification by ACAT in L-cells and enhanced intracellular cholesterol cycling in hepatoma cells [83,96,97]. SCP2 was also shown to play a role in cholesterol efflux because transfection of L-cells with *Scp2* inhibited HDL-mediated cholesterol efflux from lipid droplets to the plasma membrane through lipid rafts [95].

In addition to these studies utilizing intact cells to elucidate a role for SCP2/SCPx in cholesterol transport, studies using genetic manipulation of *Scp2* in animal models have also been done. In mice overexpressing Scp2, there was an increase in plasma LDL cholesterol, a decrease in plasma HDL cholesterol and a 70% increase in hepatic total cholesterol [98]. In *Scp2* gene-ablated mice, total hepatic cholesterol decreased 15%, likely due to decreases in cholesteryl esters [99]. In addition to these findings, animal models have also suggested a role of Scp2 in biliary cholesterol secretion. In rats with Scp2 overexpression, total hepatic cholesterol content and total bile acid content increased [98]. Conversely, in rats treated with *Scp2* antisense oligonucleotides that led to a 60% reduction in Scp2 levels in the liver, there was a delay in biliary cholesterol secretion [100]. Taken together, these data support the role of SCP2/SCPx in intracellular cholesterol transport between a variety of membranes and for a variety of critical cellular processes.

2.2.3. Role of SCP2 in Steroidogenesis

In addition to these proposed functions in cholesterol trafficking by SCP2/SCPx, much work has also been done to delineate a potential role of SCP2 in cholesterol transport to the mitochondria for steroidogenesis. While STAR is known to mediate the acute, rapid transport of cholesterol from the OMM to the IMM, thereby depleting the OMM of cholesterol, many studies have been done to investigate whether SCP2 plays a role in replenishing the OMM with cholesterol from intracellular stores. In isolated rat steroidogenic adrenal cells, SCP2 enhanced radiolabeled cholesterol transport from lipid droplets to mitochondria and also significantly increased pregnenolone production [101]. Introduction of human SCP2 and the components of the cholesterol side chain cleavage system in non-steroidogenic COS-7 cells also led to increased steroid production [102]. Furthermore, human SCP2 is most highly expressed in the steroidogenic tissues (adrenals, testis, and ovaries), and hormonal stimulation of the steroidogenic cells of these tissues leads to an increase in *SCP2* mRNA expression and protein levels by a cAMP dependent pathway and increased its association with mitochondria [103–106]. Additionally, in *Scp2*-overexpressing L-cells, it was shown that sterol transport from isolated lysosomal to mitochondrial membranes was enhanced by SCP2 [107].

However, although there is much indirect evidence that SCP2 plays a role in steroidogenesis, the function of SCP2 in cholesterol transport into the mitochondria for steroid production has been called into question. Firstly, it was shown that human SCP2 enhanced cholesterol transfer to mitochondrial membranes regardless of whether the mitochondria were isolated from MA-10 cells or from fibroblasts, showing that the cells need not be steroidogenic for SCP2 to transport cholesterol to mitochondria [12]. Secondly, it was found that *Scp2* gene-ablated mice had normal serum testosterone, progesterone, and corticosteroid levels [99]. It may be the case that, although SCP2 is dispensable for steroidogenesis, it may play a role in cholesterol transport to the mitochondria as a supplementary mechanism to STAR-mediated cholesterol transport.

2.2.4. Role of SCP2/SCPx in the Transport of Other Lipids

While SCP2/SCPx are classically known as sterol transport proteins, these proteins also possess high binding affinities for many other lipid classes and have been shown to play a role in the transport and/or metabolism of many other lipids. For example, SCP2 has high affinity for fatty acids with a reported Kd of 234 nM, similar to other fatty acid binding proteins, and previous reports show that SCP2/SCPx enhance the cellular uptake and intracellular transport of fatty acids [108–111]. Further studies have shown that SCP2/SCPx function in fatty acid transport to peroxisomes for oxidation and to ER for phospholipid incorporation [111,112]. Another lipid group for which SCP2 has high affinity is fatty acyl CoAs, with reported Kds in the range of 2–4 nM [113,114]. In vitro studies and studies in intact cells have shown that SCP2/SCPx stimulate the incorporation of microsomal fatty acyl CoA into phosphatidic acid [112,115]. Additionally, SCP2 has high affinity for phosphatidylinositol (PI) and may play a role in PI transport to the plasma membrane based on data showing that human *SCP2* overexpression results in the significant redistribution of PI from mitochondrial and ER membranes to plasma membranes [116]. Lastly, SCP2 has been shown to have nanomolar affinity to all sphingolipid classes, and in vitro studies using liver homogenates suggest that SCP2/SCPx increase sphingomyelin transport [117–119].

2.2.5. Role of SCPx in Peroxisomal β-Oxidation

In addition to being a lipid transporter, the 46 kDa SCPx protein that arises from 58 kDa SCPx exhibits 3-ketoacyl-CoA thiolase activity [120]. The 46 kDa SCPx is responsible for catalyzing the final step of the peroxisomal β-oxidation of branched-chain fatty acids and the metabolism of cholesterol for bile acid synthesis, while the classical 3-ketoacyl-CoA thiolase is specific for catalyzing the final step of peroxisomal β-oxidation of straight-chain fatty acids [86,121]. Indeed, *Scp2* gene-ablated mice had defects in the metabolism of branched-chain fatty acids with a ten-fold accumulation of phytanic acid in knockout mice [99]. Further, mice null for Scpx, but with normal levels of Scp2, had altered levels of hepatic fatty acids, suggesting that branched-chain fatty acid oxidation requires SCPx independent of SCP2 [122]. The indispensable role of SCPx in peroxisomal β-oxidation was further exemplified when a homozygous 1-nucleotide insertion in *SCP2* in a patient resulted in the complete absence of SCPx protein and led to leukoencephalopathy with dystonia and motor neuropathy, hyposmia, azoospermia, and an accumulation of branched-chain fatty acids [123]. While deficiencies in several peroxisomal enzymes and/or proteins leading to neurological diseases, such as X-linked adrenoleukodystrophy and Refsum disease, have previously been reported across many patients, this was the first report of a patient with SCPx deficiency [123]. The second report of SCPx deficiency was caused by a compound heterozygous mutation in *SCP2*, again leading to undetectable levels of SCPx and neurodegenerative symptoms [124]. Recently, our laboratory worked on the characterization of a third patient with SCPx deficiency, the first associated with a heterozygous mutation in *SCP2*, leading to low, but detectable levels of SCPx [125]. Similar to previous reports, the patient presented with severe neurological symptoms. However, in contrast to previous studies, the patient's pristanic and phytanic acid levels were normal, indicating that the patient's low levels of SCPx were sufficient for branched-chain fatty acid metabolism. Despite normal pristanic and phytanic levels, levels of many other lipid species among various lipid classes, including fatty acids, acylcarnitines, sterols, phospholipids, and sphingolipids were altered in the patient's fibroblasts, suggesting a role for SCPx in the transport and/or metabolism of these lipids [125]. Taken together, these data exemplify the critical and multifunctional roles of SCP2/SCPx as non-specific lipid transporters and SCPx as a key enzyme in peroxisomal oxidation (Figure 2).

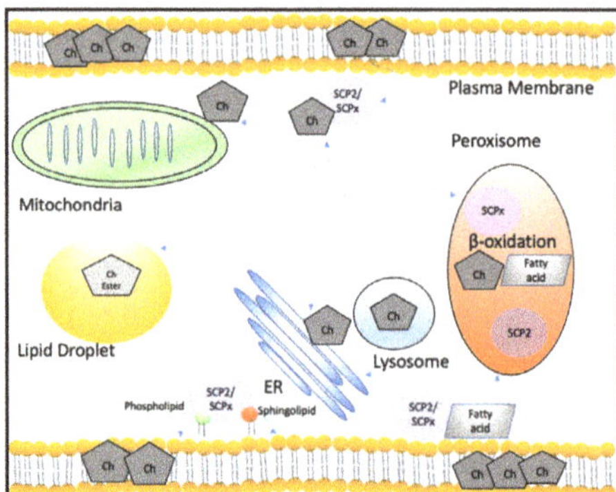

Figure 2. SCP2/SCPx function in the transport and metabolism of cholesterol and other lipids. SCP2/SCPx may transport cholesterol between various cellular membranes and organelles, such as the plasma membrane, mitochondria, lipid droplets, ER, lysosomes, and peroxisomes. Cholesterol intermediates also undergo oxidation in peroxisomes via SCPx for bile acid synthesis. In addition to cholesterol transport, SCP2/SCPx play a role in fatty acid, phospholipid, and sphingolipid transport. SCPx is also a key enzyme in the peroxisomal β-oxidation of fatty acids. Abbreviations: Ch: cholesterol; ER: endoplasmic reticulum.

3. Future Directions and Conclusions

Cholesterol transport proteins play vital roles in the intracellular distribution of cholesterol for several key cellular processes. Although STAR is classically known to function in cholesterol transport for hormone-induced steroidogenesis, recent studies suggest that it may have additional roles in the transport and metabolism of other lipids. While these studies point to additional roles of STAR, further studies should be done to clarify the mechanism by which STAR functions in these roles. For example, while current data supports a role of STAR in affecting lipid droplet content, it is still unclear how STAR may promote the transport of various lipids to lipid droplets. Additionally, it will be important to further investigate the role of STAR in mitochondrial structure and function since current data indicate that the absence of STAR leads to mitochondrial dysfunction. Furthermore, while SCP2/SCPx were first recognized as sterol transfer proteins, an accumulation of data ranging from cell-free systems to patients with SCP2 mutations have suggested a broader role for these proteins in lipid transport and metabolism. Additional work may also be done to further elucidate the multifunctional roles of these proteins, including the roles of these proteins in the transport and metabolism of the various lipids in distinct tissues, which has been supported by the current data. Thus, the data presented here suggest that STAR and SCP2/SCPx, classic examples of intracellular cholesterol transport proteins, play a more general role in lipid transport and metabolism in addition to their respective roles in cholesterol trafficking.

Author Contributions: Conceptualization, M.G., S.V. and V.P.; Writing—original draft preparation, M.G. and S.V.; Writing—review and editing, M.G., S.V. and V.P.; Visualization, M.G.; Funding acquisition, V.P. All authors have read and agreed to the published version of the manuscript.

Funding: This work was supported by funds from the John Stauffer Decanal Chair in Pharmaceutical Sciences and the School of Pharmacy (University of Southern California).

Conflicts of Interest: The authors declare no conflict of interest.

References

1. Luo, J.; Yang, H.; Song, B.L. Mechanisms and regulation of cholesterol homeostasis. *Nat. Rev. Mol. Cell Biol.* **2020**, *21*, 225–245. [CrossRef] [PubMed]
2. Shibuya, Y.; Chang, C.C.; Chang, T.Y. ACAT1/SOAT1 as a therapeutic target for Alzheimer's disease. *Future. Med. Chem.* **2015**, *7*, 2451–2467. [CrossRef]
3. Kuzu, O.F.; Noory, M.A.; Robertson, G.P. The Role of Cholesterol in Cancer. *Cancer Res.* **2016**, *76*, 2063–2070. [CrossRef]
4. Cortes, V.A.; Busso, D.; Maiz, A.; Arteaga, A.; Nervi, F.; Rigotti, A. Physiological and pathological implications of cholesterol. *Front. Biosci.* **2014**, *19*, 416–428. [CrossRef] [PubMed]
5. Maxfield, F.R.; Tabas, I. Role of cholesterol and lipid organization in disease. *Nature* **2005**, *438*, 612–621. [CrossRef] [PubMed]
6. Luu, W.; Sharpe, L.J.; Capell-Hattam, I.; Gelissen, I.C.; Brown, A.J. Oxysterols: Old Tale, New Twists. *Annu. Rev. Pharmacol. Toxicol.* **2016**, *56*, 447–467. [CrossRef] [PubMed]
7. Wong, L.H.; Gatta, A.T.; Levine, T.P. Lipid transfer proteins: The lipid commute via shuttles, bridges and tubes. *Nat. Rev. Mol. Cell Biol.* **2019**, *20*, 85–101. [CrossRef]
8. Midzak, A.; Rone, M.; Aghazadeh, Y.; Culty, M.; Papadopoulos, V. Mitochondrial protein import and the genesis of steroidogenic mitochondria. *Mol. Cell. Endocrinol.* **2011**, *336*, 70–79. [CrossRef]
9. Kriska, T.; Pilat, A.; Schmitt, J.C.; Girotti, A.W. Sterol carrier protein-2 (SCP-2) involvement in cholesterol hydroperoxide cytotoxicity as revealed by SCP-2 inhibitor effects. *J. Lipid Res.* **2010**, *51*, 3174–3184. [CrossRef]
10. Galano, M.; Li, Y.; Li, L.; Sottas, C.; Papadopoulos, V. Role of Constitutive STAR in Leydig Cells. *Int. J. Mol. Sci.* **2021**, *22*, 2021. [CrossRef] [PubMed]
11. Pfeifer, S.M.; Furth, E.E.; Ohba, T.; Chang, Y.J.; Rennert, H.; Sakuragi, N.; Billheimer, J.T.; Strauss, J.F. Sterol carrier protein 2: A role in steroid hormone synthesis? *J. Steroid Biochem. Mol. Biol.* **1993**, *47*, 167–172. [CrossRef]
12. Petrescu, A.D.; Gallegos, A.M.; Okamura, Y.; Strauss, J.F.; Schroeder, F. Steroidogenic acute regulatory protein binds cholesterol and modulates mitochondrial membrane sterol domain dynamics. *J. Biol. Chem.* **2001**, *276*, 36970–36982. [CrossRef] [PubMed]
13. Kallen, C.B.; Billheimer, J.T.; Summers, S.A.; Stayrook, S.E.; Lewis, M.; Strauss, J.F. Steroidogenic acute regulatory protein (StAR) is a sterol transfer protein. *J. Biol. Chem.* **1998**, *273*, 26285–26288. [CrossRef] [PubMed]
14. Charreau, E.H.; Calvo, J.C.; Nozu, K.; Pignataro, O.; Catt, K.J.; Dufau, M.L. Hormonal modulation of 3-hydroxy-3-methylglutaryl coenzyme A reductase activity in gonadotropin-stimulated and -desensitized testicular Leydig cells. *J. Biol. Chem.* **1981**, *256*, 12719–12724. [CrossRef]

15. Rainey, W.E.; Shay, J.W.; Mason, J.I. ACTH induction of 3-hydroxy-3-methylglutaryl coenzyme A reductase, cholesterol biosynthesis, and steroidogenesis in primary cultures of bovine adrenocortical cells. *J. Biol. Chem.* **1986**, *261*, 7322–7326. [CrossRef]
16. Goldstein, J.L.; Brown, M.S. Regulation of the mevalonate pathway. *Nature* **1990**, *343*, 425–430. [CrossRef] [PubMed]
17. Bloch, K. Biological synthesis of cholesterol. *Harvey Lect.* **1952**, *48*, 68–88. [CrossRef] [PubMed]
18. Cartwright, I.J.; Plonné, D.; Higgins, J.A. Intracellular events in the assembly of chylomicrons in rabbit enterocytes. *J. Lipid Res.* **2000**, *41*, 1728–1739. [CrossRef]
19. Chang, T.Y.; Chang, C.C.; Ohgami, N.; Yamauchi, Y. Cholesterol sensing, trafficking, and esterification. *Annu. Rev. Cell. Dev. Biol.* **2006**, *22*, 129–157. [CrossRef]
20. Hu, J.; Zhang, Z.; Shen, W.J.; Azhar, S. Cellular cholesterol delivery, intracellular processing and utilization for biosynthesis of steroid hormones. *Nutr. Metab.* **2010**, *7*, 47. [CrossRef]
21. Rigotti, A.; Miettinen, H.E.; Krieger, M. The role of the high-density lipoprotein receptor SR-BI in the lipid metabolism of endocrine and other tissues. *Endocr. Rev.* **2003**, *24*, 357–387. [CrossRef] [PubMed]
22. Leiva, A.; Verdejo, H.; Benítez, M.L.; Martínez, A.; Busso, D.; Rigotti, A. Mechanisms regulating hepatic SR-BI expression and their impact on HDL metabolism. *Atherosclerosis* **2011**, *217*, 299–307. [CrossRef]
23. Lange, Y.; Steck, T.L. Active membrane cholesterol as a physiological effector. *Chem. Phys. Lipids* **2016**, *199*, 74–93. [CrossRef]
24. Steck, T.L.; Lange, Y. Cell cholesterol homeostasis: Mediation by active cholesterol. *Trends Cell Biol.* **2010**, *20*, 680–687. [CrossRef] [PubMed]
25. Epand, R.M. Proteins and cholesterol-rich domains. *Biochim. Biophys. Acta* **2008**, *1778*, 1576–1582. [CrossRef] [PubMed]
26. Liscum, L.; Munn, N.J. Intracellular cholesterol transport. *Biochim. Biophys. Acta* **1999**, *1438*, 19–37. [CrossRef]
27. Soccio, R.E.; Breslow, J.L. Intracellular cholesterol transport. *Arter. Thromb. Vasc. Biol.* **2004**, *24*, 1150–1160. [CrossRef]
28. Mukherjee, S.; Zha, X.; Tabas, I.; Maxfield, F.R. Cholesterol distribution in living cells: Fluorescence imaging using dehydroergosterol as a fluorescent cholesterol analog. *Biophys. J.* **1998**, *75*, 1915–1925. [CrossRef]
29. Lange, Y. Disposition of intracellular cholesterol in human fibroblasts. *J. Lipid Res.* **1991**, *32*, 329–339. [CrossRef]
30. Rone, M.B.; Fan, J.; Papadopoulos, V. Cholesterol transport in steroid biosynthesis: Role of protein-protein interactions and implications in disease states. *Biochim. Biophys. Acta* **2009**, *1791*, 646–658. [CrossRef] [PubMed]
31. Papadopoulos, V.; Miller, W.L. Role of mitochondria in steroidogenesis. *Best Pract. Res. Clin. Endocrinol. Metab.* **2012**, *26*, 771–790. [CrossRef] [PubMed]
32. Lev, S. Non-vesicular lipid transport by lipid-transfer proteins and beyond. *Nat. Rev. Mol. Cell Biol.* **2010**, *11*, 739–750. [CrossRef]
33. Baumann, N.A.; Sullivan, D.P.; Ohvo-Rekilä, H.; Simonot, C.; Pottekat, A.; Klaassen, Z.; Beh, C.T.; Menon, A.K. Transport of newly synthesized sterol to the sterol-enriched plasma membrane occurs via nonvesicular equilibration. *Biochemistry* **2005**, *44*, 5816–5826. [CrossRef] [PubMed]
34. Urbani, L.; Simoni, R.D. Cholesterol and vesicular stomatitis virus G protein take separate routes from the endoplasmic reticulum to the plasma membrane. *J. Biol. Chem.* **1990**, *265*, 1919–1923. [CrossRef]
35. Kaplan, M.R.; Simoni, R.D. Transport of cholesterol from the endoplasmic reticulum to the plasma membrane. *J. Cell Biol.* **1985**, *101*, 446–453. [CrossRef]
36. DeGrella, R.F.; Simoni, R.D. Intracellular transport of cholesterol to the plasma membrane. *J. Biol. Chem.* **1982**, *257*, 14256–14262. [CrossRef]
37. Lev, S. Nonvesicular lipid transfer from the endoplasmic reticulum. *Cold Spring Harb. Perspect. Biol.* **2012**, *4*, a013300. [CrossRef] [PubMed]
38. Issop, L.; Rone, M.B.; Papadopoulos, V. Organelle plasticity and interactions in cholesterol transport and steroid biosynthesis. *Mol. Cell. Endocrinol.* **2013**, *371*, 34–46. [CrossRef]
39. Simpson, E.R.; Boyd, G.S. The cholesterol side-chain cleavage system of the adrenal cortex: A mixed-function oxidase. *Biochem. Biophys. Res. Commun.* **1966**, *24*, 10–17. [CrossRef]
40. Simpson, E.R.; Boyd, G.S. The cholesterol side-chain cleavage system of bovine adrenal cortex. *Eur. J. Biochem.* **1967**, *2*, 275–285. [CrossRef] [PubMed]
41. Jefcoate, C. High-flux mitochondrial cholesterol trafficking, a specialized function of the adrenal cortex. *J. Clin. Investig.* **2002**, *110*, 881–890. [CrossRef] [PubMed]
42. Liu, J.; Rone, M.B.; Papadopoulos, V. Protein-protein interactions mediate mitochondrial cholesterol transport and steroid biosynthesis. *J. Biol. Chem.* **2006**, *281*, 38879–38893. [CrossRef] [PubMed]
43. Hauet, T.; Yao, Z.X.; Bose, H.S.; Wall, C.T.; Han, Z.; Li, W.; Hales, D.B.; Miller, W.L.; Culty, M.; Papadopoulos, V. Peripheral-type benzodiazepine receptor-mediated action of steroidogenic acute regulatory protein on cholesterol entry into leydig cell mitochondria. *Mol. Endocrinol.* **2005**, *19*, 540–554. [CrossRef] [PubMed]
44. Li, H.; Degenhardt, B.; Tobin, D.; Yao, Z.X.; Tasken, K.; Papadopoulos, V. Identification, localization, and function in steroidogenesis of PAP7: A peripheral-type benzodiazepine receptor- and PKA (RIalpha)-associated protein. *Mol. Endocrinol.* **2001**, *15*, 2211–2228. [CrossRef]
45. Papadopoulos, V.; Amri, H.; Li, H.; Boujrad, N.; Vidic, B.; Garnier, M. Targeted disruption of the peripheral-type benzodiazepine receptor gene inhibits steroidogenesis in the R2C Leydig tumor cell line. *J. Biol. Chem.* **1997**, *272*, 32129–32135. [CrossRef]

46. Rone, M.B.; Midzak, A.S.; Issop, L.; Rammouz, G.; Jagannathan, S.; Fan, J.; Ye, X.; Blonder, J.; Veenstra, T.; Papadopoulos, V. Identification of a dynamic mitochondrial protein complex driving cholesterol import, trafficking, and metabolism to steroid hormones. *Mol. Endocrinol.* **2012**, *26*, 1868–1882. [CrossRef] [PubMed]
47. Miller, W.L. Steroidogenic acute regulatory protein (StAR), a novel mitochondrial cholesterol transporter. *Biochim. Biophys. Acta* **2007**, *1771*, 663–676. [CrossRef]
48. Bose, M.; Whittal, R.M.; Miller, W.L.; Bose, H.S. Steroidogenic activity of StAR requires contact with mitochondrial VDAC1 and phosphate carrier protein. *J. Biol. Chem.* **2008**, *283*, 8837–8845. [CrossRef] [PubMed]
49. Stocco, D.M.; Clark, B.J. Regulation of the acute production of steroids in steroidogenic cells. *Endocr. Rev.* **1996**, *17*, 221–244. [CrossRef]
50. Epstein, L.F.; Orme-Johnson, N.R. Regulation of steroid hormone biosynthesis. Identification of precursors of a phosphoprotein targeted to the mitochondrion in stimulated rat adrenal cortex cells. *J. Biol. Chem.* **1991**, *266*, 19739–19745. [CrossRef]
51. Stocco, D.M.; Sodeman, T.C. The 30-kDa mitochondrial proteins induced by hormone stimulation in MA-10 mouse Leydig tumor cells are processed from larger precursors. *J. Biol. Chem.* **1991**, *266*, 19731–19738. [CrossRef]
52. Clark, B.J.; Wells, J.; King, S.R.; Stocco, D.M. The purification, cloning, and expression of a novel luteinizing hormone-induced mitochondrial protein in MA-10 mouse Leydig tumor cells. Characterization of the steroidogenic acute regulatory protein (StAR). *J. Biol. Chem.* **1994**, *269*, 28314–28322. [CrossRef]
53. Lin, D.; Sugawara, T.; Strauss, J.F.; Clark, B.J.; Stocco, D.M.; Saenger, P.; Rogol, A.; Miller, W.L. Role of steroidogenic acute regulatory protein in adrenal and gonadal steroidogenesis. *Science* **1995**, *267*, 1828–1831. [CrossRef]
54. Sugawara, T.; Holt, J.A.; Driscoll, D.; Strauss, J.F., 3rd; Lin, D.; Miller, W.L.; Patterson, D.; Clancy, K.P.; Hart, I.M.; Clark, B.J.; et al. Human steroidogenic acute regulatory protein: Functional activity in COS-1 cells, tissue-specific expression, and mapping of the structural gene to 8p11.2 and a pseudogene to chromosome 13. *Proc. Natl. Acad. Sci. USA* **1995**, *92*, 4778–4782. [CrossRef]
55. Tee, M.K.; Lin, D.; Sugawara, T.; Holt, J.A.; Guiguen, Y.; Buckingham, B.; Strauss, J.F., 3rd; Miller, W.L. T–>A transversion 11 bp from a splice acceptor site in the human gene for steroidogenic acute regulatory protein causes congenital lipoid adrenal hyperplasia. *Hum. Mol. Genet.* **1995**, *4*, 2299–2305. [CrossRef] [PubMed]
56. Stocco, D.M. Clinical disorders associated with abnormal cholesterol transport: Mutations in the steroidogenic acute regulatory protein. *Mol. Cell. Endocrinol.* **2002**, *191*, 19–25. [CrossRef]
57. Caron, K.M.; Soo, S.C.; Wetsel, W.C.; Stocco, D.M.; Clark, B.J.; Parker, K.L. Targeted disruption of the mouse gene encoding steroidogenic acute regulatory protein provides insights into congenital lipoid adrenal hyperplasia. *Proc. Natl. Acad. Sci. USA* **1997**, *94*, 11540–11545. [CrossRef] [PubMed]
58. Strauss, J.F.; Kishida, T.; Christenson, L.K.; Fujimoto, T.; Hiroi, H. START domain proteins and the intracellular trafficking of cholesterol in steroidogenic cells. *Mol. Cell. Endocrinol.* **2003**, *202*, 59–65. [CrossRef]
59. Tsujishita, Y.; Hurley, J.H. Structure and lipid transport mechanism of a StAR-related domain. *Nat. Struct Biol.* **2000**, *7*, 408–414. [CrossRef]
60. Aghazadeh, Y.; Rone, M.B.; Blonder, J.; Ye, X.; Veenstra, T.D.; Hales, D.B.; Culty, M.; Papadopoulos, V. Hormone-induced 14-3-3γ adaptor protein regulates steroidogenic acute regulatory protein activity and steroid biosynthesis in MA-10 Leydig cells. *J. Biol. Chem.* **2012**, *287*, 15380–15394. [CrossRef] [PubMed]
61. Artemenko, I.P.; Zhao, D.; Hales, D.B.; Hales, K.H.; Jefcoate, C.R. Mitochondrial processing of newly synthesized steroidogenic acute regulatory protein (StAR), but not total StAR, mediates cholesterol transfer to cytochrome P450 side chain cleavage enzyme in adrenal cells. *J. Biol. Chem.* **2001**, *276*, 46583–46596. [CrossRef]
62. Clark, B.J.; Soo, S.C.; Caron, K.M.; Ikeda, Y.; Parker, K.L.; Stocco, D.M. Hormonal and developmental regulation of the steroidogenic acute regulatory protein. *Mol. Endocrinol.* **1995**, *9*, 1346–1355. [CrossRef]
63. Clark, B.J.; Hudson, E.A. StAR Protein Stability in Y1 and Kin-8 Mouse Adrenocortical Cells. *Biology* **2015**, *4*, 200–215. [CrossRef]
64. Granot, Z.; Geiss-Friedlander, R.; Melamed-Book, N.; Eimerl, S.; Timberg, R.; Weiss, A.M.; Hales, K.H.; Hales, D.B.; Stocco, D.M.; Orly, J. Proteolysis of normal and mutated steroidogenic acute regulatory proteins in the mitochondria: The fate of unwanted proteins. *Mol. Endocrinol.* **2003**, *17*, 2461–2476. [CrossRef]
65. Arakane, F.; Sugawara, T.; Nishino, H.; Liu, Z.; Holt, J.A.; Pain, D.; Stocco, D.M.; Miller, W.L.; Strauss, J.F. Steroidogenic acute regulatory protein (StAR) retains activity in the absence of its mitochondrial import sequence: Implications for the mechanism of StAR action. *Proc. Natl. Acad. Sci. USA* **1996**, *93*, 13731–13736. [CrossRef]
66. Bose, H.S.; Lingappa, V.R.; Miller, W.L. Rapid regulation of steroidogenesis by mitochondrial protein import. *Nature* **2002**, *417*, 87–91. [CrossRef]
67. Arakane, F.; Kallen, C.B.; Watari, H.; Stayrook, S.E.; Lewis, M.; Strauss, J.F. Steroidogenic acute regulatory protein (StAR) acts on the outside of mitochondria to stimulate steroidogenesis. *Endocr. Res.* **1998**, *24*, 463–468. [CrossRef] [PubMed]
68. Bose, H.S.; Lingappa, V.R.; Miller, W.L. The steroidogenic acute regulatory protein, StAR, works only at the outer mitochondrial membrane. *Endocr. Res.* **2002**, *28*, 295–308. [CrossRef]
69. Arakane, F.; King, S.R.; Du, Y.; Kallen, C.B.; Walsh, L.P.; Watari, H.; Stocco, D.M.; Strauss, J.F. Phosphorylation of steroidogenic acute regulatory protein (StAR) modulates its steroidogenic activity. *J. Biol. Chem.* **1997**, *272*, 32656–32662. [CrossRef]
70. Jo, Y.; King, S.R.; Khan, S.A.; Stocco, D.M. Involvement of protein kinase C and cyclic adenosine 3′,5′-monophosphate-dependent kinase in steroidogenic acute regulatory protein expression and steroid biosynthesis in Leydig cells. *Biol. Reprod.* **2005**, *73*, 244–255. [CrossRef]

71. Dyson, M.T.; Jones, J.K.; Kowalewski, M.P.; Manna, P.R.; Alonso, M.; Gottesman, M.E.; Stocco, D.M. Mitochondrial A-kinase anchoring protein 121 binds type II protein kinase A and enhances steroidogenic acute regulatory protein-mediated steroidogenesis in MA-10 mouse leydig tumor cells. *Biol. Reprod.* **2008**, *78*, 267–277. [CrossRef]

72. Baker, B.Y.; Epand, R.F.; Epand, R.M.; Miller, W.L. Cholesterol binding does not predict activity of the steroidogenic acute regulatory protein, StAR. *J. Biol. Chem.* **2007**, *282*, 10223–10232. [CrossRef] [PubMed]

73. Romanowski, M.J.; Soccio, R.E.; Breslow, J.L.; Burley, S.K. Crystal structure of the Mus musculus cholesterol-regulated START protein 4 (StarD4) containing a StAR-related lipid transfer domain. *Proc. Natl. Acad. Sci. USA* **2002**, *99*, 6949–6954. [CrossRef]

74. Mathieu, A.P.; Fleury, A.; Ducharme, L.; Lavigne, P.; LeHoux, J.G. Insights into steroidogenic acute regulatory protein (StAR)-dependent cholesterol transfer in mitochondria: Evidence from molecular modeling and structure-based thermodynamics supporting the existence of partially unfolded states of StAR. *J. Mol. Endocrinol.* **2002**, *29*, 327–345. [CrossRef]

75. Clark, B.J. The mammalian START domain protein family in lipid transport in health and disease. *J. Endocrinol.* **2012**, *212*, 257–275. [CrossRef] [PubMed]

76. Miller, W.L. StAR search–what we know about how the steroidogenic acute regulatory protein mediates mitochondrial cholesterol import. *Mol. Endocrinol.* **2007**, *21*, 589–601. [CrossRef] [PubMed]

77. Bose, H.S.; Whittal, R.M.; Baldwin, M.A.; Miller, W.L. The active form of the steroidogenic acute regulatory protein, StAR, appears to be a molten globule. *Proc. Natl. Acad. Sci. USA* **1999**, *96*, 7250–7255. [CrossRef]

78. Galano, M.; Papadopoulos, V. Role of Constitutive STAR in Mitochondrial Structure and Function in MA-10 Leydig Cells. *Endocrinology* **2022**, *163*, bqac091. [CrossRef] [PubMed]

79. Gallegos, A.M.; Atshaves, B.P.; Storey, S.M.; McIntosh, A.L.; Petrescu, A.D.; Schroeder, F. Sterol carrier protein-2 expression alters plasma membrane lipid distribution and cholesterol dynamics. *Biochemistry* **2001**, *40*, 6493–6506. [CrossRef]

80. Puglielli, L.; Rigotti, A.; Greco, A.V.; Santos, M.J.; Nervi, F. Sterol carrier protein-2 is involved in cholesterol transfer from the endoplasmic reticulum to the plasma membrane in human fibroblasts. *J. Biol. Chem.* **1995**, *270*, 18723–18726. [CrossRef]

81. Li, N.C.; Fan, J.J.; Papadopoulos, V. Sterol Carrier Protein-2, a Nonspecific Lipid-Transfer Protein, in Intracellular Cholesterol Trafficking in Testicular Leydig Cells. *PLoS ONE* **2016**, *11*, e0149728. [CrossRef] [PubMed]

82. Ohba, T.; Holt, J.A.; Billheimer, J.T.; Strauss, J.F. Human sterol carrier protein x/sterol carrier protein 2 gene has two promoters. *Biochemistry* **1995**, *34*, 10660–10668. [CrossRef]

83. Atshaves, B.P.; Petrescu, A.D.; Starodub, O.; Roths, J.B.; Kier, A.B.; Schroeder, F. Expression and intracellular processing of the 58 kDa sterol carrier protein-2/3-oxoacyl-CoA thiolase in transfected mouse L-cell fibroblasts. *J. Lipid Res.* **1999**, *40*, 610–622. [CrossRef]

84. Moncecchi, D.; Murphy, E.J.; Prows, D.R.; Schroeder, F. Sterol carrier protein-2 expression in mouse L-cell fibroblasts alters cholesterol uptake. *Biochim. Biophys. Acta* **1996**, *1302*, 110–116. [CrossRef]

85. Vahouny, G.V.; Chanderbhan, R.; Kharroubi, A.; Noland, B.J.; Pastuszyn, A.; Scallen, T.J. Sterol carrier and lipid transfer proteins. *Adv. Lipid Res.* **1987**, *22*, 83–113. [CrossRef]

86. Wanders, R.J.; Denis, S.; Wouters, F.; Wirtz, K.W.; Seedorf, U. Sterol carrier protein X (SCPx) is a peroxisomal branched-chain beta-ketothiolase specifically reacting with 3-oxo-pristanoyl-CoA: A new, unique role for SCPx in branched-chain fatty acid metabolism in peroxisomes. *Biochem. Biophys. Res. Commun.* **1997**, *236*, 565–569. [CrossRef]

87. Mendis-Handagama, S.M.; Watkins, P.A.; Gelber, S.J.; Scallen, T.J. Leydig cell peroxisomes and sterol carrier protein-2 in luteinizing hormone-deprived rats. *Endocrinology* **1992**, *131*, 2839–2845. [CrossRef]

88. Keller, G.A.; Scallen, T.J.; Clarke, D.; Maher, P.A.; Krisans, S.K.; Singer, S.J. Subcellular localization of sterol carrier protein-2 in rat hepatocytes: Its primary localization to peroxisomes. *J. Cell Biol.* **1989**, *108*, 1353–1361. [CrossRef]

89. Mori, T.; Tsukamoto, T.; Mori, H.; Tashiro, Y.; Fujiki, Y. Molecular cloning and deduced amino acid sequence of nonspecific lipid transfer protein (sterol carrier protein 2) of rat liver: A higher molecular mass (60 kDa) protein contains the primary sequence of nonspecific lipid transfer protein as its C-terminal part. *Proc. Natl. Acad. Sci. USA* **1991**, *88*, 4338–4342. [CrossRef]

90. Westerman, J.; Wirtz, K.W. The primary structure of the nonspecific lipid transfer protein (sterol carrier protein 2) from bovine liver. *Biochem. Biophys. Res. Commun.* **1985**, *127*, 333–338. [CrossRef]

91. Stolowich, N.; Frolov, A.; Petrescu, A.D.; Scott, A.I.; Billheimer, J.T.; Schroeder, F. Holo-sterol carrier protein-2. (13)C NMR investigation of cholesterol and fatty acid binding sites. *J. Biol. Chem.* **1999**, *274*, 35425–35433. [CrossRef] [PubMed]

92. Colles, S.M.; Woodford, J.K.; Moncecchi, D.; Myers-Payne, S.C.; McLean, L.R.; Billheimer, J.T.; Schroeder, F. Cholesterol interaction with recombinant human sterol carrier protein-2. *Lipids* **1995**, *30*, 795–803. [CrossRef] [PubMed]

93. Frolov, A.; Woodford, J.K.; Murphy, E.J.; Billheimer, J.T.; Schroeder, F. Spontaneous and protein-mediated sterol transfer between intracellular membranes. *J. Biol. Chem.* **1996**, *271*, 16075–16083. [CrossRef] [PubMed]

94. Gallegos, A.M.; Atshaves, B.P.; Storey, S.M.; Starodub, O.; Petrescu, A.D.; Huang, H.; McIntosh, A.L.; Martin, G.G.; Chao, H.; Kier, A.B.; et al. Gene structure, intracellular localization, and functional roles of sterol carrier protein-2. *Prog. Lipid Res.* **2001**, *40*, 498–563. [CrossRef]

95. Atshaves, B.P.; Starodub, O.; McIntosh, A.; Petrescu, A.; Roths, J.B.; Kier, A.B.; Schroeder, F. Sterol carrier protein-2 alters high density lipoprotein-mediated cholesterol efflux. *J. Biol. Chem.* **2000**, *275*, 36852–36861. [CrossRef] [PubMed]

96. Murphy, E.J.; Schroeder, F. Sterol carrier protein-2 mediated cholesterol esterification in transfected L-cell fibroblasts. *Biochim. Biophys. Acta* **1997**, *1345*, 283–292. [CrossRef]

97. Baum, C.L.; Reschly, E.J.; Gayen, A.K.; Groh, M.E.; Schadick, K. Sterol carrier protein-2 overexpression enhances sterol cycling and inhibits cholesterol ester synthesis and high density lipoprotein cholesterol secretion. *J. Biol. Chem.* **1997**, *272*, 6490–6498. [CrossRef]

98. Zanlungo, S.; Amigo, L.; Mendoza, H.; Miquel, J.F.; Vío, C.; Glick, J.M.; Rodríguez, A.; Kozarsky, K.; Quiñones, V.; Rigotti, A.; et al. Sterol carrier protein 2 gene transfer changes lipid metabolism and enterohepatic sterol circulation in mice. *Gastroenterology* **2000**, *119*, 1708–1719. [CrossRef]

99. Seedorf, U.; Raabe, M.; Ellinghaus, P.; Kannenberg, F.; Fobker, M.; Engel, T.; Denis, S.; Wouters, F.; Wirtz, K.W.A.; Wanders, R.J.A.; et al. Defective peroxisomal catabolism of branched fatty acyl coenzyme A in mice lacking the sterol carrier protein-2 sterol carrier protein-x gene function. *Genes Dev.* **1998**, *12*, 1189–1201. [CrossRef] [PubMed]

100. Puglielli, L.; Rigotti, A.; Amigo, L.; Nuñez, L.; Greco, A.V.; Santos, M.J.; Nervi, F. Modulation of intrahepatic cholesterol trafficking: Evidence by in vivo antisense treatment for the involvement of sterol carrier protein-2 in newly synthesized cholesterol transport into rat bile. *Biochem. J.* **1996**, *317 Pt 3*, 681–687. [CrossRef]

101. Chanderbhan, R.; Noland, B.J.; Scallen, T.J.; Vahouny, G.V. Sterol carrier protein2. Delivery of cholesterol from adrenal lipid droplets to mitochondria for pregnenolone synthesis. *J. Biol. Chem.* **1982**, *257*, 8928–8934. [CrossRef]

102. Yamamoto, R.; Kallen, C.B.; Babalola, G.O.; Rennert, H.; Billheimer, J.T.; Strauss, J.F. Cloning and expression of a cDNA encoding human sterol carrier protein 2. *Proc. Natl. Acad. Sci. USA* **1991**, *88*, 463–467. [CrossRef]

103. Rennert, H.; Amsterdam, A.; Billheimer, J.T.; Strauss, J.F. Regulated expression of sterol carrier protein 2 in the ovary: A key role for cyclic AMP. *Biochemistry* **1991**, *30*, 11280–11285. [CrossRef]

104. Trzeciak, W.H.; Simpson, E.R.; Scallen, T.J.; Vahouny, G.V.; Waterman, M.R. Studies on the synthesis of sterol carrier protein-2 in rat adrenocortical cells in monolayer culture. Regulation by ACTH and dibutyryl cyclic 3′,5′-AMP. *J. Biol. Chem.* **1987**, *262*, 3713–3717. [CrossRef]

105. McLean, M.P.; Puryear, T.K.; Khan, I.; Azhar, S.; Billheimer, J.T.; Orly, J.; Gibori, G. Estradiol regulation of sterol carrier protein-2 independent of cytochrome P450 side-chain cleavage expression in the rat corpus luteum. *Endocrinology* **1989**, *125*, 1337–1344. [CrossRef] [PubMed]

106. van Noort, M.; Rommerts, F.F.; van Amerongen, A.; Wirtz, K.W. Localization and hormonal regulation of the non-specific lipid transfer protein (sterol carrier protein2) in the rat testis. *J. Endocrinol.* **1986**, *109*, R13–R16. [CrossRef]

107. Gallegos, A.M.; Schoer, J.K.; Starodub, O.; Kier, A.B.; Billheimer, J.T.; Schroeder, F. A potential role for sterol carrier protein-2 in cholesterol transfer to mitochondria. *Chem. Phys. Lipids* **2000**, *105*, 9–29. [CrossRef]

108. Schroeder, F.; Myerspayne, S.C.; Billheimer, J.T.; Wood, W.G. Probing the ligand-binding sites of fatty acid and sterol carrier proteins-effects of ethanol. *Biochemistry* **1995**, *34*, 11919–11927. [CrossRef]

109. McArthur, M.J.; Atshaves, B.P.; Frolov, A.; Foxworth, W.D.; Kier, A.B.; Schroeder, F. Cellular uptake and intracellular trafficking of long chain fatty acids. *J. Lipid Res.* **1999**, *40*, 1371–1383. [CrossRef]

110. Murphy, E.J. Sterol carrier protein-2 expression increases NBD-stearate uptake and cytoplasmic diffusion in L cells. *Am. J. Physiol.* **1998**, *275*, G237–G243. [CrossRef] [PubMed]

111. Atshaves, B.P.; Storey, S.M.; Schroeder, F. Sterol carrier protein-2/sterol carrier protein-x expression differentially alters fatty acid metabolism in L cell fibroblasts. *J. Lipid Res.* **2003**, *44*, 1751–1762. [CrossRef]

112. Murphy, E.J.; Stiles, T.; Schroeder, F. Sterol carrier protein-2 expression alters phospholipid content and fatty acyl composition in L-cell fibroblasts. *J. Lipid Res.* **2000**, *41*, 788–796. [CrossRef]

113. Gossett, R.E.; Frolov, A.A.; Roths, J.B.; Behnke, W.D.; Kier, A.B.; Schroeder, F. Acyl-CoA binding proteins: Multiplicity and function. *Lipids* **1996**, *31*, 895–918. [CrossRef] [PubMed]

114. Knudsen, J.; Jensen, M.V.; Hansen, J.K.; Faergeman, N.J.; Neergaard, T.B.; Gaigg, B. Role of acylCoA binding protein in acylCoA transport, metabolism and cell signaling. *Mol. Cell. Biochem.* **1999**, *192*, 95–103. [CrossRef] [PubMed]

115. Starodub, O.; Jolly, C.A.; Atshaves, B.P.; Roths, J.B.; Murphy, E.J.; Kier, A.B.; Schroeder, F. Sterol carrier protein-2 localization in endoplasmic reticulum and role in phospholipid formation. *Am. J. Physiol. Cell. Physiol.* **2000**, *279*, C1259–C1269. [CrossRef]

116. Schroeder, F.; Zhou, M.; Swaggerty, C.L.; Atshaves, B.P.; Petrescu, A.D.; Storey, S.M.; Martin, G.G.; Huang, H.; Helmkamp, G.M.; Ball, J.M. Sterol carrier protein-2 functions in phosphatidylinositol transfer and signaling. *Biochemistry* **2003**, *42*, 3189–3202. [CrossRef] [PubMed]

117. Atshaves, B.P.; Jefferson, J.R.; McIntosh, A.L.; Gallegos, A.; McCann, B.M.; Landrock, K.K.; Kier, A.B.; Schroeder, F. Effect of sterol carrier protein-2 expression on sphingolipid distribution in plasma membrane lipid rafts/caveolae. *Lipids* **2007**, *42*, 871–884. [CrossRef]

118. Bloj, B.; Zilversmit, D.B. Accelerated transfer of neutral glycosphingolipids and ganglioside GM1 by a purified lipid transfer protein. *J. Biol. Chem.* **1981**, *256*, 5988–5991. [CrossRef]

119. Fong, T.H.; Wang, S.M.; Lin, H.S. Immunocytochemical demonstration of a lipid droplet-specific capsule in cultured Leydig cells of the golden hamsters. *J. Cell. Biochem.* **1996**, *63*, 366–373. [CrossRef]

120. Seedorf, U.; Brysch, P.; Engel, T.; Schrage, K.; Assmann, G. Sterol carrier protein X is peroxisomal 3-oxoacyl coenzyme A thiolase with intrinsic sterol carrier and lipid transfer activity. *J. Biol. Chem.* **1994**, *269*, 21277–21283. [CrossRef]

121. Ferdinandusse, S.; Denis, S.; Faust, P.L.; Wanders, R.J. Bile acids: The role of peroxisomes. *J. Lipid Res.* **2009**, *50*, 2139–2147. [CrossRef] [PubMed]

122. Atshaves, B.P.; McIntosh, A.L.; Landrock, D.; Payne, H.R.; Mackie, J.T.; Maeda, N.; Ball, J.; Schroeder, F.; Kier, A.B. Effect of SCP-x gene ablation on branched-chain fatty acid metabolism. *Am. J. Physiol. Gastrointest. Liver Physiol.* **2007**, *292*, G939–G951. [CrossRef] [PubMed]
123. Ferdinandusse, S.; Kostopoulos, P.; Denis, S.; Rusch, H.; Overmars, H.; Dillmann, U.; Reith, W.; Haas, D.; Wanders, R.J.A.; Duran, M.; et al. Mutations in the gene encoding peroxisomal sterol carrier protein X (SCPx) cause leukencephalopathy with dystonia and motor neuropathy. *Am. J. Hum. Genet.* **2006**, *78*, 1046–1052. [CrossRef] [PubMed]
124. Horvath, R.; Lewis-Smith, D.; Douroudis, K.; Duff, J.; Keogh, M.; Pyle, A.; Fletcher, N.; Chinnery, P.F. SCP2 Mutations and neurodegeneration with brain iron accumulation. *Neurology* **2015**, *85*, 1909–1911. [CrossRef] [PubMed]
125. Galano, M.; Ezzat, S.; Papadopoulos, V. SCP2 variant is associated with alterations in lipid metabolism, brainstem neurodegeneration, and testicular defects. *Hum. Genom.* **2022**, *16*, 32. [CrossRef]

International Journal of
Molecular Sciences

Article

A Short Promoter Region Containing Conserved Regulatory Motifs Is Required for Steroidogenic Acute Regulatory Protein (*Star*) Gene Expression in the Mouse Testis

Marie France Bouchard [1], Julia Picard [1], Jacques J. Tremblay [1,2] and Robert S. Viger [1,2,*]

[1] Reproduction, Mother and Child Health, Centre de Recherche du CHU de Québec-Université Laval and Centre de Recherche en Reproduction, Développement et Santé Intergénérationnelle (CRDSI), Québec, QC GIV4G2, Canada
[2] Department of Obstetrics, Gynecology, and Reproduction, Université Laval, Québec, QC G1K7P4, Canada
* Correspondence: robert.viger@crchudequebec.ulaval.ca; Tel.: +1-418-525-4444

Abstract: In the testis, Leydig cells produce steroid hormones that are needed to masculinize typical genetic males during fetal development and to initiate and maintain spermatogenesis at puberty and adulthood, respectively. Steroidogenesis is initiated by the transfer of cholesterol from the outer to the inner mitochondrial membrane through the action of steroidogenic acute regulatory protein (STAR). Given its importance for the steroidogenic process, the regulation of *STAR* gene expression has been the subject of numerous studies. These studies have involved the characterization of key promoter sequences through the identification of relevant transcription factors and the nucleotide motifs (regulatory elements) that they bind. This work has traditionally relied on in vitro studies carried out in cell cultures along with reconstructed promoter sequences. While this approach has been useful for developing models of how a gene might be transcriptionally regulated, one must ultimately validate that these modes of regulation occur in an endogenous context. We have used CRISPR/Cas9 genome editing to modify a short region of the mouse *Star* promoter (containing a subset of regulatory elements, including conserved CRE, C/EBP, AP1, and GATA motifs) that has been proposed to be critical for *Star* transcription. Analysis of the resultant mutant mice showed that this short promoter region is indeed required for maximal STAR mRNA and protein levels in the testis. Analysis also showed that both basal and hormone-activated testosterone production in mature mice was unaffected despite significant changes in *Star* expression. Our results therefore provide the first in vivo validation of regulatory sequences required for *Star* gene expression.

Keywords: steroidogenesis; testis; transcription; testosterone; Leydig cell; GATA

Citation: Bouchard, M.F.; Picard, J.; Tremblay, J.J.; Viger, R.S. A Short Promoter Region Containing Conserved Regulatory Motifs Is Required for Steroidogenic Acute Regulatory Protein (*Star*) Gene Expression in the Mouse Testis. *Int. J. Mol. Sci.* **2022**, 23, 12009. https://doi.org/10.3390/ijms231912009

Academic Editor: Walter Wahli

Received: 30 September 2022
Accepted: 7 October 2022
Published: 9 October 2022

Publisher's Note: MDPI stays neutral with regard to jurisdictional claims in published maps and institutional affiliations.

1. Introduction

Testosterone is the main sex hormone produced by the mature testis. It is essential for establishing biological maleness in typical XY males: development of the male urogenital tract during fetal development, acquisition of male secondary sex characteristics at puberty, and fertility in adults. Testosterone is synthesized and secreted by Leydig cells present in the testicular interstitium through the process of steroidogenesis. Steroidogenesis is the multistep conversion of cholesterol into steroid hormones via the sequential action of multiple proteins/enzymes. Steroidogenesis needs to be tightly regulated as too little or too much steroid hormone production can lead to incidences of differences of sex development (DSD) or pathologies such as congenital adrenal hyperplasia (CAH), osteopenia, and hormone-related cancers (reviewed in [1]). The regulation of testicular steroidogenesis is complex and involves many regulatory molecules and mechanisms such as luteinizing hormone (LH), its signaling pathways, and the factors that interpret these signals [1]. The latter include most notably transcription factors acting on the expression of genes that encode steroidogenic enzymes and other proteins involved in the biosynthesis of testosterone (reviewed in [2,3]).

Members of the GATA family of transcription factors are emerging as important regulators of steroidogenesis (reviewed in [4–6]). The GATA family is comprised of six factors (GATA1 to 6) that recognize and bind to the DNA motif (A/T)GATA(A/G) through their two conserved zinc finger domains. GATA factors are found in a broad array of tissues where they participate in cell differentiation, organogenesis, and the control of tissue-specific gene expression (reviewed in [4,7]). In both male and female gonads, GATA factors, especially GATA4 and/or GATA6, are crucial for the formation of the urogenital ridge, sex determination (gonad differentiation), fertility, and most likely steroidogenesis (reviewed in [4,7,8]). Insights into the GATA-dependence of steroidogenesis have come mainly from studies done in whole testis or immortalized Leydig cell lines. For example, GATA4 knockdown in either MA-10 or MLTC-1 Leydig cells represses the steroidogenic gene expression program and ultimately the production of sex steroid precursors [9,10]. In mice, loss of GATA4 and/or GATA6 function in the testis appears to block steroidogenic cell development and concomitant undermasculization of male embryos [11]. One of the testicular genes shown to be prominently affected by a modulation of GATA4/6 gene expression or activity is the gene encoding the steroidogenic acute regulatory protein (STAR).

Steroidogenic acute regulatory protein (STAR or STARD1) is a member of the larger START domain family, a group of transport proteins that share a common functional domain, the STAR-related lipid-transfer domain. In humans, 15 members (STARTD1-15) of the START domain family are known to transport cholesterol, ceramide, phosphatidylcholine, phosphatidylethanolamine (PE), and bile acids (reviewed in [12]). The first STARD protein discovered was the STAR or STARD1 protein which transports cholesterol from the outer to the inner mitochondrial membrane in steroidogenic cells of both the adrenals and gonads (reviewed in [13,14]). Cholesterol transport into the mitochondria is the rate-limiting step of steroidogenesis. STAR function in steroidogenesis was confirmed by both human disease (lipoid congenital adrenal hyperplasia/lipoid CAH) where the *STAR* gene is mutated [14,15], and in *Star* null mice which have a similar steroidogenic defect as seen in human lipoid CAH [16]. Being the rate-limiting step in steroidogenesis, STAR expression and activity is acutely regulated at both the transcriptional and post-transcriptional levels (reviewed in [13]).

The transcriptional control of the *STAR* gene has been intensely studied (reviewed in [13]). The proximal *STAR* promoter contains many tightly clustered regulatory motifs for the binding of different transcription factors that have been shown to modulate *STAR* promoter activity in vitro. These include motifs for the binding of NR5A factors (SF1/NR5A1 and LRH1/NR5A2), NR4A factors (NUR77/NGFI-B/NR4A1, NURR1/NR4A2, and NOR1/NR4A3), CCAAT/enhancer binding protein β (C/EBP β), CREB family factors (cAMP response element (CRE)-binding protein (CREB), CRE modulator (CREM), and sterol element binding protein (SREBP)), activator protein 1 (AP1), dosage-sensitive sex reversal, adrenal hypoplasia critical region, on chromosome X, gene 1 (DAX1), and Ying Yang 1 (YY1) (reviewed in [13]). As previously mentioned, GATA factors, mostly notably GATA4, are known to bind to the proximal *Star* promoter and enhance its transcription in vitro both basally and in response to acute hormonal stimulation [17,18]. The endogenous *Star* gene, however, has not yet been validated as a direct target for GATA binding or other transcription factors proposed to modulate its expression.

In this study, we report a novel mouse model created by CRISPR/Cas9 genome editing that modifies the mouse proximal *Star* promoter region to inactivate the critical GATA-binding motif as well as an adjacent 19-bp deletion that removes additional CRE, C/EBP, and AP1 binding motifs. Our results show that this short promoter region containing the GATA motif is required for endogenous *Star* expression in both fetal and postnatal mouse testes.

2. Results

2.1. The Integrity of a Short Promoter Region Harboring a Conserved GATA-Binding Motif Is Essential for Maximal Star Gene Expression in the Mouse Testis

Over the last two decades, the importance of GATA factors for the development and functioning of the mammalian gonads has been demonstrated through the analysis of GATA loss-of-function models in mice as well as the in vitro characterization of the GATA-dependence of the promoter regions of potential target genes (reviewed in [4,7,8]). The validation of these genes, however, as direct targets for GATA factors has been more of a challenge. CRISPR/Cas9 genome editing now offers the possibility of directly addressing this important question by allowing the precise targeting and modification of the genomic regions where GATA factors bind to their target genes. We used this strategy to define the anti-Müllerian hormone gene as a genuine target for GATA binding in Sertoli cells of the testis [19]. We now describe here the generation of a new mouse model using CRISPR/Cas9 editing to assess the importance of GATA binding for *Star* expression. As overviewed in Figure 1, the proximal mouse *Star* promoter contains a lone conserved GATA binding motif. Two sgRNAs were used in microinjections to maximize the chances of obtaining genome modifications that targeted the *Star* GATA motif. Less than 2% of founder mice (2 out of 105 births) were successfully targeted by the sgRNAs. Of these 2 founders, one was successfully repaired by the donor ssODN and contained the desired mutation of the GATA motif of the *Star* promoter. This same mouse also presented a 19-bp deletion created by nonhomologous end joining immediately upstream of the mutated GATA motif that also inactivated CRE, C/EBP, and AP1 binding motifs (Figure 1). A second founder mouse presented a longer 48-bp deletion that removed an SF1/NR5A1 motif in addition to those for CRE, AP1, and GATA. Both founders were viable and fertile. However, only the first founder (which we named p*Star*$^{\Delta19\text{-GATAmut}}$) transmitted the modified allele to its progeny. Mice heterozygous for the Δ19-GATA mutation were crossed to generate homozygotes; mice resulting from this cross were born at the expected Mendelian frequencies with a male/female ratio of 1:1. Both male and female p*Star*$^{\Delta19\text{-GATAmut}}$ heterozygous and homozygous mice were visibly undisguisable from wild-type (WT) littermates. Mature homozygous p*Star*$^{\Delta19\text{-GATAmut}}$ mice from both sexes were also fully fertile and showed no indication of adrenal insufficiency that is characteristic of *Star* null mice [16].

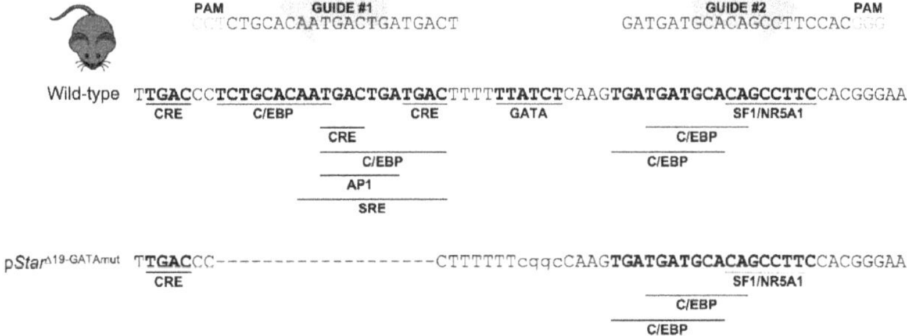

Figure 1. CRISPR/Cas9 genome editing strategy to target a conserved GATA regulatory element in the mouse *Star* promoter. The sgRNA and associated PAM sequences used are shown flanking the *Star* promoter GATA motif; regulatory motifs are underlined. Of the more than 100 founder mice screened, only 1 (named p*Star*$^{\Delta19\text{-GATAmut}}$) had a significant modification—a short 19-bp deletion generated by nonhomologous end-joining repair (NHEJ) and an adjacent mutated GATA motif introduced by the ssODN template during homology-directed repair (HDR). The 19-bp deletion removed CRE, C/EBP, AP1, and SRE binding motifs. Breeding of the p*Star*$^{\Delta19\text{-GATAmut}}$ founder with C57BL/6J mice confirmed transmission of the mutant allele.

Despite exhibiting no overt phenotype, we examined whether *Star* gene expression was nonetheless affected in p*Star*$^{\Delta19\text{-GATAmut}}$ males. *Star* mRNA levels were first assessed in male gonads at embryonic day 15 (E15.5) and E18.5, time points that cover the critical masculinization programming window and the fetal testosterone surge in the mouse. *Star* mRNA levels in homozygous p*Star*$^{\Delta19\text{-GATAmut}}$ males were significantly reduced at both E15.5 (Figure 2A; 55% decrease) and E18.5 (Figure 2B; 79% decrease) when compared to testes from age-matched WT mice. We then performed immunohistochemistry on fetal testis sections to ascertain whether the reduction in *Star* mRNA levels translated to a similar decrease at the protein level (Figure 3). Beginning at E13.5, a time point when testis differentiation is complete and when testosterone production begins, there was no difference in the pattern of STAR protein (localization and level) between WT and p*Star*$^{\Delta19\text{-GATAmut}}$ mice. No change was also observed at E18.5 when the fetal testis is actively producing testosterone. We then examined testis *Star* expression and STAR protein in WT and mutant mice at later life stages (Figure 4): (1) just after puberty (P35) when Leydig cells are ramping up testosterone production and seminiferous tubules have engaged their first cycle of spermatogenesis, and (2) at adulthood (P90) when both testosterone levels and spermatogenic output are at their peak. As we observed for the fetal testis, *Star* mRNA levels were significantly lower in p*Star*$^{\Delta19\text{-GATAmut}}$ mice than in the WT counterparts—both in P35 juvenile adults (Figure 4A; 40% less) and P90 mature adults (Figure 4B; 75% less). Western blotting on whole extracts from adult testes also showed a consistent and marked reduction in STAR protein in p*Star*$^{\Delta19\text{-GATAmut}}$ mice when compared to age-matched WT control animals (Figure 4C). Immunohistochemistry, however, revealed no change in cellular localization of STAR protein between p*Star*$^{\Delta19\text{-GATAmut}}$ and WT testes at both postnatal ages (Figure 5).

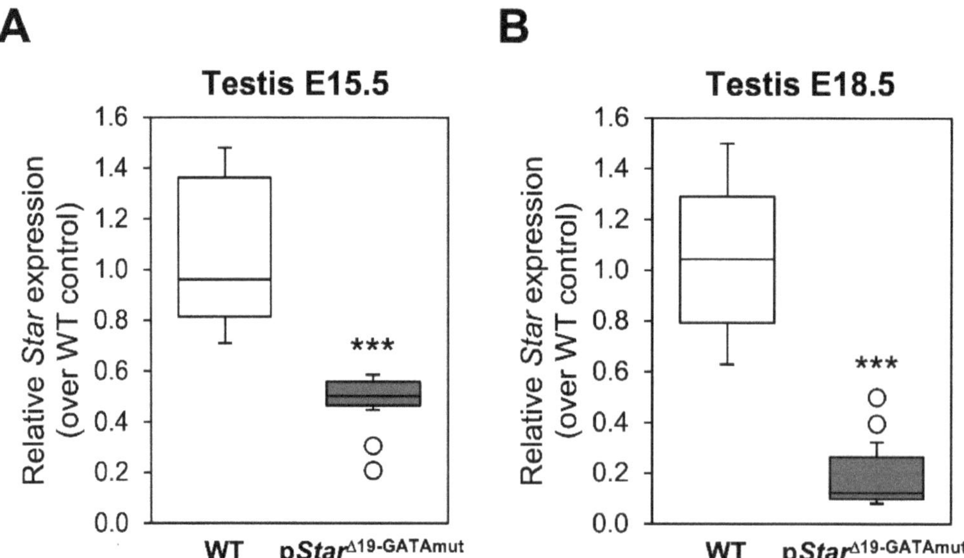

Figure 2. Comparison of *Star* gene expression in WT and p*Star*$^{\Delta19\text{-GATAmut}}$ fetal testis. *Star* expression was assessed by qPCR at (**A**) E15.5 and (**B**) E18.5. Data are reported as the value relative to age-matched WT mice (n = 6 to 11). Open circles are outliers. ***, significantly different from WT mice ($p < 0.001$).

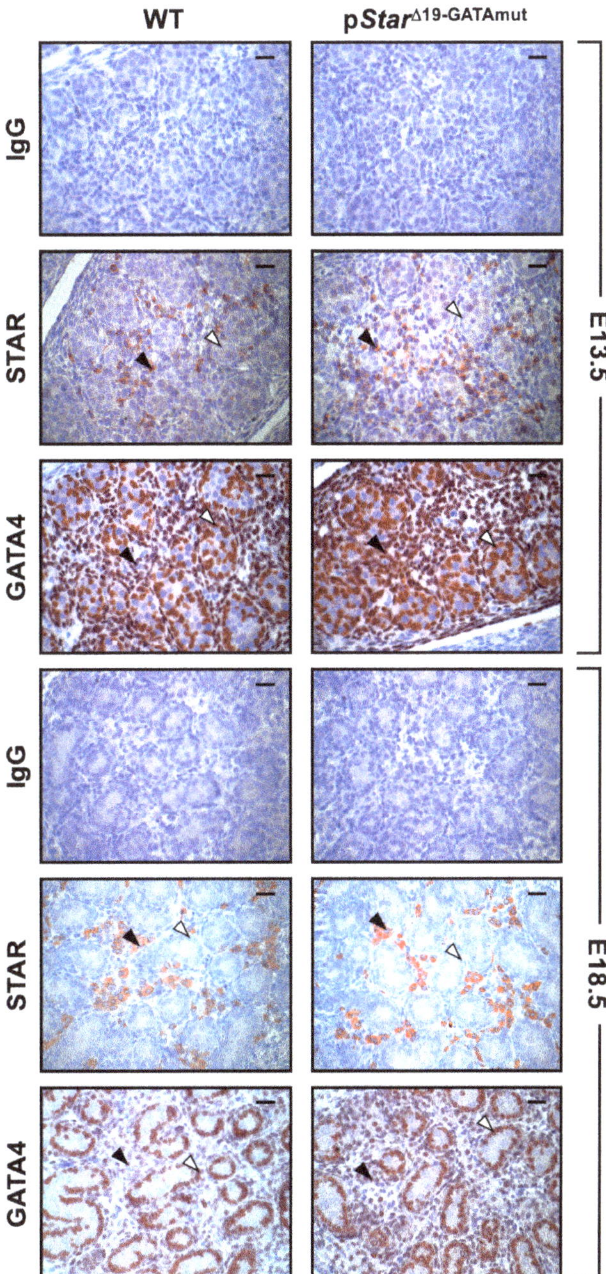

Figure 3. In situ detection of STAR and GATA4 protein in WT and p*Star*^Δ19-GATAmut fetal testes. Immunohistochemistry was performed on paraffin testis sections from both mouse genotypes at E13.5 and E18.5 using antibodies specific for STAR or GATA4; rabbit IgG was used as a negative control. Images were taken at 400× magnification; bar = 25 μm. Open arrowheads, GATA4-positive Sertoli cells; filled arrowheads, STAR-positive Leydig cells.

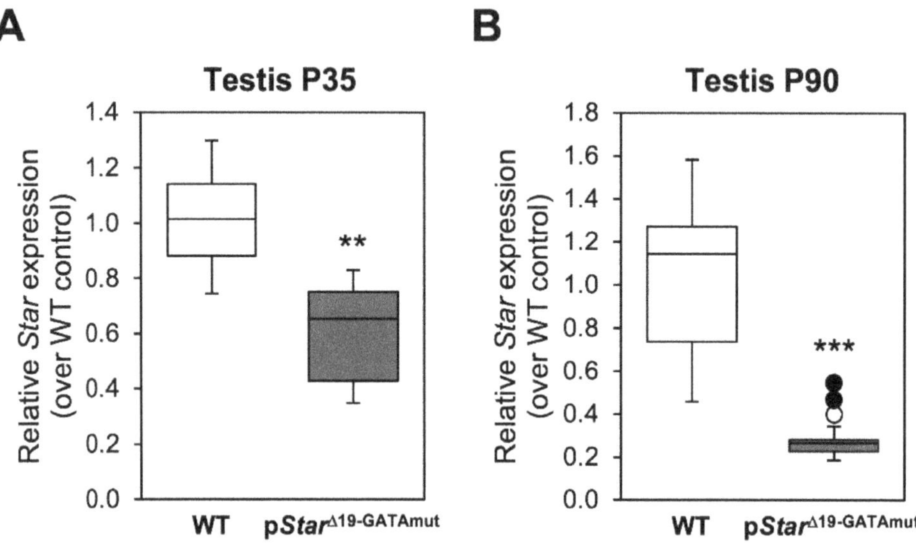

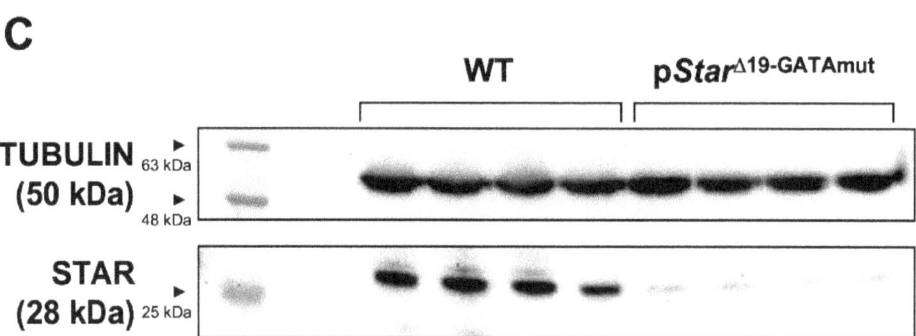

Figure 4. *Star* mRNA expression and STAR protein in postnatal testis of WT and p*Star*^Δ19-GATAmut mice. *Star* expression was assessed by qPCR in (**A**) juvenile P35 and (**B**) adult P90 testes. Data are reported as the value relative to age-matched WT mice (n = 6 to 10). The open circle is an outlier; filled circles are definitive outliers. Significantly different from WT: ** $p < 0.01$ for P35 testis and *** $p < 0.001$ for P90 testis. (**C**) Western blot detection of STAR protein levels in WT and p*Star*^Δ19-GATAmut adult mouse testis. TUBULIN was used as a loading control.

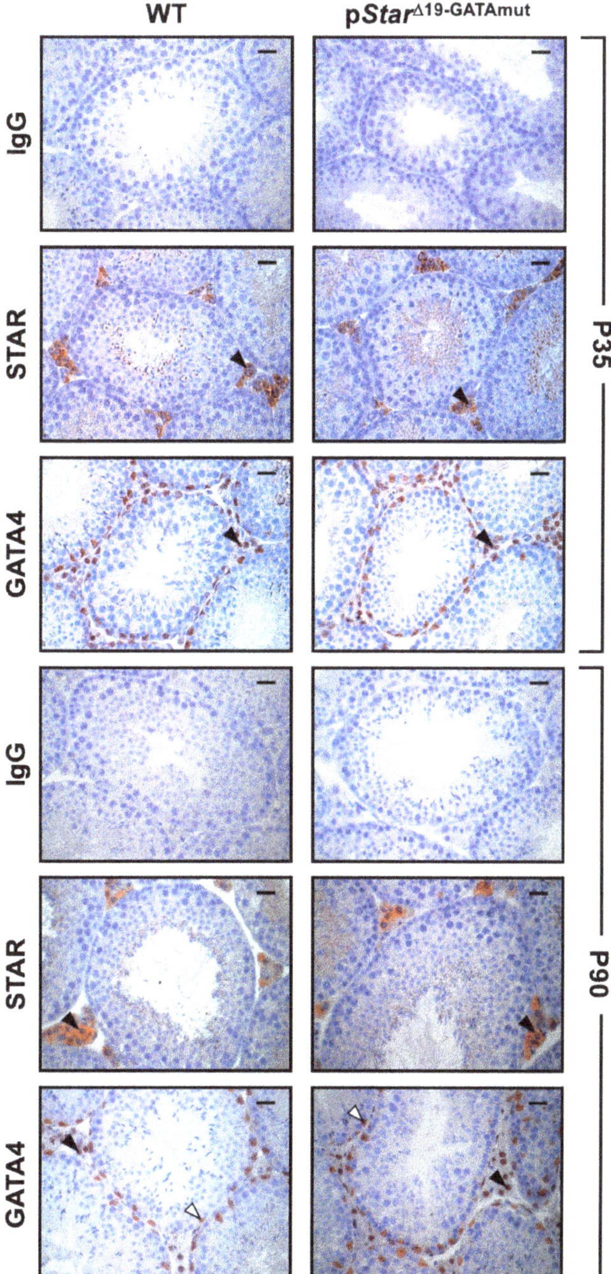

Figure 5. In situ detection of STAR and GATA4 protein in WT and p*Star*$^{\Delta19\text{-GATAmut}}$ postnatal testes. Immunohistochemistry was performed on paraffin testis sections obtained from both mouse genotypes at P35 and P90 using antibodies specific for STAR or GATA4; rabbit IgG was used as a negative control. Images were taken at 400× magnification; bar = 25 μm. Open arrowheads, GATA4-positive Sertoli cells; filled arrowheads, STAR-positive Leydig cells.

2.2. Changes in Star mRNA and STAR Protein Levels in pStar$^{\Delta 19\text{-}GATAmut}$ Mice Are Not Correlated with Alterations in Basal or Hormone-Induce Testosterone Production

In steroidogenic cells, the transfer of cholesterol from the outer to the inner membrane of the mitochondria mediated by STAR is the rate-limiting step of steroid biosynthesis. Therefore, reduced *Star* mRNA and/or STAR protein levels observed in p*Star*$^{\Delta 19\text{-}GATAmut}$ testes might compromise steroidogenesis and lead to a reduction in testosterone production. To test this hypothesis, we compared basal plasma and intratesticular testosterone levels in adult P90 WT and p*Star*$^{\Delta 19\text{-}GATAmut}$ male mice (Figure 6). Although both basal plasma and intratesticular T were slightly reduced in p*Star*$^{\Delta 19\text{-}GATAmut}$ males, the difference with WT mice was not statistically significant. Testicular *STAR* expression and activity are acutely regulated by the gonadotropin LH (reviewed in [13,20]). LH stimulation also induces steroidogenesis in the testis, which triggers numerous signaling pathways, which in turn can activate several transcription factors including GATA4 (reviewed in [2]). We therefore hypothesized that acute *Star* regulation might be impaired in testes of p*Star*$^{\Delta 19\text{-}GATAmut}$ mice. Testis explants from both WT and p*Star*$^{\Delta 19\text{-}GATAmut}$ adult P90 mice were placed in culture and exposed to a stimulating dose of hCG or left untreated. After 4 h in culture, we assessed the amount of testosterone released by the tissue ex vivo explants (Figure 7). Although testosterone production was induced after hCG treatment, there was no significant difference in hCG responsiveness between WT and p*Star*$^{\Delta 19\text{-}GATAmut}$ testes.

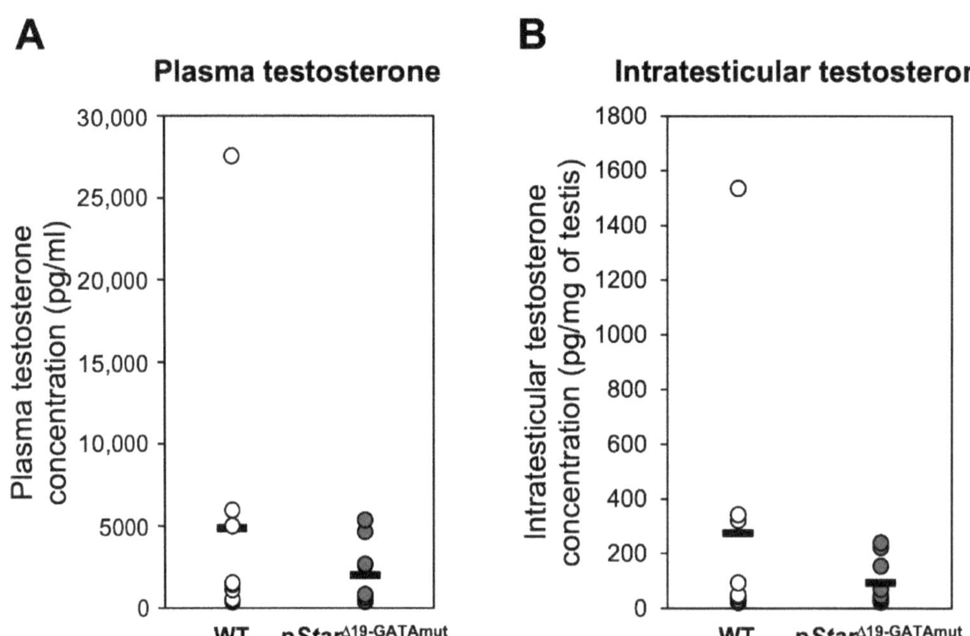

Figure 6. Testosterone levels in WT (open circles) and p*Star*$^{\Delta 19\text{-}GATAmut}$ (filled circles) adult P90 male mice. (**A**) Plasma testosterone is expressed as ng/mL (n = 9). (**B**) Intratesticular testosterone is expressed as pg/mg testis (n = 9); bar = average. No significant difference was observed between WT and p*Star*$^{\Delta 19\text{-}GATAmut}$ mice for both plasma and intratesticular testosterone levels.

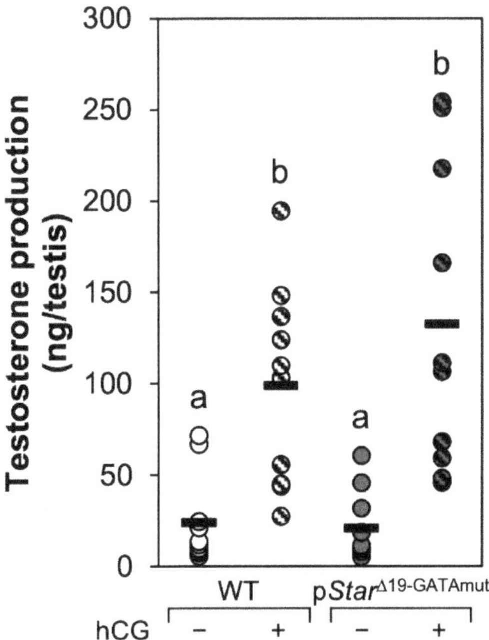

Figure 7. Ability of adult WT and p*Star*^Δ19-GATAmut testis explants to respond to hCG stimulation. Detunicated adult testis from WT or p*Star*^Δ19-GATAmut mice were placed in culture medium with antibiotics and treated with or without 1 IU/mL of hCG. After 4 h, culture media were also collected and assayed for testosterone by ELISA. Testosterone production is expressed as ng/testis (n = 10). Each individual sample is represented by a circle; bar = average. Open circles, WT in the absence of hCG; open hatched circles, WT treated with hCG; filled circles, p*Star*^Δ19-GATAmut in the absence of hCG; filled hatched circles, p*Star*^Δ19-GATAmut treated wikth hCG. Groups identified by different letters are significantly different, $p < 0.05$.

3. Discussion

Steroidogenic acute regulatory protein (STAR), identified nearly three decades ago, is an essential cholesterol transporter in all steroidogenic tissues [21]. The delivery of cholesterol from the outer to the inner mitochondrial membrane mediated by STAR is the rate-limiting step in steroidogenesis [13]. As such, many studies have been devoted to understanding how *STAR* gene expression and protein activity are regulated in steroid producing tissues such as the adrenals and gonads. At the transcriptional level, the regulation of the *STAR* gene is complex, involving the interaction of numerous transcription factors to species-conserved binding motifs that are tightly clustered, and sometimes overlapping, located within the first few hundred base pairs upstream of the transcription initiation site (reviewed in [13]). Although many transcription factors have been shown to participate in *STAR* transcription across many species, to our knowledge, this evidence has been essentially limited to in vitro studies performed in cell lines or isolated steroidogenic tissues. No studies have yet probed the importance of these transcription factor binding sites for *Star* transcription in an in vivo whole animal context. In the present study, we have generated the first such mouse model (p*Star*^Δ19-GATAmut), using CRISPR/Cas9 genome editing, to inactivate a short region of the mouse *Star* promoter containing a subset of these transcription factor binding sites. Analysis of the mutant mice confirmed that they are indeed essential for maximal expression of the endogenous *Star* gene but not basal or hormone-activated testosterone production.

The initial goal of our genome editing effort was to inactivate the lone GATA regulatory motif in the mouse *Star* promoter that has been long proposed to be critical for *Star* transcription [18]. However, even after screening more than 100 founder mice, we were unable to exclusively alter the *Star* GATA-binding motif. This was not totally unexpected given the low rate of homology directed repair (HDR) [22]. Despite this, we were successful in creating a mouse (p*Star*$^{\Delta 19\text{-GATAmut}}$) with a short 19-bp deletion in tandem with a mutated GATA motif (see Figure 1). The deleted region contains six additional regulatory motifs—many of them overlapping—for various transcription factors, including members of the C/EBP and CREB families. The sequence covered by the Δ19-GATA mutation spans one of two critical regions that largely govern the positive modulation of *STAR* expression as documented through in vitro studies (reviewed in [20,23]). The analysis of our p*Star*$^{\Delta 19\text{-GATAmut}}$ mice confirmed that this region is indeed important for maximal *Star* expression in fetal, peripubertal, and adult testes. However, we cannot at present formally attribute the decrease in *Star* expression to any one of these specific motifs since they were simultaneously disrupted in p*Star*$^{\Delta 19\text{-GATAmut}}$ mice. Contrasting our in vivo findings with existing in vitro data do suggest that the CRE motifs are likely important—MA-10 Leydig cells transfected with CREB increases *Star* mRNA levels while mutation of at least the middle CRE motif (CRE2 in ref. [24]) reduces *Star* promoter activity by approximately 50% [24], a decrease comparable to what we observed for *Star* mRNA in p*Star*$^{\Delta 19\text{-GATAmut}}$ testes. The proximal *Star* promoter also contains two C/EBP binding sites; one of these sites (site C2 identified in ref. [25]) was deleted in our p*Star*$^{\Delta 19\text{-GATAmut}}$ mice. Mutation of this C/EBP motif significantly reduces *Star* promoter activity in MA-10 Leydig cells [25]. The same can be said for the GATA motif—the *Star* promoter is potently activated by GATA factors in cells, an effect that is completely lost when motif is mutated [18]. Considering the short length of the Δ19-GATAmut region as well as the very close proximity of the regulatory motifs present therein, it is reasonable to speculate that these motifs are not simultaneously bound by different transcription factors but rather are occupied by dynamic binding. Moreover, the proximity of the regulatory motifs further implies a physical closeness of the various DNA binding proteins that would permit the proteins to directly interact. Indeed, the existence of regulatory complexes among the different transcription factors has been well-documented. For example, C/EBP and CREB proteins interact with GATA4 to increase *Star* promoter activity, at least in vitro [17,26]. Therefore, we can now conclude that the integrity of this short region is critical not only for in vitro *Star* promoter activity but also for *Star* transcription in both fetal and postnatal testis in the mouse.

Although STAR protein is plentiful in Leydig cells, it was technically difficult for us to quantify it in early fetal testes. However, in adult testes, STAR was clearly less abundant in p*Star*$^{\Delta 19\text{-GATAmut}}$ than in WT mice. Immunohistochemistry also showed that this was not accompanied by a change in cellular localization, suggesting that although less abundant, it should remain functional. Based on these observations, we expected that this would translate into reduced testosterone production either in plasma or locally within the testis itself. However, for both intratesticular and circulating (plasma) testosterone, levels in p*Star*$^{\Delta 19\text{-GATAmut}}$ adult mice were not significantly different from WT. This was somewhat surprising knowing that genetic male *Star* null fetal mice have feminized external genitalia which is suggestive of a deficit in the production of androgen precursors from fetal Leydig cells [16]. Moreover, testosterone levels are 10 times lower at 8 weeks in serum from glucocorticoid-rescued *Star* null males when compared to age-matched controls [27]. This indicates that despite a 50–75% decrease in *Star* mRNA or STAR protein observed in p*Star*$^{\Delta 19\text{-GATAmut}}$ testes, the amount of STAR remaining must be sufficient to allow for normal steroidogenesis. In steroidogenic tissues, *STAR* transcription is rapidly induced by the gonadotropin LH acting via a cAMP signaling pathway [28]. The short promoter region targeted in our p*Star*$^{\Delta 19\text{-GATAmut}}$ mice also harbors regulatory elements known to confer LH responsiveness, including the GATA-binding motif (reviewed in [20]). Based on these facts, we surmised that a steroidogenic deficit might be more easily captured under conditions of hormonal stimulation. Yet again, we observed that ex vivo testis cultures from adult p*Star*$^{\Delta19\text{-GATAmut}}$ mice still responded to hCG stimulation when compared to WT controls, therefore reinforcing the notion that STAR protein, while diminished significantly in amount, was still adequate to elicit an induction of steroidogenesis when stimulated.

Taken together, our results provide new insights into the transcriptional regulation of the *Star* gene and highlight the power of CRISPR/Cas9 genome editing for validating the importance of proposed promoter regulatory sequences for both basal and acute hormone-stimulated gene regulation.

4. Materials and Methods

4.1. Animals

All mouse experiments were carried out in accordance with the Canadian Council of Animal Care guidelines for the care and manipulation of animals used in research. Protocols were approved by the Comité de Protection des Animaux de l'Université Laval, Québec, QC, Canada (protocol nos. 2019-149 and CHU-19-046).

4.2. CRISPR/Cas9 Generation of Mouse Star Promoter Mutants

A 100-bp genomic region of the proximal murine *Star* promoter spanning a single conserved GATA binding motif was searched using the CRISPOR Web tool (www.crispor.tefor.net, accessed on 15 June 2019) for potential single-guide RNA (sgRNA) sequences [29]. Guides were selected based on their low predicted off-target potential. A pX330-U6-Chimeric_BBCBh-hSpCas9 plasmid (no. 42230) purchased from Addgene (Cambridge, MA, USA) was used to generate SpCas9/chimeric sgRNA expression plasmids [30], as previously described [19]. A single-strand oligonucleotide (ssODN) was synthesized as a template for HDR of the double-strand breaks created by the sgRNAs. The ssODN contains a mutated GATA motif of the murine *Star* promoter flanked on each side by ~90-nucleotide-long homology arms. The oligonucleotides used as primers for creating the sgRNAs as well as the ssODN used as a donor for HDR are shown in Table 1 (the GATA motif is underlined and the mutated nucleotides are in lowercase). The Sp-Cas9/chimeric sgRNA constructs were first validated in vitro and then microinjected along with the ssODN into fertilized C57BL/6J mouse eggs using the microinjection and transgenesis platform of the Institut de Recherches Cliniques de Montréal. A total of 105 founder mice were born and analyzed. Genomic DNA was isolated from tail tips collected from the founder mice using the HotSHOT method [31]. DNA screened for genomic rearrangements using genotyping primers (listed in Table 1) and Taq FroggaMix 2X master mix (FroggaBio, Concord, ON, Canada). PCR conditions were: initial denaturation for 3 min at 95 °C followed by 35 cycles of denaturation (30 s at 95 °C), annealing (30 s at 68 °C), and extension (30 s at 72 °C), and a final extension for 3 min at 72 °C. Amplicons were sequenced and analyzed using the Web tool TIDE (for tracking indels by decomposition) to evaluate the extent of Cas9-mediated rearrangements that occurred [32] and to identify the desired mutated or deleted alleles. Founder mice that targeted the GATA motif were crossed with C57BL/6J mice (stock no. 000664; The Jackson Laboratory, Bar Harbor, ME) to assess the transmission of the modified alleles. One founder mouse that possessed a combined GATA mutation and 19-bp deletion (named p*Star*$^{\Delta19\text{-GATAmut}}$) successfully transferred the modified *Star* promoter sequence to its offspring. Mice were backcrossed for a minimum of 5 generations with C57BL/6J mice to eliminate potential off-target effects. Heterozygous descendants were crossed to generate homozygous as well as wild-type (WT) control mice for experimentation as well as additional heterozygous mice that were used for colony maintenance.

Table 1. Oligonucleotide primers used for the generation and validation of genetically modified mice.

Utility	Forward Primer	Reverse Primer
Guide 1	5′-CACCGAGTCATCAGTCATTGTGCAG-3′	5′-AAACCTGCACAATGACTGATGACTC-3′
Guide 2	5′-CACCGATGATGCACAGCCTTCCAC-3′	5′-AAACGTGGAAGGCTGTGCATCATC-3′
ssODN	5′-AGTCTGCTCCCTCCCACCTTGGCCAGCACTGCAGGATGAGGCAATCATTCCATCCTTGACGCT CTGCACAATGACTGATGACTTTTTTcggcCAAGTGATGATGCACAGCCTTCCACGGCAAGCATTTA AGGCAGCGCACTTGATCTGCGCCACAGCTGCAGGACTCAGGACCTTGAAAGGCTC-3′	
Genotyping	5′-GCACCTCAGTTACTGGGCAT-3′	5′-ACACAGCTTGAACGTAGCGA-3′

In the ssODN sequence, the mutated GATA motif is underlined.

4.3. Quantitative Real-Time RT-PCR

Testes were dissected from male mice at various developmental ages: embryonic day 15.5 (E15.5), E18.5, postnatal day 35 (P35, juvenile adult), and P90 (adult). Testes from E15.5, E18.5, and P35 testes were processed for total RNA extraction using TRI Reagent solution (Sigma-Aldrich Canada, Oakville, ON, Canada) in accordance with the manufacturer's instructions. Testes from adult male mice were halved while frozen. One half was used for protein extraction (described below) and the other half was used for RNA extraction and intratesticular testosterone quantification (described below under *Hormone assay*). First-strand cDNA was synthesized from total RNA isolated from tissues using the iScript Advanced cDNA synthesis kit for quantitative real-time RT-PCR (qPCR; Bio-Rad Laboratories, Mississauga, ON, Canada). Assessment of gene expression by qPCR was done using a CFX96 plate thermal cycler and SsoAdvanced Universal SYBR Green Supermix from Bio-Rad Laboratories using their standard protocol. A panel of reference genes known for their stability in the mouse gonad [33], as well as typically used reference genes, were used for normalization. Primers for qPCR are listed in Table 2. Primer pairs were optimized beforehand for specificity and efficiency using a temperature gradient to identify the best annealing temperature and by performing a standard curve using a serial dilution of a pool of samples. PCR amplifications were run in duplicate under the following conditions: initial denaturation for 3 min at 95 °C followed by 40 cycles of denaturation (10 s at 95 °C), annealing (20 s at 62.6 °C), and extension (20 s at 72 °C) with a single acquisition of fluorescence level at the end of each extension step. Differences in mRNA levels between the mouse genotypes was determined using the ΔΔCq method [34].

Table 2. Oligonucleotide primers used for qPCR.

Gene Product	Forward Primer	Reverse Primer
Star	5′-CAACTGGAAGCAACACTCTA-3′	5′-CCTTGACATTTGGGTTCCAC-3′
Actb	5′-CTGTCGAGTCGCGTCCACC-3′	5′-ATTCCCACCATCACACCCTGG-3′
Gapdh	5′-GTCGGTGTGAACGGATTTG-3′	5′-AAGATGGTGATGGGCTTCC-3′
Polr2a	5′-ATCAACAATCAGCTGCGGCG-3′	5′-GCCAGACTTCTGCATGGCAC-3′
Ppia	5′-CGCGTCTCCTTCGAGCTGTTTG-3′	5′-TGTAAAGTCACCACCCTGGCACAT-3′
Rplp0	5′-AGATTCGGGATATGCTGTTGGC-3′	5′-TCGGGTCCTAGACCAGTGTTC-3′
Tuba1b	5′-CGCCTTCTAACCCGTTGCTA-3′	5′-CCTCCCCCAATGGTCTTGTC-3′

Abbreviations: *Actb*, actin beta; *Gapdh*, glyceraldehyde 3-phosphate dehydrogenase; *Polr2a*, RNA polymerase II subunit A; *Ppia*, peptidylprolyl isomerase A; *Rplp0*, ribosomal protein lateral stalk subunit P0; *Tuba1b*, tubulin α1B.

4.4. Immunohistochemistry

Whole male embryos were collected at E13.5. For E18.5 fetuses as well as juvenile and adult mice, gonads were harvested and prepared for histological analysis. Immunohistochemical (IHC) staining was performed using the Rabbit Specific HRP/AEC IHC Detection Kit-Micro-polymer (ab236468; Abcam, Toronto, ON, Canada) following the manufacturer's protocol. Tissue sections (4 µM) were deparaffinized and rehydrated in graded ethanols. Tissues were processed for antigen retrieval by treating them with citrate buffer (10 mM Sodium citrate, 0.05% Tween 20, pH 6.0) in a Decloaking Chamber™ NxGen (Biocare Medical, Pacheco, CA, USA) for 10 min at 110 °C. Sections were incubated overnight at 4 °C with primary antibodies for either a rabbit polyclonal anti-STAR IgG (Proteintech Cat# 12225-1-AP, RRID:AB_2115832) diluted 1:200 (1.7 µg/mL) in phosphate buffered saline (PBS) containing 1% bovine serum albumin (BSA) or a rabbit polyclonal anti-GATA4 IgG (Abcam Cat# ab84593, RRID:AB_10670538) diluted 1:500 (1.8 µg/mL) in PBS containing 1% BSA. Sections incubated with rabbit IgG isotype control (Invitrogen Cat # 02-6102, RRID:AB_2532938) diluted 1:2500 (2 µg/mL) in PBS containing 1% BSA were used as negative controls. All sections were counterstained with Harris Modified hematoxylin solution (Thermo Fisher Scientific, Nepean, ON, Canada) and mounted in MOWIOL (EMD Millipore, Gibbstown, NJ, USA). Slides were visualized with a Zeiss Axioscop II microscope (Carl Zeiss Canada, Toronto, ON, Canada) connected to a Spot RT Slider digital camera (Diagnostic Instruments, Sterling Heights, MI, USA). At least three animals per genotype were assessed.

4.5. Protein Extraction and Western Blot Analysis

For western blot analysis, half of a testis was homogenized while still frozen in ice-cold extraction buffer (50 mM Tris pH 7.4, 150 mM NaCl, 1 mM EDTA, 0.5% Igepal, 1 mM DTT, 0.5 mM phenylmethylsulfonyl fluoride (Sigma-Aldrich), and proteinase inhibitors (Sigma-Aldrich): 5 µg/mL aprotinin, 5 µg/mL leupeptin, and 5 µg/mL pepstatin. Homogenates were incubated on ice for 15 min after which the samples were briefly sonicated to break genomic DNA. Protein concentration was evaluated by Bradford assay [35], using Bio-Rad protein assay dye reagent concentrate (Bio-Rad Laboratories, Mississauga, ON, Canada) and BSA as protein standard. Aliquots (40 µg) of testicular homogenates were separated by SDS-PAGE and then electrotransferred to nitrocellulose membrane (Bio-Rad Laboratories, Mississauga, ON, Canada). Non-specific antibody binding was prevented by blocking for 1 h at RT using 5% non-fat dry milk in Tris-buffered saline (TBS: 20 mM Tris, 150 mM NaCl, pH 7.6) with 0.1% Tween 20 (TBS-T). Proteins were detected using commercially available primary antibodies for either a rabbit monoclonal anti-STAR IgG (Cell Signaling Technology Cat# 8449, RRID:AB_10889737) at a dilution of 1:5000 in 5% non-fat dry milk in TBS-T, a mouse monoclonal anti-α-tubulin IgG (used as a loading control, Sigma-Aldrich Cat# T5168, RRID:AB_477579) at a dilution of 1:10,000 in 5% non-fat dry milk in TBS-T. After washing in TBS-T, membranes were incubated with horseradish peroxidase-labeled secondary antibodies: goat anti-rabbit IgG (Vector Laboratories Cat# PI-1000, RRID:AB_2336198) diluted 1:5000 5% non-fat dry milk in TBS-T for STAR detection or horse anti-mouse IgG (Vector Laboratories Cat# PI-2000, RRID:AB_2336177) diluted 1:5000 in 5% non-fat dry milk in TBS-T for α-tubulin. After washing in TBS-T, membranes were finally incubated with Clarity Western ECL Substrate (Bio-Rad Laboratories) for 5 min. The chemiluminescent signal was detected on a ChemiDoc Imaging System (Bio-Rad Laboratories).

4.6. Hormone Assay

At the time of sacrifice, blood was drawn from adult mice by cardiac puncture. Blood was collected in EDTA-coated microtubes to prevent clotting, and plasma was isolated by centrifugation for 3 min at 2400 g and stored at −80 °C until further needed. For intratesticular testosterone quantification and RNA extraction, half a testis from each animal was homogenized while still frozen in cold PBS on ice and then centrifuged at 3000 rpm for 5 min at 4°C. Supernatant was collected and assayed for testosterone. The pellet was resuspended in TRI reagent to isolate total RNA for qPCR analysis (described above). For ex vivo stimulation of testosterone, testes were harvested and placed in 500 µL of DMEM containing 100 IU/mL penicillin and 100 µg/mL streptomycin. Testes were detunicated and treated with either 1 IU/mL of human chorionic gonadotropin (hCG, Sigma-Aldrich) or vehicle (H_2O) and incubated for 4 h at 32 °C. After incubation, tissues were centrifuged for 5 min at 2400 g and the culture medium was collected. Testosterone quantification in plasma, testicular homogenates or culture medium was performed using an ELISA kit purchased from Cayman Chemical (Cayman Chemical, Ann Arbor, MI, USA, Cat# 582701, RRID:AB_2895148) following the instructions recommended by the manufacturer. To obtain readings within the range of the standard curve, samples were diluted in EIA buffer (provided by the kit manufacturer) as follows: plasma (1:20 and 1:50), testicular homogenates (1:100 and 1:200) and culture medium (1:200 and 1:500). Microplates were read using a Tecan Spark® 10M multimode plate reader (Tecan, Morrisville, NC, USA). The assay has a cross reactivity of 100% for testosterone. Cross-reactivity is negligible for other sex steroids with the exception of 5α-dihydrotestosterone (DHT) at 27.4% (Cayman Chemical). However, since mouse intratesticular and plasma DHT levels are very low, this would not impact the testosterone measurements.

4.7. Statistical Analysis

Statistical analyses were done using JASP version 0.16.3 (https://jasp-stats.org, accessed on 10 October 2021; JASP Team (2022), Amsterdam, The Netherlands). Quantitative comparisons between wild-type and mutant mice (*Star* mRNA, plasma and intratesticular testosterone) were analyzed using a parametric Student's *t*-test. Measurement of ex vivo testosterone production was analyzed by one-way ANOVA followed by Tukey multiple comparisons tests. For all statistical analyses, $p < 0.05$ was considered significant.

Author Contributions: Conceptualization, M.F.B., J.P. and R.S.V.; methodology, M.F.B. and J.P.; formal analysis, M.F.B., J.P., J.J.T. and R.S.V.; investigation, M.F.B. and J.P.; data curation, M.F.B., J.P., J.J.T. and R.S.V.; writing—M.F.B., J.P. and R.S.V.; writing—review and editing, J.J.T. and R.S.V.; supervision, R.S.V.; project administration, R.S.V.; funding acquisition, R.S.V. All authors have read and agreed to the published version of the manuscript.

Funding: This research was funded by project grant PJT-166131 from the Canadian Institutes of Health Research (CIHR) to R.S.V.

Institutional Review Board Statement: The animal study protocols were approved by the Institutional Review Boards of Université Laval and the Centre de recherche du CHU de Québec-Université Laval (2019-149 and CHU-19-046, approved on 16 December 2019).

Informed Consent Statement: Not applicable.

Data Availability Statement: All data generated and analyzed during this study are included in this article.

Conflicts of Interest: The authors declare no conflict of interest. The funder had no role in the design of the study; in the collection, analyses, or interpretation of data; in the writing of the manuscript; or in the decision to publish the results.

References

1. Miller, W.L.; Auchus, R.J. The molecular biology, biochemistry, and physiology of human steroidogenesis and its disorders. *Endocr. Rev.* **2011**, *32*, 81–151. [CrossRef] [PubMed]
2. De Mattos, K.; Viger, R.S.; Tremblay, J.J. Transcription factors in the regulation of Leydig cell gene expression and function. *Front. Endocrinol.* **2022**, *13*, 881309. [CrossRef]
3. Tremblay, J.J. Molecular regulation of steroidogenesis in endocrine Leydig cells. *Steroids* **2015**, *103*, 3–10. [CrossRef] [PubMed]
4. Viger, R.S.; de Mattos, K.; Tremblay, J.J. Insights into the roles of GATA Factors in mammalian testis development and the control of fetal testis gene expression. *Front. Endocrinol.* **2022**, *13*, 902198. [CrossRef] [PubMed]
5. Viger, R.S.; Taniguchi, H.; Robert, N.M.; Tremblay, J.J. Role of the GATA family of transcription factors in andrology. *J. Androl.* **2004**, *25*, 441–452. [CrossRef] [PubMed]
6. Tremblay, J.J.; Viger, R.S. Novel roles for GATA transcription factors in the regulation of steroidogenesis. *J. Steroid Biochem. Mol. Biol.* **2003**, *85*, 291–298. [CrossRef]
7. Viger, R.S.; Mazaud Guittot, S.; Anttonen, M.; Wilson, D.B.; Heikinheimo, M. Role of the GATA family of transcription factors in endocrine development, function, and disease. *Mol. Endocrinol.* **2008**, *22*, 781–798. [CrossRef] [PubMed]
8. Tevosian, S.G. Transgenic mouse models in the study of reproduction: Insights into GATA protein function. *Reproduction* **2014**, *148*, R1–R14. [CrossRef]

9. Bergeron, F.; Nadeau, G.; Viger, R.S. GATA4 knockdown in MA-10 Leydig cells identifies multiple target genes in the steroidogenic pathway. *Reproduction* **2015**, *149*, 245–257. [CrossRef] [PubMed]

10. Schrade, A.; Kyronlahti, A.; Akinrinade, O.; Pihlajoki, M.; Hakkinen, M.; Fischer, S.; Alastalo, T.P.; Velagapudi, V.; Toppari, J.; Wilson, D.B.; et al. GATA4 is a key regulator of steroidogenesis and glycolysis in mouse Leydig cells. *Endocrinology* **2015**, *156*, 1860–1872. [CrossRef]

11. Padua, M.B.; Jiang, T.; Morse, D.A.; Fox, S.C.; Hatch, H.M.; Tevosian, S.G. Combined loss of the GATA4 and GATA6 transcription factors in male mice disrupts testicular development and confers adrenal-like function in the testes. *Endocrinology* **2015**, *156*, 1873–1886. [CrossRef] [PubMed]

12. Alpy, F.; Tomasetto, C. START ships lipids across interorganelle space. *Biochimie* **2014**, *96*, 85–95. [CrossRef]

13. Selvaraj, V.; Stocco, D.M.; Clark, B.J. Current knowledge on the acute regulation of steroidogenesis. *Biol. Reprod.* **2018**, *99*, 13–26. [CrossRef]

14. Manna, P.R.; Stetson, C.L.; Slominski, A.T.; Pruitt, K. Role of the steroidogenic acute regulatory protein in health and disease. *Endocrine* **2016**, *51*, 7–21. [CrossRef] [PubMed]

15. Lin, D.; Sugawara, T.; Strauss, J.F., III; Clark, B.J.; Stocco, D.M.; Saenger, P.; Rogol, A.; Miller, W.L. Role of steroidogenic acute regulatory protein in adrenal and gonadal steroidogenesis. *Science* **1995**, *267*, 1828–1831. [CrossRef]

16. Caron, K.M.; Soo, S.C.; Wetsel, W.C.; Stocco, D.M.; Clark, B.J.; Parker, K.L. Targeted disruption of the mouse gene encoding steroidogenic acute regulatory protein provides insights into congenital lipoid adrenal hyperplasia. *Proc. Natl. Acad. Sci. USA* **1997**, *94*, 11540–11545. [CrossRef]

17. Tremblay, J.J.; Hamel, F.; Viger, R.S. Protein kinase A-dependent cooperation between GATA and C/EBP transcription factors regulates StAR promoter activity. *Endocrinology* **2002**, *143*, 3935–3945. [CrossRef]

18. Tremblay, J.J.; Viger, R.S. GATA factors differentially activate multiple gonadal promoters through conserved GATA regulatory elements. *Endocrinology* **2001**, *142*, 977–986. [CrossRef]

19. Bouchard, M.F.; Bergeron, F.; Grenier Delaney, J.; Harvey, L.M.; Viger, R.S. In vivo ablation of the conserved GATA binding motif in the Amh promoter impairs Amh expression in the male mouse. *Endocrinology* **2019**, *160*, 817–826. [CrossRef]

20. Lavoie, H.A.; King, S.R. Transcriptional regulation of steroidogenic genes: STARD1, CYP11A1 and HSD3B. *Exp. Biol. Med.* **2009**, *234*, 880–907. [CrossRef] [PubMed]

21. Clark, B.J.; Wells, J.; King, S.R.; Stocco, D.M. The purification, cloning, and expression of a novel luteinizing hormone-induced mitochondrial protein in MA-10 mouse Leydig tumor cells. Characterization of the steroidogenic acute regulatory protein (StAR). *J. Biol. Chem.* **1994**, *269*, 28314–28322. [CrossRef]

22. Liu, M.; Rehman, S.; Tang, X.; Gu, K.; Fan, Q.; Chen, D.; Ma, W. Methodologies for improving HDR efficiency. *Front. Genet.* **2018**, *9*, 691. [CrossRef] [PubMed]

23. Manna, P.R.; Dyson, M.T.; Stocco, D.M. Regulation of the steroidogenic acute regulatory protein gene expression: Present and future perspectives. *Mol. Hum. Reprod.* **2009**, *15*, 321–333. [CrossRef] [PubMed]

24. Manna, P.R.; Dyson, M.T.; Eubank, D.W.; Clark, B.J.; Lalli, E.; Sassone-Corsi, P.; Zeleznik, A.J.; Stocco, D.M. Regulation of steroidogenesis and the steroidogenic acute regulatory protein by a member of the cAMP response-element binding protein family. *Mol. Endocrinol.* **2002**, *16*, 184–199. [CrossRef]

25. Reinhart, A.J.; Williams, S.C.; Clark, B.J.; Stocco, D.M. SF-1 (steroidogenic factor-1) and C/EBP beta (CCAAT/enhancer binding protein-beta) cooperate to regulate the murine StAR (steroidogenic acute regulatory) promoter. *Mol. Endocrinol.* **1999**, *13*, 729–741. [CrossRef]

26. Silverman, E.; Eimerl, S.; Orly, J. CCAAT enhancer-binding protein beta and GATA-4 binding regions within the promoter of the steroidogenic acute regulatory protein (StAR) gene are required for transcription in rat ovarian cells. *J. Biol. Chem.* **1999**, *274*, 17987–17996. [CrossRef]

27. Hasegawa, T.; Zhao, L.; Caron, K.M.; Majdic, G.; Suzuki, T.; Shizawa, S.; Sasano, H.; Parker, K.L. Developmental roles of the steroidogenic acute regulatory protein (StAR) as revealed by StAR knockout mice. *Mol. Endocrinol.* **2000**, *14*, 1462–1471. [CrossRef]

28. Reinhart, A.J.; Williams, S.C.; Stocco, D.M. Transcriptional regulation of the StAR gene. *Mol. Cell. Endocrinol.* **1999**, *151*, 161–169. [CrossRef]

29. Haeussler, M.; Schonig, K.; Eckert, H.; Eschstruth, A.; Mianne, J.; Renaud, J.B.; Schneider-Maunoury, S.; Shkumatava, A.; Teboul, L.; Kent, J.; et al. Evaluation of off-target and on-target scoring algorithms and integration into the guide RNA selection tool CRISPOR. *Genome Biol.* **2016**, *17*, 148. [CrossRef]

30. Cong, L.; Ran, F.A.; Cox, D.; Lin, S.; Barretto, R.; Habib, N.; Hsu, P.D.; Wu, X.; Jiang, W.; Marraffini, L.A.; et al. Multiplex genome engineering using CRISPR/Cas systems. *Science* **2013**, *339*, 819–823. [CrossRef]

31. Truett, G.E.; Heeger, P.; Mynatt, R.L.; Truett, A.A.; Walker, J.A.; Warman, M.L. Preparation of PCR-quality mouse genomic DNA with hot sodium hydroxide and tris (HotSHOT). *Biotechniques* **2000**, *29*, 52–54. [CrossRef] [PubMed]

32. Brinkman, E.K.; Chen, T.; Amendola, M.; van Steensel, B. Easy quantitative assessment of genome editing by sequence trace decomposition. *Nucleic Acids Res.* **2014**, *42*, e168. [CrossRef] [PubMed]

33. Yokoyama, T.; Omotehara, T.; Hirano, T.; Kubota, N.; Yanai, S.; Hasegawa, C.; Takada, T.; Mantani, Y.; Hoshi, N. Identification of reference genes for quantitative PCR analyses in developing mouse gonads. *J. Vet. Med. Sci.* **2018**, *80*, 1534–1539. [CrossRef]

34. Livak, K.J.; Schmittgen, T.D. Analysis of relative gene expression data using real-time quantitative PCR and the 2(-Delta Delta C(T)) Method. *Methods* **2001**, *25*, 402–4088. [CrossRef] [PubMed]
35. Bradford, M.M. A rapid and sensitive method for the quantitation of microgram quantities of protein utilizing the principle of protein-dye binding. *Anal. Biochem.* **1976**, *72*, 248–254. [CrossRef]

International Journal of
Molecular Sciences

Review

Basic Leucine Zipper Transcription Factors as Important Regulators of Leydig Cells' Functions

Luc J. Martin * and Ha Tuyen Nguyen

Biology Department, Université de Moncton, Moncton, NB E1A 3E9, Canada
* Correspondence: luc.martin@umoncton.ca

Abstract: Transcription factors members of the basic leucine zipper (bZIP) class play important roles in the regulation of genes and functions in testicular Leydig cells. Many of these factors, such as cAMP responsive element binding protein 1 (CREB1) and CCAAT enhancer binding protein beta (CEBPB), are regulated by the cAMP/protein kinase A (PKA) pathway, the main signaling pathway activated following the activation of the luteinizing hormone/choriogonadotropin membrane receptor LHCGR by the - hormone LH. Others, such as X-box binding protein 1 (XBP1) and members of the cAMP responsive element binding protein 3 (CREB3)-like superfamily, are implicated in the endoplasmic reticulum stress by regulating the unfolded protein response. In this review, the influences of bZIP transcription factors, including CREB1, CEBPB and activator protein 1 (AP-1) family members, on the regulation of genes important for cell proliferation, steroidogenesis and Leydig cell communication will be covered. In addition, unresolved questions regarding the mechanisms of actions of bZIP members in gene regulation will be identified.

Keywords: activator protein 1; CREB; CEBPB; testis; Leydig cells; steroidogenesis

Citation: Martin, L.J.; Nguyen, H.T. Basic Leucine Zipper Transcription Factors as Important Regulators of Leydig Cells' Functions. *Int. J. Mol. Sci.* **2022**, *23*, 12887. https://doi.org/10.3390/ijms232112887

Academic Editor: Jacques J. Tremblay

Received: 22 September 2022
Accepted: 21 October 2022
Published: 25 October 2022

Publisher's Note: MDPI stays neutral with regard to jurisdictional claims in published maps and institutional affiliations.

1. Introduction

There is increasing evidence that transcription factors of the basic leucine zipper (bZIP) class play important roles in gene regulation and function of testicular Leydig cells. This class of more than 50 members includes the cAMP responsive element binding protein/activating transcription factor (CREB/ATF), activator protein 1 (AP-1), CCAAT enhancer binding protein (CEBP) and musculoaponeurotic fibrosarcoma (MAF) families of transcription factors. The bZIP transcription factors are proteins found only in eukaryotes. They bind to specific double-stranded DNA sequences as homodimers or heterodimers to activate or repress gene transcription. Most of the genes encoding bZIP factors are classified as immediate early genes and are inducible by a multitude of endocrine and paracrine agents leading to the activation of intracellular signaling pathways. The primary stimulatory hormone acting on Leydig cells is luteinizing hormone (LH) from the anterior pituitary gland. This hormone interacts with its membrane receptor luteinizing hormone/choriogonadotropin (LHCGR) and mostly activates the cAMP/protein kinase A (PKA) pathway. Then, PKA phosphorylates a multitude of transcription factors, including bZIP members, leading to their activation and regulation of transcription of their target genes. Although the roles of certain bZIP transcription factors, such as CREB1, CEBPB, JUN and FBJ osteosarcoma (FOS), are well documented regarding the regulation of steroidogenesis in Leydig cells, other lesser-known members may play complementary roles in the regulation of proliferation, survival, and communication in these cells. Thus, this review aims to present recent advances regarding the involvement of bZIP factors in the regulation of different functions of Leydig cells from the testis.

2. Classification of bZIP Transcription Factors

Transcription factors members of the bZIP class are characterized by the presence of a bZIP domain consisting of an approximately 40 amino acid-long α-helix (Figure 1). The

N-terminal third of this helix has multiple positively charged side chains such as arginine and lysine and is defined as the basic region involved in DNA-binding. The C-terminal two-thirds of the helix is the leucine zipper and plays a role in protein dimerization. The leucine zipper is amphipathic with the hydrophobic regions on the two copies of the leucine zipper interacting to bring the two bZIP domains together in a parallel orientation. The two basic regions can then bind to DNA by passing through the major groove on opposite sides of the double helix. Importantly, the bZIP domain facilitates the dimerization of two similar transcription factors of the bZIP class. The combination of two transcription factors containing the bZIP domain enables the proper positioning of the two adjacent DNA-binding domains in the dimeric complex. For bZIP transcription factors, dimerization is required for DNA binding.

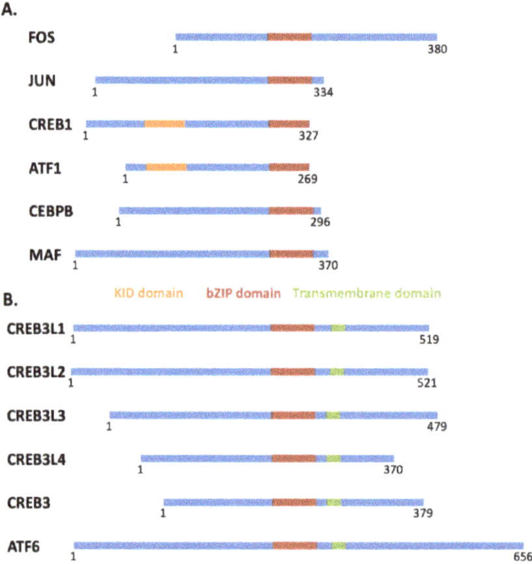

Figure 1. Schematic diagram of the primary protein structure of transcription factors members of the bZIP class. (**A**) Common nuclear bZIP transcription factors. (**B**) Transmembrane bZIP transcription factors. Abbreviations: KID, kinase A inducible activation domain; bZIP, basic leucine zipper.

According to the latest TFClass nomenclature, the class of bZIP transcription factors contains more than 50 members divided into families: JUN, FOS, MAF, B-ATF, X-box binding protein 1 (XBP1), ATF4, CREB and CEBP (Table 1) [1]. The common AP-1 family of transcription factors rather consists of members of the JUN and FOS subfamilies, whereas the common CREB/ATF family refers to CREB1, ATF1 and cAMP responsive element modulator (CREM) transcription factors.

Table 1. Classification of transcription factors members of the class of bZIP transcription factors according to the TFClass nomenclature [1].

Family	Subfamily	Transcription Factors
	JUN	JUN, JUNB, JUND
JUN-related	NFE2	NFE2, NFE2L1, NFE2L2, NFE2L3, BACH1, BACH2
	ATF2	ATF2, ATF7, CREB5

Table 1. *Cont.*

Family	Subfamily	Transcription Factors
FOS-related	FOS ATF3-like	FOS, FOSB, FOSL1, FOSL2 ATF3, JDP2
MAF-related	Large MAF Small MAF	MAF, MAFA, MAFB, NRL MAFF, MAFG, MAFK
B-ATF-related		BATF, BATF2, BATF3
XBP1-related		XBP1
ATF4-related		ATF4, ATF5
CREB-related	CREB-like CREB3-like ATF6 CREBZF-like CREBL2-like	CREB1, ATF1, CREM CREB3, CREB3L1, CREB3L2, CREB3L3, CREB3L4 ATF6, ATF6B CREBZF CREBL2
CEBP-related	CEBP PAR	CEBPA, CEBPB, CEBPG, CEBPD, CEBPE, DDIT3 DBP, HLF, NFIL3, TEF

3. Mechanisms of Gene Regulation by bZIP Transcription Factors

3.1. CREB-like Subfamily Members

CREB1 is a well-known 43 kD bZIP transcription factor that binds to an octanucleotide sequence called the cyclic AMP response element (CRE) (5′-TGACGTCA-3′) (Figure 2) [2]. This factor can bind to DNA as a homodimer or a heterodimer in association with other members of the CREB/ATF and AP-1 families [3]. CREB1 plays an essential role in the regulation of gene expression in response to a variety of extracellular signals. In response to a cytosolic increase in the second messenger cAMP, CREB can be phosphorylated by PKA at serine residue Ser^{133} in the kinase A inducible activation domain (KID) (Figure 2) [2]. This phosphorylation enhances CREB1 transcriptional activity by facilitating its interaction with the 265 kD CREB-binding protein (CBP) and subsequent recruitment of CBP to target genes [4]. This cofactor is capable of histone acetyltransferase activity and acts as an adaptor protein between CREB and the general transcription complex.

The CREB-like subfamily members CREB1, CREM and ATF1, are expressed in the MA-10 tumor Leydig cell model [5]. In testicular Leydig cells, the constitutively expressed transcription factor CREB1 represents one of the most important targets of the cAMP/PKA pathway. Among the target genes of CREB1 in Leydig cells is *Nr4a1* encoding the orphan nuclear receptor NR4A1 (NUR77). Indeed, the rapid activation of *Nr4a1* gene expression in response to LH in Leydig cells is caused, in part, by CREB1 and AP-1 transcription factors following activation of the cAMP/PKA pathway [6]. In addition to being activated by PKA, CREB1 can also be activated by the Ca^{2+}/calmodulin-dependent protein kinase (CaMK), the mitogen-activated protein kinase (MAPK) and/or AKT through phosphorylation at Ser^{133} in other cell types (Figure 2) [7–10]. By binding to a G protein-coupled receptor, the osteoblast-derived hormone osteocalcin also regulates steroidogenesis in Leydig cells following activation of CREB1 [11].

Although the importance of protein kinase C (PKC) is less than that of PKA, these two pathways appear to potentiate each other in the activation of steroidogenesis in response to LH in Leydig cells [12–14]. In addition to contributing to its activation, phosphorylation of CREB1 by PKC isoenzymes such as PKCμ (PKD) increases its ability to bind to CRE regulatory element [2,15]. Moreover, PMA (phorbol-12-myristate-13-acetate), an activator of the PKC pathway, increases the phosphorylation of CREB1 [12,16].

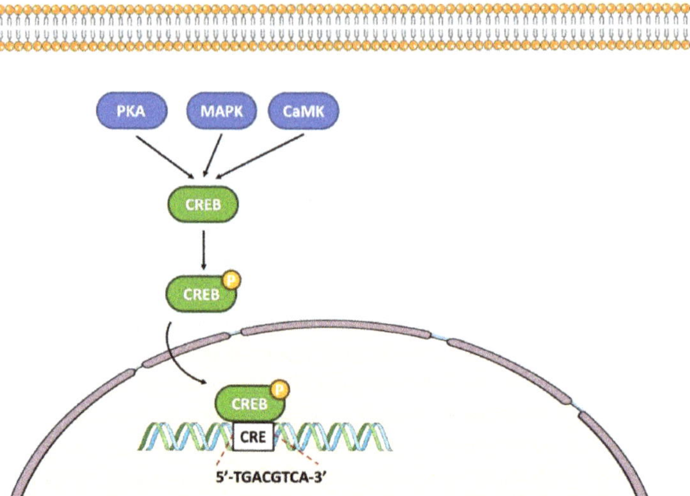

Figure 2. Regulation of the activity of CREB-like subfamily members by phosphorylation, followed by recruitment to the CRE DNA regulatory element. Abbreviations: CaMK, Ca^{2+}/calmodulin dependent protein kinase; CRE, cyclic AMP response element; CREB, cAMP responsive element binding protein; MAPK, mitogen activated protein kinase; P, phosphate; PKA, protein kinase A.

In addition to CREB1, cAMP/PKA-mediated increase in transcription can also involve activation of CREM (cAMP-response element modulator protein) by phosphorylation at Ser^{117} and its binding to CBP [17]. Alternative splicing of the *Crem* transcript can give rise to several isoforms which act as activators (τ, τ1, and τ2) or repressors (α, β, and γ) of transcription [18]. Among activators, CREMτ can interact with the steroidogenic factor 1 (NR5A1) and activate the steroidogenic acute regulatory protein (*Star*) promoter following treatment of MA-10 Leydig cells with dibutyryl-cAMP [19]. In addition, transcriptional activity of CREM, like that of CREB1, is increased post-translationally by phosphorylation following activation of PKA in response to cAMP [17]. The *Crem* gene contains a promoter that can be activated by the cAMP/PKA pathway. Such an activation results in the production of transcripts encoding a particular isoform of CREM which contains the DNA binding domain and the phosphorylatable region, but lacks the activation domain [20]. This type of isoform, called ICER (inducible cyclic AMP early repressor), binds to the CRE regulatory element and repress transcriptional activation by the active isoform of CREM, thereby limiting cAMP/PKA-dependent activation of target genes [21].

ATF1, another member of the CREB-like subfamily, can dimerize with CREB1 but not with ATF2 or ATF3, whereas ATF2 can dimerize with ATF3 but not with CREB1 [22–24]. In addition, certain CRE-binding proteins, including ATF2, ATF3 and ATF4, can heterodimerize with FOS and JUN [24,25]. Interestingly, ATF1-4 are all expressed in adult human Leydig cells [26] and may participate in the regulation of steroidogenesis. ATF1 can be phosphorylated at Ser^{63} by PKA or CAMK [27], potentially contributing to its transcriptional activity in Leydig cells.

3.2. AP-1 Members

The AP-1 family of transcription factors, including members of the JUN and FOS subfamilies, is ubiquitously expressed [28]. FOS subfamily members, including FOS, FOSB, fos-like antigen 1 (FOSL1, FRA1) and fos-like antigen 2 (FOSL2, FRA2) must heterodimerize with JUN proteins, including JUN, JUNB and JUND. However, JUN members can form homodimers or heterodimers with other AP-1 factors. Overall, JUN-FOS heterodimers bind DNA with more affinity and stability than JUN-JUN homodimers [29,30]. AP-1 members were involved in the induction of multiple cellular and viral genes through the phorbol ester tumor promoter 12-O-tetradecanoylphorbol-13-acetate response element (TRE) (Figure 3) [31]. With other bZIP transcription factors such as ATF and MAF, AP-1 members can form up to 18 different DNA-binding dimeric complexes with versatile functionality through activation or inhibition of transcription according to the composition of the dimeric complex and the cell context [32,33]. Despite having comparable DNA binding specificities, AP-1 dimers differ in their transactivation efficiencies [34,35]. Unlike FOS and FOSB, FOSL1 and FOSL2 lack transcriptional activation domains.

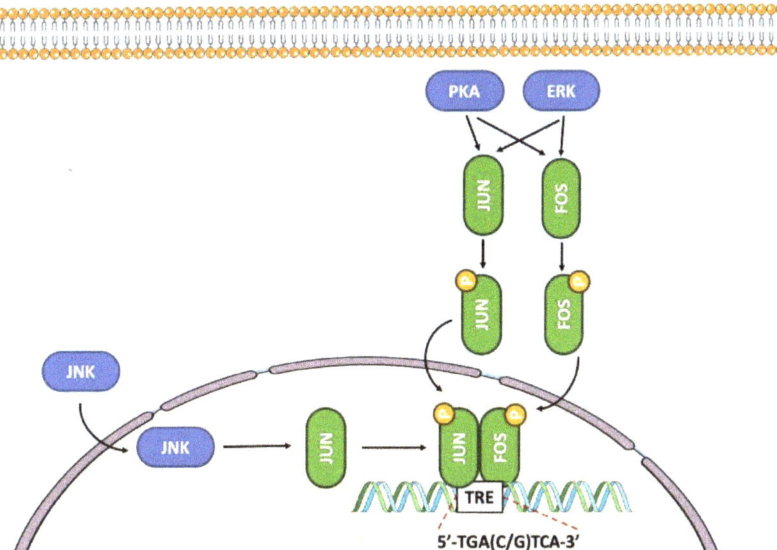

Figure 3. Regulation of the activity of AP-1 family members JUN and FOS by phosphorylation, followed by recruitment to the TRE DNA regulatory element. Abbreviations: ERK, extracellular signal-regulated kinase; JNK, JUN N-terminal kinase; P, phosphate; PKA, protein kinase A; TRE, 12-O-tetradecanoylphorbol-13-acetate response element.

Regarding their interaction with DNA regulatory elements, JUN:FOS and JUN:JUN dimers can bind to AP-1 or TRE elements with the consensus sequence 5′-TGA(C/G)TCA-3′ and to CRE [28]. Notably, JUN:FOS heterodimers have higher affinity for asymmetrical TRE motifs, whereas JUN homodimers rather bind to symmetric CRE elements [29,36]. The heterodimers formed from ATF members typically bind to CRE [3].

There may be functional redundancy among AP-1 members. Indeed, FOS, FOSB, and JUND are dispensable but JUN, JUNB, and FOSL1 are necessary for embryonic development [37]. Furthermore, substitution of *Jun* for *Junb* allows recovery of liver development but not heart development, suggesting that knock-in experiments are not equivalent for all AP-1 members [38].

There are several ways to activate and/or increase the expression of AP-1 factors. In fact, numerous signaling pathways leading to the phosphorylation of serine and threonine residues participate in the regulation of their expression and activity [39,40]. The *Jun* promoter is primarily activated by the recruitment of JUN/ATF2 heterodimer to a major Jun/TRE *cis* regulatory element [41]. Three regulatory elements—the sis-inducible enhancer (SIE), the serum response element (SRE), and the CRE—can activate the *Fos* promoter in response to a variety of hormones, growth factors, and cytokines [33,42–46]. Indeed, the expression of *Fos* will be enhanced following the recruitment of ATF or CREB factors to a CRE element in response to increased levels of second messengers cAMP and calcium [8,44].

The SRE regulatory element can be bound by a dimer consisting of the serum response factor (SRF) and ETS Like-1 (ELK1) being activated by exposure to UV light or growth factors, leading to increased *Fos* expression [46,47]. The SIE regulatory element allows the recruitment of dimers of the signal transducer and activator of transcription (STAT) family, in particular STAT1 and STAT3, to activate *Fos* gene expression [45,48]. These STAT members are mainly activated by the Janus kinase (JAK) pathway [49]. In most cell types, the SRE and CRE in the *Fos* and AP-1 (TRE) element in the *Jun* promoters are involved in basal expression of these genes [50–52]. Moreover, *Jun* is autoregulated by its own expression with the JUN/ATF2 heterodimer [41,51]. Similarly, the *Fos* gene can also self-regulate after its activation [53]. Hence, AP-1 members can regulate each other to fine-tune their expression [54].

The MAPK pathway primarily regulates the activity of AP-1 members, through phosphorylation by extracellular signal-regulated kinases (ERK), JUN N-terminal kinases (JNK), and p38 (Figure 3). In addition, other kinases such as glycogen synthase kinase-3 (GSK3), $p34^{cdc2}$ kinase (CDK1), casein kinase II (CKII), PKA and PKC can phosphorylate FOS or JUN in certain cell types. The activated JNK translocates to the nucleus [47], where it phosphorylates AP-1 members, promoting their dimerization and transcriptional activities [33]. Interestingly, the dimerization and DNA binding of AP-1 members facilitate their phosphorylation by several protein kinases [40]. Different kinases participate in the regulation of the activity of AP-1 members by phosphorylating different Ser or Thr residues. For example, the phosphorylation of JUN at Thr^{235}, Ser^{246} and Ser^{252} by GSK3, CKII or ERK1/2 decreases its transcriptional activity [39,55–57]. However, phosphorylation at Ser^{63} and Ser^{73} by JNK1 and dephosphorylation at Ser^{246} increase the stability and transcriptional activity of the JUN protein [39,55]. Such post-translational modifications of JUN potentiate its transactivation capacity by increasing its interaction with the coactivator CBP, having histone acetyltransferase activity [58]. Similar to JUN, JUNB can also be phosphorylated at $Thr^{102/104}$ by ERK2 [59,60]. However, the consequence of this post-translational modification remains to be defined. The phosphorylation of FOS at Ser^{362}, Ser^{374}, Thr^{232}, Thr^{325}, and Thr^{331} by ERK and ribosomal S6 kinase increases its protein stability and transcriptional activity [61–63]. Moreover, FOS can also be phosphorylated at Ser^{362} by PKA [64]. Compared to other AP-1 factors, JUND cannot interact directly with JNK, but can be phosphorylated by this kinase as part of a heterodimer with JUN [65]. Hence, in addition to changes in FOS and JUN protein levels, gene regulation can be influenced by their post-translational modifications (Figure 3).

In addition to LH and human chorionic gonadotrophin (hCG), several growth factors are also involved in the activation of AP-1 family members in Leydig cells. In Leydig cells from 3-week-old pigs, basic fibroblast growth factor (bFGF) and epidermal growth factor (EGF) increase the expressions of *Fos*, *Jun*, and *Junb*, whereas transforming growth factor beta (TGF) increases the expression of *Jun* only, and insulin-like growth factor 1 (IGF1) increases the expressions of *Fos* and *Junb* [66]. Interestingly, EGF and bFGF enhance the stimulatory action of hCG on *Fos* and *Junb* expressions in this model [66]. Regulation of AP-1 members in Leydig cells can also be influenced by testicular macrophages and tumor necrosis factor alpha (TNFα) production through regulation of members of the MAPK family such as ERK [67] and stress-activated protein kinases (SAPKs) [68]. In MA-10 Leydig cells, the SAPK protein level and activity are increased, whereas those of ERK are decreased, by TNFα and cAMP [69]. Such a regulation results in increases in JUN and FOS protein levels and DNA binding activities [69]. In addition, others have shown that CREB nuclear localisation is decreased in response to TNF, resulting in inhibition of steroidogenesis in MA-10 Leydig cells [70].

Although JUN and FOS are expressed in MA-10 Leydig cells, only JUN is rapidly increased in response to activation of the cAMP/PKA pathway [71,72]. However, hCG increases the expressions of *Fos*, *Fosb*, *Jun*, *Junb*, *Jund* and *Fosl2* in less than 1 h of treatment, whereas *Fosl1* is increased after 3 h of exposure of MA-10 Leydig cells [73]. In addition to the cAMP/PKA pathway, ERK1/2 and PKC are also involved in the activation of AP-1 members by LH/hCG stimulation of testicular Leydig cells [15,74,75].

3.3. CEBP Members

The CEBP subfamily includes six members: CEBPA, CEBPB, CEBPG, CEBPD, CEBPE and DNA-damage inducible transcript 3 (DDIT3, CHOP) [76]. CEBPB is the major member expressed in Leydig cells and its expression depends on Leydig cells' differentiation status [77]. Members of the CEBP subfamily can form homodimers or heterodimers with members of the CREB/ATF family via the leucine zipper dimerization domain [78], which greatly increases the diversity of DNA binding specificities and transactivation potential. CEBP members can recognize the consensus DNA-binding sequence 5′-TKNNGYAAK-3′ (Y = C or T, K = T or G) in the regulatory regions of target genes. CEBPB, whose expression is increased in response to LH and cAMP, plays an important role in hormonal regulation of gene expression in Leydig cells [77]. In human, DNA-binding of CEBPB is increased by ERK-dependent phosphorylation of Thr^{235} [79]. In addition, Ca^{2+}-mediated signals can also lead to mouse CEBPB activation through phosphorylation of Ser^{276} in the leucine zipper [74]. Acetylation and sumoylation have also been implicated in the regulation of intracellular localization and transcriptional activity of CEBP members [75,80] but have been barely investigated in the regulation of Leydig cells' functions.

3.4. Maf-Related Members

The family of MAF-related transcription factors can be divided in two sub-families: the large MAF, containing MAF, MAFA, MAFB and neural retina leucine zipper (NRL), and the small MAF, containing MAFF, MAFG and MAFK (Table 1). The heterodimers formed by MAF factors bind to MAF recognition elements (MAREs) consisting of extended sequences of the AP-1 motifs [81]. Depending on the binding site and binding partner, MAF-related members can act as transcriptional activators or repressors. By E14.5 in mouse embryo and in adult testis, *Mafb* expression is detected in Leydig cells [82,83]. However, the target genes of MAFB in fetal Leydig cells during development or in adult Leydig cells remain to be characterized. Other MAF-related members may play an important role in regulating adult Leydig cells' functions as MAF, MAFF and MAFG are highly expressed in human adult Leydig cells [26].

3.5. XBP1-Related Members

The XBP1-related members family contains only the transcription factor XBP1. This transcription factor is known to regulate the major histocompatibility complex (MHC) class II gene by binding to a promoter element known as X-Box [84]. XBP1 is implicated in cell differentiation, proliferation, apoptosis, cellular stress response and other signaling pathways. It has also been associated with the expression of genes required for membrane biogenesis and the secretory pathway [85]. Indeed, XBP1-deficient β-cells fail to secrete enough insulin to regulate blood glucose levels [86]. Several publications have evaluated ATF4 and XBP1 transcription factors as endoplasmic reticulum (ER) stress markers in Leydig cells [87,88]. With ATF4 and ATF6, XBP1 has been characterized as an unfolded protein response (UPR) transcription factor, which mediates downregulation of the 3-β-hydroxysteroid dehydrogenase (*Hsd3b1*) gene expression following high or prolonged hCG treatment in Leydig cells [89]. Such treatments result in ER stress-mediated apoptosis, possibly through binding to the ER stress response element (ERSE) found in different gene promoters [90]. Additionally, XBP1 can also form heterodimers by interacting with FOS [91] and ATF6 [92]. However, other specific target genes for XBP1 in Leydig cells remain to be further characterized.

3.6. CREB3-like Subfamily Members

The CREB3-like subfamily includes transmembrane bZIP transcription factors such as CREB3 (also known as Luman), CREB3L1, CREB3L2, CREB3L3 and CREB3L4. These transcription factors play important roles in the UPR. In addition, CREB3L1 is involved in differentiation, function, and survival of many cell types in which it is expressed. CREB3L1 can interact with other members of this subfamily such as itself and CREB3L3, as well as with CREM [78]. Targets for CREB3L1 include genes crucial for ER stress such as XBP1 and the chaperone heat shock protein 5 [93]. In addition to forming a heterodimer with CREB3L1, CREB3L3 can also heterodimerize with ATF6 and CEBPG [78,94]. Interestingly, CREB3L2 and CREB3L3 can be phosphorylated by cyclin-dependent kinases and influence cell division [63,95,96].

3.7. ATF-4 Related Members

ATF family members play essential roles in cell proliferation and differentiation, apoptosis, and inflammation. All members of this family, including ATF1, ATF2, ATF3, ATF4, ATF5, ATF6 and ATF7, can be detected in human Leydig cells [26]. However, their roles and functions in the regulation of gene expression in these cells remain to be determined. Like CREB1, ATF1 is involved in the regulation of transcription in response to variations in intracellular cAMP [97]. ATF2 and ATF3 regulate the expression of stress-response genes [98]. Interestingly, these ATF members can heterodimerize with transcription factors of the AP-1 family [3,24,99,100] to regulate gene expression. ATF4 is induced by stress signals including hypoxia, oxidative stress, ER stress and amino acid deprivation [101]. This transcription factor can bind to asymmetric CRE elements as a heterodimer and palindromic CRE elements as a homodimer. With DDIT3, ATF4 activates the transcription of the integrated stress response regulator tribbles pseudokinase 3 (TRIB3) and promotes ER stress-induced apoptosis of neuronal cells [102]. Interestingly, TRIB3 is highly expressed in different Leydig cell lines such as MA-10 and LC540 [103,104]. TRIB3 expression is rapidly modulated by xenobiotics, suggesting that its transcription may be regulated by immediate-early transcription factors like ATF4. Of interest for Leydig cells, ATF4 can heterodimerize with CEBPB [95] or JUN [3]. However, their target genes remain to be characterized. In addition, the activity of ATF4 can be modulated by phosphorylation at Ser^{69} [96,105], however the kinase involved and its functionality in Leydig cells remain to be determined.

ATF5 can bind to a ATF5-specific response element (ARE) (consensus: 5′-CYTCTYCCTTW-3′) or an amino acid response element (AARE) (consensus: 5′-WTTGCATCA-3′) present in many promoters. This transcription factor is involved in the regulation of cell survival, proliferation and differentiation [99,106]. Precisely, its transcriptional activity is enhanced by cyclin D3 and is inhibited by cyclin-dependent kinase 4 [106], suggesting that ATF5 may be an important regulator of Leydig cells' proliferation.

Similar to CREB3-like subfamily members, ATF6 is involved in the regulation of genes important for ER stress response [100]. These transcription factors contain a transmembrane domain, allowing their localization to the ER in the absence of activation of the UPR.

In general, members of the bZIP class of transcription factors are involved in tumorigenesis and normal physiological processes like cell proliferation, development, steroid synthesis, and cell-to-cell interactions.

4. bZIP Transcription Factors and Leydig Cell Proliferation and Development

Before reaching the adult stage, progenitor Leydig cells must proliferate and differentiate in the testicular interstitium to acquire their full capacity for androgen synthesis. Stem Leydig cells (SLC) proliferate and differentiate into progenitor Leydig cells (PLC) between postnatal day (P)7 and P14 in the rat. PLC are the major Leydig cells of the testis from P14 to P21 [107]. At around P35, the immature Leydig cells (ILC) population is established from the differentiation of PLC and mainly produces androstanediol (3α-Diol) as androgen. By the end of puberty (from P55 to P90), adult Leydig cells (ALC) originate from the proliferation and differentiation of ILC [108], and are characterized by high levels of LHCGR and expression of the enzyme HSD3B1 [109]. These changes contribute to the increased testosterone synthesis from ALC [110] and are essential for normal reproductive function [111] and spermatogenesis [112]. Once differentiated, ALCs are primarily involved in two processes important to support spermatogenesis: testosterone production, and INSL3 synthesis and secretion.

4.1. CREB-Related Members

Among CREB-related transcription factors, ATF1 may promote gene expression, contributing to Leydig cells' proliferation. Indeed, cyclin-dependent kinase 3 has been reported to phosphorylate ATF1 at Ser^{63}, enhancing transactivation capacity of ATF1 and promoting cell proliferation [113]. CREB3 is a transcriptional coregulator involved in ER-stress. In mouse Leydig tumor cells (MLTC-1), the shRNA-mediated knockdown of CREB3 decreased the percentage of S phase cells and increased the expressions of cyclin A1, B1 and D2, leading to reduced apoptosis and increased cell proliferation [114]. However, the use of tumor Leydig cell lines to study normal Leydig cell proliferation may not be appropriate, and the interpretation of the data should be done with caution.

4.2. AP-1 Members

AP-1 members are well known for their ability to regulate cell proliferation. Indeed, JUN and FOS regulate the expression of cyclin D1, an important activator of cyclin-dependent kinases that promote cell cycle progression [115]. JUNB and JUND, in contrast to JUN, negatively regulate cell proliferation. In fact, JUNB inhibits JUN-mediated cyclin D1 expression [116], whereas JUND activation results in decreased fibroblasts' proliferation [117]. The p21-activated protein kinase (PAK)2-dependent phosphorylation of JUN also contributes to EGF-mediated cell proliferation and transformation [118]. Since AP-1 members can regulate the expression of genes important for cell proliferation and differentiation [37,119–121], these transcription factors may influence Leydig cells' development and maturation. Interestingly, *Jun* and *Fos* are highly expressed in PLC, and like *Junb*, *Jund* and *Fosl2* show a decreased expression upon maturation of PLC to ALC [122]. In addition, *Jun* and *Fos* are highly expressed in rat mesenchymal cells involved in the regeneration of ALC [123].

Being highly expressed in PLC and ILC [124], the androgen receptor (AR) is critical for proliferation and differentiation of these cells into adult Leydig cells. Indeed, testicular feminized (*Tfm*) male mice, having a non-functional mutated androgen receptor, contain Leydig cells with reduced levels of LHCGR, are non-responsive to hCG and have decreased testosterone production due to reduced cytochrome P450 family 17 subfamily A member 1 (CYP17A1) levels [110,125]. In addition, mice with Leydig cell-specific inactivation of *Ar* by *Cre* expression under the control of the *Amhr2* promoter [126,127] produced less testosterone and were infertile due to disrupted spermatogenesis [128]. Proliferation and differentiation of PLC and ILC may require AP-1 members, since JUN can interact with AR [129,130] and can influence its transcriptional activity. However, such modulation of AR activity by AP-1 members remains to be confirmed in Leydig cells.

As suggested previously, increased ATF2 activity under pathophysiological conditions such as diabetes may increase caspase-3 activity and promote Leydig cells' apoptosis, leading to male hypogonadism [131]. ATF2, a transcription factor downstream of p38 MAPK, is a key mediator of extracellular stimulus-induced testicular injury. In addition, ATF2 can interact with AR [132,133], potentially modulating its transcriptional activity and influencing Leydig cells' proliferation.

4.3. CEBP-Related Members

During differentiation of SLC into ALC, the expression of *Alkbh5*, coding for an RNA demethylase, is increased, leading to reduced levels of N^6-methyladenosine methylation and increased autophagy in Leydig cells [134]. Interestingly, *Alkbh5* expression is increased in response to hCG by recruitment of CEBPB to its promoter region [134]. In SLCs, CEBPB also regulates the expression of *Ptgs2*, encoding the cyclooxygenase 2 enzyme which can modulate cell proliferation and apoptosis [135]. Thus, this bZIP transcription factor indirectly contributes to Leydig cell maturation by influencing mRNA methylation levels and the production of prostaglandins. DDIT3 can act as a dominant-negative inhibitor by forming a heterodimer with other CEBP members and preventing their DNA binding activity [136]. DDIT3 is activated by ER stress and promotes apoptosis [102,137]. Interestingly, decreased expression of DDIT3 is required for CEBPB to upregulate *Ptgs2* expression during SLC maturation [136].

As with other proline and acidic amino acid-rich (PAR) subfamily members, the bZIP transcription factor D site albumin promoter binding protein (DBP) is involved in the regulation of cell circadian rhythm, as reported with β-cells and the production of insulin [138]. Interestingly, the expression of DBP is inhibited by the mycotoxin zearelenone in Leydig cells [139]. However, further investigation will be required to better define its implication in regulation of proliferation and steroidogenic genes in Leydig cells.

4.4. Transmembrane bZIP Transcription Factors

In response to ER stress, cells initiate the UPR to restore protein homeostasis. The UPR is distinguished by three signaling proteins named protein kinase R (PKR)-like endoplasmic reticulum kinase (PERK), ATF6, and inositol-requiring enzyme 1 alpha (IRE1α). Activation of PERK leads to phosphorylation of eukaryotic initiation factor 2α, and then induces the bZIP transcription factor ATF4, causing increased expression of the pro-apoptotic bZIP protein DDIT3 and cell death [140]. On the other hand, the bZIP transcription factor ATF6 transits to the Golgi where it is cleaved into an activated form [140]. In addition, activated IRE1 catalyzes the splicing of a small intron from XBP1 mRNA to produce an active transcription factor. ATF6 and IRE1 pathways may also induce DDIT3 expression to trigger apoptosis [141]. Thus, ER stress can activate different bZIP transcription factors, leading Leydig cells to undergo apoptosis [89].

Overall, several transcription factors of the bZIP class influence the proliferation, maturation, and apoptosis of adult Leydig cells. However, their involvement in the differentiation of fetal Leydig cells remains to be demonstrated.

5. bZIP Transcription Factors and Steroidogenesis

Regulation of steroidogenesis in ALC is primarily mediated by the hypothalamic-pituitary-gonadal axis through LH secretion and binding to its LHCGR, resulting in de novo protein synthesis [142]. Following activation of its receptor, LH mainly stimulates cAMP production by adenylate cyclase, which leads to activation of PKA enzyme and upregulation of several genes contributing to increased androgen synthesis by Leydig cells. Members of the bZIP class being phosphorylated by PKA include CREB1, CREM, ATF1 and FOS. Hence, these transcription factors may participate in the regulation of steroidogenesis from Leydig cells.

5.1. CREB-Related Members

Different transcription factors, including bZIP members such as CREB1 [143] and AP-1 members [71,144], are involved in the regulation of *Star* expression in Leydig cells. The STAR protein participates in the import of cholesterol inside mitochondria, an essential step for the initiation of steroidogenesis. Three CRE half sites ($-96/-67$ bp) have been characterized and shown to recruit CREB1 to the cAMP-responsive region ($-151/-1$ bp) of the mouse *Star* promoter (Figure 4) [143]. CREB1 can also recruit the CREB regulated transcription coactivator 2 (CRTC2) to regulate *Star* expression in response to hormone and metabolic signals in Leydig cells [145,146]. PKA phosphorylates CREB1 on Ser^{133}, resulting in increased recruitment to DNA and target genes expression. Activation of the cAMP/PKA pathway may not be the sole inducer of CREB1 recruitment to its DNA regulatory elements. Indeed, PMA (phorbol-12-myristate-13-acetate), an activator of the PKC pathway, increases the phosphorylation of CREB1 and induces STAR expression in MA-10 Leydig cells [12,16]. Furthermore, activations of PKA and PKC signaling pathways result in an increased association of phosphorylated CREB1 to the proximal region of the *Star* promoter [147]. However, phosphorylation of the STAR protein by PKC cannot lead to its activation, unlike PKA. Thus, the contribution of PKC to the regulation of steroidogenesis in Leydig cells requires activation of PKA for the STAR protein to be phosphorylated and functional [12]. Hormone-induced recruitment of phosphorylated CREB1 and coactivator p300 to the *Nr4a1* promoter also requires activation of the Ca^{2+}/CAMKI pathway [148]. NR4A1 is an important mediator of hormone-stimulated *Star* expression [149,150]. Furthermore, silencing *Creb1* impaired NR4A1 expression and steroidogenesis in MA-10 Leydig cells [148]. NR4A1 is an orphan nuclear receptor involved in cAMP/PKA-dependent activation of steroidogenic promoters for *Star*, *HSD3B2* and *Cyp17a1* [149,151–153]. The initial substrate of steroidogenesis, cholesterol, is metabolised into pregnenolone by the cholesterol side chain cleavage enzyme CYP11A1 (cytochrome P450, family 11, subfamily a, polypeptide 1). Then, the enzyme hydroxy-delta-5-steroid dehydrogenase, 3 beta- and steroid delta-isomerase (HSD3B2 in human and HSD3B1 in rodents) catalyzes the conversion of pregnenolone into progesterone, whereas CYP17A1 converts progesterone into androstenedione. Testosterone is then synthesized from androstenedione by the enzyme HSD17B3.

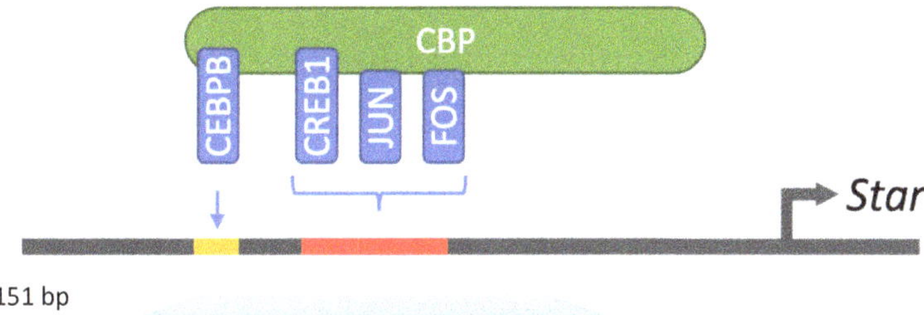

Figure 4. Overview of bZIP transcription factors recruited to the proximal region of the *Star* promoter. CEBPB is recruited to its regulatory element between −117 and −108 bp (5'-ATGAGGCAAT-3'), whereas CREB1, JUN and FOS form homo- or heterodimer competing for binding to the −96 to −67 bp region.

In addition to regulating steroidogenic genes, the CREB1 transcription factor also modulates the expression of other genes encoding enzymes in response to activation of the cAMP/PKA pathway in Leydig cells. For example, the expression of acyl-CoA synthetase 4 (*Acsl4*) is up-regulated by the recruitment of CREB1 to a DNA regulatory element at −200 bp of its promoter in MA-10 Leydig cells [154]. ACSL4 catalyzes the synthesis of acyl-CoA from arachidonic acid (AA), contributing to the regulation of AA at the cellular level. AA has been widely documented as an essential regulator of LH-dependent testosterone synthesis in rat Leydig cells [155–158].

Others have reported that CREB1 plays a critical role in activating kisspeptin (*Kiss1*) expression in mouse Leydig cells leading to autocrine regulation of steroidogenesis [159]. Indeed, CREB1-deficient Leydig cells fail to induce *Kiss1* expression in response to hCG treatment.

The expression of CREB3 is increased in mouse Leydig cells from prepubertal to adult stages [160]. The siRNA-mediated knockdown of CREB3 in Leydig cells results in the up-regulations of *Star*, *Cyp11a1*, *Hsd3b1* and *Cyp17a1* genes' expressions, leading to increased testosterone production [160]. Such CREB3-dependent regulation may involve inhibition of the nuclear receptors NR4A1 and NR5A1 [160], two nuclear receptors important for steroidogenic genes expressions. Hence, CREB3 may contribute, in part, to the decrease in testosterone production from Leydig cells according to aging by decreasing levels and activity of nuclear receptors.

The detection of CREBZF in mouse and human testis Leydig cells is debatable in vivo [26,161,162] and may be attributed to the characterization of different isoforms.

However, this transcription factor is involved in hCG-stimulated testosterone production [163]. CREBZF does not bind DNA by itself as a homodimer but rather regulates transcription by heterodimerizing with other transcription factors. In response to MAPK signaling, CREBZF can interact with ATF4 to enhance its transactivation potential [164]. CREBZF can also heterodimerize with XBP1 or CREB3 to participate in the regulation of the UPR response [165]. Silencing of *Crebzf* in primary and MLTC-1 Leydig cells decreased the expression of *Cyp17a1*, *Hsd3b1* and *Hsd17b3*, resulting in the inhibition of hCG-stimulated testosterone production [163]. Interestingly, CBEBZF activates the expression of *Nr4a1* and *Nr5a1* [163].

5.2. AP-1 Members

In addition to PKA, other signaling pathways such as PKC and MAPK may participate in the regulation of steroidogenesis in Leydig cells by phosphorylating specific AP-1 transcription factors [15,166]. Indeed, JUN can activate the *Nr4a1* promoter following its recruitment to three AP-1-like DNA regulatory elements in MA-10 Leydig cells [6,167]. In addition, JUN has additive effects with NR4A1 to activate the *Star* promoter by being recruited to the AP-1 regulatory element at −78 bp [71]. However, the binding of JUN to this regulatory element is not altered by treatment with cAMP for up to 2 h [5]. Other nuclear receptors such as NR5A1 and NR5A2 can also cooperate with JUN to activate *Star* gene expression [71]. In addition, NR5A1 cooperates with JUN to activate the human *CYP11A1* promoter in steroidogenic cells [168]. Notably, such cooperation between JUN and nuclear receptors involves protein–protein interactions. However, others have shown that JUN rather inhibits the capacity of NR4A1 to activate the *Cyp17a1* promoter in K28 Leydig cells [169]. In this regulatory process, the JNK signaling pathway is activated by reactive oxygen species, JUN and NR4A1 physically interact, and DNA binding of NR4A1 is inhibited. JUN may also interact with other nuclear receptors, such as the glucocorticoid receptor [170], the progesterone receptor [129], the estrogen receptor [129,171], the AR [129,130], the thyroid receptor [161] and the retinoic acid/retinoid X receptors [162,172], to influence the expression of steroidogenic genes. However, the implications of these protein interactions on steroidogenesis will need further investigation.

JUN also cooperates with GATA4 to activate *Star* expression [173]. This regulation is mostly dependent on the proximal GATA DNA-regulatory element in the *Star* promoter, suggesting that these transcription factors physically interact. Although a cooperation between JUN and FOS activates the *Star* promoter, JUN homodimers are rather responsible for the cooperation with GATA4 [173]. However, FOS can still be recruited to the mouse *Star* proximal promoter region [174], suggesting that JUN/FOS heterodimers may participate in the regulation of *Star* gene expression. Interestingly, *Fosl2* overexpression rather inhibits JUN-dependent activation of the *Star* promoter [174]. Thus, the combination of AP-1 factors may finely regulate *Star* expression according to the levels of androgen synthesis by Leydig cells.

Members of the AP-1 family may regulate steroidogenesis with precision through post-translational modifications. In response to PKA activation in MA-10 Leydig cells, the levels of FOS protein are increased and phosphorylated at Thr^{325}, whereas JUN is only phosphorylated at Ser^{73} after 15–30 min of dibutyryl-cAMP treatment [144]. Such post-translational modifications increase recruitment of these AP-1 members to the AP-1/CRE element located at −81 to −75 bp of the *Star* proximal promoter in mice [144]. AP-1 members JUN and FOS also promote CREB1 and CBP recruitments to the *Star* promoter. However, because AP-1 members can interact with a common DNA-regulatory element for CREB1, activation of *Star* expression by CREB1 is inhibited by FOS and JUN [144]. In addition to JUN, other AP-1 members such as FOS and JUND can also activate the *Star* promoter [71]. However, FOS rather inhibits steroidogenic gene expression in adrenal [175] and Leydig cells [144]. Precisely, AMP-activated protein kinase (AMPK)/PKC signaling pathway activation within Leydig cells represses JUN and NR4A1 but activates FOS and NR0B1 (DAX1), resulting in decreased *Star* expression and steroidogenesis [176]. Overall, JUN cooperates with several transcription factors to regulate the expression of steroidogenic genes, including specificity protein 1 (SP1) [177] and NR5A1 to regulate *Fdx1* (ferredoxin-1) [72], as well as GATA4, NR5A1 and NR4A1 to regulate *Star* [71,173], in Leydig cells. Indeed, JUN and FOS can physically interact with NR5A1, GATA4 and CEBPB as reported using the mammalian two-hybrid system [174]. Thus, AP-1 members can form major complexes contributing to the regulation of steroidogenic genes' expressions.

Several extracellular signals participate in the activation of members of the PKC family of serine/threonine kinases. In MA-10 Leydig cells, different PKC isoforms are induced by PMA and may contribute to *Star* activation and increased progesterone synthesis [15]. In addition, PMA increases JUN and FOS protein levels and their phosphorylation [15], suggesting that the activation of *Star* expression by PMA relies on activation of AP-1 members. Indeed, PKCμ inactivation decreases the expression and phosphorylation of JUN/FOS, as well as their association with the *Star* proximal promoter region, resulting in decreased *Star* expression and progesterone synthesis in MA-10 Leydig cells [15]. Interestingly, binding of JUN/FOS to the *Star* promoter in response to PMA promotes recruitment of the CBP cofactor [15], resulting in histone acetylation and increased accessibility to the promoter for the transcriptional machinery.

Once inside the mitochondria, cholesterol is converted to pregnenolone by the rate-limiting steroidogenic side chain cleavage enzyme, CYP11A1. In mouse trophoblast and granulosa cells, the *Cyp11a1* promoter activity may be regulated by AP-1 members and GATA4 [178]. Although AP-1 regulatory elements have been characterized in the *Cyp11a1* promoter [179], their implication in transcriptional regulation in Leydig cells remains to be confirmed.

After being reduced by the flavoprotein ferredoxin reductase, ferredoxin 1 (FDX1) supports steroid biosynthesis in steroidogenic cells by transferring electrons to CYP11A1. We have reported that JUN interacts with NR5A1 and these transcription factors cooperate to activate the *Fdx1* promoter in different Leydig cell lines [72]. Such activation involve regulatory elements located within the *Fdx1* proximal promoter region. Thus, AP-1 members may contribute to the activation of the CYP11A1 enzyme by increasing *Fdx1* gene expression in Leydig cells.

The gene *Tspo* encodes the translocator protein, a cholesterol-binding protein promoting the entry of cholesterol inside mitochondria [180–182]. The expression of *Tspo* in Leydig cells depends on PKCε-dependent phosphorylation of JUN through the MAPK pathway [166]. Moreover, PMA, an activator of PKCε, also promotes the binding of JUN to its AP-1 DNA regulatory element in the *Tspo* promoter [183]. Such regulatory mechanism may involve the interaction between JUN and STAT3, resulting in synergistic activation of the *Tspo* promoter [166]. However, the consequences of the regulation of *Tspo* by members of the AP-1 family on steroidogenesis by Leydig cells will require further investigations.

Different growth factors produced by Sertoli cells have paracrine actions influencing Leydig cells' functions. Among them, FGF, IGF1 and TGFα increase, whereas TGFβ decreases, androgen production from Leydig cells [184–187]. Precisely, IGF1 increases the availability of cholesterol, as well as the activities of steroidogenic enzymes, promoting the conversion of pregnenolone to testosterone [188]. Increased *Star* expression leading to progesterone synthesis has also been observed in response to stimulations of MLTC-1 cells with IGF1, FGF and TGFα [13]. Importantly, IGF1 and the cAMP/PKA pathway cooperate to activate CREB1 and JUN, leading to increased *Star* expression [189]. The expressions of the receptors for these growth factors can be regulated by AP-1 factors in Leydig cells, as shown with the upregulation of *FGFR1* by FOS in osteosarcoma cells [190]. Although several growth factors are produced by Sertoli cells, only kit ligand (KITL) and glial cell derived neurotrophic factor (GDNF) are required for male fertility. Interestingly, the transcription factors JUN and FOS may increase the expression of receptors for KITL and GDNF in Leydig cells, as indicated with the *cKIT* promoter [191].

In addition to LH, Leydig cells can also be stimulated by other hormones such as the growth hormone (GH). Indeed, GH increases *Star* gene expression and steroidogenesis by promoting cooperation between JUN and STAT5B [192]. However, it would be interesting to determine if such cooperation is also involved in the regulation of other STAT target genes such as *HSD3B2* [193] in Leydig cells.

Being part of the FOS-related family of transcription factors, the Jun dimerization protein 2 (JDP2) is highly expressed in MA-10 Leydig cells. In addition, its expression can be modulated by activation of the cAMP/PKA pathway (our unpublished observations). JDP2 binds DNA as a homodimer or as a heterodimer with ATF2 and JUN proteins, but not with FOS proteins [194]. JDP2 represses transcription either as a homo- or heterodimer [195]. However, its physiological role in regulation of steroidogenesis in testicular Leydig cells remains to be investigated further.

Overall, AP-1 members are critical regulators of the expression of genes related to steroidogenesis in Leydig cells. Precisely, the combination of expressed AP-1 factors, the post-translational modulation of their activities, and the interactions with other partners can all determine how these transcription factors influence androgen synthesis in the testis.

NFE2 Subfamily Members

The member of the nuclear factor, erythroid 2 (NFE2) subfamily of bZIP transcription factors NFE2L2 (NRF2) is a master regulator of phase 2 antioxidant genes and seems to play an important role in preventing the reduction in testosterone production from aging Leydig cells. Indeed, the knockout of *Nfe2l2* in mice results in a greater reduction in testosterone production by Leydig cells during aging compared to normal mice [196]. Although largely correlated, there is evidence that the redox imbalance occurring with aging may result in reduced steroid formation by Leydig cells [197]. Under oxidative stress, NFE2L2 dissociates from kelch-like ECH-associated protein 1 (KEAP1), translocates to the nucleus, and then activates NFE2L2-responsive genes involved in antioxidant activity and oxidant inactivation [198]. However, the target genes for NFE2L2 have not been clearly characterized in Leydig cells.

5.3. CEBP Members

CEBPB is expressed in steroidogenic cells, such as Leydig and granulosa cells [199], and plays an essential role in LH-regulated differentiation and function of Leydig cells [77]. Indeed, activations of PKA and PKC signaling pathways increase CEBPB protein levels in MA-10 Leydig cells [147]. CEBPB activates the expression of the *Star* gene, an important regulator of cholesterol import inside mitochondria [200,201]. Indeed, CEBPB can bind to the -117 to -108 bp *Star* promoter region [202–205], and can positively regulate *Star* gene basal expression [147,206]. Like other bZIP members, its transcriptional regulatory activity is enhanced by activation of the cAMP/PKA pathway [202]. In addition, CEBPB cooperates with different transcription factors, such as NR5A1 [200] and GATA4 [201], for cAMP/PKA-dependent regulation of *Star* promoter activity in MA-10 Leydig cells. It can also cooperate with nuclear factor kappa B subunit 1 (NFKB1) to regulate the expression of *Nr4a1* in Leydig cells [207].

The prostaglandin endoperoxide synthase 2 (PTGS2, COX2) enzyme, involved in the synthesis of $PGF_{2\alpha}$ and PGE_2, is highly expressed by Leydig cells from infertile patients [208] as well as during aging [209]. In MA-10 Leydig cells, CEBPB is recruited to a regulatory element in the proximal region of the *Ptgs2* promoter resulting in activation of gene expression in response to cAMP [135]. DDIT3, an inhibitor of CEBP family members by heterodimer formation [136], is also expressed in MA-10 cells [135]. In contrast to CEBPB, DDIT3 expression is rather reduced in response to hCG or cAMP [135]. Thus, induction of *Ptgs2* expression correlates with increased *Cebpb* and reduced *Ddit3* expressions in response to activation of the LH/cAMP/PKA pathway. Such activation of *Ptgs2* by CEBPB may contribute to decreased steroid production by aging Leydig cells.

5.4. Transmembrane bZIP Transcription Factors

In addition to the bZIP domain, several transcription factors also contain a transmembrane domain and participate in ER unfolded protein stress transduction. These bZIP transcription factors include ATF6 and the CREB3-like subfamily [210]. Since a major part of steroidogenesis is taking place in the ER, the UPR may have an indirect influence on gene regulation through activation of ER transmembrane bZIP transcription factors, leading to their migration in the nucleus. However, such gene regulation will require further investigation. Prolonged exposure to a high dose of hCG induces ER stress and downregulation of steroidogenic enzyme gene expression in Leydig cells [89]. Such effect may be attributed to the triggering of UPR and nuclear translocation of ATF6 in Leydig cells [89]. Interestingly, other bZIP transcription factors, such as CREB3L4, are involved in ER stress response and are crucial for proper spermatogenesis and spermiogenesis [211,212].

6. bZIP Transcription Factors and INSL3

Apart from producing steroid hormones, testicular Leydig cells are also the main source of the peptide hormone insulin-like 3 (INSL3). This hormone regulates the transabdominal phase of the descent of testes in the scrotum during fetal development as well as germ cell survival [213]. However, INSL3 is not required for spermatogenesis and germ cells' survival in adult mice [214]. The production of INSL3 is dependent on the differentiation state of Leydig cells and is correlated with the circulating levels of LH [215–218]. Intrestingly, differentiated Leydig cells produce and secrete INSL3 constitutively. The *Insl3* gene promoter contains several DNA regulatory elements for ATF3, CREB1 and MAF, suggesting a possible regulation by bZIP members. However, the regulation of *Insl3* expression by these transcription factors may be indirect and involve activation of *Nr4a1* expression, as reported for JUN [157,174,219].

7. bZIP Transcription Factors and Leydig Cell Communication

The regulation of steroidogenesis in Leydig cells depends not only on the activation of receptors associated with various signaling pathways but may also rely on the exchange of signaling molecules via gap junctions between adjacent cells. Gap junctions are formed by connexin proteins that assemble into homomeric or heteromeric connexons to create a channel for exchange of cytosolic molecules between neighboring cells. Different types of connexins are encoded by a class of genes including five groups (*Gja*, *Gjb*, *Gjc*, *Gjd* and *Gje*), each composed of several members and isoforms. Leydig cells express mainly *Gja1* (CX43), and to a lesser extent *Gja4* (CX37), *Gja6* (CX33), *Gjc1* (CX45), *Gjc2* (CX47), *Gjd2* (CX36), and *Gjd3* (CX31.9/30.2) [220–223]. Only the inactivation of GJA1 has been characterized for its effects on steroidogenesis and does not appear to be critical for Leydig cell development, differentiation and function [224,225]. This may be attributed to the functional redundancy between different types of connexins in Leydig cells [222]. However, GJA1 inactivation in Sertoli cells results in Leydig cell hyperplasia [226,227], suggesting that this connexin is involved in paracrine communication between these cell types. Interestingly, this phenotype is also associated with a decrease in *Gja1* and *Gjc1* expressions in Leydig cells, without any effect on steroidogenesis [228]. As reported in prostaglandin E2 treated cells [219], GJA1-dependent potentiation of cAMP/PKA signaling in Leydig cells is not associated with an increase in cAMP levels, but rather with an increased exchange of this second messenger between interconnected cells.

7.1. CREB Members

The expression of *Gja1* has been reported to be regulated by CREB1 in other cell types. Indeed, β-adrenoceptor stimulation upregulates cardiac *Gja1* expression via PKA and MAPK signaling pathways, possibly through activation of AP1 and CREB1 factors [229]. In addition, a complex composed of NFKB, CREB1 and CBP is formed following oxytocin stimulation and participates in the activation of *Gja1* expression in mouse embryonic stem cells [230]. It has been reported that hCG and 8-bromo-cAMP treatments, known to increase

Cyp11a1 and *Star* expression, inhibit *Gja1* expression in rat Leydig cells [231]. However, others have shown that these treatments rather increase GJA1 levels in cultured Leydig cells [232], suggesting that junctional coupling between Leydig cells is hormonally regulated. Indeed, *Gja1* expression and cell-to-cell communication are modulated by long-term (36 h) stimulation with LH or hCG [232]. Furthermore, GJA1 expression in postnatal Leydig cells has been correlated with testosterone production at different stages of postnatal development [233]. Indeed, *Gja1* expression peaks at P40 and remains elevated throughout adulthood in rat Leydig cells [231]. Thus, gap junctions may enhance testosterone production by facilitating the exchange of cAMP between neighboring cells to increase steroidogenic gene expression. However, the implication of CREB1 in the regulation of *Gja1* expression in Leydig cells remains to be confirmed.

7.2. AP-1 Members

AP-1 members JUN and FOS are major regulators of *Gja1* expression in mouse TM3 Leydig cell line [234]. Such synergy involves the recruitment of FOS to an AP-1 DNA-regulatory element located in the proximal region of the *Gja1* promoter (Figure 5). This element is highly conserved among species, suggesting that the ability of AP-1 members to regulate *Gja1* expression has been conserved throughout evolution. In addition, AP-1 members may be functionally redundant as JUN, JUNB and FOSL2 also efficiently increased *Gja1* promoter activity in TM3 Leydig cells [234]. With significant endogenous levels in Leydig cells [223,235], FOS or FOSL2 can interact with JUN, JUNB or JUND to regulate *Gja1* expression. Thus, AP-1 dimers binding to the *Gja1* promoter may consist of FOS/JUN, FOS/JUNB, FOS/JUND, FOSL2/JUN, FOSL2/JUNB or FOSL2/JUND in mouse Leydig cells. AP-1 members are therefore important for optimal *Gja1* expression in testicular Leydig cells. In addition to variations in the abundance of FOS and JUN proteins in Leydig cells, activation of the *Gja1* promoter may also involve their post-translational modifications by protein kinases and their interactions with other partners or coactivators [121].

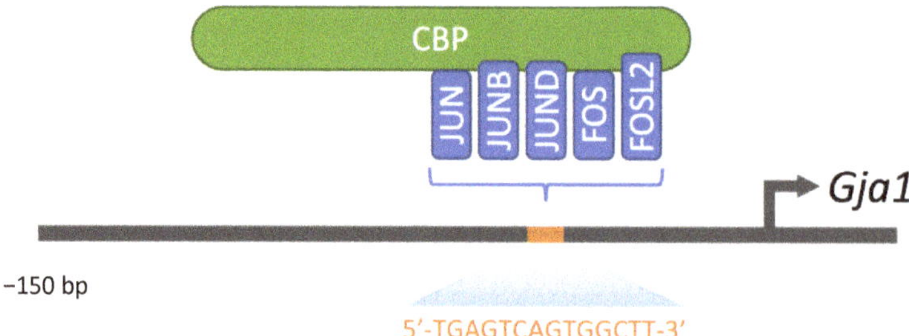

Figure 5. Overview of bZIP transcription factors recruited to the proximal region of the *Gja1* promoter. JUN, JUNB, JUND, FOS and FOSL2 can participate in the formation of homo- or heterodimers to be recruited to the −45 bp AP-1 DNA regulatory element.

7.2.1. NFE2 Subfamily Members

In response to treatments with the polyphenolic compound luteolin, the transcription factor NFE2L2 translocates into the nucleus of TM4 Sertoli cells and is associated with an increased expression of *Gja1* [236]. However, even though the *Gja1* promoter contains several NFE2L2 DNA regulatory elements in its proximal region, the recruitment of NFE2L2 to this region remains to be confirmed. In addition, such regulation of *Gja1* expression by NFE2L2 in Leydig cells should be characterized further.

7.2.2. ATF3-like Subfamily Members

As reported previously, JDP2 is highly expressed in MA-10 Leydig cells and is inducible following activation of the cAMP/PKA pathway (our unpublished observations). Interestingly, JDP2 overexpression in myocardial cells results in loss of *Gja1* expression [237], suggesting that such regulatory mechanism may be present in Leydig cells. Moreover, several JDP2 DNA regulatory elements can be found in the proximal region of the *Gja1* promoter.

7.3. MAF-Related Members

In chicken embryos, overexpression of Maf and Mafa induced ectopic expression of *Gja1* [238]. In human testis, MAF is highly expressed in Leydig cells and cells from the seminiferous tubules [26]. In addition, the *Gja1* promoter contains several MAF regulatory elements in its proximal region. However, the molecular implication of MAF transcription factors in the regulation of *Gja1* expression in Leydig cells remains to be investigated.

7.4. CEBP Members

In breast cancer cells, CEBPA regulates *GJA1* expression by binding to the DNA regulatory element 5'-AATTGTC-3' at −456 bp of the promoter region [239]. However, CEBPA expression in Leydig cells appears to be very low [26]. Thus, its role in the regulation of *Gja1* expression may be replaced by CEBPB but remains to be confirmed.

8. Conclusions

Significant progress has been made over the past 40 years in our knowledge of the signaling pathways and the underlying processes that control the activity of bZIP transcription factors in different cell types. As highlighted in this review, CREB, CEBP and AP-1 members of the bZIP family play important roles in regulating gene expression critical for adequate Leydig cell function in the testis. Their specificity of action in Leydig cells relies on bZIP-dependent homo- and heterodimer formation, as well as interactions with other types of transcription factors or cofactors. Hence, major questions remain regarding the mechanisms of action of bZIP members in Leydig cells and include: What are the contributions of signaling pathways in the activation of bZIP members and in the formation of dimers leading to the regulation of Leydig cell-specific target genes? What is the importance of redundancy between bZIP members in the formation of heterodimers? Can heterodimers between bZIP members vary depending on the target gene, cellular context, or signaling pathways activated? How is the binding of bZIP factors to their DNA regulatory elements coordinated with the recruitment of other transcription factors nearby, and what are the consequences of these recruitments to DNA on transcriptional regulation? Answers to these questions are essential to a better understanding of how transcription factors of the bZIP class influence gene expression in Leydig cells.

Author Contributions: Data curation, L.J.M. and H.T.N.; writing—original draft preparation, L.J.M.; writing—review and editing, L.J.M. and H.T.N.; funding acquisition, L.J.M. All authors have read and agreed to the published version of the manuscript.

Funding: This research was funded by the Natural Sciences and Engineering Research Council (NSERC) of Canada, grant number 386557 to L.J.M.

Institutional Review Board Statement: Not applicable.

Informed Consent Statement: Not applicable.

Data Availability Statement: Not applicable.

Conflicts of Interest: The authors declare no conflict of interest. The funders had no role in the design of the study; in the collection, analyses, or interpretation of data; in the writing of the manuscript; or in the decision to publish the results.

References

1. Wingender, E.; Schoeps, T.; Haubrock, M.; Krull, M.; Dönitz, J. TFClass: Expanding the Classification of Human Transcription Factors to Their Mammalian Orthologs. *Nucleic Acids Res.* **2018**, *46*, D343–D347. [CrossRef] [PubMed]
2. Mayr, B.; Montminy, M. Transcriptional Regulation by the Phosphorylation-Dependent Factor CREB. *Nat. Rev. Mol. Cell Biol.* **2001**, *2*, 599–609. [CrossRef] [PubMed]

3. Hai, T.; Curran, T. Cross-Family Dimerization of Transcription Factors Fos/Jun and ATF/CREB Alters DNA Binding Specificity. *Proc. Natl. Acad. Sci. USA* **1991**, *88*, 3720–3724. [CrossRef] [PubMed]
4. Radhakrishnan, I.; Pérez-Alvarado, G.C.; Parker, D.; Dyson, H.J.; Montminy, M.R.; Wright, P.E. Solution Structure of the KIX Domain of CBP Bound to the Transactivation Domain of CREB: A Model for Activator:Coactivator Interactions. *Cell* **1997**, *91*, 741–752. [CrossRef]
5. Clem, B.F.; Hudson, E.A.; Clark, B.J. Cyclic Adenosine 3′,5′-Monophosphate (CAMP) Enhances CAMP-Responsive Element Binding (CREB) Protein Phosphorylation and Phospho-CREB Interaction with the Mouse Steroidogenic Acute Regulatory Protein Gene Promoter. *Endocrinology* **2005**, *146*, 1348–1356. [CrossRef] [PubMed]
6. Inaoka, Y.; Yazawa, T.; Uesaka, M.; Mizutani, T.; Yamada, K.; Miyamoto, K. Regulation of NGFI-B/Nur77 Gene Expression in the Rat Ovary and in Leydig Tumor Cells MA-10. *Mol. Reprod. Dev.* **2008**, *75*, 931–939. [CrossRef] [PubMed]
7. Dash, P.K.; Karl, K.A.; Colicos, M.A.; Prywes, R.; Kandel, E.R. CAMP Response Element-Binding Protein Is Activated by Ca2+/Calmodulin- as Well as CAMP-Dependent Protein Kinase. *Proc. Natl. Acad. Sci. USA* **1991**, *88*, 5061–5065. [CrossRef]
8. Sheng, M.; Thompson, M.A.; Greenberg, M.E. CREB: A Ca(2+)-Regulated Transcription Factor Phosphorylated by Calmodulin-Dependent Kinases. *Science* **1991**, *252*, 1427–1430. [CrossRef] [PubMed]
9. Du, K.; Montminy, M. CREB Is a Regulatory Target for the Protein Kinase Akt/PKB. *J. Biol. Chem.* **1998**, *273*, 32377–32379. [CrossRef] [PubMed]
10. Ginty, D.D.; Bonni, A.; Greenberg, M.E. Nerve Growth Factor Activates a Ras-Dependent Protein Kinase That Stimulates c-Fos Transcription via Phosphorylation of CREB. *Cell* **1994**, *77*, 713–725. [CrossRef]
11. Oury, F.; Sumara, G.; Sumara, O.; Ferron, M.; Chang, H.; Smith, C.E.; Hermo, L.; Suarez, S.; Roth, B.L.; Ducy, P.; et al. Endocrine Regulation of Male Fertility by the Skeleton. *Cell* **2011**, *144*, 796–809. [CrossRef] [PubMed]
12. Jo, Y.; King, S.R.; Khan, S.A.; Stocco, D.M. Involvement of Protein Kinase C and Cyclic Adenosine 3′,5′-Monophosphate-Dependent Kinase in Steroidogenic Acute Regulatory Protein Expression and Steroid Biosynthesis in Leydig Cells. *Biol. Reprod.* **2005**, *73*, 244–255. [CrossRef] [PubMed]
13. Manna, P.R.; Chandrala, S.P.; Jo, Y.; Stocco, D.M. CAMP-Independent Signaling Regulates Steroidogenesis in Mouse Leydig Cells in the Absence of StAR Phosphorylation. *J. Mol. Endocrinol.* **2006**, *37*, 81–95. [CrossRef] [PubMed]
14. Stocco, D.M.; Wang, X.; Jo, Y.; Manna, P.R. Multiple Signaling Pathways Regulating Steroidogenesis and Steroidogenic Acute Regulatory Protein Expression: More Complicated than We Thought. *Mol. Endocrinol.* **2005**, *19*, 2647–2659. [CrossRef] [PubMed]
15. Manna, P.R.; Soh, J.-W.; Stocco, D.M. The Involvement of Specific PKC Isoenzymes in Phorbol Ester-Mediated Regulation of Steroidogenic Acute Regulatory Protein Expression and Steroid Synthesis in Mouse Leydig Cells. *Endocrinology* **2011**, *152*, 313–325. [CrossRef] [PubMed]
16. Manna, P.R.; Jo, Y.; Stocco, D.M. Regulation of Leydig Cell Steroidogenesis by Extracellular Signal-Regulated Kinase 1/2: Role of Protein Kinase A and Protein Kinase C Signaling. *J. Endocrinol.* **2007**, *193*, 53–63. [CrossRef] [PubMed]
17. Sassone-Corsi, P. Coupling Gene Expression to CAMP Signalling: Role of CREB and CREM. *Int. J. Biochem. Cell Biol.* **1998**, *30*, 27–38. [CrossRef]
18. Sassone-Corsi, P. Transcription Factors Responsive to CAMP. *Annu. Rev. Cell Dev. Biol.* **1995**, *11*, 355–377. [CrossRef]
19. Manna, P.R.; Eubank, D.W.; Lalli, E.; Sassone-Corsi, P.; Stocco, D.M. Transcriptional Regulation of the Mouse Steroidogenic Acute Regulatory Protein Gene by the CAMP Response-Element Binding Protein and Steroidogenic Factor 1. *J. Mol. Endocrinol.* **2003**, *30*, 381–397. [CrossRef]
20. Molina, C.A.; Foulkes, N.S.; Lalli, E.; Sassone-Corsi, P. Inducibility and Negative Autoregulation of CREM: An Alternative Promoter Directs the Expression of ICER, an Early Response Repressor. *Cell* **1993**, *75*, 875–886. [CrossRef]
21. Laoide, B.M.; Foulkes, N.S.; Schlotter, F.; Sassone-Corsi, P. The Functional Versatility of CREM Is Determined by Its Modular Structure. *EMBO J.* **1993**, *12*, 1179–1191. [CrossRef] [PubMed]
22. Hurst, H.C.; Totty, N.F.; Jones, N.C. Identification and Functional Characterisation of the Cellular Activating Transcription Factor 43 (ATF-43) Protein. *Nucleic Acids Res.* **1991**, *19*, 4601–4609. [CrossRef]
23. Hai, T.W.; Liu, F.; Coukos, W.J.; Green, M.R. Transcription Factor ATF CDNA Clones: An Extensive Family of Leucine Zipper Proteins Able to Selectively Form DNA-Binding Heterodimers. *Genes Dev.* **1989**, *3*, 2083–2090. [CrossRef] [PubMed]
24. Hoeffler, J.P.; Lustbader, J.W.; Chen, C.Y. Identification of Multiple Nuclear Factors That Interact with Cyclic Adenosine 3′,5′-Monophosphate Response Element-Binding Protein and Activating Transcription Factor-2 by Protein-Protein Interactions. *Mol. Endocrinol.* **1991**, *5*, 256–266. [CrossRef]
25. Ivashkiv, L.B.; Liou, H.C.; Kara, C.J.; Lamph, W.W.; Verma, I.M.; Glimcher, L.H. MXBP/CRE-BP2 and c-Jun Form a Complex Which Binds to the Cyclic AMP, but Not to the 12-O-Tetradecanoylphorbol-13-Acetate, Response Element. *Mol. Cell. Biol.* **1990**, *10*, 1609–1621. [CrossRef] [PubMed]
26. Uhlén, M.; Fagerberg, L.; Hallström, B.M.; Lindskog, C.; Oksvold, P.; Mardinoglu, A.; Sivertsson, Å.; Kampf, C.; Sjöstedt, E.; Asplund, A.; et al. Proteomics. Tissue-Based Map of the Human Proteome. *Science* **2015**, *347*, 1260419. [CrossRef] [PubMed]
27. Shimomura, A.; Ogawa, Y.; Kitani, T.; Fujisawa, H.; Hagiwara, M. Calmodulin-Dependent Protein Kinase II Potentiates Transcriptional Activation through Activating Transcription Factor 1 but Not CAMP Response Element-Binding Protein. *J. Biol. Chem.* **1996**, *271*, 17957–17960. [CrossRef] [PubMed]
28. O'Shea, E.K.; Rutkowski, R.; Kim, P.S. Mechanism of Specificity in the Fos-Jun Oncoprotein Heterodimer. *Cell* **1992**, *68*, 699–708. [CrossRef]

29. Ryseck, R.P.; Bravo, R. C-JUN, JUN B, and JUN D Differ in Their Binding Affinities to AP-1 and CRE Consensus Sequences: Effect of FOS Proteins. *Oncogene* **1991**, *6*, 533–542.
30. Halazonetis, T.D.; Georgopoulos, K.; Greenberg, M.E.; Leder, P. C-Jun Dimerizes with Itself and with c-Fos, Forming Complexes of Different DNA Binding Affinities. *Cell* **1988**, *55*, 917–924. [CrossRef]
31. Angel, P.; Karin, M. The Role of Jun, Fos and the AP-1 Complex in Cell-Proliferation and Transformation. *Biochim. Biophys. Acta (BBA)-Rev. Cancer* **1991**, *1072*, 129–157. [CrossRef]
32. Gazon, H.; Barbeau, B.; Mesnard, J.-M.; Peloponese, J.-M.J. Hijacking of the AP-1 Signaling Pathway during Development of ATL. *Front. Microbiol.* **2018**, *8*. [CrossRef]
33. Karin, M. The Regulation of AP-1 Activity by Mitogen-Activated Protein Kinases. *J. Biol. Chem.* **1995**, *270*, 16483–16486. [CrossRef] [PubMed]
34. Rauscher, F.J.; Voulalas, P.J.; Franza, B.R.; Curran, T. Fos and Jun Bind Cooperatively to the AP-1 Site: Reconstitution in Vitro. *Genes Dev.* **1988**, *2*, 1687–1699. [CrossRef] [PubMed]
35. Suzuki, T.; Okuno, H.; Yoshida, T.; Endo, T.; Nishina, H.; Iba, H. Difference in Transcriptional Regulatory Function between C-Fos and Fra-2. *Nucleic Acids Res.* **1991**, *19*, 5537–5542. [CrossRef]
36. Nakabeppu, Y.; Nathans, D. The Basic Region of Fos Mediates Specific DNA Binding. *EMBO J.* **1989**, *8*, 3833–3841. [CrossRef]
37. Jochum, W.; Passegué, E.; Wagner, E.F. AP-1 in Mouse Development and Tumorigenesis. *Oncogene* **2001**, *20*, 2401–2412. [CrossRef] [PubMed]
38. Passegué, E.; Jochum, W.; Behrens, A.; Ricci, R.; Wagner, E.F. JunB Can Substitute for Jun in Mouse Development and Cell Proliferation. *Nat. Genet.* **2002**, *30*, 158–166. [CrossRef] [PubMed]
39. Boyle, W.J.; Smeal, T.; Defize, L.H.; Angel, P.; Woodgett, J.R.; Karin, M.; Hunter, T. Activation of Protein Kinase C Decreases Phosphorylation of C-Jun at Sites That Negatively Regulate Its DNA-Binding Activity. *Cell* **1991**, *64*, 573–584. [CrossRef]
40. Abate, C.; Baker, S.J.; Lees-Miller, S.P.; Anderson, C.W.; Marshak, D.R.; Curran, T. Dimerization and DNA Binding Alter Phosphorylation of Fos and Jun. *Proc. Natl. Acad. Sci. USA* **1993**, *90*, 6766–6770. [CrossRef] [PubMed]
41. van Dam, H.; Duyndam, M.; Rottier, R.; Bosch, A.; de Vries-Smits, L.; Herrlich, P.; Zantema, A.; Angel, P.; van der Eb, A.J. Heterodimer Formation of CJun and ATF-2 Is Responsible for Induction of c-Jun by the 243 Amino Acid Adenovirus E1A Protein. *EMBO J.* **1993**, *12*, 479–487. [CrossRef] [PubMed]
42. Sharma, S.C.; Richards, J.S. Regulation of AP1 (Jun/Fos) Factor Expression and Activation in Ovarian Granulosa Cells: Relation of JunD and Fra2 to terminal differentiation. *J. Biol. Chem.* **2000**, *275*, 33718–33728. [CrossRef] [PubMed]
43. Robertson, L.M.; Kerppola, T.K.; Vendrell, M.; Luk, D.; Smeyne, R.J.; Morgan, J.I.; Curran, T. Regulation of C-Los Expression in Transgenic Mice Requires Multiple Interdependent Transcription Control Elements. *Neuron* **1995**, *14*, 241–252. [CrossRef]
44. Sheng, M.; McFadden, G.; Greenberg, M.E. Membrane Depolarization and Calcium Induce C-Fos Transcription via Phosphorylation of Transcription Factor CREB. *Neuron* **1990**, *4*, 571–582. [CrossRef]
45. Wagner, B.J.; Hayes, T.E.; Hoban, C.J.; Cochran, B.H. The SIF Binding Element Confers Sis/PDGF Inducibility onto the c-Fos Promoter. *EMBO J.* **1990**, *9*, 4477–4484. [CrossRef] [PubMed]
46. Sheng, M.; Dougan, S.T.; McFadden, G.; Greenberg, M.E. Calcium and Growth Factor Pathways of C-Fos Transcriptional Activation Require Distinct Upstream Regulatory Sequences. *Mol. Cell. Biol.* **1988**, *8*, 2787–2796. [CrossRef]
47. Cavigelli, M.; Dolfi, F.; Claret, F.X.; Karin, M. Induction of C-Fos Expression through JNK-Mediated TCF/Elk-1 Phosphorylation. *EMBO J.* **1995**, *14*, 5957–5964. [CrossRef]
48. Sadowski, H.B.; Shuai, K.; Darnell, J.E.; Gilman, M.Z. A Common Nuclear Signal Transduction Pathway Activated by Growth Factor and Cytokine Receptors. *Science* **1993**, *261*, 1739–1744. [CrossRef]
49. Hoey, T.; Schindler, U. STAT Structure and Function in Signaling. *Curr. Opin. Genet. Dev.* **1998**, *8*, 582–587. [CrossRef]
50. Dey, A.; Nebert, D.W.; Ozato, K. The AP-1 Site and the CAMP- and Serum Response Elements of the c-Fos Gene Are Constitutively Occupied in Vivo. *DNA Cell Biol.* **1991**, *10*, 537–544. [CrossRef]
51. Angel, P.; Hattori, K.; Smeal, T.; Karin, M. The Jun Proto-Oncogene Is Positively Autoregulated by Its Product, Jun/AP-1. *Cell* **1988**, *55*, 875–885. [CrossRef]
52. Rozek, D.; Pfeifer, G.P. In Vivo Protein-DNA Interactions at the c-Jun Promoter: Preformed Complexes Mediate the UV Response. *Mol. Cell. Biol.* **1993**, *13*, 5490–5499. [CrossRef] [PubMed]
53. Sassone-Corsi, P.; Sisson, J.C.; Verma, I.M. Transcriptional Autoregulation of the Proto-Oncogene Fos. *Nature* **1988**, *334*, 314–319. [CrossRef] [PubMed]
54. Chiu, R.; Boyle, W.J.; Meek, J.; Smeal, T.; Hunter, T.; Karin, M. The C-Fos Protein Interacts with c-Jun/AP-1 to Stimulate Transcription of AP-1 Responsive Genes. *Cell* **1988**, *54*, 541–552. [CrossRef]
55. Hibi, M.; Lin, A.; Smeal, T.; Minden, A.; Karin, M. Identification of an Oncoprotein- and UV-Responsive Protein Kinase That Binds and Potentiates the c-Jun Activation Domain. *Genes Dev.* **1993**, *7*, 2135–2148. [CrossRef]
56. Lin, A.; Frost, J.; Deng, T.; Smeal, T.; al-Alawi, N.; Kikkawa, U.; Hunter, T.; Brenner, D.; Karin, M. Casein Kinase II Is a Negative Regulator of C-Jun DNA Binding and AP-1 Activity. *Cell* **1992**, *70*, 777–789. [CrossRef]
57. Morton, S.; Davis, R.J.; McLaren, A.; Cohen, P. A Reinvestigation of the Multisite Phosphorylation of the Transcription Factor C-Jun. *EMBO J.* **2003**, *22*, 3876–3886. [CrossRef]
58. Arias, J.; Alberts, A.S.; Brindle, P.; Claret, F.X.; Smeal, T.; Karin, M.; Feramisco, J.; Montminy, M. Activation of CAMP and Mitogen Responsive Genes Relies on a Common Nuclear Factor. *Nature* **1994**, *370*, 226–229. [CrossRef]

59. Mendelson, K.G.; Contois, L.R.; Tevosian, S.G.; Davis, R.J.; Paulson, K.E. Independent Regulation of JNK/P38 Mitogen-Activated Protein Kinases by Metabolic Oxidative Stress in the Liver. *Proc. Natl. Acad. Sci. USA* **1996**, *93*, 12908–12913. [CrossRef]

60. Dephoure, N.; Zhou, C.; Villén, J.; Beausoleil, S.A.; Bakalarski, C.E.; Elledge, S.J.; Gygi, S.P. A Quantitative Atlas of Mitotic Phosphorylation. *Proc. Natl. Acad. Sci. USA* **2008**, *105*, 10762–10767. [CrossRef]

61. Deng, T.; Karin, M. C-Fos Transcriptional Activity Stimulated by H-Ras-Activated Protein Kinase Distinct from JNK and ERK. *Nature* **1994**, *371*, 171–175. [CrossRef] [PubMed]

62. Murphy, L.O.; Smith, S.; Chen, R.-H.; Fingar, D.C.; Blenis, J. Molecular Interpretation of ERK Signal Duration by Immediate Early Gene Products. *Nat. Cell Biol.* **2002**, *4*, 556–564. [CrossRef]

63. Pellegrino, M.J.; Stork, P.J.S. Sustained Activation of Extracellular Signal-Regulated Kinase by Nerve Growth Factor Regulates c-Fos Protein Stabilization and Transactivation in PC12 Cells. *J Neurochem* **2006**, *99*, 1480–1493. [CrossRef] [PubMed]

64. Tratner, I.; Ofir, R.; Verma, I.M. Alteration of a Cyclic AMP-Dependent Protein Kinase Phosphorylation Site in the c-Fos Protein Augments Its Transforming Potential. *Mol. Cell. Biol.* **1992**, *12*, 998–1006. [CrossRef] [PubMed]

65. Kallunki, T.; Deng, T.; Hibi, M.; Karin, M. C-Jun Can Recruit JNK to Phosphorylate Dimerization Partners via Specific Docking Interactions. *Cell* **1996**, *87*, 929–939. [CrossRef]

66. Hall, S.H.; Berthelon, M.C.; Avallet, O.; Saez, J.M. Regulation of C-Fos, c-Jun, Jun-B, and c-Myc Messenger Ribonucleic Acids by Gonadotropin and Growth Factors in Cultured Pig Leydig Cell. *Endocrinology* **1991**, *129*, 1243–1249. [CrossRef]

67. Vietor, I.; Schwenger, P.; Li, W.; Schlessinger, J.; Vilcek, J. Tumor Necrosis Factor-Induced Activation and Increased Tyrosine Phosphorylation of Mitogen-Activated Protein (MAP) Kinase in Human Fibroblasts. *J. Biol. Chem.* **1993**, *268*, 18994–18999. [CrossRef]

68. Sánchez, I.; Hughes, R.T.; Mayer, B.J.; Yee, K.; Woodgett, J.R.; Avruch, J.; Kyriakis, J.M.; Zon, L.I. Role of SAPK/ERK Kinase-1 in the Stress-Activated Pathway Regulating Transcription Factor c-Jun. *Nature* **1994**, *372*, 794–798. [CrossRef]

69. Li, X.; Hales, K.H.; Watanabe, G.; Lee, R.J.; Pestell, R.G.; Hales, D.B. The Effect of Tumor Necrosis Factor-Alpha and CAMP on Induction of AP-1 Activity in MA-10 Tumor Leydig Cells. *Endocrine* **1997**, *6*, 317–324. [CrossRef]

70. Arai, K.Y.; Roby, K.F.; Terranova, P.F. Tumor Necrosis Factor Alpha (TNF) Suppresses CAMP Response Element (CRE) Activity and Nuclear CRE Binding Protein in MA-10 Mouse Leydig Tumor Cells. *Endocrine* **2005**, *27*, 17–24. [CrossRef]

71. Martin, L.J.; Tremblay, J.J. The Nuclear Receptors NUR77 and SF1 Play Additive Roles with C-JUN through Distinct Elements on the Mouse Star Promoter. *J. Mol. Endocrinol.* **2009**, *42*, 119–129. [CrossRef]

72. Roumaud, P.; Rwigemera, A.; Martin, L.J. Transcription Factors SF1 and CJUN Cooperate to Activate the Fdx1 Promoter in MA-10 Leydig Cells. *J. Steroid Biochem. Mol. Biol.* **2017**, *171*, 121–132. [CrossRef] [PubMed]

73. Suzuki, S.; Nagaya, T.; Suganuma, N.; Tomoda, Y.; Seo, H. Inductions of Immediate Early Genes (IEGS) and Ref-1 by Human Chorionic Gonadotropin in Murine Leydig Cell Line (MA-10). *Biochem. Mol. Biol. Int.* **1998**, *44*, 217–224. [CrossRef] [PubMed]

74. Wegner, M.; Cao, Z.; Rosenfeld, M.G. Calcium-Regulated Phosphorylation within the Leucine Zipper of C/EBP Beta. *Science* **1992**, *256*, 370–373. [CrossRef] [PubMed]

75. Xu, M.; Nie, L.; Kim, S.-H.; Sun, X.-H. STAT5-Induced Id-1 Transcription Involves Recruitment of HDAC1 and Deacetylation of C/EBPbeta. *EMBO J.* **2003**, *22*, 893–904. [CrossRef] [PubMed]

76. Ramji, D.P.; Foka, P. CCAAT/Enhancer-Binding Proteins: Structure, Function and Regulation. *Biochem. J.* **2002**, *365*, 561–575. [CrossRef] [PubMed]

77. Nalbant, D.; Williams, S.C.; Stocco, D.M.; Khan, S.A. Luteinizing Hormone-Dependent Gene Regulation in Leydig Cells May Be Mediated by CCAAT/Enhancer-Binding Protein-Beta. *Endocrinology* **1998**, *139*, 272–279. [CrossRef] [PubMed]

78. Newman, J.R.S.; Keating, A.E. Comprehensive Identification of Human BZIP Interactions with Coiled-Coil Arrays. *Science* **2003**, *300*, 2097–2101. [CrossRef]

79. Nakajima, T.; Kinoshita, S.; Sasagawa, T.; Sasaki, K.; Naruto, M.; Kishimoto, T.; Akira, S. Phosphorylation at Threonine-235 by a Ras-Dependent Mitogen-Activated Protein Kinase Cascade Is Essential for Transcription Factor NF-IL6. *Proc. Natl. Acad. Sci. USA* **1993**, *90*, 2207–2211. [CrossRef]

80. Eaton, E.M.; Sealy, L. Modification of CCAAT/Enhancer-Binding Protein-Beta by the Small Ubiquitin-like Modifier (SUMO) Family Members, SUMO-2 and SUMO-3. *J. Biol. Chem.* **2003**, *278*, 33416–33421. [CrossRef]

81. Kataoka, K.; Noda, M.; Nishizawa, M. Maf Nuclear Oncoprotein Recognizes Sequences Related to an AP-1 Site and Forms Heterodimers with Both Fos and Jun. *Mol. Cell. Biol.* **1994**, *14*, 700–712. [CrossRef]

82. DeFalco, T.; Takahashi, S.; Capel, B. Two Distinct Origins for Leydig Cell Progenitors in the Fetal Testis. *Dev. Biol.* **2011**, *352*, 14–26. [CrossRef]

83. Shawki, H.H.; Oishi, H.; Usui, T.; Kitadate, Y.; Basha, W.A.; Abdellatif, A.M.; Hasegawa, K.; Okada, R.; Mochida, K.; El-Shemy, H.A.; et al. MAFB Is Dispensable for the Fetal Testis Morphogenesis and the Maintenance of Spermatogenesis in Adult Mice. *PLoS ONE* **2018**, *13*, e0190800. [CrossRef]

84. Liou, H.C.; Boothby, M.R.; Glimcher, L.H. Distinct Cloned Class II MHC DNA Binding Proteins Recognize the X Box Transcription Element. *Science* **1988**, *242*, 69–71. [CrossRef] [PubMed]

85. Sriburi, R.; Jackowski, S.; Mori, K.; Brewer, J.W. XBP1: A Link between the Unfolded Protein Response, Lipid Biosynthesis, and Biogenesis of the Endoplasmic Reticulum. *J. Cell Biol.* **2004**, *167*, 35–41. [CrossRef] [PubMed]

86. Lee, A.-H.; Heidtman, K.; Hotamisligil, G.S.; Glimcher, L.H. Dual and Opposing Roles of the Unfolded Protein Response Regulated by IRE1alpha and XBP1 in Proinsulin Processing and Insulin Secretion. *Proc. Natl. Acad. Sci. USA* **2011**, *108*, 8885–8890. [CrossRef]

87. Gao, Y.; Wu, X.; Zhao, S.; Zhang, Y.; Ma, H.; Yang, Z.; Yang, W.; Zhao, C.; Wang, L.; Zhang, Q. Melatonin Receptor Depletion Suppressed HCG-Induced Testosterone Expression in Mouse Leydig Cells. *Cell. Mol. Biol. Lett.* **2019**, *24*, 21. [CrossRef]

88. Kang, F.-C.; Wang, S.-C.; Chang, M.-M.; Pan, B.-S.; Wong, K.-L.; Cheng, K.-S.; So, E.C.; Huang, B.-M. Midazolam Activates Caspase, MAPKs and Endoplasmic Reticulum Stress Pathways, and Inhibits Cell Cycle and Akt Pathway, to Induce Apoptosis in TM3 Mouse Leydig Progenitor Cells. *Onco Targets Ther.* **2018**, *11*, 1475–1490. [CrossRef]

89. Park, S.-J.; Kim, T.-S.; Park, C.-K.; Lee, S.-H.; Kim, J.-M.; Lee, K.-S.; Lee, I.-K.; Park, J.-W.; Lawson, M.A.; Lee, D.-S. HCG-Induced Endoplasmic Reticulum Stress Triggers Apoptosis and Reduces Steroidogenic Enzyme Expression through Activating Transcription Factor 6 in Leydig Cells of the Testis. *J. Mol. Endocrinol.* **2013**, *50*, 151–166. [CrossRef]

90. Yoshida, H.; Matsui, T.; Yamamoto, A.; Okada, T.; Mori, K. XBP1 MRNA Is Induced by ATF6 and Spliced by IRE1 in Response to ER Stress to Produce a Highly Active Transcription Factor. *Cell* **2001**, *107*, 881–891. [CrossRef]

91. Ono, S.J.; Liou, H.C.; Davidon, R.; Strominger, J.L.; Glimcher, L.H. Human X-Box-Binding Protein 1 Is Required for the Transcription of a Subset of Human Class II Major Histocompatibility Genes and Forms a Heterodimer with c-Fos. *Proc. Natl. Acad. Sci. USA* **1991**, *88*, 4309–4312. [CrossRef]

92. Yamamoto, K.; Sato, T.; Matsui, T.; Sato, M.; Okada, T.; Yoshida, H.; Harada, A.; Mori, K. Transcriptional Induction of Mammalian ER Quality Control Proteins Is Mediated by Single or Combined Action of ATF6alpha and XBP1. *Dev. Cell* **2007**, *13*, 365–376. [CrossRef] [PubMed]

93. Murakami, T.; Saito, A.; Hino, S.; Kondo, S.; Kanemoto, S.; Chihara, K.; Sekiya, H.; Tsumagari, K.; Ochiai, K.; Yoshinaga, K.; et al. Signalling Mediated by the Endoplasmic Reticulum Stress Transducer OASIS Is Involved in Bone Formation. *Nat. Cell Biol.* **2009**, *11*, 1205–1211. [CrossRef] [PubMed]

94. Zhang, K.; Shen, X.; Wu, J.; Sakaki, K.; Saunders, T.; Rutkowski, D.T.; Back, S.H.; Kaufman, R.J. Endoplasmic Reticulum Stress Activates Cleavage of CREBH to Induce a Systemic Inflammatory Response. *Cell* **2006**, *124*, 587–599. [CrossRef] [PubMed]

95. Podust, L.M.; Krezel, A.M.; Kim, Y. Crystal Structure of the CCAAT Box/Enhancer-Binding Protein Beta Activating Transcription Factor-4 Basic Leucine Zipper Heterodimer in the Absence of DNA. *J. Biol. Chem.* **2001**, *276*, 505–513. [CrossRef]

96. Olsen, J.V.; Vermeulen, M.; Santamaria, A.; Kumar, C.; Miller, M.L.; Jensen, L.J.; Gnad, F.; Cox, J.; Jensen, T.S.; Nigg, E.A.; et al. Quantitative Phosphoproteomics Reveals Widespread Full Phosphorylation Site Occupancy during Mitosis. *Sci. Signal.* **2010**, *3*, ra3. [CrossRef] [PubMed]

97. De Cesare, D.; Sassone-Corsi, P. Transcriptional Regulation by Cyclic AMP-Responsive Factors. *Prog. Nucleic Acid Res. Mol. Biol.* **2000**, *64*, 343–369. [CrossRef] [PubMed]

98. Hai, T.; Wolfgang, C.D.; Marsee, D.K.; Allen, A.E.; Sivaprasad, U. ATF3 and Stress Responses. *Gene Expr.* **1999**, *7*, 321–335. [PubMed]

99. Dluzen, D.; Li, G.; Tacelosky, D.; Moreau, M.; Liu, D.X. BCL-2 Is a Downstream Target of ATF5 That Mediates the Prosurvival Function of ATF5 in a Cell Type-Dependent Manner. *J. Biol. Chem.* **2011**, *286*, 7705–7713. [CrossRef] [PubMed]

100. Bailey, D.; O'Hare, P. Transmembrane BZIP Transcription Factors in ER Stress Signaling and the Unfolded Protein Response. *Antioxid. Redox Signal.* **2007**, *9*, 2305–2321. [CrossRef] [PubMed]

101. Guo, X.; Aviles, G.; Liu, Y.; Tian, R.; Unger, B.A.; Lin, Y.-H.T.; Wiita, A.P.; Xu, K.; Correia, M.A.; Kampmann, M. Mitochondrial Stress Is Relayed to the Cytosol by an OMA1-DELE1-HRI Pathway. *Nature* **2020**, *579*, 427–432. [CrossRef]

102. Ohoka, N.; Yoshii, S.; Hattori, T.; Onozaki, K.; Hayashi, H. TRB3, a Novel ER Stress-Inducible Gene, Is Induced via ATF4-CHOP Pathway and Is Involved in Cell Death. *EMBO J.* **2005**, *24*, 1243–1255. [CrossRef] [PubMed]

103. Basque, A.; Nguyen, H.T.; Touaibia, M.; Martin, L.J. Gigantol Improves Cholesterol Metabolism and Progesterone Biosynthesis in MA-10 Leydig Cells. *Curr. Issues Mol. Biol.* **2022**, *44*, 73–93. [CrossRef] [PubMed]

104. Nguyen, H.T.; Couture, R.; Touaibia, M.; Martin, L.J. Transcriptome Modulation Following Administration of Luteolin to Bleomycin-Etoposide-Cisplatin Chemotherapy on Rat LC540 Tumor Leydig Cells. *Andrologia* **2021**, *53*, e13960. [CrossRef] [PubMed]

105. Gauci, S.; Helbig, A.O.; Slijper, M.; Krijgsveld, J.; Heck, A.J.R.; Mohammed, S. Lys-N and Trypsin Cover Complementary Parts of the Phosphoproteome in a Refined SCX-Based Approach. *Anal. Chem.* **2009**, *81*, 4493–4501. [CrossRef] [PubMed]

106. Liu, W.; Sun, M.; Jiang, J.; Shen, X.; Sun, Q.; Liu, W.; Shen, H.; Gu, J. Cyclin D3 Interacts with Human Activating Transcription Factor 5 and Potentiates Its Transcription Activity. *Biochem. Biophys. Res. Commun.* **2004**, *321*, 954–960. [CrossRef]

107. Chen, H.; Ge, R.-S.; Zirkin, B.R. Leydig Cells: From Stem Cells to Aging. *Mol. Cell. Endocrinol.* **2009**, *306*, 9–16. [CrossRef] [PubMed]

108. Teerds, K.J.; Rijntjes, E.; Veldhuizen-Tsoerkan, M.B.; Rommerts, F.F.G.; de Boer-Brouwer, M. The Development of Rat Leydig Cell Progenitors in Vitro: How Essential Is Luteinising Hormone? *J. Endocrinol.* **2007**, *194*, 579–593. [CrossRef] [PubMed]

109. Ge, R.S.; Hardy, D.O.; Catterall, J.F.; Hardy, M.P. Opposing Changes in 3alpha-Hydroxysteroid Dehydrogenase Oxidative and Reductive Activities in Rat Leydig Cells during Pubertal Development. *Biol. Reprod.* **1999**, *60*, 855–860. [CrossRef] [PubMed]

110. Benton, L.; Shan, L.X.; Hardy, M.P. Differentiation of Adult Leydig Cells. *J. Steroid Biochem. Mol. Biol.* **1995**, *53*, 61–68. [CrossRef]

111. Kaukua, J.; Pekkarinen, T.; Sane, T.; Mustajoki, P. Sex Hormones and Sexual Function in Obese Men Losing Weight. *Obes. Res.* **2003**, *11*, 689–694. [CrossRef] [PubMed]

112. Walker, W.H. Molecular Mechanisms of Testosterone Action in Spermatogenesis. *Steroids* **2009**, *74*, 602–607. [CrossRef] [PubMed]
113. Zheng, D.; Cho, Y.-Y.; Lau, A.T.Y.; Zhang, J.; Ma, W.-Y.; Bode, A.M.; Dong, Z. Cyclin-Dependent Kinase 3-Mediated Activating Transcription Factor 1 Phosphorylation Enhances Cell Transformation. *Cancer Res.* **2008**, *68*, 7650–7660. [CrossRef] [PubMed]
114. Wang, L.; Meng, Q.; Yang, L.; Yang, D.; Guo, W.; Lin, P.; Chen, H.; Tang, K.; Wang, A.; Jin, Y. Luman/CREB3 Knock-down Inhibit HCG Induced MLTC-1 Apoptosis. *Theriogenology* **2021**, *161*, 140–150. [CrossRef]
115. Wisdom, R.; Johnson, R.S.; Moore, C. C-Jun Regulates Cell Cycle Progression and Apoptosis by Distinct Mechanisms. *EMBO J.* **1999**, *18*, 188–197. [CrossRef] [PubMed]
116. Bakiri, L.; Lallemand, D.; Bossy-Wetzel, E.; Yaniv, M. Cell Cycle-Dependent Variations in c-Jun and JunB Phosphorylation: A Role in the Control of Cyclin D1 Expression. *EMBO J.* **2000**, *19*, 2056–2068. [CrossRef] [PubMed]
117. Weitzman, J.B.; Fiette, L.; Matsuo, K.; Yaniv, M. JunD Protects Cells from P53-Dependent Senescence and Apoptosis. *Mol Cell* **2000**, *6*, 1109–1119. [CrossRef]
118. Li, T.; Zhang, J.; Zhu, F.; Wen, W.; Zykova, T.; Li, X.; Liu, K.; Peng, C.; Ma, W.; Shi, G.; et al. P21-Activated Protein Kinase (PAK2)-Mediated c-Jun Phosphorylation at 5 Threonine Sites Promotes Cell Transformation. *Carcinogenesis* **2011**, *32*, 659–666. [CrossRef]
119. Eferl, R.; Wagner, E.F. AP-1: A Double-Edged Sword in Tumorigenesis. *Nat. Rev. Cancer* **2003**, *3*, 859–868. [CrossRef]
120. Piechaczyk, M.; Farràs, R. Regulation and Function of JunB in Cell Proliferation. *Biochem Soc Trans* **2008**, *36*, 864–867. [CrossRef]
121. Karin, M.; Liu, Z.; Zandi, E. AP-1 Function and Regulation. *Curr. Opin. Cell Biol.* **1997**, *9*, 240–246. [CrossRef]
122. Ge, R.-S.; Dong, Q.; Sottas, C.M.; Chen, H.; Zirkin, B.R.; Hardy, M.P. Gene Expression in Rat Leydig Cells during Development from the Progenitor to Adult Stage: A Cluster Analysis. *Biol. Reprod.* **2005**, *72*, 1405–1415. [CrossRef] [PubMed]
123. Guan, X.; Chen, P.; Ji, M.; Wen, X.; Chen, D.; Zhao, X.; Huang, F.; Wang, J.; Shao, J.; Xie, J.; et al. Identification of Rat Testicular Leydig Precursor Cells by Single-Cell-RNA-Sequence Analysis. *Front. Cell Dev Biol.* **2022**, *10*, 805249. [CrossRef] [PubMed]
124. Shan, L.X.; Hardy, M.P. Developmental Changes in Levels of Luteinizing Hormone Receptor and Androgen Receptor in Rat Leydig Cells. *Endocrinology* **1992**, *131*, 1107–1114. [CrossRef] [PubMed]
125. Murphy, L.; O'Shaughnessy, P.J. Testicular Steroidogenesis in the Testicular Feminized (Tfm) Mouse: Loss of 17 Alpha-Hydroxylase Activity. *J. Endocrinol.* **1991**, *131*, 443–449. [CrossRef] [PubMed]
126. Tsai, M.-Y.; Yeh, S.-D.; Wang, R.-S.; Yeh, S.; Zhang, C.; Lin, H.-Y.; Tzeng, C.-R.; Chang, C. Differential Effects of Spermatogenesis and Fertility in Mice Lacking Androgen Receptor in Individual Testis Cells. *Proc. Natl. Acad. Sci. USA* **2006**, *103*, 18975–18980. [CrossRef] [PubMed]
127. Xu, Q.; Lin, H.-Y.; Yeh, S.-D.; Yu, I.-C.; Wang, R.-S.; Chen, Y.-T.; Zhang, C.; Altuwaijri, S.; Chen, L.-M.; Chuang, K.-H.; et al. Infertility with Defective Spermatogenesis and Steroidogenesis in Male Mice Lacking Androgen Receptor in Leydig Cells. *Endocrine* **2007**, *32*, 96–106. [CrossRef] [PubMed]
128. Arango, N.A.; Kobayashi, A.; Wang, Y.; Jamin, S.P.; Lee, H.-H.; Orvis, G.D.; Behringer, R.R. A Mesenchymal Perspective of Müllerian Duct Differentiation and Regression in Amhr2-LacZ Mice. *Mol. Reprod. Dev.* **2008**, *75*, 1154–1162. [CrossRef] [PubMed]
129. Shemshedini, L.; Knauthe, R.; Sassone-Corsi, P.; Pornon, A.; Gronemeyer, H. Cell-Specific Inhibitory and Stimulatory Effects of Fos and Jun on Transcription Activation by Nuclear Receptors. *EMBO J.* **1991**, *10*, 3839–3849. [CrossRef] [PubMed]
130. Bubulya, A.; Wise, S.C.; Shen, X.Q.; Burmeister, L.A.; Shemshedini, L. C-Jun Can Mediate Androgen Receptor-Induced Transactivation. *J. Biol. Chem.* **1996**, *271*, 24583–24589. [CrossRef] [PubMed]
131. Faid, I.; Al-Hussaini, H.; Kilarkaje, N. Resveratrol Alleviates Diabetes-Induced Testicular Dysfunction by Inhibiting Oxidative Stress and c-Jun N-Terminal Kinase Signaling in Rats. *Toxicol. Appl. Pharmacol.* **2015**, *289*, 482–494. [CrossRef] [PubMed]
132. Ngan, E.S.W.; Hashimoto, Y.; Ma, Z.-Q.; Tsai, M.-J.; Tsai, S.Y. Overexpression of Cdc25B, an Androgen Receptor Coactivator, in Prostate Cancer. *Oncogene* **2003**, *22*, 734–739. [CrossRef] [PubMed]
133. Moilanen, A.M.; Karvonen, U.; Poukka, H.; Jänne, O.A.; Palvimo, J.J. Activation of Androgen Receptor Function by a Novel Nuclear Protein Kinase. *Mol. Biol. Cell* **1998**, *9*, 2527–2543. [CrossRef]
134. Chen, Y.; Wang, J.; Xu, D.; Xiang, Z.; Ding, J.; Yang, X.; Li, D.; Han, X. M6A MRNA Methylation Regulates Testosterone Synthesis through Modulating Autophagy in Leydig Cells. *Autophagy* **2021**, *17*, 457–475. [CrossRef]
135. Yazawa, T.; Imamichi, Y.; Yuhki, K.-I.; Uwada, J.; Mikami, D.; Shimada, M.; Miyamoto, K.; Kitano, T.; Takahashi, S.; Sekiguchi, T.; et al. Cyclooxygenase-2 Is Acutely Induced by CCAAT/Enhancer-Binding Protein β to Produce Prostaglandin E 2 and F 2α Following Gonadotropin Stimulation in Leydig Cells. *Mol. Reprod. Dev.* **2019**, *86*, 786–797. [CrossRef]
136. Ron, D.; Habener, J.F. CHOP, a Novel Developmentally Regulated Nuclear Protein That Dimerizes with Transcription Factors C/EBP and LAP and Functions as a Dominant-Negative Inhibitor of Gene Transcription. *Genes Dev.* **1992**, *6*, 439–453. [CrossRef] [PubMed]
137. Yamaguchi, H.; Wang, H.-G. CHOP Is Involved in Endoplasmic Reticulum Stress-Induced Apoptosis by Enhancing DR5 Expression in Human Carcinoma Cells. *J. Biol. Chem.* **2004**, *279*, 45495–45502. [CrossRef]
138. Allaman-Pillet, N.; Roduit, R.; Oberson, A.; Abdelli, S.; Ruiz, J.; Beckmann, J.S.; Schorderet, D.F.; Bonny, C. Circadian Regulation of Islet Genes Involved in Insulin Production and Secretion. *Mol. Cell. Endocrinol.* **2004**, *226*, 59–66. [CrossRef] [PubMed]
139. Zhao, L.; Xiao, Y.; Li, C.; Zhang, J.; Zhang, Y.; Wu, M.; Ma, T.; Yang, L.; Wang, X.; Jiang, H.; et al. Zearalenone Perturbs the Circadian Clock and Inhibits Testosterone Synthesis in Mouse Leydig Cells. *J. Toxicol. Environ. Health A* **2021**, *84*, 112–124. [CrossRef] [PubMed]

140. Logue, S.E.; Cleary, P.; Saveljeva, S.; Samali, A. New Directions in ER Stress-Induced Cell Death. *Apoptosis* **2013**, *18*, 537–546. [CrossRef] [PubMed]

141. Kania, E.; Pająk, B.; Orzechowski, A. Calcium Homeostasis and ER Stress in Control of Autophagy in Cancer Cells. *Biomed. Res. Int.* **2015**, *2015*, 352794. [CrossRef] [PubMed]

142. Clark, B.J.; Wells, J.; King, S.R.; Stocco, D.M. The Purification, Cloning, and Expression of a Novel Luteinizing Hormone-Induced Mitochondrial Protein in MA-10 Mouse Leydig Tumor Cells. Characterization of the Steroidogenic Acute Regulatory Protein (StAR). *J. Biol. Chem.* **1994**, *269*, 28314–28322. [CrossRef]

143. Manna, P.R.; Dyson, M.T.; Eubank, D.W.; Clark, B.J.; Lalli, E.; Sassone-Corsi, P.; Zeleznik, A.J.; Stocco, D.M. Regulation of Steroidogenesis and the Steroidogenic Acute Regulatory Protein by a Member of the CAMP Response-Element Binding Protein Family. *Mol. Endocrinol.* **2002**, *16*, 184–199. [CrossRef]

144. Manna, P.R.; Stocco, D.M. Crosstalk of CREB and Fos/Jun on a Single Cis-Element: Transcriptional Repression of the Steroidogenic Acute Regulatory Protein Gene. *J. Mol. Endocrinol.* **2007**, *39*, 261–277. [CrossRef]

145. Qiu, L.; Wang, H.; Dong, T.; Huang, J.; Li, T.; Ren, H.; Wang, X.; Qu, J.; Wang, S. Perfluorooctane Sulfonate (PFOS) Disrupts Testosterone Biosynthesis via CREB/CRTC2/StAR Signaling Pathway in Leydig Cells. *Toxicology* **2021**, *449*, 152663. [CrossRef]

146. Altarejos, J.Y.; Montminy, M. CREB and the CRTC Co-Activators: Sensors for Hormonal and Metabolic Signals. *Nat. Rev. Mol. Cell Biol.* **2011**, *12*, 141–151. [CrossRef]

147. Manna, P.R.; Huhtaniemi, I.T.; Stocco, D.M. Mechanisms of Protein Kinase C Signaling in the Modulation of 3′,5′-Cyclic Adenosine Monophosphate-Mediated Steroidogenesis in Mouse Gonadal Cells. *Endocrinology* **2009**, *150*, 3308–3317. [CrossRef]

148. Abdou, H.S.; Robert, N.M.; Tremblay, J.J. Calcium-Dependent Nr4a1 Expression in Mouse Leydig Cells Requires Distinct AP1/CRE and MEF2 Elements. *J. Mol. Endocrinol.* **2016**, *56*, 151–161. [CrossRef]

149. Martin, L.J.; Boucher, N.; Brousseau, C.; Tremblay, J.J. The Orphan Nuclear Receptor NUR77 Regulates Hormone-Induced StAR Transcription in Leydig Cells through Cooperation with Ca2+/Calmodulin-Dependent Protein Kinase I. *Mol. Endocrinol.* **2008**, *22*, 2021–2037. [CrossRef]

150. Abdou, H.S.; Villeneuve, G.; Tremblay, J.J. The Calcium Signaling Pathway Regulates Leydig Cell Steroidogenesis through a Transcriptional Cascade Involving the Nuclear Receptor NR4A1 and the Steroidogenic Acute Regulatory Protein. *Endocrinology* **2013**, *154*, 511–520. [CrossRef]

151. Martin, L.J.; Tremblay, J.J. The Human 3beta-Hydroxysteroid Dehydrogenase/Delta5-Delta4 Isomerase Type 2 Promoter Is a Novel Target for the Immediate Early Orphan Nuclear Receptor Nur77 in Steroidogenic Cells. *Endocrinology* **2005**, *146*, 861–869. [CrossRef] [PubMed]

152. Zhang, P.; Mellon, S.H. Multiple Orphan Nuclear Receptors Converge to Regulate Rat P450c17 Gene Transcription: Novel Mechanisms for Orphan Nuclear Receptor Action. *Mol. Endocrinol.* **1997**, *11*, 891–904. [CrossRef] [PubMed]

153. Bassett, M.H.; Suzuki, T.; Sasano, H.; De Vries, C.J.M.; Jimenez, P.T.; Carr, B.R.; Rainey, W.E. The Orphan Nuclear Receptor NGFIB Regulates Transcription of 3beta-Hydroxysteroid Dehydrogenase. Implications for the Control of Adrenal Functional Zonation. *J. Biol. Chem.* **2004**, *279*, 37622–37630. [CrossRef] [PubMed]

154. Orlando, U.; Cooke, M.; Cornejo Maciel, F.; Papadopoulos, V.; Podestá, E.J.; Maloberti, P. Characterization of the Mouse Promoter Region of the Acyl-CoA Synthetase 4 Gene: Role of Sp1 and CREB. *Mol. Cell. Endocrinol.* **2013**, *369*, 15–26. [CrossRef]

155. Marinero, M.J.; Colas, B.; Prieto, J.C.; López-Ruiz, M.P. Different Sites of Action of Arachidonic Acid on Steroidogenesis in Rat Leydig Cells. *Mol. Cell. Endocrinol.* **1996**, *118*, 193–200. [CrossRef]

156. Abayasekara, D.R.; Band, A.M.; Cooke, B.A. Evidence for the Involvement of Phospholipase A2 in the Regulation of Luteinizing Hormone-Stimulated Steroidogenesis in Rat Testis Leydig Cells. *Mol. Cell. Endocrinol.* **1990**, *70*, 147–153. [CrossRef]

157. Ronco, A.M.; Moraga, P.F.; Llanos, M.N. Arachidonic Acid Release from Rat Leydig Cells: The Involvement of G Protein, Phospholipase A2 and Regulation of CAMP Production. *J. Endocrinol.* **2002**, *172*, 95–104. [CrossRef]

158. Cooke, B.A.; Dirami, G.; Chaudry, L.; Choi, M.S.; Abayasekara, D.R.; Phipp, L. Release of Arachidonic Acid and the Effects of Corticosteroids on Steroidogenesis in Rat Testis Leydig Cells. *J. Steroid Biochem. Mol. Biol.* **1991**, *40*, 465–471. [CrossRef]

159. Hsu, M.-C.; Wu, L.-S.; Jong, D.-S.; Chiu, C.-H. KISS1R Signaling Modulates Gonadotropin Sensitivity in Mouse Leydig Cell. *Reproduction* **2020**, *160*, 843–852. [CrossRef]

160. Wang, L.; Lu, M.; Zhang, R.; Guo, W.; Lin, P.; Yang, D.; Chen, H.; Tang, K.; Zhou, D.; Wang, A.; et al. Inhibition of Luman/CREB3 Expression Leads to the Upregulation of Testosterone Synthesis in Mouse Leydig Cells. *J. Cell. Physiol.* **2019**. [CrossRef]

161. Zhang, X.K.; Wills, K.N.; Husmann, M.; Hermann, T.; Pfahl, M. Novel Pathway for Thyroid Hormone Receptor Action through Interaction with Jun and Fos Oncogene Activities. *Mol. Cell. Biol.* **1991**, *11*, 6016–6025. [CrossRef] [PubMed]

162. Schüle, R.; Rangarajan, P.; Yang, N.; Kliewer, S.; Ransone, L.J.; Bolado, J.; Verma, I.M.; Evans, R.M. Retinoic Acid Is a Negative Regulator of AP-1-Responsive Genes. *Proc. Natl. Acad. Sci. USA* **1991**, *88*, 6092–6096. [CrossRef] [PubMed]

163. Lu, M.; Zhang, R.; Yu, T.; Wang, L.; Liu, S.; Cai, R.; Guo, X.; Jia, Y.; Wang, A.; Jin, Y.; et al. CREBZF Regulates Testosterone Production in Mouse Leydig Cells. *J. Cell. Physiol.* **2019**, *234*, 22819–22832. [CrossRef] [PubMed]

164. Hogan, M.R.; Cockram, G.P.; Lu, R. Cooperative Interaction of Zhangfei and ATF4 in Transactivation of the Cyclic AMP Response Element. *FEBS Lett.* **2006**, *580*, 58–62. [CrossRef]

165. Zhang, R.; Rapin, N.; Ying, Z.; Shklanka, E.; Bodnarchuk, T.W.; Verge, V.M.K.; Misra, V. Zhangfei/CREB-ZF - a Potential Regulator of the Unfolded Protein Response. *PLoS ONE* **2013**, *8*, e77256. [CrossRef] [PubMed]

166. Batarseh, A.; Li, J.; Papadopoulos, V. Protein Kinase C Epsilon Regulation of Translocator Protein (18 KDa) Tspo Gene Expression Is Mediated through a MAPK Pathway Targeting STAT3 and c-Jun Transcription Factors. *Biochemistry* **2010**, *49*, 4766–4778. [CrossRef]
167. Martin, L.J.; Boucher, N.; El-Asmar, B.; Tremblay, J.J. CAMP-Induced Expression of the Orphan Nuclear Receptor Nur77 in MA-10 Leydig Cells Involves a CaMKI Pathway. *J. Androl.* **2009**, *30*, 134–145. [CrossRef]
168. Guo, I.-C.; Huang, C.-Y.; Wang, C.-K.L.; Chung, B. Activating Protein-1 Cooperates with Steroidogenic Factor-1 to Regulate 3′,5′-Cyclic Adenosine 5′-Monophosphate-Dependent Human CYP11A1 Transcription in Vitro and in Vivo. *Endocrinology* **2007**, *148*, 1804–1812. [CrossRef]
169. Lee, S.-Y.; Gong, E.-Y.; Hong, C.Y.; Kim, K.-H.; Han, J.-S.; Ryu, J.C.; Chae, H.Z.; Yun, C.-H.; Lee, K. ROS Inhibit the Expression of Testicular Steroidogenic Enzyme Genes via the Suppression of Nur77 Transactivation. *Free Radic. Biol. Med.* **2009**, *47*, 1591–1600. [CrossRef]
170. Göttlicher, M.; Heck, S.; Doucas, V.; Wade, E.; Kullmann, M.; Cato, A.C.; Evans, R.M.; Herrlich, P. Interaction of the Ubc9 Human Homologue with C-Jun and with the Glucocorticoid Receptor. *Steroids* **1996**, *61*, 257–262. [CrossRef]
171. Teyssier, C.; Belguise, K.; Galtier, F.; Chalbos, D. Characterization of the Physical Interaction between Estrogen Receptor Alpha and JUN Proteins. *J. Biol. Chem.* **2001**, *276*, 36361–36369. [CrossRef]
172. Yang-Yen, H.F.; Zhang, X.K.; Graupner, G.; Tzukerman, M.; Sakamoto, B.; Karin, M.; Pfahl, M. Antagonism between Retinoic Acid Receptors and AP-1: Implications for Tumor Promotion and Inflammation. *New Biol.* **1991**, *3*, 1206–1219. [PubMed]
173. Martin, L.J.; Bergeron, F.; Viger, R.S.; Tremblay, J.J. Functional Cooperation between GATA Factors and CJUN on the Star Promoter in MA-10 Leydig Cells. *J. Androl.* **2012**, *33*, 81–87. [CrossRef] [PubMed]
174. Manna, P.R.; Eubank, D.W.; Stocco, D.M. Assessment of the Role of Activator Protein-1 on Transcription of the Mouse Steroidogenic Acute Regulatory Protein Gene. *Mol. Endocrinol.* **2004**, *18*, 558–573. [CrossRef]
175. Shea-Eaton, W.; Sandhoff, T.W.; Lopez, D.; Hales, D.B.; McLean, M.P. Transcriptional Repression of the Rat Steroidogenic Acute Regulatory (StAR) Protein Gene by the AP-1 Family Member c-Fos. *Mol. Cell. Endocrinol.* **2002**, *188*, 161–170. [CrossRef]
176. Abdou, H.S.; Bergeron, F.; Tremblay, J.J. A Cell-Autonomous Molecular Cascade Initiated by AMP-Activated Protein Kinase Represses Steroidogenesis. *Mol. Cell. Biol.* **2014**, *34*, 4257–4271. [CrossRef]
177. Kardassis, D.; Papakosta, P.; Pardali, K.; Moustakas, A. C-Jun Transactivates the Promoter of the Human P21(WAF1/Cip1) Gene by Acting as a Superactivator of the Ubiquitous Transcription Factor Sp1. *J. Biol. Chem.* **1999**, *274*, 29572–29581. [CrossRef]
178. Sher, N.; Yivgi-Ohana, N.; Orly, J. Transcriptional Regulation of the Cholesterol Side Chain Cleavage Cytochrome P450 Gene (CYP11A1) Revisited: Binding of GATA, Cyclic Adenosine 3′,5′-Monophosphate Response Element-Binding Protein and Activating Protein (AP)-1 Proteins to a Distal Novel Cluster of Cis-Regulatory Elements Potentiates AP-2 and Steroidogenic Factor-1-Dependent Gene Expression in the Rodent Placenta and Ovary. *Mol. Endocrinol.* **2007**, *21*, 948–962. [CrossRef]
179. Rice, D.A.; Kirkman, M.S.; Aitken, L.D.; Mouw, A.R.; Schimmer, B.P.; Parker, K.L. Analysis of the Promoter Region of the Gene Encoding Mouse Cholesterol Side-Chain Cleavage Enzyme. *J. Biol. Chem.* **1990**, *265*, 11713–11720. [CrossRef]
180. Selvaraj, V.; Stocco, D.M.; Clark, B.J. Current Knowledge on the Acute Regulation of Steroidogenesis. *Biol. Reprod.* **2018**, *99*, 13–26. [CrossRef] [PubMed]
181. Morohaku, K.; Pelton, S.H.; Daugherty, D.J.; Butler, W.R.; Deng, W.; Selvaraj, V. Translocator Protein/Peripheral Benzodiazepine Receptor Is Not Required for Steroid Hormone Biosynthesis. *Endocrinology* **2014**, *155*, 89–97. [CrossRef] [PubMed]
182. Fan, J.; Wang, K.; Zirkin, B.; Papadopoulos, V. CRISPR/Cas9–Mediated Tspo Gene Mutations Lead to Reduced Mitochondrial Membrane Potential and Steroid Formation in MA-10 Mouse Tumor Leydig Cells. *Endocrinology* **2018**, *159*, 1130–1146. [CrossRef] [PubMed]
183. Batarseh, A.; Giatzakis, C.; Papadopoulos, V. Phorbol-12-Myristate 13-Acetate Acting through Protein Kinase Cepsilon Induces Translocator Protein (18-KDa) TSPO Gene Expression. *Biochemistry* **2008**, *47*, 12886–12899. [CrossRef] [PubMed]
184. Lin, T.; Haskell, J.; Vinson, N.; Terracio, L. Characterization of Insulin and Insulin-like Growth Factor I Receptors of Purified Leydig Cells and Their Role in Steroidogenesis in Primary Culture: A Comparative Study. *Endocrinology* **1986**, *119*, 1641–1647. [CrossRef]
185. Lin, T.; Blaisdell, J.; Haskell, J.F. Transforming Growth Factor-Beta Inhibits Leydig Cell Steroidogenesis in Primary Culture. *Biochem. Biophys. Res. Commun.* **1987**, *146*, 387–394. [CrossRef]
186. Sordoillet, C.; Chauvin, M.A.; Revol, A.; Morera, A.M.; Benahmed, M. Fibroblast Growth Factor Is a Regulator of Testosterone Secretion in Cultured Immature Leydig Cells. *Mol. Cell. Endocrinol.* **1988**, *58*, 283–286. [CrossRef]
187. Millena, A.C.; Reddy, S.C.; Bowling, G.H.; Khan, S.A. Autocrine Regulation of Steroidogenic Function of Leydig Cells by Transforming Growth Factor-Alpha. *Mol. Cell. Endocrinol.* **2004**, *224*, 29–39. [CrossRef]
188. Saez, J.M.; Avallet, O.; Naville, D.; Perrard-Sapori, M.H.; Chatelain, P.G. Sertoli-Leydig Cell Communications. *Ann. N. Y. Acad. Sci.* **1989**, *564*, 210–231. [CrossRef]
189. Manna, P.R.; Chandrala, S.P.; King, S.R.; Jo, Y.; Counis, R.; Huhtaniemi, I.T.; Stocco, D.M. Molecular Mechanisms of Insulin-like Growth Factor-I Mediated Regulation of the Steroidogenic Acute Regulatory Protein in Mouse Leydig Cells. *Mol. Endocrinol.* **2006**, *20*, 362–378. [CrossRef]
190. Weekes, D.; Kashima, T.G.; Zandueta, C.; Perurena, N.; Thomas, D.P.; Sunters, A.; Vuillier, C.; Bozec, A.; El-Emir, E.; Miletich, I.; et al. Regulation of Osteosarcoma Cell Lung Metastasis by the C-Fos/AP-1 Target FGFR1. *Oncogene* **2016**, *35*, 2852–2861. [CrossRef] [PubMed]

191. Tian, Y.; Wang, G.; Hu, Q.; Xiao, X.; Chen, S. AML1/ETO Trans-Activates c-KIT Expression through the Long Range Interaction between Promoter and Intronic Enhancer. *J. Cell. Biochem.* **2018**, *119*, 3706–3715. [CrossRef] [PubMed]

192. Hébert-Mercier, P.-O.; Bergeron, F.; Robert, N.M.; Mehanovic, S.; Pierre, K.J.; Mendoza-Villarroel, R.E.; de Mattos, K.; Brousseau, C.; Tremblay, J.J. Growth Hormone-Induced STAT5B Regulates Star Gene Expression Through a Cooperation With CJUN in Mouse MA-10 Leydig Cells. *Endocrinology* **2022**, *163*, bqab267. [CrossRef] [PubMed]

193. Simard, J.; Ricketts, M.-L.; Gingras, S.; Soucy, P.; Feltus, F.A.; Melner, M.H. Molecular Biology of the 3beta-Hydroxysteroid Dehydrogenase/Delta5-Delta4 Isomerase Gene Family. *Endocr. Rev* **2005**, *26*, 525–582. [CrossRef] [PubMed]

194. Katz, S.; Heinrich, R.; Aronheim, A. The AP-1 Repressor, JDP2, Is a Bona Fide Substrate for the c-Jun N-Terminal Kinase. *FEBS Lett.* **2001**, *506*, 196–200. [CrossRef]

195. Jin, C.; Ugai, H.; Song, J.; Murata, T.; Nili, F.; Sun, K.; Horikoshi, M.; Yokoyama, K.K. Identification of Mouse Jun Dimerization Protein 2 as a Novel Repressor of ATF-2. *FEBS Lett.* **2001**, *489*, 34–41. [CrossRef]

196. Chen, H.; Jin, S.; Guo, J.; Kombairaju, P.; Biswal, S.; Zirkin, B.R. Knockout of the Transcription Factor Nrf2: Effects on Testosterone Production by Aging Mouse Leydig Cells. *Mol. Cell. Endocrinol.* **2015**, *409*, 113–120. [CrossRef] [PubMed]

197. Beattie, M.C.; Adekola, L.; Papadopoulos, V.; Chen, H.; Zirkin, B.R. Leydig Cell Aging and Hypogonadism. *Exp. Gerontol.* **2015**, *68*, 87–91. [CrossRef] [PubMed]

198. Li, W.; Kong, A.-N. Molecular Mechanisms of Nrf2-Mediated Antioxidant Response. *Mol. Carcinog.* **2009**, *48*, 91–104. [CrossRef] [PubMed]

199. Fan, H.-Y.; Liu, Z.; Johnson, P.F.; Richards, J.S. CCAAT/Enhancer-Binding Proteins (C/EBP)-α and -β Are Essential for Ovulation, Luteinization, and the Expression of Key Target Genes. *Mol. Endocrinol.* **2011**, *25*, 253–268. [CrossRef] [PubMed]

200. Reinhart, A.J.; Williams, S.C.; Clark, B.J.; Stocco, D.M. SF-1 (Steroidogenic Factor-1) and C/EBP Beta (CCAAT/Enhancer Binding Protein-Beta) Cooperate to Regulate the Murine StAR (Steroidogenic Acute Regulatory) Promoter. *Mol. Endocrinol.* **1999**, *13*, 729–741. [CrossRef]

201. Tremblay, J.J.; Hamel, F.; Viger, R.S. Protein Kinase A-Dependent Cooperation between GATA and CCAAT/Enhancer-Binding Protein Transcription Factors Regulates Steroidogenic Acute Regulatory Protein Promoter Activity. *Endocrinology* **2002**, *143*, 3935–3945. [CrossRef] [PubMed]

202. Hiroi, H.; Christenson, L.K.; Chang, L.; Sammel, M.D.; Berger, S.L.; Strauss, J.F. Temporal and Spatial Changes in Transcription Factor Binding and Histone Modifications at the Steroidogenic Acute Regulatory Protein (StAR) Locus Associated with StAR Transcription. *Mol. Endocrinol.* **2004**, *18*, 791–806. [CrossRef] [PubMed]

203. Kuhl, A.J.; Ross, S.M.; Gaido, K.W. CCAAT/Enhancer Binding Protein Beta, but Not Steroidogenic Factor-1, Modulates the Phthalate-Induced Dysregulation of Rat Fetal Testicular Steroidogenesis. *Endocrinology* **2007**, *148*, 5851–5864. [CrossRef] [PubMed]

204. Hu, Y.; Dong, C.; Chen, M.; Lu, J.; Han, X.; Qiu, L.; Chen, Y.; Qin, J.; Li, X.; Gu, A.; et al. Low-Dose Monobutyl Phthalate Stimulates Steroidogenesis through Steroidogenic Acute Regulatory Protein Regulated by SF-1, GATA-4 and C/EBP-Beta in Mouse Leydig Tumor Cells. *Reprod. Biol. Endocrinol.* **2013**, *11*, 72. [CrossRef] [PubMed]

205. Manna, P.R.; Wang, X.-J.; Stocco, D.M. Involvement of Multiple Transcription Factors in the Regulation of Steroidogenic Acute Regulatory Protein Gene Expression. *Steroids* **2003**, *68*, 1125–1134. [CrossRef] [PubMed]

206. Christenson, L.K.; Strauss, J.F. Steroidogenic Acute Regulatory Protein (StAR) and the Intramitochondrial Translocation of Cholesterol. *Biochim. Biophys. Acta* **2000**, *1529*, 175–187. [CrossRef]

207. El-Asmar, B.; Giner, X.C.; Tremblay, J.J. Transcriptional Cooperation between NF-KappaB P50 and CCAAT/Enhancer Binding Protein Beta Regulates Nur77 Transcription in Leydig Cells. *J. Mol. Endocrinol.* **2009**, *42*, 131–138. [CrossRef] [PubMed]

208. Frungieri, M.B.; Calandra, R.S.; Mayerhofer, A.; Matzkin, M.E. Cyclooxygenase and Prostaglandins in Somatic Cell Populations of the Testis. *Reproduction* **2015**, *149*, R169–R180. [CrossRef] [PubMed]

209. Chen, H.; Luo, L.; Liu, J.; Zirkin, B.R. Cyclooxygenases in Rat Leydig Cells: Effects of Luteinizing Hormone and Aging. *Endocrinology* **2007**, *148*, 735–742. [CrossRef] [PubMed]

210. Kondo, S.; Saito, A.; Asada, R.; Kanemoto, S.; Imaizumi, K. Physiological Unfolded Protein Response Regulated by OASIS Family Members, Transmembrane BZIP Transcription Factors. *IUBMB Life* **2011**, *63*, 233–239. [CrossRef]

211. Adham, I.M.; Eck, T.J.; Mierau, K.; Müller, N.; Sallam, M.A.; Paprotta, I.; Schubert, S.; Hoyer-Fender, S.; Engel, W. Reduction of Spermatogenesis but Not Fertility in Creb3l4-Deficient Mice. *Mol. Cell. Biol.* **2005**, *25*, 7657–7664. [CrossRef]

212. Nagamori, I.; Yomogida, K.; Ikawa, M.; Okabe, M.; Yabuta, N.; Nojima, H. The Testes-Specific BZip Type Transcription Factor Tisp40 Plays a Role in ER Stress Responses and Chromatin Packaging during Spermiogenesis. *Genes Cells* **2006**, *11*, 1161–1171. [CrossRef] [PubMed]

213. Ivell, R.; Anand-Ivell, R. Biology of Insulin-like Factor 3 in Human Reproduction. *Hum. Reprod. Update* **2009**, *15*, 463–476. [CrossRef] [PubMed]

214. Huang, Z.; Rivas, B.; Agoulnik, A.I. Insulin-like 3 Signaling Is Important for Testicular Descent but Dispensable for Spermatogenesis and Germ Cell Survival in Adult Mice. *Biol. Reprod.* **2012**, *87*, 143. [CrossRef]

215. Zimmermann, S.; Schöttler, P.; Engel, W.; Adham, I.M. Mouse Leydig Insulin-like (Ley I-L) Gene: Structure and Expression during Testis and Ovary Development. *Mol. Reprod. Dev.* **1997**, *47*, 30–38. [CrossRef]

216. Ivell, R.; Balvers, M.; Domagalski, R.; Ungefroren, H.; Hunt, N.; Schulze, W. Relaxin-like Factor: A Highly Specific and Constitutive New Marker for Leydig Cells in the Human Testis. *Mol. Hum. Reprod.* **1997**, *3*, 459–466. [CrossRef] [PubMed]

217. Balvers, M.; Spiess, A.N.; Domagalski, R.; Hunt, N.; Kilic, E.; Mukhopadhyay, A.K.; Hanks, E.; Charlton, H.M.; Ivell, R. Relaxin-like Factor Expression as a Marker of Differentiation in the Mouse Testis and Ovary. *Endocrinology* **1998**, *139*, 2960–2970. [CrossRef]
218. Bay, K.; Andersson, A.-M. Human Testicular Insulin-like Factor 3: In Relation to Development, Reproductive Hormones and Andrological Disorders. *Int. J. Androl.* **2011**, *34*, 97–109. [CrossRef]
219. Gupta, A.; Anderson, H.; Buo, A.M.; Moorer, M.C.; Ren, M.; Stains, J.P. Communication of CAMP by Connexin43 Gap Junctions Regulates Osteoblast Signaling and Gene Expression. *Cell Signal* **2016**, *28*, 1048–1057. [CrossRef] [PubMed]
220. Nielsen, P.A.; Kumar, N.M. Differences in Expression Patterns between Mouse Connexin-30.2 (Cx30.2) and Its Putative Human Orthologue, Connexin-31.9. *FEBS Lett.* **2003**, *540*, 151–156. [CrossRef]
221. Risley, M.S.; Tan, I.P.; Roy, C.; Sáez, J.C. Cell-, Age- and Stage-Dependent Distribution of Connexin43 Gap Junctions in Testes. *J. Cell. Sci.* **1992**, *103 Pt 1*, 81–96. [CrossRef]
222. Li, D.; Sekhon, P.; Barr, K.J.; Márquez-Rosado, L.; Lampe, P.D.; Kidder, G.M. Connexins and Steroidogenesis in Mouse Leydig Cells. *Can. J. Physiol. Pharmacol.* **2013**, *91*, 157–164. [CrossRef] [PubMed]
223. Najih, M.; Nguyen, H.T.; Martin, L.J. Involvement of Calmodulin-Dependent Protein Kinase I in the Regulation of the Expression of Connexin 43 in MA-10 Tumor Leydig Cells. *Mol. Cell. Biochem.* **2022**. [CrossRef] [PubMed]
224. Roscoe, W.A.; Barr, K.J.; Mhawi, A.A.; Pomerantz, D.K.; Kidder, G.M. Failure of Spermatogenesis in Mice Lacking Connexin43. *Biol. Reprod.* **2001**, *65*, 829–838. [CrossRef] [PubMed]
225. Kahiri, C.N.; Khalil, M.W.; Tekpetey, F.; Kidder, G.M. Leydig Cell Function in Mice Lacking Connexin43. *Reproduction* **2006**, *132*, 607–616. [CrossRef] [PubMed]
226. Sridharan, S.; Simon, L.; Meling, D.D.; Cyr, D.G.; Gutstein, D.E.; Fishman, G.I.; Guillou, F.; Cooke, P.S. Proliferation of Adult Sertoli Cells Following Conditional Knockout of the Gap Junctional Protein GJA1 (Connexin 43) in Mice. *Biol. Reprod.* **2007**, *76*, 804–812. [CrossRef] [PubMed]
227. Brehm, R.; Zeiler, M.; Rüttinger, C.; Herde, K.; Kibschull, M.; Winterhager, E.; Willecke, K.; Guillou, F.; Lécureuil, C.; Steger, K.; et al. A Sertoli Cell-Specific Knockout of Connexin43 Prevents Initiation of Spermatogenesis. *Am. J. Pathol.* **2007**, *171*, 19–31. [CrossRef] [PubMed]
228. Noelke, J.; Wistuba, J.; Damm, O.S.; Fietz, D.; Gerber, J.; Gaehle, M.; Brehm, R. A Sertoli Cell-Specific Connexin43 Knockout Leads to Altered Interstitial Connexin Expression and Increased Leydig Cell Numbers. *Cell Tissue Res.* **2015**, *361*, 633–644. [CrossRef] [PubMed]
229. Salameh, A.; Krautblatter, S.; Karl, S.; Blanke, K.; Gomez, D.R.; Dhein, S.; Pfeiffer, D.; Janousek, J. The Signal Transduction Cascade Regulating the Expression of the Gap Junction Protein Connexin43 by Beta-Adrenoceptors. *Br. J. Pharmacol.* **2009**, *158*, 198–208. [CrossRef] [PubMed]
230. Yun, S.P.; Park, S.S.; Ryu, J.M.; Park, J.H.; Kim, M.O.; Lee, J.-H.; Han, H.J. Mechanism of PKA-Dependent and Lipid-Raft Independent Stimulation of Connexin43 Expression by Oxytoxin in Mouse Embryonic Stem Cells. *Mol. Endocrinol.* **2012**, *26*, 1144–1157. [CrossRef] [PubMed]
231. You, S.; Li, W.; Lin, T. Expression and Regulation of Connexin43 in Rat Leydig Cells. *J. Endocrinol.* **2000**, *166*, 447–453. [CrossRef] [PubMed]
232. Pérez-Armendariz, E.M.; Luna, J.; Miranda, C.; Talavera, D.; Romano, M.C. Luteinizing and Human Chorionic Gonadotropin Hormones Increase Intercellular Communication and Gap Junctions in Cultured Mouse Leydig Cells. *Endocrine* **1996**, *4*, 141–157. [CrossRef] [PubMed]
233. Bravo-Moreno, J.F.; Díaz-Sánchez, V.; Montoya-Flores, J.G.; Lamoyi, E.; Saéz, J.C.; Pérez-Armendariz, E.M. Expression of Connexin43 in Mouse Leydig, Sertoli, and Germinal Cells at Different Stages of Postnatal Development. *Anat. Rec.* **2001**, *264*, 13–24. [CrossRef]
234. Ghouili, F.; Martin, L.J. Cooperative Regulation of Gja1 Expression by Members of the AP-1 Family CJun and CFos in TM3 Leydig and TM4 Sertoli Cells. *Gene* **2017**, *635*, 24–32. [CrossRef]
235. Landry, D.A.; Sormany, F.; Haché, J.; Roumaud, P.; Martin, L.J. Steroidogenic Genes Expressions Are Repressed by High Levels of Leptin and the JAK/STAT Signaling Pathway in MA-10 Leydig Cells. *Mol. Cell. Biochem.* **2017**, *433*, 79–95. [CrossRef] [PubMed]
236. Ma, B.; Zhang, J.; Zhu, Z.; Zhao, A.; Zhou, Y.; Ying, H.; Zhang, Q. Luteolin Ameliorates Testis Injury and Blood-Testis Barrier Disruption through the Nrf2 Signaling Pathway and by Upregulating Cx43. *Mol. Nutr. Food Res.* **2019**, *63*, e1800843. [CrossRef]
237. Kehat, I.; Hasin, T.; Aronheim, A. The Role of Basic Leucine Zipper Protein-Mediated Transcription in Physiological and Pathological Myocardial Hypertrophy. *Ann. N. Y. Acad. Sci.* **2006**, *1080*, 97–109. [CrossRef]
238. Reza, H.M.; Urano, A.; Shimada, N.; Yasuda, K. Sequential and Combinatorial Roles of Maf Family Genes Define Proper Lens Development. *Mol. Vis.* **2007**, *13*, 18–30.
239. Ming, J.; Zhou, Y.; Du, J.; Fan, S.; Pan, B.; Wang, Y.; Fan, L.; Jiang, J. MiR-381 Suppresses C/EBPα-Dependent Cx43 Expression in Breast Cancer Cells. *Biosci. Rep.* **2015**, *35*, e00266. [CrossRef]

International Journal of
Molecular Sciences

MDPI

Article

Comprehensive and Quantitative Analysis of the Changes in Proteomic and Phosphoproteomic Profiles during Stimulation and Repression of Steroidogenesis in MA-10 Leydig Cells [†]

Zoheir B. Demmouche [1] and Jacques J. Tremblay [1,2,*]

[1] Reproduction, Mother and Child Health, Room T3-67, CHU de Québec—Université Laval Research Centre, Québec, QC G1V 4G2, Canada
[2] Centre for Research in Reproduction, Development and Intergenerational Health, Department of Obstetrics, Gynecology and Reproduction, Faculty of Medicine, Université Laval, Québec, QC G1V 0A6, Canada
* Correspondence: jacques-j.tremblay@crchudequebec.ulaval.ca; Tel.: +1-418-525-4444 (ext. 46254)
† This research was conducted as part of the requirements for a PhD degree.

Abstract: Leydig cells produce testosterone, a hormone essential for male sex differentiation and spermatogenesis. The pituitary hormone, LH, stimulates testosterone production in Leydig cells by increasing the intracellular cAMP levels, which leads to the activation of various kinases and transcription factors, ultimately stimulating the expression of the genes involved in steroidogenesis. The second messenger, cAMP, is subsequently degraded to AMP, and the increase in the intracellular AMP levels activates AMP-dependent protein kinase (AMPK). Activated AMPK potently represses steroidogenesis. Despite the key roles played by the various stimulatory and inhibitory kinases, the proteins phosphorylated by these kinases during steroidogenesis remain poorly characterized. In the present study, we have used a quantitative LC-MS/MS approach, using total and phosphopeptide-enriched proteins to identify the global changes that occur in the proteome and phosphoproteome of MA-10 Leydig cells during both the stimulatory phase (Fsk/cAMP treatment) and inhibitory phase (AICAR-mediated activation of AMPK) of steroidogenesis. The phosphorylation levels of several proteins, including some never before described in Leydig cells, were significantly altered during the stimulation and inhibition of steroidogenesis. Our data also provide new key insights into the finely tuned and dynamic processes that ensure adequate steroid hormone production.

Keywords: testis; Leydig cells; steroidogenesis; star; AMPK; phosphoproteomics; proteomics

Citation: Demmouche, Z.B.; Tremblay, J.J. Comprehensive and Quantitative Analysis of the Changes in Proteomic and Phosphoproteomic Profiles during Stimulation and Repression of Steroidogenesis in MA-10 Leydig Cells. *Int. J. Mol. Sci.* 2022, 23, 12846. https://doi.org/10.3390/ijms232112846

Academic Editor: Antonio Lucacchini

Received: 29 September 2022
Accepted: 21 October 2022
Published: 25 October 2022

Publisher's Note: MDPI stays neutral with regard to jurisdictional claims in published maps and institutional affiliations.

1. Introduction

Leydig cells are located in the interstitial space between the seminiferous tubules of the mammalian testis [1,2]. These cells are the source of androgens, the main one being testosterone [3]. Steroidogenesis is the biological process of converting cholesterol into steroid hormones, which involves cholesterol transport into the mitochondria, where steroidogenesis is initiated. In males, androgens are essential for male sex differentiation during fetal life and for initiating and maintaining spermatogenesis from puberty onwards. Androgens are also needed to acquire male secondary sex characteristics during puberty. Inadequate androgen production is associated with some cases of differences/disorders of sex development (DSD) in males [4].

Since androgens have pleiotropic roles in male reproductive function and overall health, their synthesis is regulated tightly. Steroidogenesis in Leydig cells is stimulated mainly by the pituitary luteinizing hormone (LH) [5]. The binding of LH to its G protein-coupled receptor (LHCGR) on Leydig cells activates adenylate cyclase, which leads to an increase in intracellular cAMP levels [6]. LH/cAMP-induced steroidogenesis then triggers the activation of several kinases, including protein kinase A (PKA), mitogen-activated protein kinase (MAPK), and calcium/calmodulin-dependent protein kinase I (CAMKI), which in turn phosphorylates several proteins required for increased steroid hormone synthesis (reviewed in [7]). This includes several transcription factors, such as the cAMP response element-binding protein (CREB)/CRE modulator (CREM), GATA4, and SF1 (reviewed in [8]). Furthermore, the de novo synthesis of the NR4A1/NUR77 (nuclear receptor subfamily 4, group A, member 1) transcription factor is required for maximal hormone-induced steroidogenesis [9]. The induction of *Nr4a1* expression and the stimulation of steroidogenesis in Leydig cells also requires the release of Ca^{2+} from internal stores through the ryanodine receptors, leading to the activation of CAMKI [10]. Once adequate testosterone levels are reached, stimulation of steroidogenesis is blunted by two mechanisms. The first is the classic negative feedback loop, where testosterone acts at the level of the hypothalamus and pituitary to inhibit LH production [11]. The second is at the level of the Leydig cell itself. In Leydig cells, cAMP is degraded into AMP by phosphodiesterase (PDE) 8A, PDE8B, and PDE4 [12]. The ensuing increase in the intracellular AMP levels activates AMP-activated protein kinase (AMPK), a ubiquitous serine/threonine kinase best known as an energy balance sensor. AMPK is a heterotrimeric complex containing one catalytic subunit (alpha) and two regulatory subunits (beta and gamma) [13]. Once activated, AMPK potently blunts the LH/cAMP-induced steroidogenesis in Leydig cells [14].

We previously compared the transcriptome of Leydig cells that were either untreated, stimulated with Forskolin (Fsk, an agonist of adenylate cyclase), or co-treated with Fsk+AICAR (an agonist of AMPK). This led to the identification and characterization of several genes that were upregulated by the LH/cAMP stimulatory pathway and subsequently downregulated by the AMPK repressive pathway [14]. In addition to the changes in the transcriptome, the protein levels are also affected by the stimulation/repression of steroidogenesis in Leydig cells. However, global changes in protein and phosphoprotein levels have never been reported in this context. In the present work, we have used a quantitative mass spectrometry approach to elucidate the global and dynamic changes in the phosphoproteome of Leydig cells in response to stimulatory (Fsk/cAMP) and inhibitory (AICAR/AMPK) treatments.

2. Results

2.1. Validation of MA-10 Leydig Cell Responsiveness

Before analyzing the samples by quantitative LC-MS/MS, we first validated the responsiveness of the MA-10 Leydig cells to the different treatments. MA-10 Leydig cells were treated for 1 h with the vehicle (DMSO), Forskolin alone (Fsk), or Fsk+AICAR. Fsk is an agonist of adenylate cyclase, leading to increased intracellular cAMP levels and stimulation of steroidogenesis. AICAR is an agonist of the AMPK kinase, which we have identified as a potent repressor of hormone-activated steroidogenesis [14]. Although 1 h is sufficient to detect changes in the protein phosphorylation levels, it is usually too short to detect changes in the total protein levels by Western blot. We, therefore, isolated total RNA, which was used in a qPCR to quantify the mRNA levels for the steroidogenic acute regulatory (*Star*) protein. The *Star* gene codes for the steroidogenic acute regulatory (STAR) protein, a protein essential for hormone-induced cholesterol transport into mitochondria and, consequently, steroidogenesis [15]. The *Star* gene is an excellent marker of the dynamic steroidogenic process in Leydig cells, as its expression is strongly induced by LH/Fsk/cAMP and repressed by AICAR/AMPK ([14] and reviewed in [8]). As shown in Figure 1, the *Star* mRNA levels from the three samples used in LC-MS/MS (described below) were increased six- to ten-fold in the presence of Fsk. As expected [14], this increase was potently repressed when the MA-10 Leydig cells were co-treated with AICAR in addition to Fsk (Figure 1). These results confirm the responsiveness of the MA-10 Leydig cells to Fsk and AICAR and validate that the protein samples isolated from these cells are suitable for LC-MS/MS analysis.

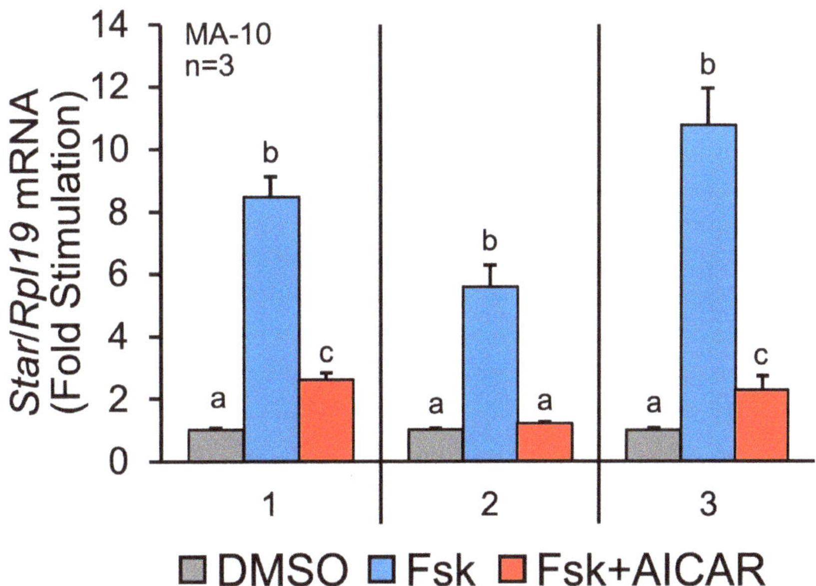

Figure 1. Validation of treatments on MA-10 Leydig cells by assessing *Star* mRNA levels. MA-10 Leydig cells were treated with either DMSO (control, grey bars), Forskolin alone (Fsk, 10 μM, blue bars) or Forskolin and AICAR (Fsk+AICAR, 10 μM, and 1 mM, respectively, red bars) for 1 h. Total RNA was extracted and reverse-transcribed, and qPCR was performed to quantify *Star* mRNA levels. *Rpl19* was used to normalize the data. The numbers on the x-axis refer to the three different samples used in the LC-MS/MS analysis. Results are displayed as the mean of three individual experiments, each performed in duplicate. For a given experiment, different letters indicate a statistically significant difference between groups ($p < 0.05$).

2.2. Treatment of MA-10 Leydig Cells with Fsk or Fsk+AICAR Significantly Affects the Levels of 20 Proteins

Total proteins from the MA-10 Leydig cells treated for 1 h with either the vehicle (DMSO), Fsk alone, or Fsk+AICAR were extracted, digested, and quantitively analyzed by LC-MS/MS. A total of 5887 proteins were identified, and 4819 were quantified (data not shown). The level of the majority of these proteins was not significantly changed by the treatments due to the short treatment time. Only 20 proteins (listed in Table 1) were significantly affected between treatments (DMSO vs. Fsk, Fsk vs. Fsk+AICAR, and DMSO vs. Fsk+AICAR). The protein levels either increased (shown in red in Table 1) or decreased (shown in blue in Table 1). In addition, the same protein was sometimes found in different comparison groups. For instance, the protein levels of the orphan nuclear receptor NR4A1 (NUR77), a known regulator of the hormone-induced steroidogenic gene expression in Leydig cells [10,16], was increased by 2.4-fold by the Fsk treatment (DMSO vs. Fsk group), and this increase was blunted after the AICAR treatment (down by 2.34-fold; Fsk vs. Fsk+AICAR group), as previously reported [14]. Since the levels of the NR4A1 protein are back down to the control levels after treatment with Fsk+AICAR, the NR4A1 protein levels were not significantly changed in the DMSO vs. Fsk+AICAR group. These changes in NR4A1 protein levels (increased by Fsk/cAMP and reduced upon activation of AMPK by AICAR) serve as a positive control and validate the quantitative LC MS/MS approach used.

Table 1. Variation in protein levels from MA-10 Leydig cells treated for 1 h with vehicle (DMSO), Fsk, or Fsk+AICAR.

DMSO vs. Fsk							
Upregulated by Forskolin							
Protein name	Protein	Gene name	Fsk	DMSO	Fsk/DMSO	q-Value	Fsk/DMSO
Dymeclin	Q8CHY3	*Dym*	3.64×10^7	1.57×10^6	23.2	5.08×10^{-6}	Up
Homologous-pairing protein 2 homolog	O35047	*Psmc3ip*	7.93×10^6	1.57×10^6	5.06	1.07×10^{-4}	Up
Acyl-Coenzyme A dehydrogenase family member 12	D3Z7X0	*Acad12*	8.76×10^6	2.22×10^6	3.94	2.56×10^{-2}	Up
N-acetyltransferase domain containing 1	Q9DBW3	*Natd1*	5.61×10^6	1.57×10^6	3.58	1.58×10^{-4}	Up
Nuclear receptor subfamily 4 group A member 1	P12813	*Nr4a1*	3.76×10^6	1.57×10^6	2.4	1.41×10^{-3}	Up
Downregulated by Forskolin							
Protein name	Protein	Gene name	Fsk	DMSO	DMSO/Fsk	q-Value	Fsk/DMSO
H/ACA ribonucleoprotein complex subunit 3	Q9CQS2	*Nop10*	1.21×10^6	7.91×10^6	6.55	1.44×10^{-2}	Down
Mpv17-like protein 2	Q8VIK2	*Mpv17l2*	1.21×10^6	6.46×10^6	5.35	8.13×10^{-3}	Down
Overexpressed in colon carcinoma 1 protein homolog	P0C913	*Occ1*	1.21×10^6	6.36×10^6	5.27	8.32×10^{-3}	Down
Glutathione peroxidase 7	Q99LJ6	*Gpx7*	1.21×10^6	4.79×10^6	3.97	3.60×10^{-2}	Down
Fsk vs. Fsk+AICAR							
Upregulated by AICAR							
Protein name	Protein	Gene name	Fsk+AICAR	Fsk	Fsk+AICAR/Fsk	q-Value	Fsk+AICAR/Fsk
Cyclic AMP-responsive element-binding protein 1	Q01147	*Creb1*	2.19×10^7	1.21×10^6	18.15	3.83×10^{-4}	Up
Mpv17-like protein 2	Q8VIK2	*Mpv17l2*	6.32×10^6	1.21×10^6	5.23	5.72×10^{-3}	Up
Overexpressed in colon carcinoma 1 protein homolog	P0C913	*Occ1*	5.95×10^6	1.21×10^6	4.93	7.06×10^{-3}	Up
HAUS augmin-like complex subunit 3	Q8QZX2	*Haus3*	3.98×10^6	1.21×10^6	3.3	3.10×10^{-2}	Up
Downregulated by AICAR							
Protein name	Protein	Gene name	Fsk+AICAR	Fsk	Fsk/Fsk+AICAR	q-Value	Fsk+AICAR/Fsk
Dymeclin	Q8CHY3	*Dym*	1.61×10^6	3.64×10^7	22.64	8.11×10^{-6}	Down
Ubinuclein-2	Q80WC1	*Ubn2*	1.61×10^6	1.56×10^7	9.7	1.71×10^{-5}	Down
Ferrochelatase	P22315	*Fech*	1.61×10^6	6.69×10^6	4.16	4.02×10^{-4}	Down
Serine/threonine-protein kinase LMTK2	Q3TYD6	*Lmtk2*	1.61×10^6	4.56×10^6	2.84	6.67×10^{-3}	Down
Armadillo-like helical domain-containing protein 3	Q6PD19	*Armh3*	1.61×10^6	4.52×10^6	2.81	8.76×10^{-4}	Down
Ribokinase	Q8R1Q9	*Rbks*	1.61×10^6	4.05×10^6	2.52	1.13×10^{-3}	Down
Nuclear receptor subfamily 4 group A member 1	P12813	*Nr4a1*	1.61×10^6	3.76×10^6	2.34	1.48×10^{-3}	Down

Table 1. *Cont.*

DMSO vs. Fsk+AICAR							
Upregulated by Forskolin + AICAR							
Protein name	Protein	Gene name	Fsk+AICAR	DMSO	Fsk+AICAR/DMSO	*q*-Value	Fsk+AICAR/DMSO
Glutamyl-tRNA(Gln) amidotransferase subunit B. mitochondrial	Q99JT1	*Gatb*	3.56×10^6	1.57×10^6	2.27	8.86×10^{-4}	Up
Decapping and exoribonuclease protein	O70348	*Dxo*	2.89×10^6	1.57×10^6	1.84	3.70×10^{-2}	Up
Downregulated by Forskolin + AICAR							
Protein name	Protein	Gene name	Fsk+AICAR	DMSO	DMSO/Fsk+AICAR	*q*-Value	Fsk+AICAR/DMSO
SAGA-associated factor 29	Q9DA08	*Sgf29*	1.61×10^6	1.39×10^9	864.22	3.00×10^{-5}	Down
Coiled-coil domain-containing protein 187	Q8C5V8	*Ccdc187*	1.61×10^6	1.40×10^7	8.71	3.87×10^{-4}	Down
Ferrochelatase	P22315	*Fech*	1.61×10^6	6.73×10^6	4.19	1.34×10^{-4}	Down
Armadillo-like helical domain-containing protein 3	Q6PD19	*Armh3*	1.61×10^6	5.59×10^6	3.48	3.67×10^{-2}	Down

The level of the protein Dymeclin, a Golgi-associated protein believed to be involved in intracellular trafficking [17–19], was the most increased after 1 h of Fsk treatment (23-fold; Table 1). Similar to NR4A1, the Fsk-dependent increase of Dymeclin was prevented when AMPK was activated by AICAR (decreased by 23-fold and back to control levels). Another protein affected by both Fsk and AICAR is the Mpv17-like protein 2 (MPV17L2). MPV17L2 is required for the assembly and stability of the mitochondrial ribosome, and in its absence, protein synthesis in mitochondria is impaired, and mitochondrial DNA aggregates [20]. Contrary to NR4A1 and Dymeclin, the MPV17L2 levels were reduced by 5.35-fold in the presence of Fsk and increased by 5.23-fold by AICAR-mediated AMPK activation (Table 1).

As summarized in Table 1, in addition to Dymeclin (up by 23-fold) and NR4A1 (up by 2.4-fold), the top Fsk-induced proteins included the homologous-pairing protein 2 homolog (PSMC3IP; up by 5-fold), acyl-coenzyme A dehydrogenase family member 12 (ACAD12; up by 4-fold), and N-acetyltransferase domain containing 1 (NATD1; up by 3.6-fold). In addition to MPV17L2, the most induced proteins in the presence of AICAR (activation of AMPK) included the cyclic AMP-responsive element-binding protein 1 (CREB1; up by 18-fold), while decreased proteins, in addition to Dymeclin and NR4A1, included the Lemur serine/threonine-protein kinase 2 (LMTK2; down by 2.8-fold). These results showed that after a short 1-h treatment, the level of some proteins involved in mitochondrial function and in steroidogenesis was already altered.

2.3. Identification of Proteins Differentially Phosphorylated in MA-10 Leydig Cells in Response to Fsk and AICAR

Our quantitative analysis of the phosphoproteome of MA-10 Leydig cells revealed significant changes in the phosphorylation levels of several phosphopeptides corresponding to various proteins in response to the treatments. The treatments included DMSO (the vehicle used as the control), Fsk to increase the cAMP levels mimicking LH-stimulated steroidogenesis, and Fsk+AICAR simultaneously, where AICAR activates AMPK, a kinase previously reported to potently inhibit hormone-induced steroidogenesis [14]. The top 10 differentially phosphorylated (up and down) phosphopeptides for each pairwise comparison (Fsk vs. DMSO, Fsk+AICAR vs. Fsk, Fsk+AICAR vs. DMSO) are presented in Table 2.

Table 2. Top 10 phosphopeptides differentially phosphorylated in MA-10 Leydig cells treated with vehicle (DMS), Fsk, or Fsk+AICAR (pair comparisons).

Fsk vs. DMSO							
Phosphorylation decreased by Forskolin							
Phosphopeptide Sequence	Protein	Protein Name	Gene Name	Fsk	DMSO	DMSO/ Fsk	q-Value	Fsk/ DMSO
IQTSNVTNKNDPK	Q64012	RNA-binding protein Raly	*Raly*	2.12×10^6	3.47×10^8	163.6	4.54×10^{-4}	Down
DLPPFEDESEGLLGTEGPMEEEEDGEELIGDGMER	P97310	DNA replication licensing factor MCM2	*Mcm2*	2.12×10^6	5.46×10^7	25.7	1.63×10^{-3}	Down
IDEPNTPYHNMIGDDEDAYSDSEGNEVMTPDILAK	Q9DCL8	Protein phosphatase inhibitor 2	*Ppp1r2*	2.12×10^6	5.14×10^7	24.19	1.53×10^{-3}	Down
HLSNVSSTGSIDMVDSPQLATLADEVSASLAK	P10637	Microtubule-associated protein tau	*Mapt*	2.12×10^6	4.06×10^7	19.14	1.53×10^{-3}	Down
SSSPSASLTEHEVSDSPGDEPSESPYESADETQTEASVSSK	Q8BVE8	Histone-lysine N-methyltransferase NSD2	*Whsc1*	2.12×10^6	3.51×10^7	16.54	8.63×10^{-3}	Down
ETGGTYPPSPPPHSSPTPAATVAATVSTAVPGEPLLPR	Q80U72	Protein scribble homolog	*Scrib*	2.12×10^6	2.36×10^7	11.13	1.55×10^{-2}	Down
RISHSLYSGIEGLDESPTR	Q99NH2	Partitioning defective 3 homolog	*Pard3*	2.12×10^6	2.19×10^7	10.33	2.37×10^{-2}	Down
KSMYSRVPECQVTTYYYVGFAYLMMR	Q8QZY1	Eukaryotic translation initiation factor 3 subunit L	*Eif3l*	2.12×10^6	2.12×10^7	9.99	3.82×10^{-3}	Down
ALEETPPDSPAAEQENSVNCVDPLR	Q8K1K3	TERF1-interacting nuclear factor 2	*Tinf2*	2.12×10^6	2.03×10^7	9.54	3.36×10^{-3}	Down
LAAQESSEAEDVTVDR	Q6PGL7	WASH complex subunit FAM21	*Fam21*	2.12×10^6	1.99×10^7	9.36	4.73×10^{-2}	Down
Phosphorylation increased by Forskolin							
Phosphopeptide Sequence	Protein	Protein Name	Gene Name	Fsk	DMSO	Fsk/ DMSO	q-Value	Fsk/ DMSO
KSSVEGLEPAENK	P16254	Signal recognition particle 14 kDa protein	*Srp14*	2.63×10^8	2.90×10^6	90.6	4.54×10^{-4}	Up
STSQGSINSPVYSR	Q8K4G5	Actin-binding LIM protein 1	*Ablim1*	7.08×10^7	2.12×10^6	33.48	1.63×10^{-3}	Up
QNPEQSADEDAEKNEEDSEGSSDEDEDEDGVGNTTFLK	Q8R1B4	Eukaryotic translation initiation factor 3 subunit C	*Eif3c*	7.06×10^7	2.12×10^6	33.39	1.53×10^{-3}	Up
PCIQAQYGTPATSPGPR	P12813	Nuclear receptor subfamily 4 group A member 1	*Nr4a1*	4.52×10^7	2.12×10^6	21.36	1.53×10^{-3}	Up
NTFTAWSEEDSDYEIDDR	Q6ZQ58	La-related protein 1	*Larp1*	1.66×10^8	1.02×10^7	16.28	9.61×10^{-3}	Up
EKVESAGPGGDSEPTGSTGALAHTPR	O08550	Histone-lysine N-methyltransferase 2B	*Kmt2b*	2.42×10^7	2.12×10^6	11.44	3.17×10^{-3}	Up
SPDLSNQNSDQANEEWETASESSDFASER	Q7TSC1	Protein PRRC2A	*Prrc2a*	1.77×10^7	2.12×10^6	8.35	8.52×10^{-3}	Up
KVSVEPQDSHQDAQPR	Q8BXB6	Solute carrier organic anion transporter family member 2B1	*Slco2b1*	1.97×10^8	2.47×10^7	7.98	1.30×10^{-2}	Up

Table 2. *Cont.*

Fsk vs. DMSO								
Phosphopeptide Sequence	Protein	Protein Name	Gene Name	Fsk	DMSO	DMSO/ Fsk	q-Value	Fsk/ DMSO
Phosphorylation increased by Forskolin								
KSSAAAAAAAAAEGALLPQTPPSPR	Q9ERD6	Ras-specific guanine nucleotide-releasing factor RalGPS2	*Ralgps2*	1.64×10^7	2.12×10^6	7.77	9.61×10^{-3}	Up
SHSESASPSALSSSPNNLSPTGWSQPK	Q99N57	RAF proto-oncogene serine/threonine-protein kinase	*Raf1*	1.61×10^7	2.12×10^6	7.63	1.30×10^{-2}	Up

Fsk+AICAR vs. Fsk								
Phosphorylation decreased by AICAR								
Phosphopeptide Sequence	Protein	Protein Name	Gene Name	Fsk + AICAR	Fsk	Fsk/ Fsk+AICAR	q-Value	Fsk+AICAR/Fsk
YDEKTGLALQTEEFIPNYYCTDERR	K7N6T2	Vomeronasal 2 receptor 67	*Vmn2r67*	2.15×10^6	1.56×10^9	727.23	1.81×10^{-4}	Down
GASAATGIPLESDEDSNDNDNDLENENCMHTN	Q8R5H1	Ubiquitin carboxyl-terminal hydrolase 15	*Usp15*	2.15×10^6	6.62×10^7	30.83	2.50×10^{-3}	Down
KGSDDDGGDSPVQDIDTPEVDLYQLQVNTLR	O88574	Histone deacetylase complex subunit SAP30	*Sap30*	2.15×10^6	3.75×10^7	17.43	1.12×10^{-2}	Down
WGQPPSPTPVPRPPDADPNTPSPK	Q6PDQ2	Chromodomain-helicase-DNA-binding protein 4	*Chd4*	2.15×10^6	3.33×10^7	15.52	8.64×10^{-3}	Down
TNSMGSATGPLPGTK	Q8BZ47	Zinc finger protein 609	*Znf609*	2.15×10^6	3.21×10^7	14.94	4.99×10^{-3}	Down
NLATSADTPPSTIPGTGK	Q8BKX6	Serine/threonine-protein kinase SMG1	*Smg1*	2.15×10^6	1.63×10^7	7.57	4.56×10^{-3}	Down
SPHDSKSPLDHRSPLER	E9PZM4	Chromodomain-helicase-DNA-binding protein 2	*Chd2*	2.15×10^6	1.58×10^7	7.35	4.63×10^{-3}	Down
IASSSSENNFLSGSPSSPMGDILQTPQFQMR	E9Q7G0	Nuclear mitotic apparatus protein 1	*Numa1*	2.15×10^6	1.40×10^7	6.52	1.14×10^{-2}	Down
NQENVSHLSVSSASPTSSVASAAGSVTSSSLQK	O54826	Protein AF-10	*Mllt10*	2.15×10^6	1.39×10^7	6.45	2.00×10^{-2}	Down
GSSGEGLPFAEEGNLTIK	Q8VHK1	Caskin-2	*Caskin2*	2.15×10^6	1.23×10^7	5.75	1.02×10^{-2}	Down
Phosphorylation increased by AICAR								
Phosphopeptide Sequence	Protein	Protein Name	Gene Name	Fsk + AICAR	Fsk	Fsk+AICAR/ Fsk	q-Value	Fsk+AICAR/Fsk
ARPAQAPVSEELPPSPKPGK	Q6PGL7	WASH complex subunit FAM21	*Fam21*	6.22×10^7	2.12×10^6	29.29	1.54×10^{-2}	Up
HLSNVSSTGSIDMVDSPQLATLADEVSASLAK	P10637	Microtubule-associated protein tau	*Mapt*	4.88×10^7	2.12×10^6	22.99	2.50×10^{-3}	Up

Table 2. *Cont.*

				Fsk vs. DMSO				
Phosphopeptide Sequence	Protein	Protein Name	Gene Name	Fsk	DMSO	DMSO/ Fsk	q-Value	Fsk/ DMSO
Phosphorylation increased by AICAR								
APSPEPPTEEVAAETNSTPDDLEAQDALSPETTEEK	Q9ESJ4	NCK-interacting protein with SH3 domain	*Nckipsd*	4.73×10^7	2.12×10^6	22.26	2.50×10^{-3}	Up
SSSPSASLTEHEVSDSPGDEPSESPYESADETQTEASVSSK	Q8BVE8	Histone-lysine N-methyltransferase NSD2	*Whsc1*	3.40×10^7	2.12×10^6	16.00	4.63×10^{-3}	Up
VLLTHEVMCSR	O08792	Transcription factor COE2	*Ebf2*	3.04×10^7	2.12×10^6	14.33	2.50×10^{-3}	Up
VTETEDDSDSDSDDDEDDVHVTIGDIK	Q9D824	Pre-mRNA 3'-end-processing factor FIP1	*Fip1l1*	2.40×10^7	2.12×10^6	11.31	2.50×10^{-3}	Up
SRSVEDDEEGHLICQSGDVLSAR	P22518	Dual specificity protein kinase CLK1	*Clk1*	2.04×10^7	2.12×10^6	9.63	8.25×10^{-3}	Up
RISHSLYSGIEGLDESPTR	Q99NH2	Partitioning defective 3 homolog	*Pard3*	1.94×10^7	2.12×10^6	9.16	3.87×10^{-3}	Up
DAEDLSPCLPSSSQEDTAVPSSPGPSDEVSNTEAEAR	G5E8P0	Gamma-tubulin complex component 6	*Tubgcp6*	1.92×10^7	2.12×10^6	9.06	3.47×10^{-2}	Up
SVDLKTASPESGRSGFQDEESFR	E9Q4Y4	Centrosomal protein 192	*Cep192*	1.89×10^7	2.12×10^6	8.91	4.99×10^{-3}	Up
				Fsk+AICAR vs. DMSO				
Phosphorylation decreased by Forskolin + AICAR								
Phosphopeptide Sequence	Protein	Protein Name	Gene Name	Fsk + AICAR	DMSO	DMSO/ Fsk+AICAR	q-Value	Fsk+AICAR/ DMSO
AADPPAENSSAPEAEQGGAE	P62960	Nuclease-sensitive element-binding protein 1	*Ybx1*	2.15×10^6	1.60×10^8	74.33	2.97×10^{-3}	Down
DLPPFEDESEGLLGTEGPMEEEEDGEELIGDGMER	P97310	DNA replication licensing factor MCM2	*Mcm2*	2.15×10^6	5.46×10^7	25.4	2.27×10^{-3}	Down
SLTISVDSASTSR	Q99MZ6	Unconventional myosin-VIIb	*Myo7b*	2.15×10^6	4.54×10^7	21.12	3.65×10^{-2}	Down
LDGESDKEQFDDDQK	Q8C1B1	Calmodulin-regulated spectrin-associated protein 2	*Camsap2*	2.15×10^6	3.27×10^7	15.23	6.41×10^{-3}	Down
ALEETPPDSPAAEQENSVNCVDPLR	Q8K1K3	TERF1-interacting nuclear factor 2	*Tinf2*	2.15×10^6	2.03×10^7	9.43	4.79×10^{-3}	Down
QATESPAYGIPLKDGSEQTDEEAEGPFSDDEMVTHK	Q99KK1	Receptor expression-enhancing protein 3	*Reep3*	2.15×10^6	1.89×10^7	8.8	2.18×10^{-2}	Down
HSPIKEEPCGSISETVCK	Q6PDM1	Male-specific lethal 1 homolog	*Msl1*	2.15×10^6	1.88×10^7	8.75	2.63×10^{-2}	Down

Table 2. *Cont.*

Fsk vs. DMSO							
Phosphorylation decreased by Forskolin + AICAR							
Phosphopeptide Sequence	Protein	Protein Name	Gene Name	Fsk + AICAR	DMSO	DMSO/Fsk+AICAR *q*-Value	Fsk+AICAR/DMSO
QDVETCRPSSPFGR	Q6PG16	Holliday junction recognition protein	*Hjurp*	2.15×10^6	1.78×10^7	8.27	1.96×10^{-2}	Down
SSPDPAVNPVPK	Q924W7	Suppression of tumorigenicity 5 protein	*St5*	2.15×10^6	1.73×10^7	8.03	6.41×10^{-3}	Down
SEVNSEDSDIQEVLPVPK	Q7TQI8	Testis-specific Y-encoded-like protein 2	*Tspyl2*	2.83×10^6	2.20×10^7	7.8	1.83×10^{-2}	Down
Phosphorylation increased by Forskolin + AICAR							
Sequence	Protein	Protein Name	Gene Name	Fsk + AICAR	DMSO	Fsk+AICAR/ DMSO	*q*-Value	Fsk+AICAR/DMSO
KSSVEGLEPAENK	P16254	Signal recognition particle 14 kDa protein	*Srp14*	2.76×10^8	2.90×10^6	95.06	1.47×10^{-3}	Up
TPLGASLDEQSSGTPK	Q922B9	Sperm-specific antigen 2 homolog	*Ssfa2*	1.37×10^8	2.35×10^6	58.37	1.47×10^{-3}	Up
TVSTQHSTESQDNDQPDYDSVASDEDTDVETR	Q9JLQ2	ARF GTPase-activating protein GIT2	*Git2*	1.84×10^8	3.32×10^6	55.33	6.87×10^{-3}	Up
QNPEQSADEDAEKNEEDSEGSSDEDEDEDGVGNTTFLK	Q8R1B4	Eukaryotic translation initiation factor 3 subunit C	*Eif3c*	8.66×10^7	2.12×10^6	40.93	1.74×10^{-3}	Up
STSQGSINSPVYSR	Q8K4G5	Actin-binding LIM protein 1	*Ablim1*	8.20×10^7	2.12×10^6	38.77	3.41×10^{-3}	Up
LKNDSDLFGLGLEEMGPKESSDEDR	Q5U3K5	Rab-like protein 6	*Rabl6*	4.16×10^7	2.12×10^6	19.67	1.76×10^{-3}	Up
NTFTAWSEEDSDYEIDDR	Q6ZQ58	La-related protein 1	*Larp1*	1.88×10^8	1.02×10^7	18.39	6.41×10^{-3}	Up
AELEEMEEVHPSDEEEEETKAESFYQK	Q9QYR6	Microtubule-associated protein 1A	*Map1a*	3.86×10^7	2.12×10^6	18.26	1.76×10^{-3}	Up
QSNASSDVEVEEKETNVSKEDTDQEEK	Q99JF8	PC4 and SFRS1-interacting protein	*Psip1*	3.81×10^7	2.12×10^6	18.01	6.41×10^{-3}	Up
PCIQAQYGTPATSPGPR	P12813	Nuclear receptor subfamily 4 group A member 1	*Nr4a1*	2.46×10^7	2.12×10^6	11.62	2.39×10^{-2}	Up

2.3.1. Proteins Differentially Phosphorylated in Response to Fsk

We first compared the phosphopeptides between the control and Fsk-treated MA-10 Leydig cells to identify changes in protein phosphorylation during the stimulatory phase of steroidogenesis. This led to the identification of 12,125 phosphopeptides, of which 8471 were quantified. As shown in Figure 2A, the quantified phosphopeptides were mapped on a volcano plot based on the significance and the ratio between the Fsk-treated and control (DMSO) MA-10 Leydig cells. A total of 61 phosphopeptides had significantly different phosphorylation levels between these two conditions; the phosphorylation of 35 phosphopeptides was increased, while the phosphorylation of 26 phosphopeptides was decreased following the Fsk treatment. These 61 phosphopeptides correspond to 60 unique phosphoproteins and are shown along with their relative phosphorylation levels between the two conditions on the heatmap in Figure 2B.

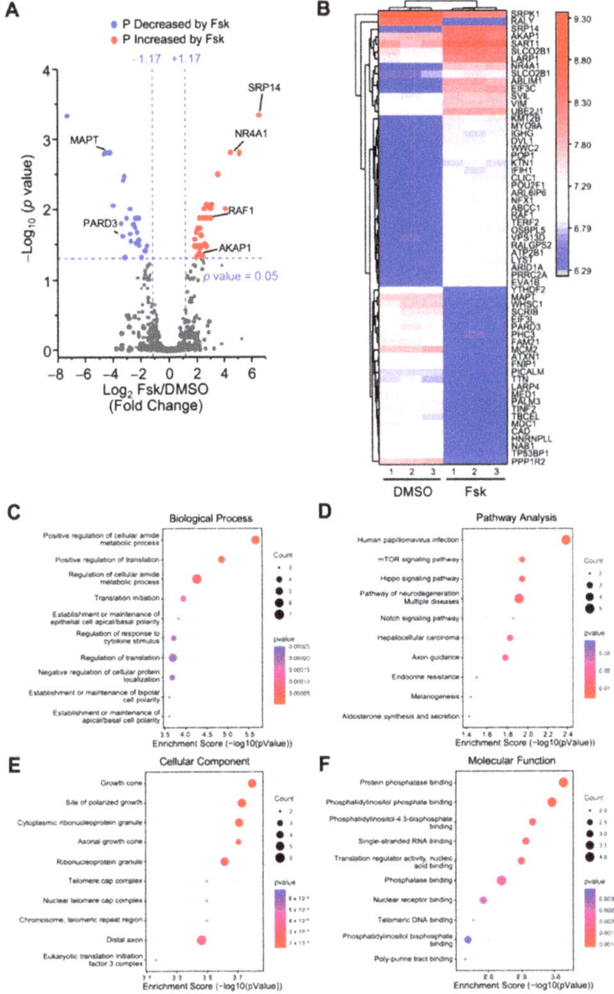

Figure 2. Analysis of differentially phosphorylated proteins in MA-10 Leydig cells in response Forskolin. LC-MS/MS was performed on enriched phosphopeptides extracted and digested from MA-10 Leydig cells treated with either DMSO (control) or Forskolin (Fsk, 10 μm) for 1 h. Phosphorylated proteins were then quantified and compared between treatments. Results are the mean of three individual experiments,

each performed in triplicate. (**A**) Volcano plot comparing the level of phosphorylated proteins detected by LC-MS/MS in control and Fsk-treated MA-10 Leydig cells. Each dot represents a distinct phosphoprotein. A Log2 fold change of $+/-1.17$ (corresponds to 2.25-fold) with a *p*-value of 0.05 was used and is represented by the blue dotted lines. In response to Fsk, blue dots depict a decrease in phosphorylation level (two proteins are identified in that group: MAPT and PARD3), while red dots correspond to an increase in phosphorylation level (four proteins are identified in that group: SRP14, NR4A1, RAF1, AKAP1). Grey dots correspond to proteins with no significant changes in phosphorylation levels. (**B**) Heatmap of protein phosphorylation levels for differentially phosphorylated proteins between treatment groups (control DMSO vs. Fsk). The name of the protein is indicated. The scale on the right represents expression levels (Log10). Biological Process (**C**), Pathway Analysis (**D**), Cellular Component (**E**), and Molecular Function (**F**) of the differentially phosphorylated proteins are shown.

The proteins exhibiting increased phosphorylation after the Fsk treatment included the nuclear receptor NR4A1, the signal recognition particle SRP14, the protein kinase A anchoring protein AKAP1, and the serine/threonine kinase RAF1. In comparison, the proteins showing a reduction in their phosphorylation level in response to Fsk included the microtubule-associated protein tau (MAPT) and the Par-3 family cell polarity regulator protein PARD3 (Figure 2A,B). In response to Fsk in MA-10 Leydig cells, the differentially phosphorylated proteins are mostly implicated in the metabolic processes and the regulation of translation (Figure 2C). A pathway analysis revealed that these proteins are implicated mainly in the mTOR and Hippo signaling pathways, two pathways involved in metabolism (Figure 2D). As shown in Figure 2E, the differentially phosphorylated proteins are associated with various cellular components. Finally, the 60 differentially phosphorylated proteins are associated with various molecular functions, including binding to protein phosphatase, phosphatidyl inositol phosphate, and nucleic acid (Figure 2F).

2.3.2. Proteins Differentially Phosphorylated in Response to AMPK Activation

The activation of AMPK with AICAR represses LH/cAMP-induced steroidogenesis but does not significantly affect the unstimulated steroidogenic cells [14]. Therefore, to identify potential target proteins of AMPK, we compared all phosphopeptides between the Fsk- and Fsk+AICAR-treated MA-10 Leydig cells. The differences between those two conditions should reflect mainly the proteins phosphorylated by AMPK. This comparison identified 12,125 phosphopeptides, of which 8581 phosphopeptides were quantified. It also revealed that the phosphorylated amino acid in cells exposed to AICAR was, by far, a serine, as shown by the sequence motif presented in Figure 3A. Quantified phosphopeptides were then mapped on a volcano plot based on the significance and the ratio between the Fsk+AICAR- and Fsk-treated cells (Figure 3B). A total of 46 phosphopeptides had significantly different phosphorylation levels between these two conditions; in the Fsk+AICAR group, the phosphorylation of 27 phosphopeptides was increased, while the phosphorylation of 19 phosphopeptides was decreased compared to the Fsk group (Figure 3B,C). These 46 phosphopeptides correspond to 44 unique phosphoproteins, as some proteins have more than one phosphorylated residue (phosphopeptide sequence). These 44 phosphoproteins are presented along with their relative phosphorylation levels between the two conditions on the heatmap presented in Figure 3C.

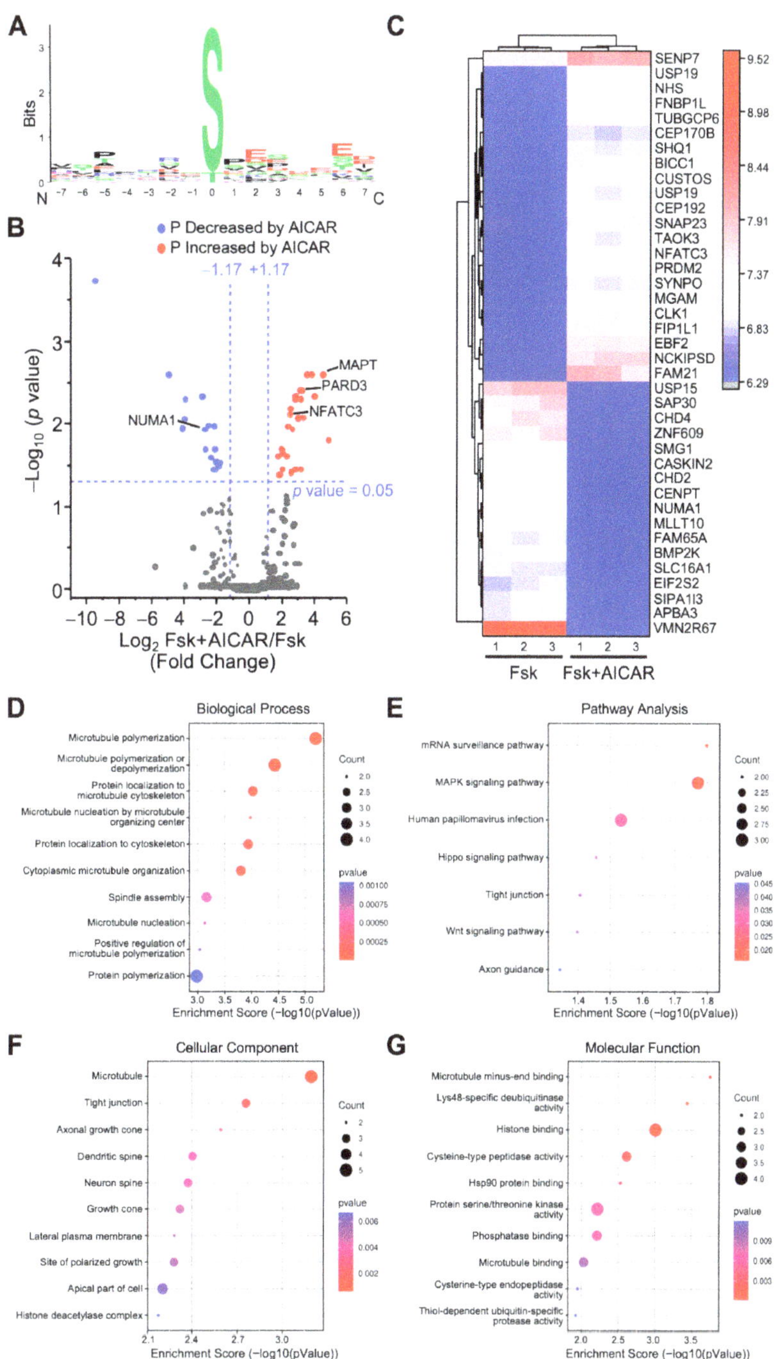

Figure 3. Analysis of differentially phosphorylated proteins in MA-10 Leydig cells treated with Forskolin or Forskolin+AICAR. LC-MS/MS was performed on enriched phosphopeptides extracted and digested

from MA-10 Leydig cells treated with Forskolin (Fsk, 10 μm) or Fsk+AICAR (10 μM and 1 mM, respectively) for 1 h. Phosphorylated proteins were then quantified and compared between treatments. Results are the mean of three individual experiments, each performed in triplicate. (**A**) Consensus sequence motif of phosphorylated peptides. The amino acid chosen is the one with the highest probability of being phosphorylated. (**B**) Volcano plot comparing the level of phosphorylated proteins detected by LC-MS/MS in Fsk-treated vs. Fsk+AICAR-treated MA-10 Leydig cells (to observe the effects of AICAR). Each dot represents a distinct phosphoprotein. A Log2 fold change of +/−1.17 (corresponds to 2.25-fold) with a *p*-value of 0.05 was used and is represented by the blue dotted lines. In response to AICAR, blue dots depict a decrease in phosphorylation level (one protein is identified in that group: NUMA1), while red dots correspond to an increase in phosphorylation level (three proteins are identified in that group: MAPT, PARD3, NFATC3). Grey dots correspond to proteins with no significant changes in phosphorylation levels. (**C**) Heatmap of protein phosphorylation levels for differentially phosphorylated proteins between treatments (Fsk vs. Fsk+AICAR). The name of the protein in indicated. The scale on the right represents expression levels (Log10). Biological Process (**D**), Pathway Analysis (**E**), Cellular Component (**F**), and Molecular Function (**G**) of the differentially phosphorylated proteins are shown.

The proteins with increased phosphorylation in the AICAR group included NFATC3 (transcription factor), PARD3, and MAPT; an example of a protein with reduced phosphorylation in the AICAR group is the nuclear mitotic apparatus protein NUMA1 (Figure 3A,B). The proteins differentially phosphorylated between the Fsk+AICAR and Fsk groups are mostly implicated in the microtubule polymerization or its organization (Figure 3D). This is supported by the localization of these proteins, mostly with microtubules (Figure 3F), with their main molecular functions being microtubule binding, histone binding, and deubiquitinase activity (Figure 3G). The pathway analysis revealed that the 44 proteins are mainly associated with the MAPK signaling pathway and mRNA surveillance pathway (Figure 3E).

2.3.3. Proteins Differentially Phosphorylated in Response to Combined Fsk+AICAR

Comparisons were made between the Fsk+AICAR and the control (DMSO) groups to identify all the changes in the protein phosphorylation triggered by the simultaneous treatment of the MA-10 Leydig cells with Fsk (stimulatory) and AICAR (inhibitory). A total of 12,125 phosphopeptides were identified, of which 8711 were quantified. As shown in Figure 4A, the quantified phosphopeptides were mapped on a volcano plot based on the significance and the ratio between the Fsk+AICAR and control (DMSO) cells. A total of 73 phosphopeptides exhibited significantly different phosphorylation levels between these two conditions; the phosphorylation of 38 phosphopeptides was increased in the MA-10 Leydig cells treated with Fsk+AICAR, while the phosphorylation of 35 phosphopeptides was decreased. These 73 phosphopeptides correspond to 68 unique phosphoproteins, as some proteins harbor more than one phosphorylation site, such as MAP1A. The 68 phosphoproteins identified, along with their relative phosphorylation levels between the two conditions, are shown on the heatmap presented in Figure 4B.

Some of the proteins with increased phosphorylation in the MA-10 Leydig cells in response to Fsk+AICAR included NR4A1, NUMA1, the protein GOLGA5 involved in maintaining the Golgi structure, and the microtubule-associated proteins MAP1A and MAP1B. MAP1A was also present amongst the proteins with reduced phosphorylation. This is explained by the fact that MAP1A contains two distinct phosphopeptides: one showed increased phosphorylation, while the other showed reduced phosphorylation in the presence of Fsk+AICAR (Figure 4A). In terms of the biological processes, differentially phosphorylated proteins were mainly associated with microtubule polymerization, depolymerization, and organization (Figure 4C). Consistent with this, these proteins are mostly localized with microtubules, and their main functions are to bind to the microtubules, actin, and tubulin (Figure 4F). The pathway analysis associated the differentially phosphorylated proteins with the axon guidance pathway (Figure 4D).

Taken together, these results provide the first description of the global changes in the phosphoproteome of MA-10 steroidogenic Leydig cells after a stimulatory treatment (Fsk) and an inhibitory treatment (activation of AMPK by AICAR).

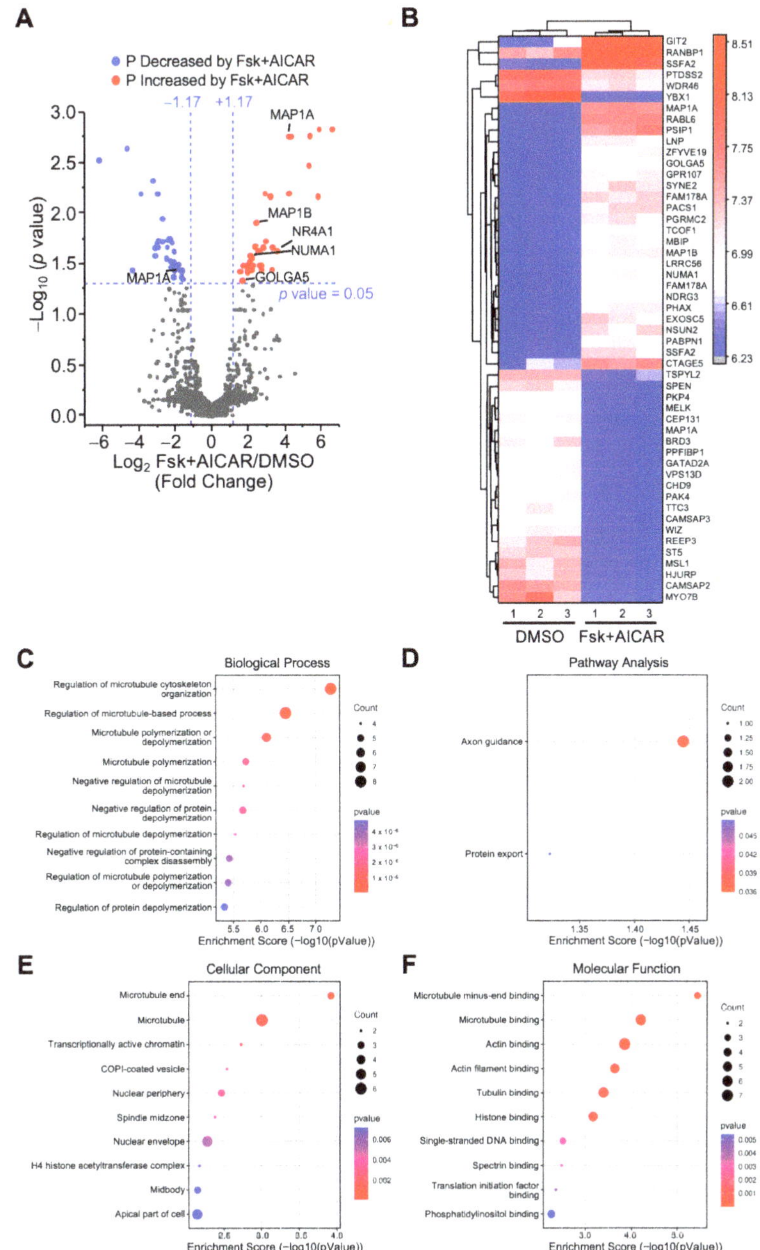

Figure 4. Analysis of differentially phosphorylated proteins in MA-10 Leydig cells in response to Forskolin+AICAR. LC-MS/MS was performed on enriched phosphopeptides extracted and digested

from MA-10 Leydig cells treated with DMSO (control) or Fsk+AICAR (10 μM and 1 mM, respectively) for 1 h. Phosphorylated proteins were then quantified and compared between treatments. Results are the mean of three individual experiments, each performed in triplicate. (**A**) Volcano plot comparing the level of phosphorylated proteins detected by LC-MS/MS in control (DMSO) vs. Fsk+AICAR-treated MA-10 Leydig cells. Each dot represents a distinct phosphoprotein. A Log2 fold change of +/−1.17 (corresponds to 2.25-fold) with a *p*-value of 0.05 was used and is represented by the blue dotted lines. In response to Fsk+AICAR, blue dots depict a decrease in phosphorylation level (one protein is identified in that group: MAP1A), while red dots correspond to an increase in phosphorylation level (five proteins are identified in that group: MAP1A, MAP1B, NR4A1, NUMA1, GOLGA5). Grey dots correspond to proteins with no significant changes in phosphorylation levels. (**B**) Heatmap of protein phosphorylation levels for differentially phosphorylated proteins between treatments (control DMSO vs. Fsk+AICAR). The name of the protein is indicated. The scale on the right represents expression levels (Log10). Biological Process (**C**), Pathway Analysis (**D**), Cellular Component (**E**) and Molecular Function (**F**) of the differentially phosphorylated proteins are shown.

2.4. Phosphorylation Trajectory of Representative Proteins

Several proteins exhibited an altered phosphorylation status in response to Fsk and to Fsk+AICAR in the MA-10 Leydig cells (Figures 2–4). Although the protein levels remained unchanged for the majority of proteins (Table 1), changes in the phosphorylation level in response to Fsk and to Fsk+AICAR could still be the result of a concomitant change in the protein level. Therefore, the phosphorylation status of the representative proteins was plotted and compared to their protein levels, allowing for better visualization of the phosphorylation trajectory of a given protein.

The phosphorylation level of NR4A1, an orphan nuclear receptor known to regulate the expression of several steroidogenic genes, was increased by more than 20-fold after the Fsk treatment, while the total NR4A1 protein level was only increased by 2.4-fold. This increase tended to be attenuated when cells were treated with AICAR in addition to Fsk (Figure 5A). Interestingly, the phosphorylation levels of the progesterone membrane receptor components 2 (PGRMC2), a protein belonging to the membrane-associated progesterone receptor (MAPR) protein family, was significantly increased when cells were treated with Fsk and with Fsk+AICAR (Figure 5B). No change in the total PGRMC2 protein level was noted. The NUMA1 exhibited the same phosphorylation pattern as PGRMC2 (Figure 5C), with an increase in phosphorylation of the phosphopeptide QAASSQEPSELEELR. On the other hand, a different phosphopeptide (IASSSSENNFLSGSPSSPMGDILQTPQFQMR) of the NUMA1 protein remained unchanged in response to Fsk, whereas AICAR caused a significant reduction in its phosphorylation level (Figure 5D). Finally, the phosphorylation level of two proteins, known to be downstream of AMPK, MAPT, and PARD3, was significantly reduced when the MA-10 Leydig cells were treated with Fsk. However, this decrease in phosphorylation was no longer apparent when the cells were co-treated with Fsk+AICAR, suggesting that AMPK phosphorylates these proteins (Figure 5E,F). The total protein level of PGRMC2, NUMA1, MAPT, and PARD3 was not affected by the treatments. These results validate AMPK activation in the MA-10 Leydig cells treated with AICAR.

Altogether, these results reveal that protein phosphorylation, in response to Fsk and Fsk+AICAR in MA-10 Leydig cells, is a very dynamic process and is mainly independent of the changes in protein levels.

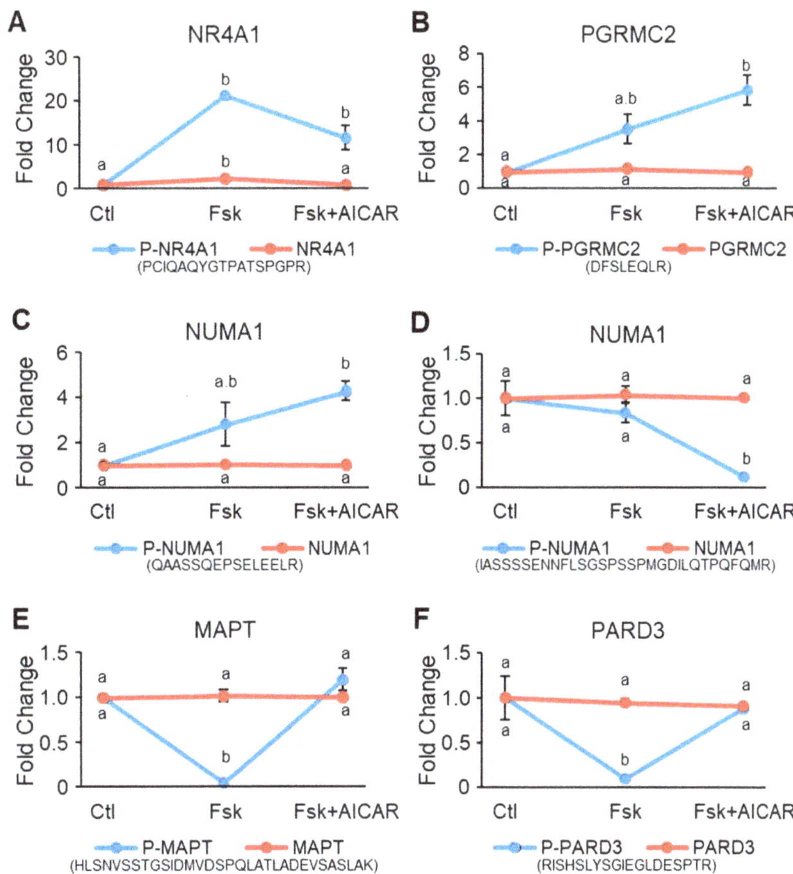

Figure 5. Representative proteins that are differentially phosphorylated in MA-10 Leydig cells in response to Forskolin and to Forskolin+AICAR. Intensity plots of representative phosphoproteins (blue lines) and total proteins (red lines). Ctl, DMSO-treated cells; Fsk, Forskolin-treated cells; Fsk+AICAR, cells treated with Forskolin and AICAR. Results are shown as Fold Change compared to the control. Results are the mean of three individual experiments +/−SEM. The phosphopeptide sequence is shown beneath the graph. For a given protein, different letters indicate a statistically significant difference between treatment groups ($p < 0.05$).

3. Discussion

Several studies have reported changes in the expression of a single gene or its protein levels in response to hormone-induced steroidogenesis. However, Leydig cell steroidogenesis is a strictly regulated process that requires changes in gene expression, protein levels, and protein phosphorylation of many genes and proteins. Therefore, the main objective of this work was to elucidate the global phosphoproteomic changes that occur in MA-10 Leydig cells in response to Forskolin-induced steroidogenesis, followed by AICAR-activated AMPK, leading to repression of steroidogenesis.

3.1. Global Variation in Protein Levels in MA-10 Leydig Cells 1 h after Fsk or Fsk+AICAR Treatment

We first determined the changes in the global protein levels and found that the level of only 20 proteins was significantly modified in the Leydig cells treated for 1 h with either Forskolin alone (stimulatory) or Forskolin+AICAR (inhibitory). One of those proteins was the nuclear receptor NR4A1. Our global proteomic analysis revealed that the NR4A1 protein levels were increased by 2.4-fold after the Forskolin treatment, and this increase was blunted when cells were co-treated with AICAR. This is consistent with previous studies from our group that reported the pattern of *Nr4a1* mRNA levels [14], as well as a time course of NR4A1 protein levels [10], in response to $(Bu)_2$-cAMP treatment of mouse MA-10 Leydig cells and primary Leydig cells from rats. Therefore, the observed changes in the NR4A1 protein levels in our current study validate our experimental approach.

Another protein affected by the treatments was Dymeclin, which was induced 23-fold in response to Fsk and then lost upon AMPK activation. Although it has never been reported in Leydig cells, Dymeclin is believed to be implicated in vesicle trafficking, as it shuttles between the cytosol and the Golgi apparatus in a highly dynamic manner [18,19]. Vesicle trafficking is an important process for steroidogenesis. For instance, the life cycle of lipid droplets, which are a source of cholesterol substrate for steroid hormones [21], involves intracellular trafficking (reviewed in [22]). The significant and rapid changes in Dymeclin protein levels in response to Fsk and Fsk+AICAR suggest that Dymeclin might play an active role in hormone-regulated steroidogenesis in Leydig cells.

The transcription factor, CREB, is known to activate steroidogenic gene expression in Leydig cells [23–28]. Despite this, the CREB protein levels remained unchanged in response to the Fsk/cAMP treatment, which is identical to what we observed in our current global proteomic analysis. Unexpectedly, the CREB protein levels were increased by 18-fold in the MA-10 Leydig cells treated with AICAR to turn on the repressive kinase, AMPK (Fsk+AICAR-treated MA-10 Leydig cells compared to the Fsk-treated cells). This would appear to suggest a repressive role for CREB in steroidogenesis. Although mainly known as an activator, CREB has also been reported to act as a repressor. For instance, CREB represses the expression of the *cFos* gene [29]. Furthermore, alternate exon usage is also known to switch CREB from an activator to a repressor [30]. It is, therefore, possible that activated AMPK influences CREB exon usage and that CREB would, therefore, repress steroidogenic gene expression in that context.

The MPV17L2 protein levels were also modulated by the treatments. MPV17L2 is an integral mitochondrial inner membrane protein [20], and as such, it is implicated in mitochondrial ribosome assembly [20] and in reactive oxygen species (ROS) production [31], two processes known to influence steroidogenesis in Leydig cells [32]. We found that the MPV17L2 protein levels increased 5-fold in the presence of AICAR, while it was reduced by 5-fold in the Fsk-treated cells. Our data are consistent with transcriptomic data from the pancreas, where the *Mpv17l2* mRNA levels are decreased by 1.6-fold in AMPK-deficient mice [33], indicating that activated AMPK increases its expression.

Overall, our data show that globally, the levels of 20 proteins, including some never before reported in Leydig cells, are rapidly and differentially modulated in response to Fsk (stimulatory) and to Fsk+AICAR (inhibitory).

3.2. Comprehensive Analysis of the Phosphoproteome during Stimulation and Inhibition of Steroidogenesis in MA-10 Leydig Cells

In Leydig cells, the response to hormone stimulation involves the activation of various signaling pathways, leading to increased activity of various kinases, including PKA, MAPK, CAMKI, and AMPK. These kinases then phosphorylate various proteins, thus ensuring a proper cellular response to the stimulus. Although some phosphorylated proteins have been identified, the majority of studies have focused on a single protein at the time. One study reported the changes in the phosphoproteome of Leydig cells in which the cAMP levels were maintained artificially high by inhibiting all phosphodiesterase activity [34]. This study identified alterations in several phosphosites, some of which were also detected in our current study. However, constitutively high cAMP levels might trigger non-physiological responses. In our present work, we used Fsk, an agonist of adenylate cyclase that leads to a typical increase in cAMP levels, triggering a stimulatory response, and AICAR, an agonist of AMPK, leading to an inhibitory response. In both cases, the levels of cAMP, the main second messenger in Leydig cells, followed a normal increase–decrease rate typical of a more classic response [14]. Using this treatment scheme and a quantitative phosphoproteomics approach, we now report the global changes in the phosphoproteome that occur in Leydig cells in response to both a stimulatory and an inhibitory signal. We found that the phosphorylation level of a large number of proteins was modified in response to either Fsk (stimulatory response) or Fsk+AICAR (inhibitory response).

The proteins displaying increased phosphorylation levels in the MA-10 Leydig cells treated with Fsk include AKAP1 (also known as AKAP121), RAF1, NR4A1, and SRP14. The phosphorylation of both AKAP1 and RAF1 was previously reported as being altered in the MA-10 Leydig cells depleted of PDE4 and PDE8 activity [34]. PDE activity is required for normal Leydig cell function. Indeed, PDEs regulate steroidogenesis by their ability to degrade cAMP into AMP, thus inactivating several cAMP-dependent pathways and kinases, such as PKA, while simultaneously activating the repressive kinase AMPK [12]. AKAP1 is a member of the AKAP family of scaffold proteins. AKAP1 is known to interact with PKA and therefore regulates its intracellular localization, especially at the outer mitochondrial membrane (OMM) (reviewed in [35]), where PKA can phosphorylate its target proteins [36]. A key target of PKA at the OMM for the stimulation of steroidogenesis is the STAR protein [36,37]. PKA-dependent phosphorylation of STAR increases its cholesterol shuttling activity, leading to enhanced steroidogenesis [36,37]. AKAP1 is, therefore, considered an important mitochondrial signaling hub, and the mitochondrion is an essential organelle for steroidogenesis in Leydig cells. To date, AKAP1 is not known to be phosphorylated. Our current work demonstrates that AKAP1 is phosphorylated in response to Fsk. The physiological implication of AKAP1 phosphorylation remains to be established, although a previous study suggested that phosphorylation of AKAP1 can modulate the AKAP1-PKA association [38].

Another protein exhibiting increased phosphorylation in response to Fsk is SRP14. SRP14 is involved in protein translation and, more specifically, in the elongation arrest and efficient translocation of proteins into the endoplasmic reticulum [39]. This is consistent with the requirement of a de novo protein synthesis in hormone-induced steroidogenic cells [10]. Interestingly, the level of SRP14 protein in glioblastoma cells is increased by CYP17A1 [40], an enzyme implicated in steroidogenesis. As we observed in the Fsk-treated MA-10 Leydig cells, SRP14 was also found to be phosphorylated during a large screen of phosphorylated proteins in Ras-transformed cells [41]. The significance of SRP14 phosphorylation, however, remains unknown.

The proteins displaying reduced phosphorylation levels after Fsk treatment include MAPT and PARD3. Phosphorylation of both proteins was restored when AMPK was activated (Fsk+AICAR-treated cells). MAPT, a protein involved in microtubule assembly and stabilization, is a known target of AMPK [42]. Furthermore, phosphorylation of MAPT is reduced in the MA-10 Leydig cells deficient in PDE4 and PDE8 activity to maintain high cAMP levels [34]. These data are consistent with our present data showing a reduction in

MAPT phosphorylation in the context of high cAMP levels (Fsk-treated cells) and increased phosphorylation in AMPK-activated cells. Phosphorylation of MAPT is known to influence protein stability and degradation (reviewed in [43]).

The protein PARD3 is part of the PAR3/PAR6/aPKC complex implicated in tight-junction assembly [44]. Activation of AMPK facilitates tight-junction assembly [45]. Whether activated AMPK interacts with and directly phosphorylates the PAR3/PAR6/aPKC complex remains uncertain [45]. However, our current data now indicate that PARD3 is phosphorylated upon AMPK activation. Furthermore, PARD3 is known to be phosphorylated by the PAR1 kinase, leading to PARD3 translocation from the tight junction to the cytosol [46]. Once in the cytosol, PARD3 promotes the interaction between PP1A phosphatase and LATS1, a mediator of the Hippo signaling pathway [46]. This results in the dephosphorylation and inactivation of LATS1 and, consequently, dephosphorylation, activation, and nuclear translocation of the TAZ transcriptional co-activator [46]. Once in the nucleus, TAZ represses Leydig cell steroidogenesis by suppressing the expression of several steroidogenic genes via an interaction with the nuclear receptor NR4A1, preventing it from binding to DNA and activating its target genes [47]. It is tempting to speculate that the phosphorylation of PARD3, which we observed in the AMPK-activated Leydig cells, might lead to the same events, therefore contributing to a repression of steroidogenesis.

The transcription factors belonging to the nuclear factor of activated T cells (NFAT) family are well-known transcriptional activators (reviewed in [48,49]). In our phosphoproteomic screen, the phosphorylation of the transcription factor NFATC3 remained unchanged in the presence of Fsk alone. However, it was significantly increased in the MA-10 Leydig cells treated with Fsk+AICAR, indicating that NFATC3 becomes phosphorylated upon activation of AMPK. Interestingly, phosphorylated NFAT transcription factors are known to be sequestered in the cytoplasm since phosphorylation masks the nuclear localization sequence [50,51]. The phosphorylation of NFATC3 that occurs upon activation of AMPK in MA-10 Leydig cells would be consistent with the exclusion of NFATC3 from the nucleus and a reduction in its ability to activate gene expression, therefore contributing to reduced steroidogenesis.

Another group of differentially phosphorylated proteins includes those where the phosphorylation level was significantly different only in the Fsk+AICAR-treated MA-10 Leydig cells vs. control cells (vehicle-treated). This included the NUMA1 protein, which was differentially phosphorylated on two different peptides; one exhibited increased phosphorylation, while the phosphorylation of another residue of the protein was significantly decreased. NUMA1 is a component of the nuclear matrix (reviewed in [52]) and is implicated in ciliogenesis and autophagy, as is AMPK [53]. NUMA1 was indeed proposed as a potential target of AMPK, although this remains to be validated [54]. Our current results support direct phosphorylation of NUMA1 by AMPK.

The PGRMC2 protein also exhibited increased phosphorylation in the MA-10 Leydig cells treated with Fsk+AICAR vs. controls. PGRMC2 is a membrane-associated progesterone receptor (MAPR) present in Leydig cells and is believed to play a role in the progesterone auto/paracrine action on Leydig cells [55,56] in addition to potentially regulating the activity of some cytochrome P450 enzymes [57]. We showed that PGRMC2 is present in MA-10 Leydig cells and that it is phosphorylated after AMPK activation.

The MA-10 Leydig cell line is the gold standard for studying Leydig cells, although they are not identical to the cultured primary Leydig cells. The MA-10 Leydig cell line was originally established from a Leydig cell tumor in a mouse (M5480P) [58] and has since been meticulously characterized. Treatment of MA-10 cells with the luteinizing hormone, Forskolin, or cAMP analogs, increases steroid hormone production in the same way as in normal Leydig cells ([58] and reviewed in [59]). The rate-limiting step in hormone-induced steroidogenesis is the transport of cholesterol into the mitochondria, where steroidogenesis is initiated (reviewed in [60]); this is also true for MA-10 Leydig cells [10,15,61]. MA-10 cells mainly produce progesterone because of a mutation in the *Cyp17a1* coding sequence [58]. MA-10 Leydig cells are, therefore, suitable for the study of hormone-induced steroidogenesis and, more specifically, the early steps of steroidogenesis [62]. Since our current work focuses on the modulation of protein phosphorylation that occurs early following hormone treatment, the MA-10 Leydig cell line is a suitable and convenient model to use.

In conclusion, using a quantitative approach, our present study determined the global phosphoproteomic profile of MA-10 Leydig cells in response to Fsk (stimulation of steroidogenesis) and AICAR-mediated AMPK activation (repression of steroidogenesis). Our results indicate that steroidogenesis is a very dynamic process that involves differential protein phosphorylation by various kinases, altogether contributing to the fine-tuned regulation that is needed to achieve the appropriate production of steroid hormones.

4. Materials and Methods

4.1. Cell Culture

The MA-10 cell line was obtained from ATCC (Cat# CRL-3050, RRID:CVCL_D789. ATCC, Manassas, VA, USA). The MA-10 cells were grown in a DMEM/F12 medium supplemented with 2.438 g/L sodium bicarbonate, 3.57 g/L HEPES, and 15% horse serum on gelatin-coated plates. Penicillin and streptomycin sulphate were added to the cell culture media to a final concentration of 50 mg/L, and the cells were kept at 37 °C, 5% CO_2, in a humidified incubator. The MA-10 Leydig cell line was validated by morphology and by quantifying the progesterone output, as previously described.

4.2. Chemicals

The AMPK agonist, AICAR, was obtained from Tocris Bioscience (Minneapolis, MN, USA). Forskolin (Fsk) was purchased from Sigma-Aldrich, Canada (Oakville, ON, Canada).

4.3. RNA Isolation, Reverse Transcription, and Quantitative PCR

The MA-10 Leydig cells were cultured in the presence of either DMSO (vehicle), Fsk (10 µM), or Fsk+AICAR (10 µM + 1 mM) for 60 min. Isolation of RNA, cDNA synthesis, and reverse transcription-quantitative PCR (RT-qPCR) were performed as previously described [63,64]. Briefly, the total RNA from the MA-10 Leydig cells grown and treated, as described above, was isolated using TRIZOL (Life Technologies, Burlington, ON, Canada) and reverse-transcribed using the iScript Advanced cDNA Synthesis Kit (Bio-Rad Laboratories, ON, Canada). A quantitative real-time PCR was performed using a CFX96™ Real-Time PCR Detection System (Bio-Rad Laboratories, ON, Canada) along with the SsoAdvanced Universal SYBR Green Supermix kit (Bio-Rad Laboratories, ON, Canada) according to the manufacturer's protocols. Relative expression was normalized to the expression of *Rpl19*, which was used as an internal control. The mouse *Star* primers were: forward 5′-GTT CCT CGC TAC GTT CAA GC-3′ and reverse 5′-GAA ACA CCT TGC CCA CAT CT-3′. The mouse *Rpl19* primers were: forward 5′-CTG AAG GTC AAA GGG AAT GTG-3′ and reverse 5′-GGA CAG AGT CTT GAT GAT CTC-3′. The melting temperature of both primer couples was 62.6 °C.

4.4. Sample Preparation for LC-MS/MS

The MA-10 Leydig cells (5.8 million) were plated in 15-cm plates and then treated at 80% confluence with either the vehicle (DMSO), Fsk (10 µM) or Fsk+AICAR (10 µM + 1 mM) for 60 min at 37 °C and 5% CO_2. This experiment was repeated three times in triplicate.

Protein extraction, digestion, phosphopeptide enrichment, and mass spectrometry analyses were performed by the proteomics platform of the CHU de Quebec Research Centre (Quebec City, QC, Canada). Cell pellets were resuspended in an extraction buffer (50 mM ammonium bicarbonate, 0.5% sodium deoxycholate (SDC), 50 mM DTT, protease inhibitor cocktail, phosphatase inhibitor PhosSTOP, and 1 µM pepstatin) and vortexed for 1 min. They were then sonicated with a microprobe (Sonic Dismembrator 550, Fisher Scientific) 20 times for 1 s (on/off). The extract was centrifuged at 16,000× g for 15 min, the supernatant was collected, and then acetone-precipitated. The pellet was resuspended in ammonium bicarbonate 50 mM/SDC 1% and treated with bioruptor 15 times for 30 s (on/off). The protein concentration was determined by performing a Bradford assay.

Protein aliquots of 10 μg for the global protein analysis or 2 mg for the phosphopeptide analysis for each condition (vehicle, Fsk, Fsk+AICAR) were used. The proteins were first heated at 95 °C for 5 min and reduced with 0.2 mM DTT at 37 °C 30 min. They were then alkylated with 0.8 mM IAA (iodoacetamide) for 30 min at 37 °C in the dark. This was followed by digestion with trypsin (1:50) at 37 °C overnight. Digestion was stopped with the addition of formic acid; the sodium deoxycholate was removed after centrifugation. The peptides were desalted using a stage-tip column (for the global protein analysis) or HLB column (for the phosphopeptide analysis).

Phosphopeptide enrichment was performed using the High Select Phosphopeptide Enrichment Kits & Reagents kit (Fisher Scientific Canada, Catalog number A32993) according to the manufacturer's protocol. Two (2) mg of dry protein from each condition (vehicle, Fsk, Fsk+AICAR) were resuspended in 150 μL of binding/equilibration buffer (pH < 3). The TiO_2 spin-tip columns were washed and then equilibrated before adding 150 μL of the resuspended samples. The TiO_2 spin-tip columns loaded with the samples were centrifuged twice at $1000\times g$ for 5 min to bind the phosphopeptides. The TiO_2 spin-tip columns were then washed twice with the binding/equilibration buffer, once with the wash buffer and then with HPLC water. The phosphopeptides were finally eluted with 50 μL of phosphopeptide elution buffer twice.

4.5. Quantitative Sample Analysis by LC-MS/MS

Aliquots of 1 μg (for global protein analysis) and 1 mg of enriched phosphopeptide were analyzed by nanoLC-MS/MS using a Dionex UltiMate 3000 nanoRSLC chromatography system (Thermo Fisher Scientific, Ontario, Canada), connected to an Orbitrap Fusion mass spectrometer (Thermo Fisher Scientific Ontario, Canada). Peptides were trapped at 20 μL/min in a loading solvent (2% acetonitrile, 0.05% TFA) on a 5 mm × 300 μm C18 Pepmap cartridge pre-column (Thermo Fisher Scientific, Ontario, Canada) for 5 min. The pre-column was then switched online with a Pepmap Acclaim column (Thermo Fisher Scientific, Ontario, Canada) 50 cm × 75 μm internal diameter separation column. The peptides were eluted with a linear gradient of 5–40% solvent B (A: 0.1% formic acid, B: 80% acetonitrile, 0.1% formic acid) over 270 min for a total of a 300-min run at 300 nL/min for the global protein analysis. For the phosphopeptide analysis, elution with the solvent gradient was completed in 90 min for a total run of 120 min. Mass spectra were acquired using a data-dependent acquisition mode using Thermo Xcalibur software, version 4.1.50. Full-scan mass spectra (350 to 1800 m/z) were acquired in the orbitrap using an AGC target of 4e5, a maximum injection time of 50 ms, and a resolution of 120,000. An internal calibration using lock mass on the m/z 445.12003 siloxane ion was used. Each MS scan was followed by the acquisition of fragmentation involving MS/MS spectra of the most intense ions for a total cycle time of 3 s (top speed mode). The selected ions were isolated using the quadrupole analyzer in a window of 1.6 m/z and fragmented by higher energy collision-induced dissociation (HCD) with 35% of collision energy. The resulting fragments were detected by the linear ion trap at a rapid scan rate with an AGC target of 1×10^4 and a maximum injection time of 50 ms. The dynamic exclusion of previously fragmented peptides was set for a period of 30 s and a tolerance of 10 ppm.

4.6. Quantitative Data Analyses

Spectra were searched against the Uniprot Reference mus musculus (61,295 entries) using the Andromeda module of the MaxQuant software, v. 1.6.10.43. The trypsin/P enzyme parameter was selected with two possible missed cleavages. The carbamidomethylation of cysteines was set as a fixed modification, while methionine oxidation and phosphorylation (Ser, Thr, Tyr) were set as variable modifications. The mass search tolerances were 5 ppm and 0.5 Da for MS and MS/MS, respectively. For the protein validation, a maximum false discovery rate of 1% at the peptide and protein levels was used based on a target/decoy search. MaxQuant was also used for label-free quantifications. The 'match between runs' option was used with a 20 min value as the alignment time window and 0.7 min as the match time window. For the global protein analysis, the protein group files were used, while for the phosphopeptide analysis, the modificationSpecifPeptides file was used. RStudio 1.2.5019 was used for data processing. A normalization step was performed using the median of the median intensities of each condition. The missing peptide intensity values were replaced by a noise value corresponding to the 1% percentile of the normalized value for each condition. A peptide was considered quantifiable only if at least three intensity values in one of the two conditions were present and with a minimum of two peptides (for the global protein analysis).

4.7. Statistical Analysis

For the RT-qPCR, statistical analyses were carried out using a nonparametric one-way ANOVA on ranks via the Kruskal–Wallis test, followed by a post-hoc Mann–Whitney U test. A p-value of < 0.05 was considered significant. For the LC-MS/MS analysis, a p-value limma test, a limma q-value (Benjamin–Hochberg correction), and a z-score were calculated. Phosphopeptides or proteins were considered variants if the q-value was <0.05 and the z-score was $+/- 1.96$. All statistical analyses were performed using the OriginPro Version 2021 software (www.originlab.com, accessed on 10 August 2022) (OriginLab Corporation, Northampton, MA, USA).

Author Contributions: Z.B.D. performed all the experiments. Z.B.D. and J.J.T. analyzed and interpreted the data. J.J.T. conceived the study and coordinated and supervised the project. Z.B.D. drafted the manuscript with the assistance of J.J.T. All authors have read and agreed to the published version of the manuscript.

Funding: This research was funded by a grant from the Canadian Institutes of Health Research (CIHR) (funding reference number PJT-148738) to JJT.

Institutional Review Board Statement: Not applicable.

Informed Consent Statement: Not applicable.

Data Availability Statement: All mass spectrometry data (raw files and MaxQuant search result files) are publicly available on the ProteomeXchange repository (www.proteomexchange.org) with the identifier PXD037514.

Acknowledgments: We would like to thank Sylvie Bourassa (proteomics platform of the CHU de Quebec Research Centre, Quebec City) for her assistance with the quantitative protein and phosphoprotein mass spectrometry approach and Lisa Lahens for her assistance with the OriginPro software.

Conflicts of Interest: The authors declare no conflict of interest.

References

1. Leydig, F. Zur anatomie der mannlichen geschlechtsorgane und analdrusen der saugethiere. *Z. Wiss. Zool.* **1850**, *2*, 1–57.
2. Bouin, P.; Ancel, P. Recherches sur les cellules interstitielles du testicule des mammiferes. *Arch. Zool. Exp. Gen.* **1903**, *1*, 437–523.
3. Hall, P.F.; Irby, D.C.; De Kretser, D.M. Conversion of cholesterol to androgens by rat testes: Comparison of interstitial cells and seminiferous tubules. *Endocrinology* **1969**, *84*, 488–496. [CrossRef]
4. Hughes, I.A. Consensus statement on management of intersex disorders. *Arch. Dis. Child.* **2005**, *91*, 554–563. [CrossRef]
5. Ewing, L.L.; Eik-Nes, K.B. On the formation of testosterone by the perfused rabbit testis. *Can. J. Biochem.* **1966**, *44*, 1327–1344. [CrossRef] [PubMed]
6. Dufau, M.L.; Tsuruhara, T.; Horner, K.A.; Podesta, E.; Catt, K.J. Intermediate role of adenosine 3′:5′-cyclic monophosphate and protein kinase during gonadotropin-induced steroidogenesis in testicular interstitial cells. *Proc. Natl. Acad. Sci. USA* **1977**, *74*, 3419–3423. [CrossRef] [PubMed]
7. Tremblay, J.J. Molecular regulation of steroidogenesis in endocrine Leydig cells. *Steroids* **2015**, *103*, 3–10. [CrossRef]
8. de Mattos, K.; Viger, R.S.; Tremblay, J.J. Transcription factors in the regulation of Leydig cell gene expression and function. *Front. Endocrinol.* **2022**, *13*, 881309. [CrossRef]
9. Martin, L.J.; Tremblay, J.J. The human 3β-hydroxysteroid dehydrogenase/Δ5-Δ4 isomerase type 2 promoter is a novel target for the immediate early orphan nuclear receptor Nur77 in steroidogenic cells. *Endocrinology* **2005**, *146*, 861–869. [CrossRef]

10. Martin, L.J.; Boucher, N.; Brousseau, C.; Tremblay, J.J. The orphan nuclear receptor NUR77 regulates hormone-induced StAR transcription in Leydig cells through cooperation with Ca^{2+}/calmodulin-dependent protein kinase I. *Mol. Endocrinol.* **2008**, *22*, 2021–2037. [CrossRef]

11. Darney, K.J.; Ewing, L. Autoregulation of testosterone secretion in perfused rat testes. *Endocrinology* **1981**, *109*, 993–995. [CrossRef] [PubMed]

12. Shimizu-Albergine, M.; Tsai, L.-C.L.; Patrucco, E.; Beavo, J.A. cAMP-specific phosphodiesterases 8A and 8B, essential regulators of Leydig cell steroidogenesis. *Mol. Pharmacol.* **2012**, *81*, 556–566. [CrossRef] [PubMed]

13. Xiao, B.; Sanders, M.J.; Underwood, E.; Heath, R.; Mayer, F.V.; Carmena, D.; Jing, C.; Walker, P.A.; Eccleston, J.F.; Haire, L.F.; et al. Structure of mammalian AMPK and its regulation by ADP. *Nature* **2011**, *472*, 230–233. [CrossRef] [PubMed]

14. Abdou, H.S.; Bergeron, F.; Tremblay, J.J. A cell-autonomous molecular cascade initiated by AMP-activated protein kinase represses steroidogenesis. *Mol. Cell. Biol.* **2014**, *34*, 4257–4271. [CrossRef] [PubMed]

15. Clark, B.J.; Wells, J.; King, S.R.; Stocco, D.M. The purification, cloning, and expression of a novel luteinizing hormone-induced mitochondrial protein in MA-10 mouse Leydig tumor cells. Characterization of the steroidogenic acute regulatory protein (StAR). *J. Biol. Chem.* **1994**, *269*, 28314–28322. [CrossRef]

16. Abdou, H.S.; Villeneuve, G.; Tremblay, J.J. The calcium signaling pathway regulates Leydig cell steroidogenesis through a transcriptional cascade involving the nuclear receptor NR4A1 and the steroidogenic acute regulatory protein. *Endocrinology* **2013**, *154*, 511–520. [CrossRef]

17. Dupuis, N.; Fafouri, A.; Bayot, A.; Kumar, M.; Lecharpentier, T.; Ball, G.; Edwards, D.; Bernard, V.; Dournaud, P.; Drunat, S.; et al. Dymeclin deficiency causes postnatal microcephaly, hypomyelination and reticulum-to-Golgi trafficking defects in mice and humans. *Hum. Mol. Genet.* **2015**, *24*, 2771–2783. [CrossRef]

18. Dimitrov, A.; Paupe, V.; Gueudry, C.; Sibarita, J.-B.; Raposo, G.; Vielemeyer, O.; Gilbert, T.; Csaba, Z.; Attie-Bitach, T.; Cormier-Daire, V.; et al. The gene responsible for Dyggve-Melchior-Clausen syndrome encodes a novel peripheral membrane protein dynamically associated with the Golgi apparatus. *Hum. Mol. Genet.* **2009**, *18*, 440–453. [CrossRef]

19. Osipovich, A.B.; Jennings, J.L.; Lin, Q.; Link, A.J.; Ruley, H.E. Dyggve–Melchior–Clausen syndrome: Chondrodysplasia resulting from defects in intracellular vesicle traffic. *Proc. Natl. Acad. Sci. USA* **2008**, *105*, 16171–16176. [CrossRef]

20. Dalla Rosa, I.; Durigon, R.; Pearce, S.F.; Rorbach, J.; Hirst, E.M.A.; Vidoni, S.; Reyes, A.; Brea-Calvo, G.; Minczuk, M.; Woellhaf, M.W.; et al. MPV17L2 is required for ribosome assembly in mitochondria. *Nucleic Acids Res.* **2014**, *42*, 8500–8515. [CrossRef]

21. Miller, W.L.; Bose, H.S. Early steps in steroidogenesis: Intracellular cholesterol trafficking. *J. Lipid Res.* **2011**, *52*, 2111–2135. [CrossRef]

22. Dejgaard, S.; Presley, J. Interactions of lipid droplets with the intracellular transport machinery. *Int. J. Mol. Sci.* **2021**, *22*, 2776. [CrossRef] [PubMed]

23. Kumar, S.; Kang, H.; Park, E.; Park, H.-S.; Lee, K. The expression of CKLFSF2B is regulated by GATA1 and CREB in the Leydig cells, which modulates testicular steroidogenesis. *Biochim. Biophys. Acta Gene Regul. Mech.* **2018**, *1861*, 1063–1075. [CrossRef] [PubMed]

24. Orlando, U.; Cooke, M.; Maciel, F.C.; Papadopoulos, V.; Podestá, E.J.; Maloberti, P. Characterization of the mouse promoter region of the acyl-CoA synthetase 4 gene: Role of Sp1 and CREB. *Mol. Cell. Endocrinol.* **2013**, *369*, 15–26. [CrossRef]

25. Manna, P.R.; Stocco, D.M. Crosstalk of CREB and Fos/Jun on a single cis-element: Transcriptional repression of the steroidogenic acute regulatory protein gene. *J. Mol. Endocrinol.* **2007**, *39*, 261–277. [CrossRef] [PubMed]

26. Clem, B.F.; Hudson, E.A.; Clark, B.J. Cyclic adenosine $3',5'$-monophosphate (cAMP) enhances cAMP-responsive element binding (CREB) protein phosphorylation and phospho-CREB interaction with the mouse steroidogenic acute regulatory protein gene promoter. *Endocrinology* **2005**, *146*, 1348–1356. [CrossRef] [PubMed]

27. Manna, P.; Eubank, D.; Lalli, E.; Sassone-Corsi, P.; Stocco, D. Transcriptional regulation of the mouse steroidogenic acute regulatory protein gene by the cAMP response-element binding protein and steroidogenic factor 1. *J. Mol. Endocrinol.* **2003**, *30*, 381–397. [CrossRef] [PubMed]

28. Manna, P.R.; Dyson, M.T.; Eubank, D.W.; Clark, B.J.; Lalli, E.; Sassone-Corsi, P.; Zeleznik, A.J.; Stocco, D.M. Regulation of Steroidogenesis and the Steroidogenic Acute Regulatory Protein by a Member of the cAMP Response-Element Binding Protein Family. *Mol. Endocrinol.* **2002**, *16*, 184–199. [CrossRef]

29. Ofir, R.; Dwarki, V.J.; Rashid, D.; Verma, I.M. CREB represses transcription of fos promoter: Role of phosphorylation. *Gene Expr.* **1991**, *1*, 55–60.

30. Walker, W.H.; Girardet, C.; Habener, J.F. Alternative exon splicing controls a translational switch from activator to repressor isoforms of transcription factor CREB during spermatogenesis. *J. Biol. Chem.* **1996**, *271*, 20145–21050. [CrossRef]

31. Yi, W.-R.; Tu, M.-J.; Yu, A.-X.; Lin, J.; Yu, A.-M. Bioengineered miR-34a modulates mitochondrial inner membrane protein 17 like 2 (MPV17L2) expression toward the control of cancer cell mitochondrial functions. *Bioengineered* **2022**, *13*, 12489–12503. [CrossRef] [PubMed]

32. Li, J.; Gao, L.; Chen, J.; Zhang, W.-W.; Zhang, X.-Y.; Wang, B.; Zhang, C.; Wang, Y.; Huang, Y.-C.; Wang, H.; et al. Mitochondrial ROS-mediated ribosome stalling and GCN2 activation are partially involved in 1-nitropyrene-induced steroidogenic inhibition in testes. *Environ. Int.* **2022**, *167*, 107393. [CrossRef] [PubMed]

33. Kone, M.; Pullen, T.J.; Sun, G.; Ibberson, M.; Martinez-Sanchez, A.; Sayers, S.; Nguyen-Tu, M.S.; Kantor, C.; Swisa, A.; Dor, Y.; et al. LKB1 and AMPK differentially regulate pancreatic β-cell identity. *FASEB J.* **2014**, *28*, 4972–4985. [CrossRef] [PubMed]

34. Golkowski, M.; Shimizu-Albergine, M.; Suh, H.W.; Beavo, J.A.; Ong, S.-E. Studying mechanisms of cAMP and cyclic nucleotide phosphodiesterase signaling in Leydig cell function with phosphoproteomics. *Cell. Signal.* **2016**, *28*, 764. [CrossRef]
35. Liu, Y.; Merrill, R.A.; Strack, S. A-kinase anchoring protein 1: Emerging roles in regulating mitochondrial form and function in health and disease. *Cells* **2020**, *9*, 298. [CrossRef] [PubMed]
36. Dyson, M.T.; Jones, J.K.; Kowalewski, M.P.; Manna, P.R.; Alonso, M.; Gottesman, M.E.; Stocco, U.M. Mitochondrial A-kinase anchoring protein 121 binds type II protein kinase A and enhances steroidogenic acute regulatory protein-mediated steroidogenesis in MA-10 mouse Leydig tumor cells. *Biol. Reprod.* **2008**, *78*, 267–277. [CrossRef] [PubMed]
37. Arakane, F.; King, S.R.; Du, Y.; Kallen, C.B.; Walsh, L.P.; Watari, H.; Stocco, D.M.; Strauss, J.F. Phosphorylation of steroidogenic acute regulatory protein (StAR) modulates its steroidogenic activity. *J. Biol. Chem.* **1997**, *272*, 32656–32662. [CrossRef] [PubMed]
38. Pryde, K.R.; Smith, H.L.; Chau, K.-Y.; Schapira, A.H. PINK1 disables the anti-fission machinery to segregate damaged mitochondria for mitophagy. *J. Cell Biol.* **2016**, *213*, 163–171. [CrossRef] [PubMed]
39. Lakkaraju, A.K.; Mary, C.; Scherrer, A.; Johnson, A.E.; Strub, K. SRP keeps polypeptides translocation-competent by slowing translation to match limiting ER-targeting sites. *Cell* **2008**, *133*, 440–451. [CrossRef]
40. Lin, H.-Y.; Ko, C.-Y.; Kao, T.-J.; Yang, W.-B.; Tsai, Y.-T.; Chuang, J.-Y.; Hu, S.-L.; Yang, P.-Y.; Lo, W.-L.; Hsu, T.-I. CYP17A1 maintains the survival of glioblastomas by regulating SAR1-mediated endoplasmic reticulum health and redox homeostasis. *Cancers* **2019**, *11*, 1378. [CrossRef]
41. Anton, K.; Sinclair, J.; Ohoka, A.; Kajita, M.; Ishikawa, S.; Benz, P.M.; Renne, T.; Balda, M.; Jorgensen, C.; Matter, K.; et al. PKA-regulated VASP phosphorylation promotes extrusion of transformed cells from the epithelium. *J. Cell Sci.* **2014**, *127*, 3425–3433. [CrossRef] [PubMed]
42. Thornton, C.; Bright, N.J.; Sastre, M.; Muckett, P.J.; Carling, D. AMP-activated protein kinase (AMPK) is a tau kinase, activated in response to amyloid β-peptide exposure. *Biochem. J.* **2011**, *434*, 503–512. [CrossRef]
43. Alquezar, C.; Arya, S.; Kao, A.W. Tau Post-translational modifications: Dynamic transformers of Tau function, degradation, and aggregation. *Front. Neurol.* **2020**, *11*, 595532. [CrossRef] [PubMed]
44. Chen, X.; Macara, I.G. Par-3 controls tight junction assembly through the Rac exchange factor Tiam1. *Nat. Cell Biol.* **2005**, *7*, 262–269. [CrossRef]
45. Zhang, L.; Li, J.; Young, L.H.; Caplan, M.J. AMP-activated protein kinase regulates the assembly of epithelial tight junctions. *Proc. Natl. Acad. Sci. USA* **2006**, *103*, 17272–17277. [CrossRef] [PubMed]
46. Lv, X.; Liu, C.; Wang, Z.; Sun, Y.; Xiong, Y.; Lei, Q.; Guan, K. PARD3 induces TAZ activation and cell growth by promotingLATS1 andPP1 interaction. *EMBO Rep.* **2015**, *16*, 975–985. [CrossRef] [PubMed]
47. Shin, J.H.; Lee, G.; Jeong, M.G.; Kim, H.K.; Won, H.Y.; Choi, Y.; Lee, J.; Nam, M.; Choi, C.S.; Hwang, G.; et al. Transcriptional coactivator with PDZ-binding motif suppresses the expression of steroidogenic enzymes by nuclear receptor 4 A1 in Leydig cells. *FASEB J.* **2020**, *34*, 5332–5347. [CrossRef]
48. Horsley, V.; Pavlath, G.K. Nfat: Ubiquitous regulator of cell differentiation and adaptation. *J. Cell Biol.* **2002**, *156*, 771–774. [CrossRef]
49. Rao, A.; Luo, C.; Hogan, P.G. Transcription factors of the NFAT family: Regulation and Function. *Annu. Rev. Immunol.* **1997**, *15*, 707–747. [CrossRef]
50. Okamura, H.; Aramburu, J.; Garcia-Rodriguez, C.; Viola, J.; Raghavan, A.; Tahiliani, M.; Zhang, X.; Qin, J.; Hogan, P.G.; Rao, A. concerted dephosphorylation of the transcription factor NFAT1 induces a conformational switch that regulates transcriptional activity. *Mol. Cell* **2000**, *6*, 539–550. [CrossRef]
51. Okamura, H.; Garcia-Rodriguez, C.; Martinson, H.; Qin, J.; Virshup, D.M.; Rao, A. A conserved docking motif for CK1 binding controls the nuclear localization of NFAT1. *Mol. Cell. Biol.* **2004**, *24*, 4184–4195. [CrossRef] [PubMed]
52. Radulescu, A.E.; Cleveland, D.W. NuMA after 30 years: The matrix revisited. *Trends Cell Biol.* **2010**, *20*, 214–222. [CrossRef] [PubMed]
53. Akhshi, T.; Trimble, W.S. A non-canonical Hedgehog pathway initiates ciliogenesis and autophagy. *J. Cell Biol.* **2021**, *220*, e202004179. [CrossRef]
54. Schaffer, B.E.; Levin, R.S.; Hertz, N.T.; Maures, T.J.; Schoof, M.L.; Hollstein, P.E.; Benayoun, B.A.; Banko, M.R.; Shaw, R.J.; Shokat, K.M.; et al. Identification of AMPK phosphorylation sites reveals a network of proteins involved in cell invasion and facilitates large-scale substrate prediction. *Cell Metab.* **2015**, *22*, 907–921. [CrossRef] [PubMed]
55. Braun, B.C.; Okuyama, M.W.; Müller, K.; Dehnhard, M.; Jewgenow, K. Steroidogenic enzymes, their products and sex steroid receptors during testis development and spermatogenesis in the domestic cat (Felis catus). *J. Steroid Biochem. Mol. Biol.* **2018**, *178*, 135–149. [CrossRef]
56. Ponikwicka-Tyszko, D.; Chrusciel, M.; Pulawska, K.; Bernaczyk, P.; Sztachelska, M.; Guo, P.; Li, X.; Toppari, J.; Huhtaniemi, I.T.; Wolczynski, S.; et al. Mifepristone treatment promotes testicular Leydig cell tumor progression in transgenic mice. *Cancers* **2020**, *12*, E3263. [CrossRef]
57. Wendler, A.; Wehling, M. PGRMC2, a yet uncharacterized protein with potential as tumor suppressor, migration inhibitor, and regulator of cytochrome P450 enzyme activity. *Steroids* **2013**, *78*, 555–558. [CrossRef]
58. Ascoli, M. Characterization of several clonal lines of cultured Leydig tumor cells: Gonadotropin receptors and steroidogenic responses. *Endocrinology* **1981**, *108*, 88–95. [CrossRef]
59. Rahman, N.A.; Huhtaniemi, I.T. Testicular cell lines. *Mol. Cell. Endocrinol.* **2004**, *228*, 53–65. [CrossRef]

60. Selvaraj, V.; Stocco, D.M.; Clark, B.J. Current knowledge on the acute regulation of steroidogenesis. *Biol. Reprod.* **2018**, *99*, 13–26. [CrossRef]

61. King, S.R.; Ronen-Fuhrmann, T.; Timberg, R.; Clark, B.J.; Orly, J.; Stocco, D.M. Steroid production after in vitro transcription, translation, and mitochondrial processing of protein products of complementary deoxyribonucleic acid for steroidogenic acute regulatory protein. *Endocrinology* **1995**, *136*, 5165–5176. [CrossRef] [PubMed]

62. Engeli, R.T.; Fürstenberger, C.; Kratschmar, D.V.; Odermatt, A. Currently available murine Leydig cell lines can be applied to study early steps of steroidogenesis but not testosterone synthesis. *Heliyon* **2018**, *4*, e00527. [CrossRef] [PubMed]

63. Mehanovic, S.; E Mendoza-Villarroel, R.; de Mattos, K.; Talbot, P.; Viger, R.S.; Tremblay, J.J. Identification of novel genes and pathways regulated by the orphan nuclear receptor COUP-TFII in mouse MA-10 Leydig cells. *Biol. Reprod.* **2021**, *105*, 1283–1306. [CrossRef]

64. Hébert-Mercier, P.-O.; Bergeron, F.; Robert, N.M.; Mehanovic, S.; Pierre, K.J.; Mendoza-Villarroel, R.E.; de Mattos, K.; Brousseau, C.; Tremblay, J.J. Growth hormone-induced STAT5B regulates *Star* gene expression through a cooperation with cJUN in mouse MA-10 Leydig cells. *Endocrinology* **2022**, *163*, bqab267. [CrossRef] [PubMed]

International Journal of
Molecular Sciences

Article

A Novel Model Using AAV9-Cre to Knockout Adult Leydig Cell Gene Expression Reveals a Physiological Role of Glucocorticoid Receptor Signalling in Leydig Cell Function

Anne-Louise Gannon [1], Annalucia L. Darbey [1], Grace Chensee [1], Ben M. Lawrence [1], Liza O'Donnell [1], Joanna Kelso [1], Natalie Reed [1], Shanmathi Parameswaran [1], Sarah Smith [1], Lee B. Smith [1,2,3] and Diane Rebourcet [1,*]

[1] College of Engineering, Science and Environment, The University of Newcastle, Callaghan, NSW 2308, Australia
[2] MRC Centre for Reproductive Health, The Queen's Medical Research Institute, University of Edinburgh, 47 Little France Crescent, Edinburgh EH16 4TJ, UK
[3] Office for Research, Griffith University, Southport, QLD 4222, Australia
* Correspondence: diane.rebourcet@newcastle.edu.au

Citation: Gannon, A.-L.; Darbey, A.L.; Chensee, G.; Lawrence, B.M.; O'Donnell, L.; Kelso, J.; Reed, N.; Parameswaran, S.; Smith, S.; Smith, L.B.; et al. A Novel Model Using AAV9-Cre to Knockout Adult Leydig Cell Gene Expression Reveals a Physiological Role of Glucocorticoid Receptor Signalling in Leydig Cell Function. *Int. J. Mol. Sci.* **2022**, *23*, 15015. https://doi.org/10.3390/ijms232315015

Academic Editor: Jacques J. Tremblay

Received: 28 October 2022
Accepted: 26 November 2022
Published: 30 November 2022

Abstract: Glucocorticoids are steroids involved in key physiological processes such as development, metabolism, inflammatory and stress responses and are mostly used exogenously as medications to treat various inflammation-based conditions. They act via the glucocorticoid receptor (GR) expressed in most cells. Exogenous glucocorticoids can negatively impact the function of the Leydig cells in the testis, leading to decreased androgen production. However, endogenous glucocorticoids are produced by the adrenal and within the testis, but whether their action on GR in Leydig cells regulates steroidogenesis is unknown. This study aimed to define the role of endogenous GR signalling in adult Leydig cells. We developed and compared two models; an inducible Cre transgene driven by expression of the *Cyp17a1* steroidogenic gene (*Cyp17*-iCre) that depletes GR during development and a viral vector-driven Cre (AAV9-Cre) to deplete GR in adulthood. The delivery of AAV9-Cre ablated GR in adult mouse Leydig cells depleted Leydig cell GR more efficiently than the *Cyp17*-iCre model. Importantly, adult depletion of GR in Leydig cells caused reduced expression of luteinising hormone receptor (*Lhcgr*) and of steroidogenic enzymes required for normal androgen production. These findings reveal that Leydig cell GR signalling plays a physiological role in the testis and highlight that a normal balance of glucocorticoid activity in the testis is important for steroidogenesis.

Keywords: glucocorticoid receptor; androgens; leydig cells; AAV9; steroidogenesis

1. Introduction

Stress and illness are known to reduce testosterone production by the testis. Although the mechanisms underpinning this observation remain unclear, it is widely accepted that, under stress conditions, the hypothalamus-pituitary-adrenal axis acts to suppress gonadal function as fertility is a secondary consideration to survival [1]. The production of androgens by the testis is essential for men's health and fertility. Disruption to the production or action of androgens is associated with lifelong chronic pathologies, including infertility [2–5].

It is well established that elevated circulating levels of glucocorticoids, such as those induced during inflammation, reduces androgen production and fertility in men [6,7]. Exogenous glucocorticoids are commonly used to treat a wide variety of inflammatory conditions, such as arthritis, asthma, and allergies. Conversely, long term medication or prolonged stress can lead to glucocorticoid resistance, which can also impact on androgen production [6]. Therefore, before such therapies are considered, it is essential to establish the role of GR in the steroidogenic Leydig cells.

Glucocorticoids act via a ligand activated nuclear receptor GR (encoded by the nuclear receptor subfamily 3 group member 1 gene *Nr3c1*). GR signalling regulates the homeostasis between basal and stress-related conditions of key processes such as metabolism and immune functions [8–10]. Glucocorticoids can also act at the level of the hypothalamus-gonad axis and stress, or exogenous glucocorticoids can shut down the reproductive function and inhibit LH pulsatile secretion which can lead to a reduction of testosterone production [11]. In the testis, GR is expressed in both somatic and germ cells [12–15]. GR signalling in Sertoli cells is required for normal Sertoli cell maturation, spermatogenesis and for Leydig cell steroidogenesis [16], suggesting a physiological role of GR in the testis. Whilst most glucocorticoids are produced by the adrenal, resident interstitial macrophages produce glucocorticoids within the testis [17] and the intratesticular concentration of glucocorticoids is tightly regulated by specific metabolizing enzymes (11β-hydroxysteroid dehydrogenase enzymes, or HSD11Bs) [14,18]. Previous studies have shown that elevated glucocorticoids can regulate testicular LH receptor expression [19,20] and decrease testosterone production by Leydig cells via a reduction in the expression of steroidogenic enzymes [21–25]. However, whether GR signalling in Leydig cells supports normal steroidogenesis is unknown.

To investigate the role of GR signalling in Leydig cells, we developed and compared two inducible transgenic mouse models to specifically ablate GR signalling in Leydig cells along testis development; an inducible Cre transgene driven by expression of the *Cyp17a1* steroidogenic gene (*Cyp17*-iCre) and a viral vector-driven Adeno Associated Virus 9 Cre (AAV9-Cre). The results demonstrate that viral-mediated vectors can be used to selectively knockdown gene expression in adult Leydig cells and reveal that GR signalling regulates Leydig cell steroidogenesis.

2. Results

2.1. Confirmation of Cyp17-iCre Leydig Cell Cre Recombinase Activity

GR is widely expressed in the testis, including in Leydig cells, Sertoli cells and germ cells from fetal life through to adulthood [12,15,16]. The transgenic mouse line expressing an inducible Cre recombinase (iCre) under the control of the *Cyp17a1* (cytochrome P450 17αhydroxylase/17, 20-lyase) promoter (referred to as *Cyp17*-iCre) [26] was used to direct gene deletions to testicular Leydig cells and ovarian theca cells. The *Cyp17a1*-iCre line was bred to a Rosa-26:RFP reporter floxed line (CYPTR) (Figure 1A). Previous studies using an independently derived line have shown that the transgene is active in E16.5 testes through to adulthood [26]. Epifluorescence analysis of freshly dissected CYPTR mice at day (d) 80 confirms RFP expression in the testis (Figure 1B), but expression was also observed in the adrenal, epididymis and liver (Supplementary Figure S1). Immunohistochemical localisation of RFP and the Leydig cell-specific marker protein 3βHSD in CYPTR testis sections demonstrated co-localisation of RFP and 3βHSD in Leydig cells, with no off-target expression in other testicular cells (Figure 1C). To determine targeting efficiency, we quantified the percentage of Leydig cells (3βHSD-positive cells) that were RFP-positive and showed that 99% of Leydig cells expressed the CYPTR transgene (Figure 1D). Finally, we determined that the *Cyp17*-iCre transgene co-localised with GR positive Leydig cells (Figure 1E). Taken together, these data confirm that the *Cyp17*-iCre transgene is specific to Leydig cells in the testis and co-localises with Leydig cell GR, making it a suitable model for the deletion of GR in Leydig cells.

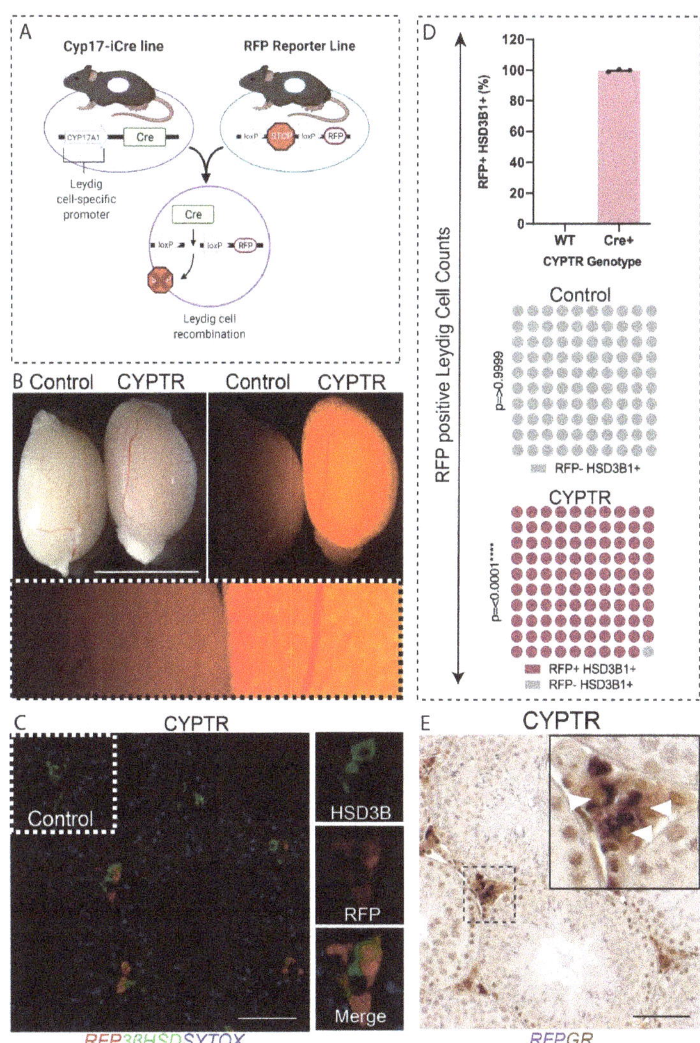

Figure 1. Confirmation of RFP expression in testes from Cyp17a1-iCre:RT26 RFP (CYPTR) mice. (**A**) Diagram of the breeding strategy to generate Cyp17-iCre;RT26 RFP reporter (CYPTR) mice. (**B**) Representative epifluorescence microscopy images of freshly dissected CYPTR testis at d80 confirming RFP expression in adult testis. Organs were imaged under an RFP filter and show a lack of RFP in Controls but positive RFP expression in CYPTR testis (fluorescent images taken after 5.5 s of exposure). Scale bars 0.5 cm. (**C**) Dual-label immunofluorescence of RFP and the testicular Leydig cell marker protein 3βHSD (green), counterstained with the nuclear stain Sytox (blue). Insets demonstrate 40× magnification of single channel 3βHSD, RFP and merged channels. (**D**) Percentage of 3βHSD+ interstitial cells that co-express RFP in wildtype and CYPTR mice, indicating that CYPTR mice express RFP in Leydig cells, whereas wildtype mice do not ($n = 3$ per group, Chi Square $p = 0.0001$). (**E**) Double immunostaining of RFP (purple) and GR (brown) in testis from adult CYPTR mice demonstrated co-localisation of RFP with GR in Leydig cells. Insets demonstrate 40× magnification of single channel 3βHSD, RFP and merged channels. Scale Bars 50 μm.

2.2. Assessment of Glucocorticoid Receptor Ablation in Leydig Cells Using a Traditional Cre/LoxP Model

We next utilised the *Cyp17*-iCre line to generate a mouse model with Leydig cell-specific GR ablation. We bred the *Cyp17*-iCre females to homozygous (Hom) $GR^{flox/flox}$ males ('GRGR' mice) [27], resulting in an offspring heterozygous (Het) for $Cre^{+/}$;GR^{flox}/$^{+}$ or $Cre^{-/}$;GR^{flox}/$^{+}$. For total Leydig cell GR ablation, $Cre^{+/}$; GR^{flox}/$^{+}$ males were again bred to $Gr^{flox/flox}$ female resulting in the following offspring: $Cre^{-/}$;GR^{flox}/^{flox} (Cre- Hom) termed 'littermate control', $Cre^{-/}$;GR^{flox}/$^{+}$ (Cre-Het), termed 'Het littermate control', $Cre^{+/}$;GR^{flox}/^{flox} (Cre+ Hom), termed 'CYP GR knockout' (CYPGRKO), $Cre^{+/}$;GR^{flox}/$^{+}$ (Cre+ Het), termed 'Het CYPGR knockout' (Het CYPGRKO).

Co-immunohistochemical localisation of GR and 3βHSD in adult (d80) control mice demonstrated that GR is expressed in the nuclei of most interstitial cells; Leydig cells, endothelial cells, and macrophages; as well in the tubules in Sertoli and peritubular myoid cells and early stage germ cells. In contrast, in the CYPGRKO testis, GR depletion was observed in some, but not all, Leydig cells (black and white arrowheads) (Figure 2A). Stereological quantification of GR positive and negative Leydig cells (3βHSD positive cells) in control and CYPGRKO revealed that an average of 28% Leydig cells had lost their GR expression (Figure 2B). *Nr3c1* transcripts and direct downstream targets of GR, *Stc1* and *Tsc22d3* are unperturbed in the CYPGRKO testis in all cohorts analysed (Figure 2C). In adulthood (d80), there was no difference in body weight between the genotype and the gross reproductive system of CYPGRKO males appeared unchanged compared to control (Figure 2D,E). Testis weight remained unchanged, and the histology of the testis showed no differences between control and CYPGRKO males (Figure 2F,G).

Despite the significant decrease in the number of GR-positive, 3βHSD-positive Leydig cells in CYPGRKO mice, no changes were observed in the expression of Leydig cell-enriched steroidogenic enzyme transcripts (Supplementary Figure S2). Leydig cells are well known for their ability to show steroidogenic compensation following physiological challenge or manipulation, and normal steroidogenesis can be observed in mice with as little as 30% functional Leydig cells [28,29]. The results suggest that the targeting efficiency of the *Cyp17*-iCre in Leydig cells is only able to ablate GR in 28% of Leydig cells and this is not sufficient to impact testis function. Thus, CYPGRKO mice are an inadequate model to study the role of GR in Leydig cell development and regulation.

2.3. Validation of the AAV9 Inducible Cre/loxP System

We next investigated the physiological role of GR by acute ablation in adult Leydig cells (mice > 60 days of age) to dissect the effects of developmental vs. adult GR function in Leydig cells. We have previously demonstrated that AAV9 specifically and efficiently targets adult Leydig cells [30]. Thus, we utilised AAV9 carrying Cre recombinase to generate a GR knockout in adult Leydig cells (Figure 3A).

First, to validate the targeting capability of AAV9-Cre in the adult testis, we used a Cre inducible Tomato Red Floxed reporter line (TRTR) [31] to verify the site of expression of the AAV9-Cre transgene (via expression of RFP). TRTR adult (postnatal day (d) 80) males were injected within the testicular interstitium with either vehicle, AAV9-GFP and AAV9-Cre-GFP (both carrying transgenes downstream of a CMV promoter) and the testes were collected 7 days post injection (dpi) (Figure 3A). The fluorescence was assessed on freshly dissected testis. While both AAV9-GFP and AAV9-Cre-GFP induced expression of GFP in the interstitium compared to the vehicle injected testis, only AAV9-Cre-GFP generated RFP expression within the testis (Figure 3B). To evaluate potential off-target effects following delivery of the Cre into the interstitial compartment, several organs were collected and assessed under for RFP expression. RFP was only detected in adrenal and liver (Supplementary Figure S3). To confirm that the downstream results are correlated with a direct consequence of GR depletion within the Leydig cell and not the off-target effects, we assessed the colocalization of GR and RFP proteins in both liver and adrenal. These results showed that cells expressing GR in the adrenal are not targeted by AAV9, suggesting that disruption of glucocorticoid signalling in the adrenal would be unlikely. There was a small degree of RFP and GR co-localisation observed in the liver (Supplementary Figure S4) but no changes in liver size, gross histology or health of animals were noted (not shown). We conclude that the AAV9-Cre-GFP is a suitable model to acutely target the testis with minimal off-target effects.

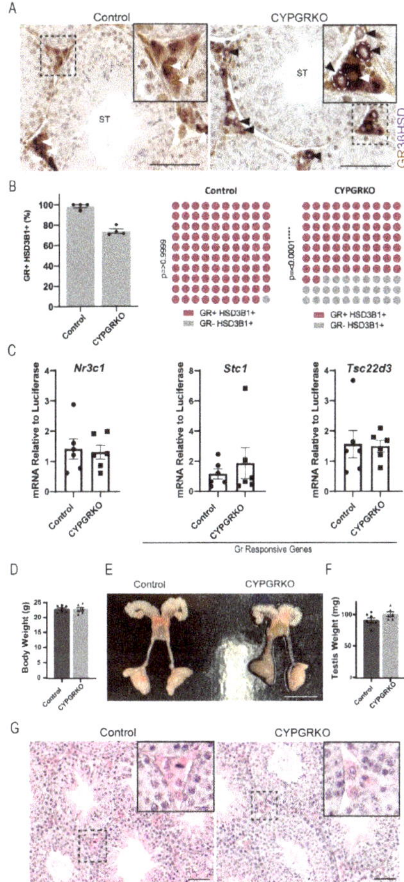

Figure 2. Validation of glucocorticoid receptor ablation in CYPGRKO Leydig cells. (**A**) Immunohisto-chemical localisation of testicular GR demonstrates ablation in 3βHSD+ Leydig cells but not in other testicular cells. Black arrowheads denote Leydig cells with GR ablation, white arrowheads denote Leydig cells with GR intact. (**B**) Percentage of cells co-expressing GR protein and the Leydig cell marker 3βHSD demonstrates a significant reduction in the number of Leydig cells expressing GR (n = 3, Chi Square p = 0.0001). (**C**) Comparative analysis of testicular Nr3c1 gene expression shows no changes in CYPGRKO when compared with littermate controls. Direct GR response gene transcripts Stc1 and Tsc22d3 do not show any changes in CYPGRKO when compared with littermate controls (one-way ANOVA; n = 6–8, Tukey's post hoc analysis, error bars SEM). Scale Bars 50 μm. Annotations; I, Interstitium; ST, seminiferous tubule). (**D**) Adult body weight of CYPGRKO compared to control did not change. (**E**) The gross morphology of the reproductive system in adult (d80) was comparable across the different groups (n = 6–7, t test). (**F**) No difference was noted in adult testis weight between groups (n = 6–7, T Test), and (**G**) testicular histology was normal. Scale Bars 50 μm.

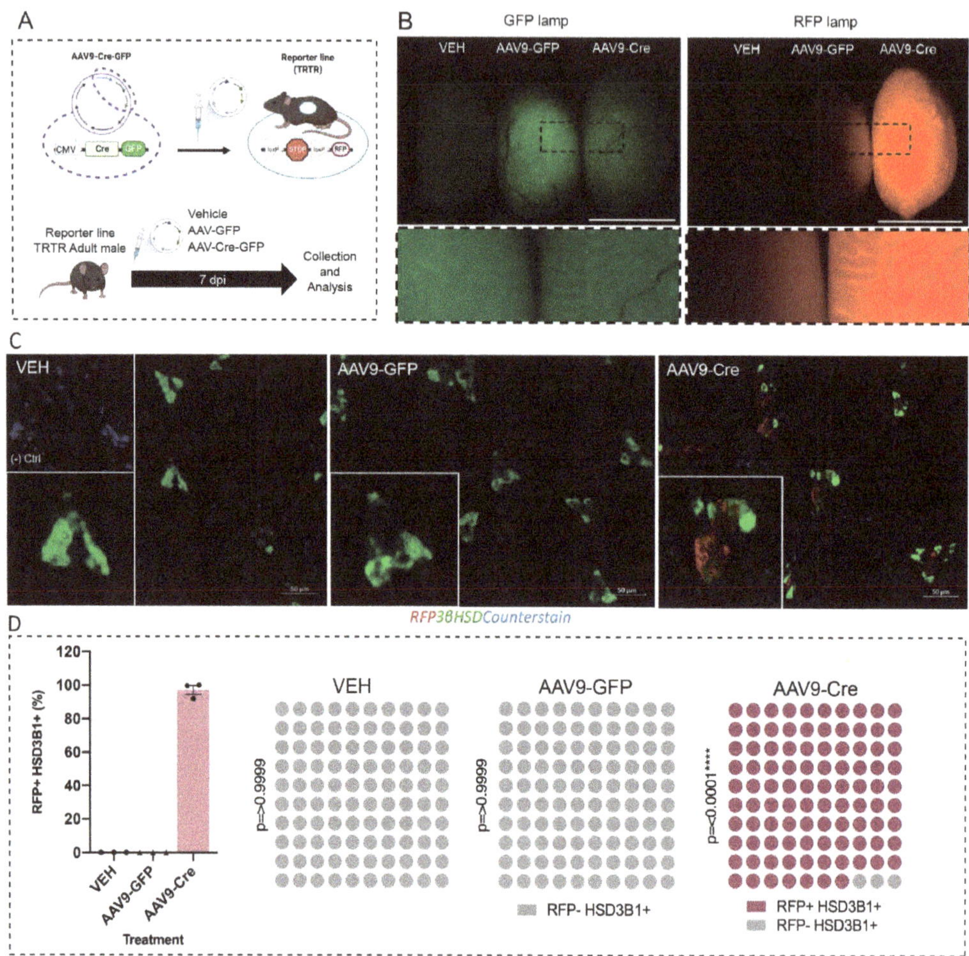

Figure 3. RFP expression in Leydig cells following delivery of the Cre Recombinase using adeno-associated virus serotype 9 (AAV9-Cre). (**A**) Schematic diagram of the generation of the inducible model (**B**) Representative images of freshly dissected Tomato Red (TRTR) testis at 7 days post injection (dpi) following injection of vehicle, AAV-GFP or AAV9-Cre-GFP; higher magnification of the dashed boxes is shown in the panel below. (**C**) Co-immunostaining of RFP and 3βHSD demonstrates RFP expression localised within Leydig cells (**D**) Quantification of cells co-expressing 3βHSD protein and RFP indicates only the AAV9-Cre-GFP induced expression of RFP in ~97% of Leydig cells (*n* = 3). (Chi~2 **** *p* = 0.0001). Scale Bars 100 µm.

Next, to confirm that the AAV9-inducible Cre/*LoxP* system specifically targets Leydig cells within the testis, we co-localised the Leydig cell marker 3βHSD and RFP [32]. RFP localisation was restricted to 3βHSD-positive Leydig cells in the testis (Figure 3C). The number of RFP-positive versus RFP-negative Leydig cells was quantified in vehicle, AAV9-GFP and AAV9-Cre-GFP injected testes (7 dpi). In accordance with the immunohistochemistry (Figure 3C), only the AAV9-Cre-GFP induced expression of RFP in ~97% of Leydig cells (Figure 3D). These data demonstrate that AAV9 inducible Cre/*loxP* can target Leydig cells within 7dpi and thus is a suitable model for the inducible knockdown of GR expression in adult mice.

2.4. Validation of the Inducible Model of GR Depletion in Adult Testis (AAV9-LCGR Mice)

Adult (>d60) GR$^{flox/flox}$ males (GRGR mice) (Figure 4A) were injected with either vehicle, AAV9-GFP or AAV9-Cre-GFP in their testicular interstitium and sacrificed 35 days later (the duration of mouse spermatogenesis) (Figure 4A). Fluorescence was assessed on freshly dissected testis and confirmed GFP expression in the interstitial compartment in both AAV9-GFP and AAV9-Cre-GFP compared the vehicle-injected testis (Figure 4B). Consistent with the inducible AAV9-Cre-GFP recombinase expression and action in the Tomato Red reporter line (Figure 3), GRGR mice injected with AAV9-Cre-GFP (hereafter referred to as AAV9-LCGR mice) displayed an absence of GR protein specifically in 3βHSD+ Leydig cells but not in other GR-expressing cells including Sertoli cells, germ cells, peritubular myoid, endothelial cells and interstitial macrophages (Figure 4C). Quantification of the number of 3βHSD-positive Leydig cells with or without GR expression revealed GR was depleted in ~48% of Leydig cells (Figure 4D). The increased number of GR-ablated Leydig cells in AAV9-LCGR mice, compared to the CYPGRKO mice, was reflected by decreased testicular expression of *Nr3c1* expression and of its direct downstream target, *Stc1* (Figure 4D), indicating that even a halving of the number of adult Leydig cells expressing GR leads to a reduced read out of GR-mediated transcription in the testis.

In summary, we have developed a model that permits the acute and specific ablation of GR in approximately half of adult Leydig cells, allowing the investigation of the effects of reduced GR signalling in Leydig cells separate from developmental impacts. This model also provides a unique opportunity to determine the roles of autocrine GR signalling within Leydig cells, as has been previously shown for AR [28].

2.5. Reduced GR Signalling in Adult Leydig Cells Does Not Impact the Reproductive System

We assessed the impact of reduced GR signalling in AAV9-LCGR adult Leydig cells on the gross morphology of the reproductive system in males 35d post injection, however no differences were observed between the cohorts (Figure 5A). The body and reproductive organs (testis, seminal vesicle, and epididymis) weights between the cohorts were also unchanged (Figure 5B–D). As seminal vesicle weights serve as a key biomarker of circulating androgen action, this suggest that a reduction in GR signalling in adult Leydig cells does not have a major influence on circulating androgen levels. We next determined whether the loss of GR signalling in approximately half of the adult Leydig cells alters adult testicular architecture, however gross testis morphology in AAV9-LCGR mice was similar to controls (Figure 5E).

2.6. Reduced GR Signalling in Adult Leydig Cells Impairs Their Function

We further investigated the impact of reduced GR signalling on specific functional endpoints. Analysis of steroid enzyme transcripts and circulating hormones were carried out in uman Chorionic Gonadotrophin (hCG) stimulated mice. This is due to the intra-variability of the AAV9 targeting, and natural variation observed in Leydig cell function in micc, which is observed in unstimulated cohorts (Supplementary Figure S5). We stimulated Leydig cells of AAV9-LCGR and control mice), to induce maximum output from Leydig cells [28]. Steroidogenic enzymes transcripts (*StAR*, *Cyp11a1*, *Hsd3b6* and *Cyp17a1*) were significantly decreased (Figure 6A–F), however, the expression of *Hsd3b1* and *Hsd17b3*, the enzyme responsible for testosterone synthesis, was unchanged (Figure 6G). Circulating LH, and testosterone were not significantly altered (Figure 6H–I), though we do note an increase in serum DHT between AAV9-LCGR and vehicle-treated animals (Figure 6J), when analysed without the hCG-treated cohort; this could be physiologically important as DHT is a particularly potent androgen. Despite reduced levels of *Lhcgr*, there was no change in hCG-stimulated circulating androgen levels (Figure 6I,J). Taken together, the data suggest that a 50% reduction in GR signalling reduces Leydig cell expression level of key steroidogenic enzymes, indicating that GR signalling in Leydig cells supports their function.

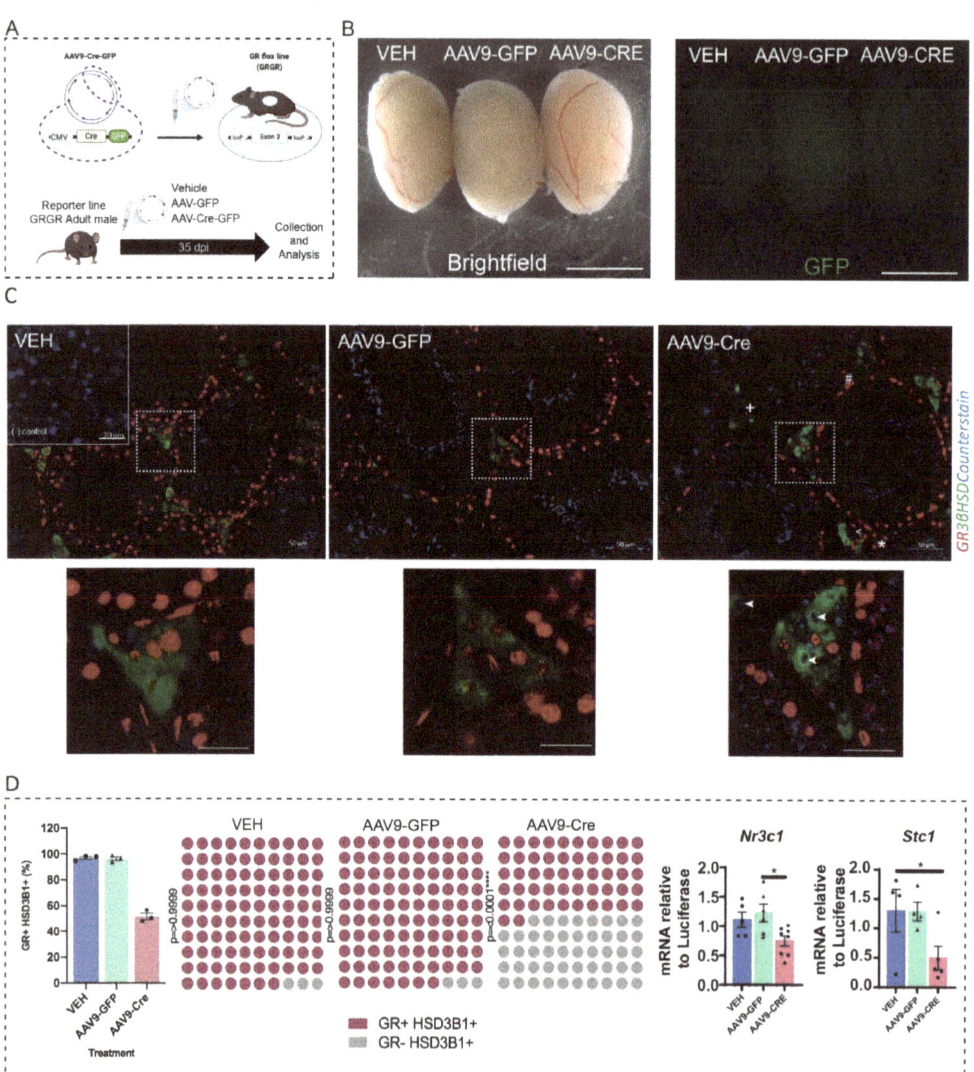

Figure 4. GR depletion in Leydig cells following delivery of the Cre Recombinase using AAV9-Cre-GFP. (**A**) Diagram of the model to ablate GR in adult Leydig cells (**B**) Representative fluorescent images of freshly dissected GRGR testis at 35 dpi injection of vehicle, AAV-GFP or AAV9-Cre. Scale bars 0.5 cm. (**C**) Co-immunostaining of GR and 3βHSD demonstrates depletion of GR protein in Leydig cells at 35 dpi AAV9-Cre. The micrographs in the bottom Insets demonstrate 40× magnification. Scale Bars = 50 μm. GR+ Sertoli cells, spermatogonia and spermatocytes (*), peritubular myoid cells (#), endothelial cells (<) and interstitial macrophages (+) are indicated. Arrowheads denote GR negative Leydig cells. (**D**) Quantification of cells co-expressing 3βHSD protein and GR indicates only the AAV9-Cre-GFP induced GR ablation, and this was observed in 48% of Leydig cells (*n* = 3, Chi-square **** *p* = 0.0001). Analysis of the testicular expression of Nr3c1 and Stc1 by qPCR demonstrates significantly reduced transcript levels (one-way ANOVA; *n* = 6–8, * *p* < 0.05, Tukey's post hoc analysis, error bars SEM).

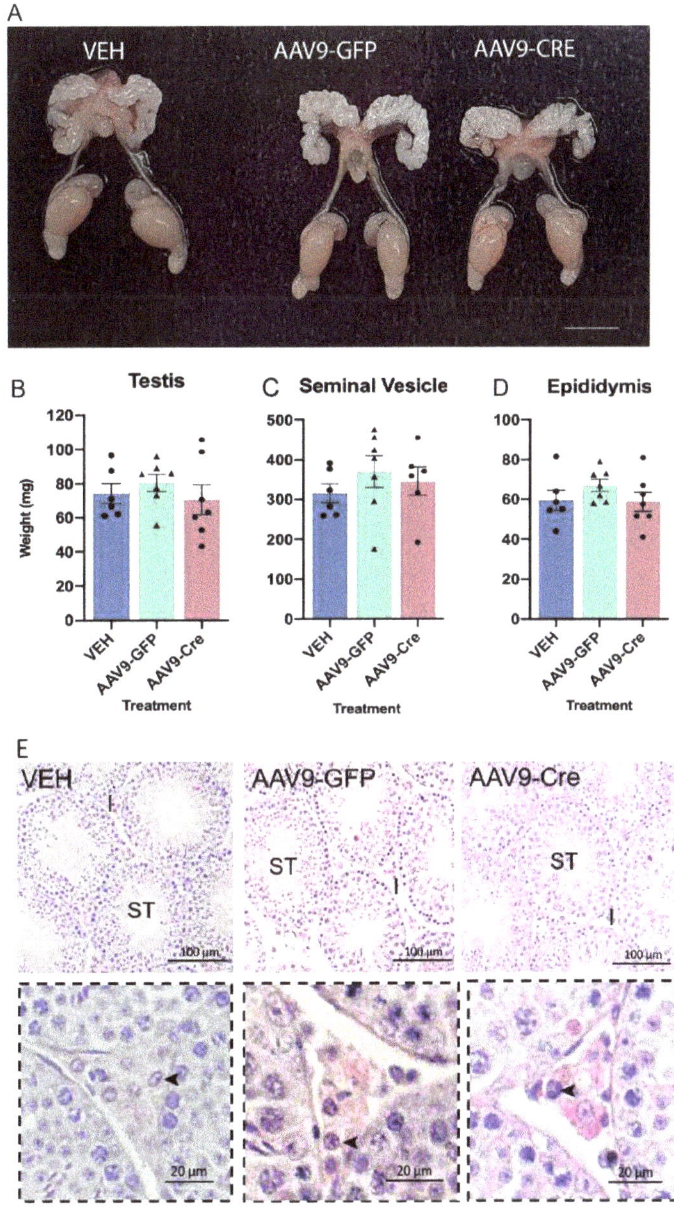

Figure 5. Reduced GR signalling in adult Leydig cells does not alter the gross morphology of the male reproductive system. (**A**) The gross morphology of the reproductive system in adult (d80) was comparable across the different groups. In adult (d80) no difference was noted in the (**B**) testis (**C**) seminal vesicles or (**D**) epididymis between the groups (n = 6–7, ANOVA). (**E**) Gross testicular morphology appeared normal. Insets in the panel below demonstrate $40\times$ magnification of the testicular interstitium and arrowheads indicate the normal appearance of Leydig cells. Scale Bars = 100 µm, Abbreviations; I, Interstitium; ST, seminiferous tubule.

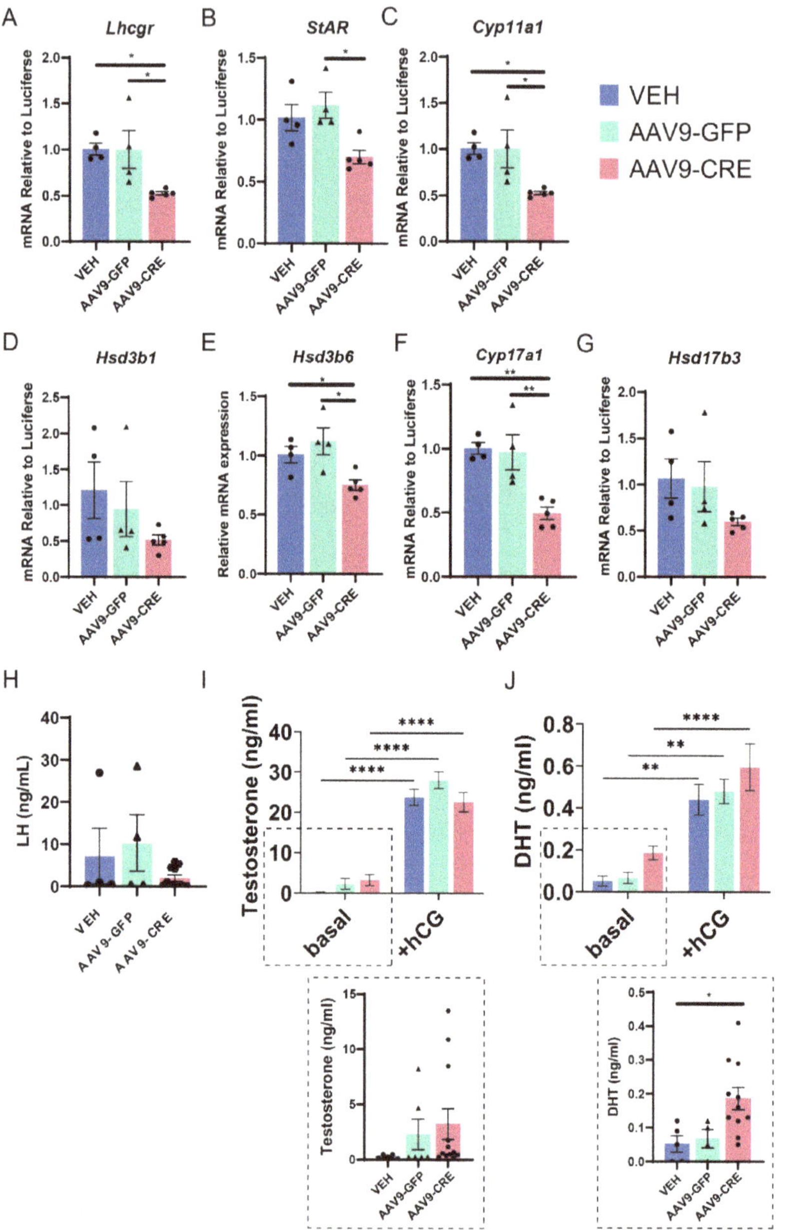

Figure 6. Reduced GR signalling impacts on Leydig cell function. (**A–I**) Comparative testicular expression of Leydig cell steroidogenic transcripts in adults (d80) in human chorionic gonadotrophin (hCG)-stimulated conditions (n = 5–7 per group ANOVA (* $p < 0.05$, ** $p < 0.01$ Tukey's post hoc analysis, error bars SEM). (**H**) Circulating LH and circulating (**I**) testosterone, (**J**) DHT were assessed in basal and hCG-stimulated conditions. Insets show statistical analysis of testosterone and DHT in non hCG-stimulated mice, n = 5–7 per group ANOVA (* $p < 0.05$, ** $p < 0.01$ **** $p < 0.0001$, Tukey's post hoc analysis, error bars SEM).

3. Discussion

Glucocorticoids control fundamental processes in the human body including cell metabolism [33], growth [34], differentiation [35], apoptosis and inflammation [36]). Glucocorticoid signalling is a common target for pharmaceutical intervention in certain conditions such as in the treatment of chronic inflammation [37]) or anxiety and depression [38]) and GR agonists and antagonists are commonly used as treatments in a wide range of clinical settings. Given the widespread use of GR agonists and antagonists by males at different stages of the life course, it is important to determine the role of GR signalling in Leydig cell development and adult function, particularly in terms of steroidogenic capacity of the Leydig cells in adulthood.

GR is expressed in the Leydig cells from fetal life onwards [15,39]. We first utilised the *Cyp17a1*-iCre model to investigate the role of GR in supporting the development and function of the adult Leydig cell population [26]. While the *Cyp17*-iCre model has previously been shown to efficiently target estrogen receptor alpha (*Esr1*) in Leydig cells, we found that only 28% of adult Leydig cells had depleted GR. Because this model could deplete *Nr3c1* expression in Leydig cells from late fetal life [26], it is possible that there could be some compensation during development that could lead to the majority of adult Leydig cells retaining GR. Depletion of GR signalling in Leydig cells from e16.5 encompasses the masculinisation programming window [26,40]. The normal development of the reproductive system in the adult CYPGRKO males suggests that the Leydig cells can maintain normal androgen levels across fetal life. The low level of GR ablation (28%) in adult CYPGRKO mice could be due to reduced targeting efficiency and/or developmental compensation during perinatal and pubertal Leydig cell proliferation and development. In adult CYPGRKO mice, ~70% of adult Leydig cells retain GR expression with no alterations in testicular morphology and steroid enzyme transcripts. demonstrating that the 70% of Leydig cells retaining GR expression can adequately compensate for any physiological loss of GR. This study is in in line with our previous androgen receptor knock out model in Leydig cells which has shown that targeting of approximately 50% is needed to have a substantial developmental and functional impact in Leydig cells [28]. The ability for nuclear receptors to be able to compensate for and activate reciprocal response elements, can add additional challenges when trying to define the role of a single nuclear receptor. Further consideration should be given to mineralocorticoid receptor, which has been shown to also have a high binding affinity for glucocorticoids and are present in Leydig cells [41], which may have also contributed to the lack of phenotype observed. Given the lack of suitability of the CYPGRKO model to effectively knockdown GR signalling in adult Leydig cells, and the lack of a phenotype in these mice, we next chose to develop a model of adult Leydig cell GR ablation.

We next focused on using a viral-mediated delivery system to knockdown Leydig cell GR signalling in adult mice. The recent characterisation of AAV-9 as tools to target Leydig cells provided an opportunity to develop an inducible Cre/loxP system [30] for this purpose. Utilising AAV-9. we generated a model that depletes GR in adult Leydig cells and found that this approach had improved targeting efficiency for Leydig cell gene knockdown compared to the Cyp17-iCre line. Furthermore, we demonstrated that this model had minimal off target GR ablation effects. Thus, our study offers the first evidence of a novel cost-effective method to investigate adult Leydig cell function and demonstrates that viral-mediated delivery of Cre recombinase to the interstitial compartment of the testis permits the knockdown of a gene of interest (in this instance GR) in adult Leydig cells.

Use of the AAV9 system enabled refined assessment of the effects of reduced GR signalling in approximately half of the adult Leydig cells. It is known that supra- or subphysiological levels of glucocorticoids can impact Leydig cell function and survival [21,42] and down-regulate the expression of *Lhcgr* [43] and steroidogenic transcripts [44]. However, there is little information regarding a requirement for GR-signalling in adult Leydig cell function. GR signalling in Sertoli cells is required to support normal testis development and Leydig cell differentiation and steroidogenesis [16]. Results from the present study demonstrate that GR signalling within Leydig cells is also important for Leydig cell steroidogenic function, as a significant decrease in Leydig cell steroidogenic enzyme expression was observed even when half of the Leydig cell population retained GR expression. We have previously shown that Sertoli cells support adult Leydig cell function and steroidogenesis [45] and thus the data from Hazra and colleagues [16] suggests that some of the Sertoli cell effects on Leydig cells are mediated by GR signalling in Sertoli cells. The mechanisms underlying the cross talk between Sertoli, and Leydig cell GR regulation warrants further investigation.

Whilst our inducible Leydig cell knockout model did not impact the reproductive or testicular morphology, we observed a significant downregulation in steroidogenic enzyme transcripts, including *StAR, Cyp11a1*, and *Cyp17a1*. This suggests that GR-signalling is required to support Leydig cell steroidogenesis and confirms that GR has a physiological role across the somatic cells of the testis [16]. These variations were, however, not reflected in terms of circulating LH and testosterone levels, although DHT did show a significant increase. Taken together, the data suggests that depleting only 50% of GR signalling can induce a biological effect in Leydig cells, with a reduction in LHCGR expression and reduced transcription of key steroidogenic enzymes. The normal levels of circulating LH and testosterone may be associated with an ability of Leydig cells to maintain steroidogenic output in a setting of reduced enzyme expression [28] combined with the AAV9 targeting variations, across the cohorts, may explain the overall sampling variation.

Both low and high levels of circulating glucocorticoids (as reported in Cushing Syndrome or Addison disease) suppress male steroidogenesis and fertility, whilst normal physiological levels regulate testis function [46,47]. This study developed two distinct Cre/*LoxP* models to target GR signalling in Leydig cells and showed that there is a threshold of GR inactivation to induce a Leydig cell defect (approximately 30% versus 50% GR targeting). This data highlights an intricate homeostatic balance of GR signalling in reproductive function, and whether it involves compensation mechanisms via the mineralocorticoid or other nuclear receptor remains to be investigated [48,49].

To conclude, our findings suggest that GR-signalling plays a physiological role in normal adult Leydig cell function. Our results also demonstrate the development of a novel transgenic mouse model that provides new opportunities to elucidate the roles of glucocorticoid signalling in Leydig cells, while simultaneously validating a new way to rapidly generate a model of adult Leydig cell gene knockdown. A considerable advantage of deliverable transgenics is that the panel of AAV isotypes permits targeting of different organs and cells, and thus our model is applicable to dissecting the genetic regulation of many physiological processes.

4. Materials and Methods

4.1. Ethics Statement

The research animals used in this study were monitored, handled, and euthanized in accordance with the NSW Animal Research Act 1998, NSW Animal Research Regulation 2010 and the Australian Code for the Care and Use of Animals for Scientific Purposes 8th Edition as approved by the University of Newcastle Animal Care and Ethics Committee (approval number A-2018-827 and A-2018-823). Mice used for experiments were housed at the institute's Central Animal House under a 12 h light/12 h dark cycle at a constant temperature of 21–22 °C with food and water ad libitum. Animals were euthanized immediately before use via carbon dioxide asphyxiation.

4.2. Generation of CYPTR Reporter and CYPGRKO Knockout Mice Using Cyp17a1-iCre

To generate Leydig cell reporter mice, female C57BL/6 mice carrying a random insertion of the *Cyp17a1-i* Cre [50] were mated to C57BL/6 male mice homozygous (Hom) for a floxed *loxP*-flanked STOP cassette preventing transcription of a CAG promoter-driven red fluorescent protein variant (tdTomato) [31]. These matings resulted in Cre-Heterozygous (Het) 'Controls' and Cre+ Het 'CYPTR' mice. To generate Leydig cell GR knockout mice, female C57BL/6 mice carrying a random insertion of the *Cyp17a1-i* Cre [50] were mated to C57BL/6 male mice homozygous (Hom) for floxed GR [27]. The first generation resulted in offspring heterozygous (Het) for GRflox that were either Cre+ or Cre-. For total Leydig cell GR ablation Cre+ GR heterozygous males were again bred to C57BL/6 female mice homozygous for floxed GR resulting in the following offspring: Cre-Hom termed 'littermate control', Cre- Het, termed 'Het littermate control', Cre+ Hom, termed 'CYP GR knockout' (CYPGRKO), and Cre+ Het, termed 'Het CYPGR knockout' (Het CYPGRKO).

4.3. PCR Genotyping of Mice

Mice were genotyped for the inheritance of Cre recombinase as previously described [51]. PCR amplification products were resolved using QIAxcel capillary system (QIAGEN, Sydney, NSW, Australia). An amplicon of 102 bp indicated the inheritance of the Cre recombinase transgene. Mice were also genotyped for the inheritance of floxed GR using primer sequences forward GGCATGCACATTACTGGCCTTCT, reverse 1 GTGTAGCAG CCAGCTTACAGGA and reverse 2 CCTTCTCATTCCATGTCAGCATGT. Expected band sizes are 2.5 kb for wild type GR and 500 bp for recombined GR.

4.4. Viral Vectors

Adeno-Associated viral particle 9 (AAV9) containing GFP and CRE expressing transgenes downstream of a CMV promoter were obtained from GeneCopoeia (via United Bioresearch, Maroota NSW, Australia, Cat. No: AC001). Viral particles were supplied at a titre of $\geq 5 \times 10^{12}$ GC/mL. Utilising viral vectors that express fluorescent reporters downstream of the powerful CMV promoter enabled confirmation and identification of all transduced cells carrying delivered transgenes (GFP and/or Cre recombinase).

4.5. Testicular Delivery of Adeno-Associated Viral Particle 9

Viral particles were introduced into the interstitial compartment of adult (>60 days post-natal) mouse testes using an Ultra-Fine 23 gauge 0.3-mL insulin needle, injecting close to the rete testis as previously described [30]. A maximum volume of 20 µL was delivered. Successful delivery of the particles was monitored via the addition of Trypan Blue dye to the viral particles (0.04%). Animals were culled 7 days or 35 days post injection.

4.6. Inducible Model Using AAV9 Viral Vector

For assessment of the viral vector Cre recombinase delivery, we utilised Gt(ROSA) 26Sor(tdTomato-WPRE) termed 'TRTR' mice carrying an insertion of a Cre reporter allele inserted into the Gt(ROSA)26Sor locus [31]. When TRTR mice are injected with the Cre-encoding viral construct directly into the interstitial area, RFP will be expressed specifically in cells where Cre was active. These mice were obtained from the Jackson laboratory (JAX stock #007914). To generate Leydig cell GR knockout mice, GR floxed ('GRGR', JAX stock #007909) mice which possess *loxP* sites flanking exon 3 of the *Nr3c1* gene, were injected as previously described.

4.7. Tissue Collection and Processing

Mice were culled between d80 and d100 by inhalation of carbon dioxide and subsequent cervical dislocation Body weight was measured and organs were removed and weighed. Tissues were fixed in Bouin's fixative for 4–24 h depending on tissue size. Bouin's-fixed tissues were processed and embedded in paraffin wax, and 5 µm sections were used for histological analysis. Tissues were stained with haematoxylin and eosin using standard protocols and examined for histological abnormalities.

4.8. Quantitative RT-PCR

RNA was obtained from frozen tissues using the RNeasy Mini extraction kit with RNase-free DNase on the column digestion kit (Qiagen, Sydney, NSW, Australia) according to the manufacturer's protocol. RNA yield was quantified using a NanoDrop 1000 spectrophotometer (Thermo Fisher Scientific, Waltham, MA, USA). Random hexamer primed cDNA was prepared using the SuperScript VILO cDNA synthesis kit (Life Technologies) according to manufacturers' protocols. Quantitative PCR was performed on the genes of interest listed in Table 1 using an Lightcycler 96 instrument (Roche, Sydney, NSW, Australia) and the Roche Universal Probe Library (Roche, AU). The expression of each gene was related to external housekeeping gene assay Luciferase (Roche, AU).

Table 1. Details of Antibodies and detection methods used.

Primary Antibody (AbI) Name	References	Dilution AbI	RRID	Detection System
GFP	Abcam ab6556	1/1500	AB_305564	IHC/IF
3 b-HSD	Elabscience E-AB15112	1/1000	ND	IHC/IF
RFP	Evrogen #AB233	1/1500	AB_2571743	IHC/IF
GR	Abcam ab183127	1/1000	AB_2833234	IHC/IF

IHC: Immunohistochemistry IF: immunofluorescence.

4.9. Immunohistochemistry

Immunolocalization was performed either by a single antibody colourimetric (DAB) immunostaining method, as described previously [52] a single or double antibody tyramide fluorescent immunostaining method, as described previously [50,53], or automated Bond immunostaining method, as described previously [52]. Antibodies used are listed in Table 2. A minimum of five individual sections for each genotype were immunostained in each experiment. For whole organ fluorescence, freshly dissected organs were visualised with a Zeiss Discovery V.12 Fluoroscope under either brightfield (transmitted light) or an epifluorescent GFP/RFP filter.

4.10. Extraction and Analysis of Steroid Hormones from Plasma

Immediately after culling (after CO_2 and before cervical dislocation), blood was collected from mice via cardiac puncture with a syringe and needle with a wide bevel to reduce lysis, blood was collected in EDTA coated tubes to prevent coagulation. Plasma was separated by centrifugation and stored at $-80\,°C$. LH was measured by ELISA according to manufacturer instructions (ab235648). The inter-assay coefficient of variance (CV) was <5.2% and the intra-assay CV was <5.4%. Steroid hormones were assessed using LC-MS/MS at the ANZAC Research Institute (University of Sydney, Sydney, NSW, Australia) as previously described [54].

4.11. Assessment of Cre Efficiency in Cre/LoxP and AAV9 Inducible Mice

Analysis of Cre efficiency in reporter mice and GR knockout mice was achieved via quantitation of cells positive for the Leydig cell specific protein 3βHSD and the reporter gene RFP, and the presence/absence of glucocorticoid receptor immunohistochemical localisation. Testis sections were imaged and then counted in Zen lite (ZEISS)). Each cohort had n = 3 and 5 sections from each animal was counted. Each section was scanned on the Axioscan to provide a whole section view.

Table 2. Details of primer sequences used for genotyping and qRT-PCR.

Gene	Forward Primer	Reverse Primer	Probe
GRGR-FLOX	Atgcctgctaggcaa atgat	Ttccagggctataggaagca	Genomic
CYP17A1 ICRE	CaggttttggtgcacagtCa	GctgtagcttctccactcCac	Genomic
TRTR WT	Aagggagctgcagtggagta	Ccgaaaatctgtgggaagtc	Genomic
TRTR MUTANT	Ggcattaaagcagcgtatcc	Ctgttcctgtacggcatgg	Genomic
Lhcgr	Gggacgacgctaatctcg	Cctggaaggtgccactgt	Upl #80
Star	Aaactcacttggctgctcagta	Tgcgataggacctggttgat	Upl #83
Cyp11a1	Cccattggggtcctgttta	Tggtagacagcattgatgaacc	Upl #67
Cyp17a1	Catcccacacaaggctaaca	Cagtgcccagagattgatga	Upl #67
Hsd3b1	Gaactgcaggaggtcagagc	Gcactgggcatccagaat	Upl #12
Hsd3b6	Accatccttccacagttctagc	Acagtgaccctggagatggt	Upl #95
Hsd17b3	Gagttggccagacatggact	Agcttccagtggtcctctca	Upl #47
Srd5a1	Gggaaactggatacaaaataccc	Ccacgagctccccaaaata	Upl #41
Srd5a2	Ggtcatctacaggatcccaca	Tcaataatctcgcccaggaa	Upl #50
NR3C1	Caaagattgcaggtatcctatgaa	Cttggctcttcagaccttcc	Upl #81
STC1	Gaggcggaacaaaatgattc	Gcagcgaaccacttcagc	Upl #45
TSC22D3	Ggtggccctagacaacaaga	Tcaagcagctcacgaatctg	Upl #10
Luciferase	Gcacatatcgaggtgaacatcac	Gccaaccgaacggacattt	5′ned- tacgcggaatacttc

4.12. Statistical Analysis

Power calculations based on previous cell quantitation experiments determined that a sample size of 3 is appropriate for quantitative end points for cell counting and immunohistochemistry [55]. Statistical analysis is performed via GraphPad Prism (version 8; GraphPad Software Inc., San Diego, CA, USA). Statistical tests include a one-way ANOVA with Tukey's post hoc test (if comparing multiple groups), a two-way ANOVA with Tukey's post hoc test (if comparing multiple groups and variables), and Chi Squared test for determining proportion targeting/ablation. Values are expressed as mean $\pm$ S.E.M.

Supplementary Materials: The following supporting information can be downloaded at: https://www.mdpi.com/article/10.3390/ijms232315015/s1.

Author Contributions: A.-L.G. and D.R. conceived and designed the study. A.-L.G., D.R., A.L.D., B.M.L., J.K., S.S., S.P., N.R. and G.C. carried out the experiments. A.-L.G., L.O., L.B.S. and D.R. analysed the results. A.-L.G. and D.R. wrote the paper. All authors have read and agreed to the published version of the manuscript.

Funding: This work was funded by the Hunter Medical Research Institute (HRMI) Infertility fund awarded to CIA Dr Rebourcet and CIB Dr Anne-Louise Gannon (G2101102) and from the University of Newcastle internal Research and Innovation funding (awarded to LBS).

Institutional Review Board Statement: The study was conducted in accordance with the Institutional Biosafety Committee (IBC), approval number 2018.77 held by A-LG. The research animals used in this study were monitored, handled, and euthanized in accordance with the NSW Animal Research Act 1998, NSW Animal Research Regulation 2010 and the Australian Code for the Care and Use of Animals for Scientific Purposes 8th Edition as approved by the University of Newcastle Animal Care and Ethics Committee (approval number A-2018-827 and A-2018-823).

Data Availability Statement: All data relating to this study has been included in the main manuscript or in the supplementary document. Schematics of breeding method Created with BioRender.com, agreement number BV24KRQ930.

Acknowledgments: We thank Animal Support Unit (ASU) staff Adrian Bernard and Annemaree Probert for their technical assistance.

Conflicts of Interest: The authors declare no conflict of interest.

References

1. Mooij, C.F.; van Herwaarden, A.E.; Claahsen-van der Grinten, H.L. Disorders of Adrenal Steroidogenesis: Impact on Gonadal Function and Sex Development. *Pediatr. Endocrinol. Rev.* **2016**, *14*, 109–128. [CrossRef] [PubMed]
2. Chu, K.Y.; Achua, J.K.; Ramasamy, R. Strategies to increase testosterone in men seeking fertility. *Turk. J. Urol.* **2020**. [CrossRef] [PubMed]
3. Farrell, J.B.; Deshmukh, A.; Baghaie, A.A. Low testosterone and the association with type 2 diabetes. *Diabetes Educ.* **2008**, *34*, 799–806. [CrossRef] [PubMed]
4. Ohlander, S.J.; Lindgren, M.C.; Lipshultz, L.I. Testosterone and Male Infertility. *Urol. Clin. N. Am.* **2016**, *43*, 195–202. [CrossRef]
5. Sidhom, K.; Panchendrabose, K.; Mann, U.; Patel, P. An update on male infertility and intratesticular testosterone-insight into novel serum biomarkers. *Int. J. Impot. Res.* **2022**, *34*, 673–678. [CrossRef]
6. Whirledge, S.; Cidlowski, J.A. Glucocorticoids, stress, and fertility. *Minerva Endocrinol.* **2010**, *35*, 109–125.
7. Silva, E.J.; Vendramini, V.; Restelli, A.; Bertolla, R.P.; Kempinas, W.G.; Avellar, M.C. Impact of adrenalectomy and dexamethasone treatment on testicular morphology and sperm parameters in rats: Insights into the adrenal control of male reproduction. *Andrology* **2014**, *2*, 835–846. [CrossRef]
8. Rook, G.A. Glucocorticoids and immune function. *Baillieres Best Pract. Res. Clin. Endocrinol. Metab.* **1999**, *13*, 567–581. [CrossRef]
9. Mir, N.; Chin, S.A.; Riddell, M.C.; Beaudry, J.L. Genomic and Non-Genomic Actions of Glucocorticoids on Adipose Tissue Lipid Metabolism. *Int. J. Mol. Sci.* **2021**, *22*, 8503. [CrossRef]
10. Weger, B.D.; Weger, M.; Gorling, B.; Schink, A.; Gobet, C.; Keime, C.; Poschet, G.; Jost, B.; Krone, N.; Hell, R.; et al. Extensive Regulation of Diurnal Transcription and Metabolism by Glucocorticoids. *PLoS Genet.* **2016**, *12*, e1006512. [CrossRef]
11. Matsuwaki, T.; Kayasuga, Y.; Yamanouchi, K.; Nishihara, M. Maintenance of gonadotropin secretion by glucocorticoids under stress conditions through the inhibition of prostaglandin synthesis in the brain. *Endocrinology* **2006**, *147*, 1087–1093. [CrossRef] [PubMed]
12. Schultz, R.; Isola, J.; Parvinen, M.; Honkaniemi, J.; Wikstrom, A.C.; Gustafsson, J.A.; Pelto-Huikko, M. Localization of the glucocorticoid receptor in testis and accessory sexual organs of male rat. *Mol. Cell. Endocrinol.* **1993**, *95*, 115–120. [CrossRef] [PubMed]
13. Weber, M.A.; Groos, S.; Hopfl, U.; Spielmann, M.; Aumuller, G.; Konrad, L. Glucocorticoid receptor distribution in rat testis during postnatal development and effects of dexamethasone on immature peritubular cells in vitro. *Andrologia* **2000**, *32*, 23–30. [CrossRef] [PubMed]
14. Ge, R.S.; Hardy, D.O.; Catterall, J.F.; Hardy, M.P. Developmental changes in glucocorticoid receptor and 11beta-hydroxysteroid dehydrogenase oxidative and reductive activities in rat Leydig cells. *Endocrinology* **1997**, *138*, 5089–5095. [CrossRef] [PubMed]
15. Nordkap, L.; Almstrup, K.; Nielsen, J.E.; Bang, A.K.; Priskorn, L.; Krause, M.; Holmboe, S.A.; Winge, S.B.; Egeberg Palme, D.L.; Morup, N.; et al. Possible involvement of the glucocorticoid receptor (NR3C1) and selected NR3C1 gene variants in regulation of human testicular function. *Andrology* **2017**, *5*, 1105–1114. [CrossRef]
16. Hazra, R.; Upton, D.; Jimenez, M.; Desai, R.; Handelsman, D.J.; Allan, C.M. In vivo actions of the Sertoli cell glucocorticoid receptor. *Endocrinology* **2014**, *155*, 1120–1130. [CrossRef]
17. Wang, M.; Fijak, M.; Hossain, H.; Markmann, M.; Nusing, R.M.; Lochnit, G.; Hartmann, M.F.; Wudy, S.A.; Zhang, L.; Gu, H.; et al. Characterization of the Micro-Environment of the Testis that Shapes the Phenotype and Function of Testicular Macrophages. *J. Immunol.* **2017**, *198*, 4327–4340. [CrossRef]
18. Ge, R.S.; Dong, Q.; Niu, E.M.; Sottas, C.M.; Hardy, D.O.; Catterall, J.F.; Latif, S.A.; Morris, D.J.; Hardy, M.P. 11{beta}-Hydroxysteroid dehydrogenase 2 in rat leydig cells: Its role in blunting glucocorticoid action at physiological levels of substrate. *Endocrinology* **2005**, *146*, 2657–2664. [CrossRef]
19. Bambino, T.H.; Hsueh, A.J. Direct inhibitory effect of glucocorticoids upon testicular luteinizing hormone receptor and steroidogenesis in vivo and in vitro. *Endocrinology* **1981**, *108*, 2142–2148. [CrossRef]
20. Saez, J.M.; Morera, A.M.; Haour, F.; Evain, D. Effects of in vivo administration of dexamethasone, corticotropin and human chorionic gonadotropin on steroidogenesis and protein and DNA synthesis of testicular interstitial cells in prepuberal rats. *Endocrinology* **1977**, *101*, 1256–1263. [CrossRef]
21. Gao, H.B.; Tong, M.H.; Hu, Y.Q.; You, H.Y.; Guo, Q.S.; Ge, R.S.; Hardy, M.P. Mechanisms of glucocorticoid-induced Leydig cell apoptosis. *Mol. Cell. Endocrinol.* **2003**, *199*, 153–163. [CrossRef] [PubMed]
22. Hales, D.B.; Payne, A.H. Glucocorticoid-mediated repression of P450scc mRNA and de novo synthesis in cultured Leydig cells. *Endocrinology* **1989**, *124*, 2099–2104. [CrossRef] [PubMed]
23. Hales, D.B.; Sha, L.; Payne, A.H. Glucocorticoid and cyclic adenosine 3'5'-monophosphate-mediated induction of cholesterol side-chain cleavage cytochrome P450 (P450scc) in MA-10 tumor Leydig cells. Increases in mRNA are cycloheximide sensitive. *Endocrinology* **1990**, *126*, 2800–2808. [CrossRef] [PubMed]

24. Hu, G.X.; Lian, Q.Q.; Lin, H.; Latif, S.A.; Morris, D.J.; Hardy, M.P.; Ge, R.S. Rapid mechanisms of glucocorticoid signaling in the Leydig cell. *Steroids* **2008**, *73*, 1018–1024. [CrossRef]

25. Xiao, Y.C.; Huang, Y.D.; Hardy, D.O.; Li, X.K.; Ge, R.S. Glucocorticoid suppresses steroidogenesis in rat progenitor Leydig cells. *J. Androl.* **2010**, *31*, 365–371. [CrossRef]

26. Bridges, P.J.; Koo, Y.; Kang, D.W.; Hudgins-Spivey, S.; Lan, Z.J.; Xu, X.; DeMayo, F.; Cooney, A.; Ko, C. Generation of Cyp17iCre transgenic mice and their application to conditionally delete estrogen receptor alpha (Esr1) from the ovary and testis. *Genesis* **2008**, *46*, 499–505. [CrossRef]

27. De Gendt, K.; Swinnen, J.V.; Saunders, P.T.; Schoonjans, L.; Dewerchin, M.; Devos, A.; Tan, K.; Atanassova, N.; Claessens, F.; Lécureuil, C. A Sertoli cell-selective knockout of the androgen receptor causes spermatogenic arrest in meiosis. *Proc. Natl. Acad. Sci. USA* **2004**, *101*, 1327–1332. [CrossRef]

28. O'Hara, L.; McInnes, K.; Simitsidellis, I.; Morgan, S.; Atanassova, N.; Slowikowska-Hilczer, J.; Kula, K.; Szarras-Czapnik, M.; Milne, L.; Mitchell, R.T. Autocrine androgen action is essential for Leydig cell maturation and function, and protects against late-onset Leydig cell apoptosis in both mice and men. *FASEB J.* **2015**, *29*, 894–910. [CrossRef]

29. O'Hara, L.; Smith, L.B. Androgen receptor roles in spermatogenesis and infertility. *Best Pract. Res. Clin. Endocrinol. Metab.* **2015**, *29*, 595–605. [CrossRef]

30. Darbey, A.; Rebourcet, D.; Curley, M.; Kilcoyne, K.; Jeffery, N.; Reed, N.; Milne, L.; Roesl, C.; Brown, P.; Smith, L.B. A comparison of in vivo viral targeting systems identifies adeno-associated virus serotype 9 (AAV9) as an effective vector for genetic manipulation of Leydig cells in adult mice. *Andrology* **2021**, *9*, 460–473. [CrossRef]

31. Madisen, L.; Zwingman, T.A.; Sunkin, S.M.; Oh, S.W.; Zariwala, H.A.; Gu, H.; Ng, L.L.; Palmiter, R.D.; Hawrylycz, M.J.; Jones, A.R. A robust and high-throughput Cre reporting and characterization system for the whole mouse brain. *Nat. Neurosci.* **2010**, *13*, 133–140. [CrossRef] [PubMed]

32. Kotula-Balak, M.; Hejmej, A.; Bilińska, B. Hydroxysteroid Dehydrogenases–localization, function and regulation in the testis. *Intech Chapter* **2012**, *11*, 265–288. [CrossRef]

33. Vegiopoulos, A.; Herzig, S. Glucocorticoids, metabolism and metabolic diseases. *Mol. Cell. Endocrinol.* **2007**, *275*, 43–61. [CrossRef] [PubMed]

34. Ren, R.; Oakley, R.H.; Cruz-Topete, D.; Cidlowski, J.A. Dual role for glucocorticoids in cardiomyocyte hypertrophy and apoptosis. *Endocrinology* **2012**, *153*, 5346–5360. [CrossRef]

35. Croxtall, J.D.; Choudhury, Q.; Flower, R.J. Glucocorticoids act within minutes to inhibit recruitment of signalling factors to activated EGF receptors through a receptor-dependent, transcription-independent mechanism. *Br. J. Pharmacol.* **2000**, *130*, 289–298. [CrossRef]

36. Oakley, R.H.; Cidlowski, J.A. Cellular processing of the glucocorticoid receptor gene and protein: New mechanisms for generating tissue-specific actions of glucocorticoids. *J. Biol. Chem.* **2011**, *286*, 3177–3184. [CrossRef]

37. Al-Hejjaj, W.K.; Numan, I.T.; Al-Sa'ad, R.Z.; Hussain, S.A. Anti-inflammatory activity of telmisartan in rat models of experimentally-induced chronic inflammation: Comparative study with dexamethasone. *Saudi Pharm. J.* **2011**, *19*, 29–34. [CrossRef]

38. Anacker, C.; Zunszain, P.A.; Carvalho, L.A.; Pariante, C.M. The glucocorticoid receptor: Pivot of depression and of antidepressant treatment? *Psychoneuroendocrinology* **2011**, *36*, 415–425. [CrossRef]

39. Welter, H.; Herrmann, C.; Dellweg, N.; Missel, A.; Thanisch, C.; Urbanski, H.F.; Kohn, F.M.; Schwarzer, J.U.; Muller-Taubenberger, A.; Mayerhofer, A. The Glucocorticoid Receptor NR3C1 in Testicular Peritubular Cells is Developmentally Regulated and Linked to the Smooth Muscle-Like Cellular Phenotype. *J. Clin. Med.* **2020**, *9*, 961. [CrossRef]

40. Welsh, M.; Saunders, P.T.; Fisken, M.; Scott, H.M.; Hutchison, G.R.; Smith, L.B.; Sharpe, R.M. Identification in rats of a programming window for reproductive tract masculinization, disruption of which leads to hypospadias and cryptorchidism. *J. Clin. Investig.* **2008**, *118*, 1479–1490. [CrossRef]

41. Ge, R.-S.; Dong, Q.; Sottas, C.M.; Latif, S.A.; Morris, D.J.; Hardy, M.P. Stimulation of testosterone production in rat Leydig cells by aldosterone is mineralocorticoid receptor mediated. *Mol. Cell. Endocrinol.* **2005**, *243*, 35–42. [CrossRef] [PubMed]

42. Chen, Y.; Wang, Q.; Wang, F.F.; Gao, H.B.; Zhang, P. Stress induces glucocorticoid-mediated apoptosis of rat Leydig cells in vivo. *Stress* **2012**, *15*, 74–84. [CrossRef] [PubMed]

43. Zhang, J.; Hu, G.; Huang, B.; Zhuo, D.; Xu, Y.; Li, H.; Zhan, X.; Ge, R.-S.; Xu, Y. Dexamethasone suppresses the differentiation of stem Leydig cells in rats in vitro. *BMC Pharmacol. Toxicol.* **2019**, *20*, 32. [CrossRef] [PubMed]

44. Fon, W.P.; Li, P.H. Dexamethasone-induced suppression of steroidogenic acute regulatory protein gene expression in mouse Y-1 adrenocortical cells is associated with reduced histone H3 acetylation. *Endocrine* **2007**, *32*, 155–165. [CrossRef] [PubMed]

45. Rebourcet, D.; Mackay, R.; Darbey, A.; Curley, M.K.; Jørgensen, A.; Frederiksen, H.; Mitchell, R.T.; O'Shaughnessy, P.J.; Nef, S.; Smith, L.B. Ablation of the canonical testosterone production pathway via knockout of the steroidogenic enzyme HSD17B3, reveals a novel mechanism of testicular testosterone production. *FASEB J.* **2020**, *34*, 10373–10386. [CrossRef] [PubMed]

46. Urban, M.D.; Lee, P.A.; Gutai, J.P.; Migeon, C.J. Androgens in pubertal males with Addison's disease. *J. Clin. Endocrinol. Metab.* **1980**, *51*, 925–929. [CrossRef]

47. Zheng, H.; Wang, Q.; Cui, Q.; Sun, Q.; Wu, W.; Ji, L.; He, M.; Lu, B.; Zhang, Z.; Ma, Z.; et al. The hypothalamic-pituitary-gonad axis in male Cushing's disease before and after curative surgery. *Endocrine* **2022**, *77*, 357–362. [CrossRef]

48. Daskalakis, N.P.; Meijer, O.C.; de Kloet, E.R. Mineralocorticoid receptor and glucocorticoid receptor work alone and together in cell-type-specific manner: Implications for resilience prediction and targeted therapy. *Neurobiol. Stress* **2022**, *18*, 100455. [CrossRef]
49. Arora, V.K.; Schenkein, E.; Murali, R.; Subudhi, S.K.; Wongvipat, J.; Balbas, M.D.; Shah, N.; Cai, L.; Efstathiou, E.; Logothetis, C.; et al. Glucocorticoid receptor confers resistance to antiandrogens by bypassing androgen receptor blockade. *Cell* **2013**, *155*, 1309–1322. [CrossRef]
50. O'Hara, L.; York, J.P.; Zhang, P.; Smith, L.B. Targeting of GFP-Cre to the mouse Cyp11a1 locus both drives cre recombinase expression in steroidogenic cells and permits generation of Cyp11a1 knock out mice. *PLoS ONE* **2014**, *9*, e84541. [CrossRef]
51. Welsh, M.; Saunders, P.T.; Atanassova, N.; Sharpe, R.M.; Smith, L.B. Androgen action via testicular peritubular myoid cells is essential for male fertility. *FASEB J.* **2009**, *23*, 4218–4230. [CrossRef] [PubMed]
52. O'Hara, L.; Welsh, M.; Saunders, P.T.; Smith, L.B. Androgen receptor expression in the caput epididymal epithelium is essential for development of the initial segment and epididymal spermatozoa transit. *Endocrinology* **2010**, *152*, 718–729. [CrossRef] [PubMed]
53. O'Hara, L.; Smith, L.B. Androgen receptor signalling in Vascular Endothelial cells is dispensable for spermatogenesis and male fertility. *BMC Res. Notes* **2012**, *5*, 16. [CrossRef] [PubMed]
54. Skiba, M.A.; Bell, R.J.; Islam, R.M.; Handelsman, D.J.; Desai, R.; Davis, S.R. Androgens During the Reproductive Years: What Is Normal for Women? *J. Clin. Endocrinol. Metab.* **2019**, *104*, 5382–5392. [CrossRef] [PubMed]
55. Gannon, A.-L.; O'Hara, L.; Mason, I.J.; Jørgensen, A.; Frederiksen, H.; Curley, M.; Milne, L.; Smith, S.; Mitchell, R.T.; Smith, L.B. Androgen Receptor Is Dispensable for X-Zone Regression in the Female Adrenal but Regulates Post-Partum Corticosterone Levels and Protects Cortex Integrity. *Front. Endocrinol.* **2021**, *11*, 1026. [CrossRef]

 International Journal of
Molecular Sciences

Article

A 35-bp Conserved Region Is Crucial for *Insl3* Promoter Activity in Mouse MA-10 Leydig Cells

Xavier C. Giner [1], Kenley Joule Pierre [1], Nicholas M. Robert [1] and Jacques J. Tremblay [1,2,*]

[1] Reproduction, Mother and Child Health, Room T3-67, CHU de Québec—Université Laval Research Centre, Québec, QC G1V 4G2, Canada
[2] Centre for Research in Reproduction, Development and Intergenerational Health, Department of Obstetrics, Gynecology and Reproduction, Faculty of Medicine, Université Laval, Québec, QC G1V 0A6, Canada
* Correspondence: jacques-j.tremblay@crchudequebec.ulaval.ca

Abstract: The peptide hormone insulin-like 3 (INSL3) is produced almost exclusively by Leydig cells of the male gonad. INSL3 has several functions such as fetal testis descent and bone metabolism in adults. *Insl3* gene expression in Leydig cells is not hormonally regulated but rather is constitutively expressed. The regulatory region of the *Insl3* gene has been described in various species; moreover, functional studies have revealed that the *Insl3* promoter is regulated by various transcription factors that include the nuclear receptors AR, NUR77, COUP-TFII, LRH1, and SF1, as well as the Krüppel-like factor KLF6. However, these transcription factors are also found in several tissues that do not express *Insl3*, indicating that other, yet unidentified factors, must be involved to drive *Insl3* expression specifically in Leydig cells. Through a fine functional promoter analysis, we have identified a 35-bp region that is responsible for conferring 70% of the activity of the mouse *Insl3* promoter in Leydig cells. All tri- and dinucleotide mutations introduced dramatically reduced *Insl3* promoter activity, indicating that the entire 35-bp sequence is required. Nuclear proteins from MA-10 Leydig cells bound specifically to the 35-bp region. The 35-bp sequence contains GC- and GA-rich motifs as well as potential binding elements for members of the CREB, C/EBP, AP1, AP2, and NF-κB families. The *Insl3* promoter was indeed activated 2-fold by NF-κB p50 but not by other transcription factors tested. These results help to further define the regulation of *Insl3* gene transcription in Leydig cells.

Keywords: Leydig cells; insulin-like 3; transcription; transcription factors; linker-scanning mutagenesis; DNA–protein interactions

Citation: Giner, X.C.; Pierre, K.J.; Robert, N.M.; Tremblay, J.J. A 35-bp Conserved Region Is Crucial for *Insl3* Promoter Activity in Mouse MA-10 Leydig Cells. *Int. J. Mol. Sci.* **2022**, 23, 15060. https://doi.org/10.3390/ijms232315060

Academic Editor: Jacek Z. Kubiak

Received: 1 November 2022
Accepted: 28 November 2022
Published: 1 December 2022

Publisher's Note: MDPI stays neutral with regard to jurisdictional claims in published maps and institutional affiliations.

1. Introduction

Leydig cells located in the testis produce testosterone and insulin-like 3 (INSL3), two hormones essential for male sex differentiation and reproductive function. More specifically, testosterone is required to masculinize the male embryo during fetal life, and postnatally to complete internal and external male organ development as well as for the initiation and maintenance of spermatogenesis [1]. INSL3 regulates the inguino-scrotal phase of testicular descent during fetal life [2,3] and bone metabolism in adults [4]. INSL3 has also been proposed to reduce germ cell apoptosis [5,6], modulate skeletal muscle metabolism and function [7], regulate motor and sensory brain functions [8], and contribute to corneal wound healing [9]. INSL3 is an exclusive marker for the differentiation and functional status of Leydig cells in the testis [10,11].

The *Insl3* gene is composed of two exons (219 and 404 bp) separated by a 739-bp intron. The *Insl3* gene is atypically located entirely within the last intron of the *Jak3* gene [12–14]. Consequently, the regulatory elements controlling *Insl3* gene expression in Leydig cells are believed to be located within a short promoter region of about 1000 bp. The genomic region corresponding to the *Insl3* promoter has been isolated from rat [13], mouse [12,15], human [16], dog [17], and pig [16]. Functional studies have identified several transcription factors regulating *Insl3* promoter activity in Leydig cells. These include the Krüppel-like factor KLF6 [18] as well as the nuclear receptors SF1 (Ad4BP, NR5A1) [13,17,19,20], LRH1 (NR5A2) [21], NUR77 (NGFI-B, NR4A1) [21,22], testosterone-activated androgen receptor (AR, NR3C4) [23,24], COUP-TFII (NR2F2) [25], and DAX1 (NR0B1) [19]. Transcriptional cooperation between COUP-TFII and SF1 [25,26], and between KLF6 and NUR77 and SF1 [18] on the *Insl3* promoter has also been reported. However, most of these transcription factors are also found in cell types that do not express the *Insl3* gene, such as Sertoli and adrenal cells, indicating that additional transcription factors are likely involved in directing *Insl3* expression in Leydig cells. In the present study, we have identified a 35-bp regulatory region bound by nuclear proteins that is essential for mouse *Insl3* promoter activity in MA-10 Leydig cells.

2. Results

2.1. A 35-bp Region Is Responsible for 70% of Mouse Insl3 Promoter Activity

To identify the regions responsible for the activity of the mouse *Insl3* promoter in Leydig cells, several 5′ progressive deletions of the promoter were generated and transfected in MA-10 Leydig cells. The MA-10 cell line expresses the *Insl3* gene and has been validated as an appropriate model to study *Insl3* gene expression [20,22,27]. As shown in Figure 1, *Insl3* promoter deletions from −1082 bp to −186 bp had no significant effect on *Insl3* promoter activity. However, further deletion to −111 bp led to a 70% reduction in *Insl3* promoter activity (Figure 1). Finally, a further reduction in *Insl3* promoter activity to 15% was observed with a deletion to −79 bp, which is considered a minimal promoter (Figure 1). These data, therefore, identify a critical region located between −186 and −111 bp that is responsible for 70% of *Insl3* promoter activity.

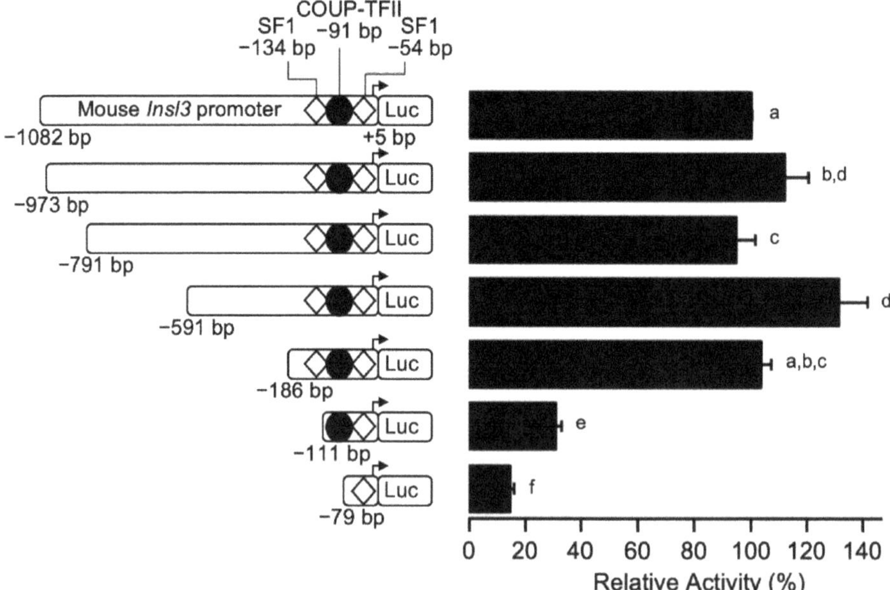

Figure 1. Identification of a critical regulatory region within the proximal mouse *Insl3* promoter. MA-10 Leydig cells were transfected with various 5′ deletion constructs of the mouse *Insl3* promoter; the 5′ end point of each construct is indicated on the left of the graph. The position of previously identified binding sites for SF1 (white diamonds) and COUP-TFII (black circle) is indicated. Results are shown as % Relative Activity (±SEM) relative to the activity of the −1132 bp reporter, which was set to 100%. Different letters indicate a statistically significant difference between groups ($p < 0.05$).

To more precisely locate the region conferring 70% of *Insl3* promoter activity, additional 5′ deletion constructs to −151, −141, −131, and −120 bp were generated and transfected in MA-10 Leydig cells. As shown in Figure 2, deletion of a 35-bp region between −186 and −151 bp led to a 70% reduction in *Insl3* promoter activity. Promoter activity remained similar with all other deletion constructs, except for the minimal −79 bp which was reduced to 15% (Figure 2). Together, these results indicate that a 35-bp region located between −186 and −151 bp is responsible for conferring 70% of activity to the mouse *Insl3* promoter in MA-10 Leydig cells.

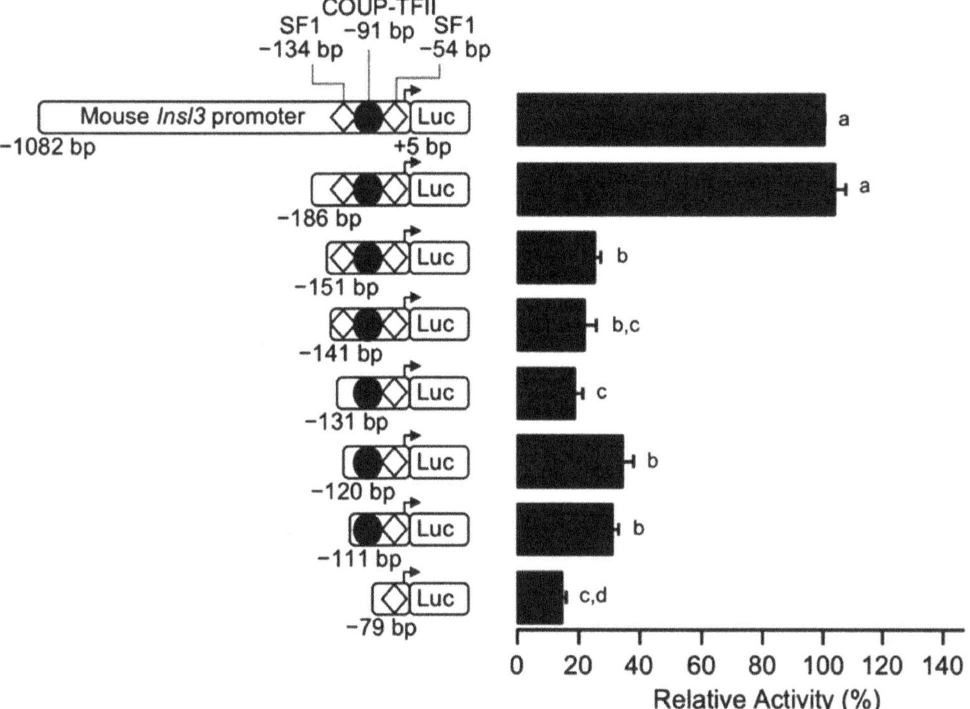

Figure 2. The region between −186 and −151 bp confers 70% of mouse *Insl3* promoter activity in Leydig cells. MA-10 Leydig cells were transfected with a series of fine 5′ deletion constructs of the mouse *Insl3* promoter; the 5′ end point of each construct is indicated on the left of the graph. The position of previously identified binding sites for SF1 (white diamonds) and COUP-TFII (black circle) is indicated. Results are shown as % Relative Activity ($\pm$ SEM) relative to the activity of the −1132 bp reporter, which was set to 100%. Different letters indicate a statistically significant difference between groups ($p < 0.05$).

2.2. Fine Mapping of the 35-bp Region Conferring 70% of Insl3 Promoter Activity

To identify the sequences responsible for conferring 70% of mouse *Insl3* promoter activity within the −186 and −151 bp region, a linker-scanning approach was used. Eight sequential mutations were introduced in the 35-bp region in the context of the −1182 bp *Insl3* promoter (Figure 3) and the reporter constructs were transfected in MA-10 Leydig cells. As shown in Figure 3, all the mutations (M1 to M8) led to a loss of 60–65% in *Insl3* promoter activity compared to the wild type −1182 bp promoter. This indicates that the entire 35-bp region is necessary for full mouse *Insl3* promoter activity.

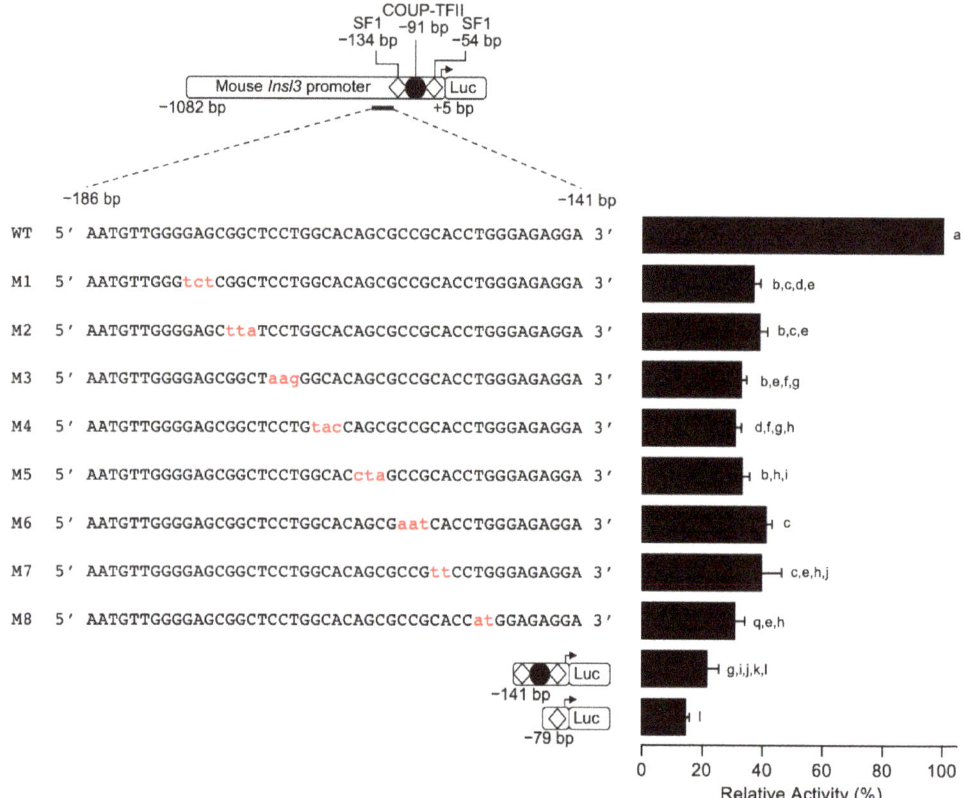

Figure 3. Fine mapping of the 35-bp region in the proximal mouse *Insl3* promoter. MA-10 Leydig cells were transfected with various −1132 bp mouse *Insl3* promoter constructs: a wild-type promoter and a series of trinucleotide or dinucleotide mutated constructs (M1 to M8; the mutations are in red lowercase). The position of previously identified binding sites for SF1 (white diamonds) and COUP-TFII (black circle) is indicated. Results are shown as % Relative Activity (± SEM) relative to the activity of the −1132 bp wild-type reporter, which was set to 100%. Different letters indicate a statistically significant difference between groups ($p < 0.05$).

2.3. Binding of Nuclear Proteins to the 35-bp Regulatory Region

The linker-scanning site-directed mutagenesis data (Figure 3) established that the full 35-bp region is important and that the activity of the *Insl3* promoter does not depend on any specific motif within this region. This suggests that binding of more than one transcription factor may be involved. A DNA–protein interaction approach (electromobility shift assay (EMSA)) was used to determine if protein(s) from Leydig cell nuclear extracts can bind to the 35-bp region. As shown in Figure 4, binding was detected (Figure 4, lane 2) that could be competed with increasing molar excess (5- and 25-fold) of unlabelled wild-type (WT) oligonucleotides (Figure 4, lanes 3 and 4). Due to the size of the band, it is possible that more than one protein binds to this region. To assess the specificity of the protein–DNA interaction, competition experiments were performed using unlabelled oligonucleotides (5× and 25× molar excess) containing the same mutations as those described in Figure 3 for promoter activity. As shown in Figure 4, oligonucleotides corresponding to mutants M1 (lanes 5 and 6), M3 (lanes 9 and 10), and M4 (lanes 11 and 12) were unable to compete the binding complex. Oligonucleotides corresponding to mutants M2 (Figure 4, lanes 7 and 8) and M5 (Figure 4, lanes 13 and 14) were as efficient as the WT oligonucleotide (Figure 4, lanes 3 and 4) at displacing the binding complex. On the other hand, oligonucleotides for mutants M6 only partially competed the binding complex (Figure 4, lanes 15 and 16). Together, these data indicate that protein(s) do bind to the 35-bp *Insl3* promoter region and that the nucleotides mutated in mutants M1, M3, M4, and to a lesser extent M6, are important for this binding.

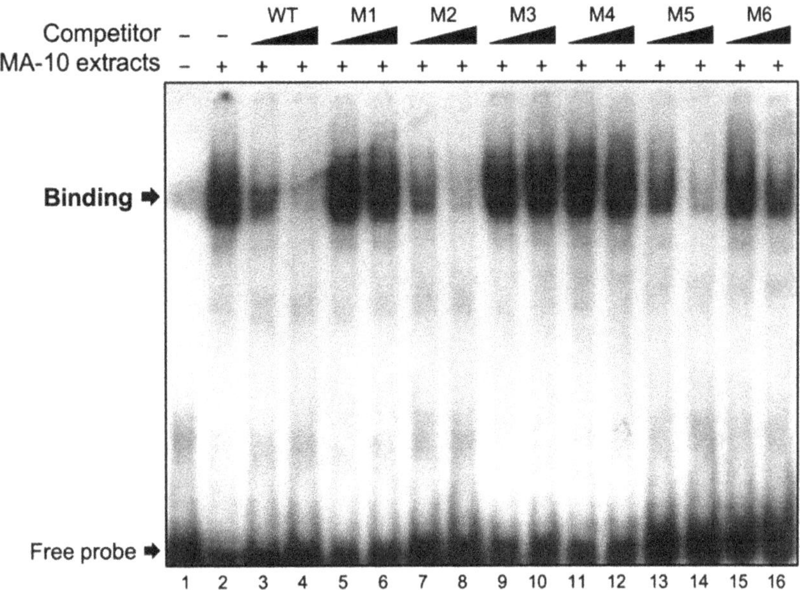

Figure 4. Nuclear proteins from MA-10 cells bind specifically to the 35-bp region. EMSA was used to determine the binding of nuclear extracts from MA-10 Leydig cells (MA-10 extracts) to a double-stranded ^{32}P-labelled oligonucleotide corresponding to the 35-bp region (-186 to -151 bp) of the *Insl3* promoter. Protein binding was challenged by increasing concentrations (black triangles; molar excesses of $5\times$ and $25\times$) of unlabelled oligonucleotides corresponding to the wild-type 35-bp region (WT) or oligonucleotides containing linker-scanning mutations (M1, M2, M3, M4, M5, M6; defined in Figure 3) in the 35-bp region.

2.4. Binding Assessment of Various Transcription Factors to the 35-bp Region

The 35-bp nucleotide sequence (5′- AAT GTT GGG GAG CGG CTC CTG GCA CAG CGC CGC AC) contains potential binding sites for zinc finger-containing transcription factors known to bind GC- and GA-rich motifs. These include RBP2 (retinoblastoma-binding protein 2), GLI (GLI-Krüppel family zinc finger), IKZF1 (IKAROS family zinc finger 1), SP1 (specificity protein 1), KLF6 (Krüppel-like factor 6), HLTF (helicase-like transcription factor), ZEB1 (zinc finger E-box binding homeobox 1), and HAND1 (heart and neural crest derivatives-expressed protein 1). We, therefore, used oligonucleotides containing a consensus binding sequence for each of these transcription factors in EMSAs. As shown in Figure 5, binding of proteins from MA-10 Leydig cell nuclear extract was detected on the 35-bp region used as a probe (Figure 5A,B, lane 2). As expected, this binding was competed with unlabelled oligonucleotides corresponding the WT sequence and M1 mutant, but not M2 mutant (Figure 5A,B, lanes 3–8). Surprisingly, oligonucleotides containing a consensus binding site for RBP2, GLI, IKZF, SP1 (Figure 5A, lanes 9–16) and KLF6, HLTF, ZEB1, and HAND1 (Figure 5B, lanes 9–16) were unable to compete the binding complex. This indicates that these transcription factors do not bind to the 35-bp region of the mouse *Insl3* promoter.

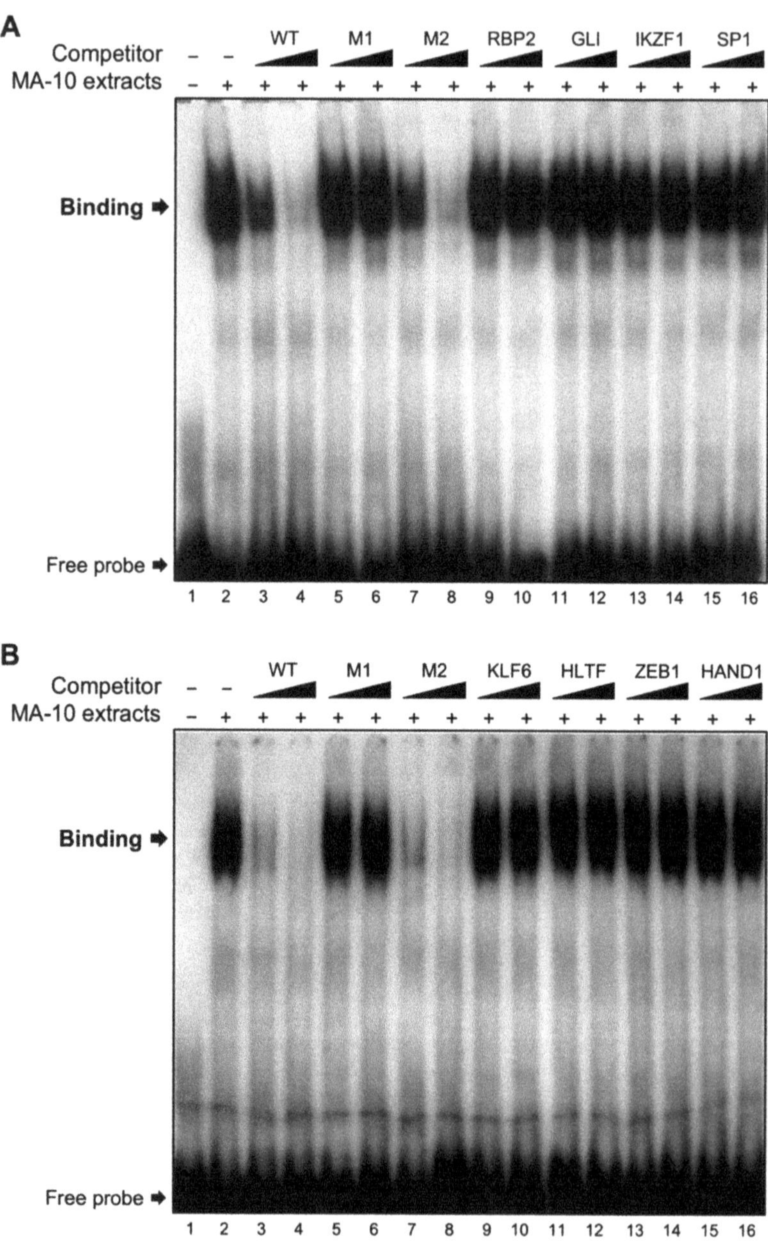

Figure 5. Assessment of potential transcription factors binding to the 35-bp region. EMSA was used to challenge the binding of nuclear protein(s) from MA-10 Leydig cells (MA-10 extracts) to a double-stranded ^{32}P-labelled oligonucleotide corresponding to the 35-bp region (-186 to -151 bp) of the *Insl3* promoter. Protein binding was challenged by increasing concentrations (black triangles; molar excesses of $5\times$ and $25\times$) of unlabelled oligonucleotides corresponding to the wild-type 35-bp region (WT), mutants M1 and M2, or oligonucleotides containing a consensus binding sequence for transcription factors (**A**) RBP2, GLI, IKZF1, SP1, and (**B**) KLF6, HLTF, ZEB1, HAND1.

2.5. NF-κB p50 Activates the Mouse Insl3 Promoter via the 35-bp Region

A closer analysis of the M1 sequence suggested that this might represent a potential binding site for NF-κB p50, or for members of the CREB or C/EBP families of transcription factors. In silico analysis of the sequences in mutants M3 and M4 predicts that they could be recognized by members of the AP1 and AP2 families. Several AP1 members are known to be present in Leydig cells, including cJUN, a well-characterized activator of several genes (reviewed in [28]). We, therefore, tested whether members of these families could activate the *Insl3* promoter in MA-10 Leydig cells. As shown in Figure 6A,B, the −1082 bp *Insl3* promoter was activated by SF1 and COUP-TFII as previously reported [13,17,19,20,25,26]. On the other hand, cJUN, CREB, and C/EBPβ failed to activate the *Insl3* promoter construct (Figure 6A,B). In the presence of NF-κB p50, a 2-fold activation of the −1082 bp *Insl3* promoter was observed (Figure 6B). Combining two transcription factors did not result in any functional cooperation (Figure 6A,B). However, in the presence of CREB or C/EBPβ, activation by the nuclear receptors SF1 and COUP-TFII (Figure 6A), and by NF-κB p50 (Figure 6B), was abolished, a phenomenon that occurs when transcription factors compete for a limited amount of common co-factors.

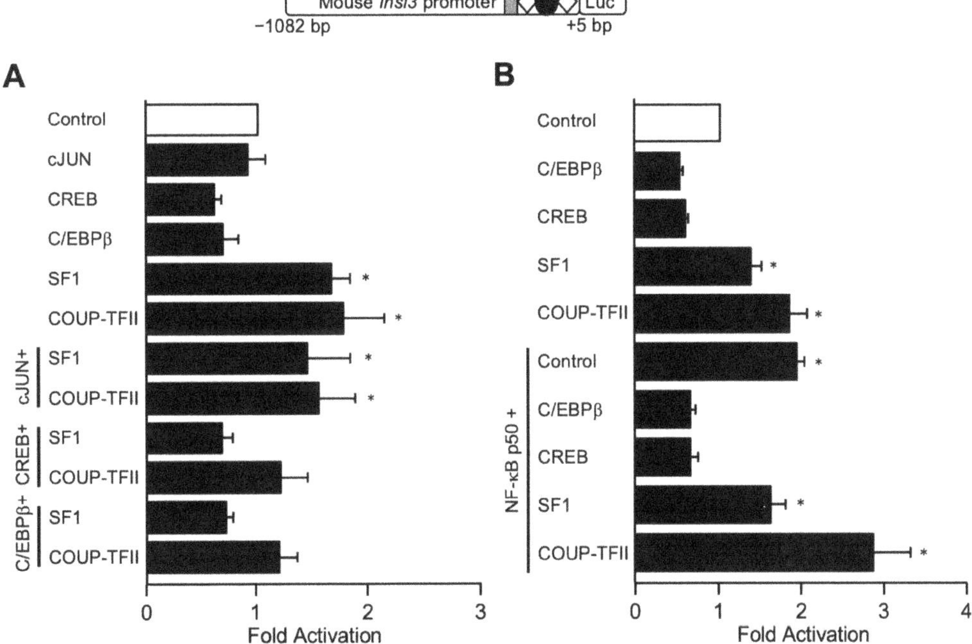

Figure 6. The mouse *Insl3* promoter is activated by NF-κB p50 but not CREB, cJUN, or C/EBPβ. MA-10 Leydig cells were transfected with a −1132 bp mouse *Insl3* promoter along with an empty expression vector (open bar) or expression vectors for cJUN, CREB, C/EBPβ, NF-κB p50, SF1, and COUP-TFII either alone or in combination as indicated in (**A,B**) (black bars). The position of previously identified binding sites for SF1 (white diamonds) and COUP-TFII (black circle) is indicated. The grey box represents the 35-bp sequence responsible for 70% of *Insl3* promoter activity. Results are shown as Fold Activation over control ± SEM. An asterisk (*) represents a statistically significant activation compared to control (empty expression vector, value set at 1, *p* < 0.05).

Next, to locate the NF-κB p50 responsive region, MA-10 Leydig cells were transfected with *Insl3* promoter deletion constructs. As shown in Figure 7, a deletion to −186 bp that retains the 35-bp region was still activated ~2-fold by NF-κB p50. However, a deletion construct to −151 bp that removes the 35-bp region was no longer activated by NF-κB p50 (Figure 7).

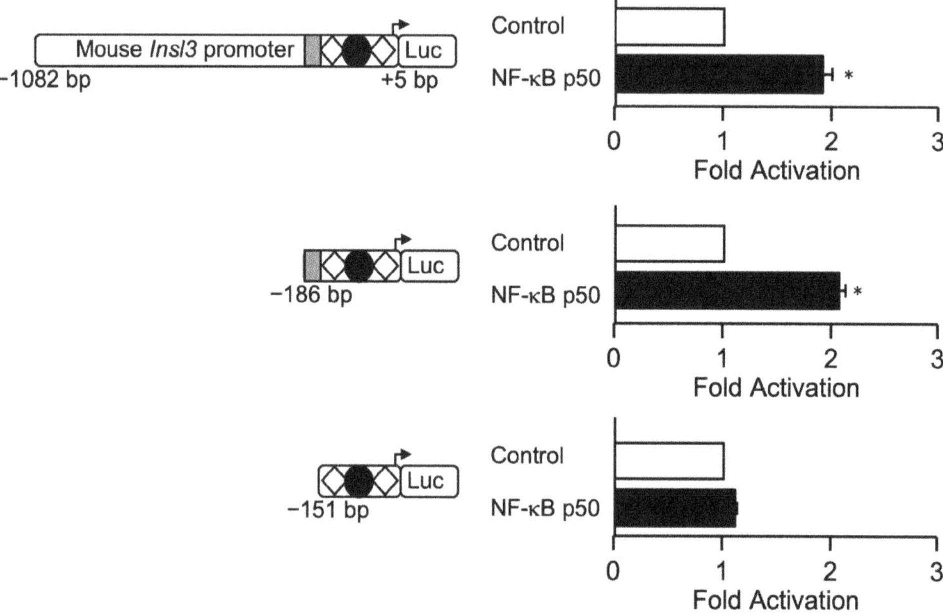

Figure 7. NF-κB p50-dependent activation of the mouse *Insl3* promoter requires the 35-bp region. MA-10 Leydig cells were transfected with three *Insl3* promoter constructs (−1132 bp, −186 bp, −151 bp) along with an empty expression vector (open bar) or an expression vector for NF-κB p50 (black bars). The position of previously identified binding sites for SF1 (white diamonds) and COUP-TFII (black circle) is indicated. The grey box represents the 35-bp sequence responsible for 70% of *Insl3* promoter activity. Results are shown as Fold Activation over control ± SEM. An asterisk (*) represents a statistically significant activation compared to control (empty expression vector, value set at 1, $p < 0.05$).

3. Discussion

The peptide hormone INSL3 produced by Leydig cells plays important roles in male reproductive development and function (reviewed in [29]). Although some transcription factors have been implicated in the regulation of *Insl3* gene expression in Leydig cells, most are also present in cells that do not express *Insl3* indicating that additional, yet unidentified transcription factors are also required to direct *Insl3* expression in Leydig cells. In the present study, we have located and characterized new regulatory elements important for *Insl3* promoter activity in Leydig cells.

In our study, we made use of the mouse MA-10 Leydig cell line [30], which are immortalized cells corresponding to immature Leydig cells from the adult population [30,31]. Contrary to primary Leydig cells, MA-10 cells proliferate and contain an aberrant chromosome number [32]. Despite these issues, MA-10 cells nonetheless respond to hormonal stimulation, like primary Leydig cells, with an increase in steroid hormone production [30,33–36], indicating that the signalling cascades, kinases, and transcription factors required for this response are present in MA-10 Leydig cells. More relevant to our current work, MA-10 Leydig cells constitutively express *Insl3* [13,19,22,23,27,37–39] and have been validated as a suitable model to study *Insl3* gene transcription [27].

3.1. Identification of an Essential 35-bp Region within the Proximal Insl3 Promoter

For the promoter functional assays, we used a genomic fragment of −1082 bp upstream of the mouse *Insl3* transcription start site. Because the entire *Insl3* gene is located within the last intron of the *Jak3* gene [12–14], the −1082 bp fragment corresponds to the longest sequence that can be used before entering the coding region of the *Jak3* gene. Our 5′ progressive deletion analysis of the mouse *Insl3* −1082 bp regulatory region revealed that deletion of up to −186 bp did not significantly affect *Insl3* promoter activity in MA-10 Leydig cells. This is similar to a previous study performed in another Leydig cell line, the MLTC-1 cells, where a construct of −188 bp retained all mouse *Insl3* promoter activity [19]. Another study also reported that a short proximal region of the mouse *Insl3* promoter contained within −157 bp was required for Leydig-specific transcription [20]. Deletion analysis of the human *INSL3* promoter also showed that truncation to −132 bp still maintained full promoter activity in both MA-10 and MLTC-1 Leydig cells but not non-steroidogenic cells [18]. Although the minimal required region appears shorter in the human promoter, the *Insl3* promoter from various rodents contains an additional ~50 bp compared to the *INSL3* promoter from primates, as revealed by the alignment of the *INSL3* promoter sequence from various species (Figure 8). Therefore, −132 bp in the human *INSL3* promoter is equivalent to −178 bp in the mouse *Insl3* promoter (Figure 8). Although the *Insl3* promoter has been isolated from rat, pig, and dog [13,16,17], a 5′ progressive promoter deletion approach to locate species-specific regulatory elements required for its activity in Leydig cells was not performed.

Through fine promoter deletions, we identified a 35-bp region located between −186 and −151 bp that is essential for maximal *Insl3* promoter activity in Leydig cells. Site-directed mutagenesis in the context of the −1082 bp *Insl3* promoter further confirmed the importance of the entire 35-bp region since any trinucleotide or dinucleotide mutation within this region reduced promoter activity. This suggests that more than one transcription factor may bind to this region. The 35-bp region contains several GC- and GA-rich boxes, some of which have been conserved across species (Figure 8). In addition, a potential E-box motif (CAnnTG) for the binding of bHLH transcription factors is also present. The significant reduction in *Insl3* promoter activity that occurred when mutations were introduced in either half-site of this palindromic E-box motif (mutations M7 and M8) strongly supports the involvement of bHLH family members. In addition, mutations that target GC/GA-rich motifs reduce the activity of both the mouse (present work) and human [18] *INSL3* promoter in Leydig cells. This prompted us to assess whether proteins known to bind to GC/GA-rich motifs could bind to the 35-bp region of the mouse *Insl3* promoter. Although a protein complex from MA-10 Leydig cells specifically bound to the 35-bp region as identified by EMSA, competition experiments using several oligonucleotides containing binding sites for various transcription factors known to bind GC/GA-rich sequences such as RBP2, GLI, IKZF1, SP1, KLF6, HLTF, ZEB1, and HAND1, failed to displace the binding. This indicates that the protein complex binding to the 35-bp region does not contain members of these transcription factor families.

3.2. Potential Transcription Factors Acting via the 35-bp Element

Our protein–DNA interaction assays indicated that the protein(s) responsible for the activity of the 35-bp region might not be zinc finger-containing transcription factors recognizing GC/GA-rich motifs. Of all the mutations introduced in the 35-bp region and tested by EMSA, mutants M1, M3, M4, and M6 failed displace the binding complex. This indicates that the integrity of these sequences is essential for protein binding, and thus, provides clues regarding the nature of the transcription factor that could bind to these sequences.

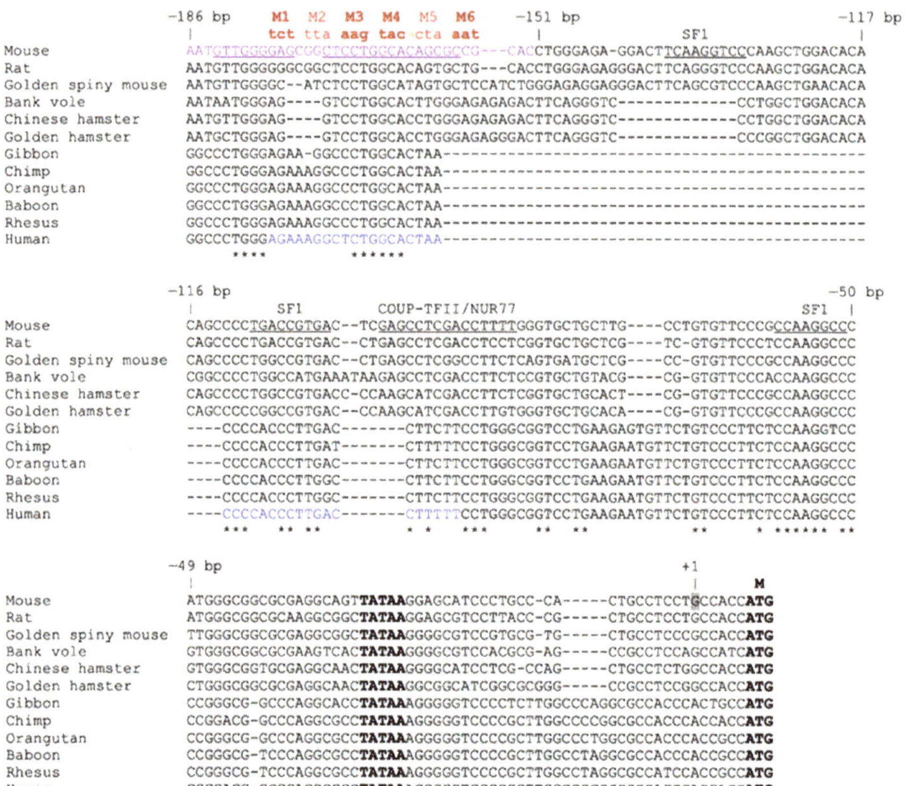

Figure 8. Alignment of the proximal *Insl3* gene promoter from different species. The sequences of the proximal *Insl3* promoter (−186 bp to the ATG of the mouse promoter) from various species indicated on the left were aligned using Clustal Omega multiple sequence alignment tool. The numbering corresponds to the position in the mouse *Insl3* promoter. The 35-bp region important for mouse *Insl3* promoter activity identified in this study is shown in purple. The mutations generated in the 35-bp region are shown in red (M1–M6); the bolded ones (M1, M3, M4, M6) were unable to compete the binding in EMSA. The underlined sequences correspond to potential binding sites for CREB, NF-κB, AP1, AP2, and C/EBP family members. The sequence shown in blue corresponds to an approximately 40-bp region (−132 and −93 bp) previously identified as important for the activity of the human *INSL3* promoter [18]. The positions of previously identified binding sites for transcription factors SF1, COUP-TFII, and NUR77 are shown. The TATA-box, transcription start site (+1), and ATG (M) are also indicated. An asterisk indicates a nucleotide conserved across all 12 species.

In silico analysis of the sequence surrounding mutant M1 identified potential binding sites for members of the CREB, C/EBP, and NF-κB families of transcription factors, which are expressed in Leydig cells where they regulate gene expression [40–45]. However, when assayed in functional promoter assays, CREB and C/EBPβ failed to activate the *Insl3* promoter, either by themselves or in combination with other transcription factors. On the other hand, activity of the −1082 bp *Insl3* promoter was increased by 2-fold in the presence of NF-κB p50, supporting a role for this transcription factor in *Insl3* gene transcription in Leydig cells. Furthermore, the NF-κB p50-dependent activation of the *Insl3* promoter required the 35-bp region as a deletion construct lacking this region was no longer activated by NF-κB p50. Additional work such as performing EMSAs and competition assays using a consensus NF-κB p50 motif, using mutated *Insl3* promoter constructs, as well as testing other potential partners, would be needed to fully understand the mechanism of NF-κB p50 action in the regulation of *Insl3* promoter activity in Leydig cells.

Another sequence we identified within the 35-bp element as important for the binding of nuclear proteins and for the activity of the mouse *Insl3* promoter in MA-10 Leydig cells was the sequence 5′-TCCTGGCACA-3′ located between −170 and −161 bp. Mutation of this sequence (mutants M3 and M4) prevented protein binding and reduced *Insl3* promoter activity by 70%. In silico analysis of this sequence predicted that it could be recognized by members of the AP1 and AP2 families. Several AP1 members are known to be expressed in Leydig cells, including cJUN, a well-characterized activator of several genes (reviewed in [28]). AP2 factors are present in Leydig cells and regulate the activity of the luteinizing hormone receptor promoter [46]. Despite the fact that cJUN failed to activate the *Insl3* promoter in our assays, we cannot exclude the possibility that other AP1 or AP2 family members might contribute to *Insl3* promoter activity. At this time, the nature of the factor binding to this sequence remains to be established. This could be determined using a DNA–protein precipitation assay where a double-stranded oligonucleotide containing the sequence of interest is biotinylated and incubated with nuclear extracts. DNA-bound proteins are then isolated using avidin beads, eluted, and analyzed by LC-MS/MS.

In conclusion, although our present work has identified a key 35-bp regulatory region, additional work is needed to fully decipher the transcription factors acting via this 35-bp element responsible for 70% of *Insl3* promoter activity in Leydig cells.

4. Materials and Methods

4.1. Plasmids

The mouse *Insl3* promoter constructs −1082 to +5 bp and −186 to +5 bp were previously described [22,25]. Mouse *Insl3* promoter deletions to −973 bp, −791 bp, −591 bp, −151 bp, −141 bp, −131 bp, −120 bp, −111 bp, and −79 bp were obtained by PCR using the −1082 bp *Insl3* promoter as a template, along the primers listed in Table 1. All promoter fragments were cloned into a modified pXP1 luciferase reporter plasmid [47,48].

Table 1. Sequence of the primers used to generate the various mouse *Insl3* promoter deletion constructs.

−1082 bp	Forward	5′- GCG GAT CCT GGT TCC TAT GAT CTG GCT G -3′
−973 bp	Forward	5′- GCG GAT CCG AAT GGG GAT ATT AAA TAT GTG -3′
−791 bp	Forward	5′- GCG GAT CCC CCT TGC TCC CCT GAC TGT G -3′
−591 bp	Forward	5′- GCG GAT CCC TGG GAG AGT AGA GGT CTT G -3′
−186 bp	Forward	5′- CGG GAT CCA ATG TTG GGG AGC GGC TCC TG -3′
−151 bp	Forward	5′- GCG GAT CCC TGG GAG AGG ACT TCA AGG T -3′
−141 bp	Forward	5′- GCG GAT CCA CTT CAA GGT CCC AAG CTG G -3′
−131 bp	Forward	5′- GCG GAT CCC CCA AGC TGG ACA CAC AGC C -3′
−120 bp	Forward	5′- GGG GAT CCA CAC AGC CCC TGA CCG TG -3′
−111 bp	Forward	5′- GCG GAT CCC CTG ACC GTG ACT CGA GCC T -3′
−79 bp	Forward	5′- GGG GAT CCT GCT GCT TGC CTG TGT TC -3′
+5 bp	Reverse	5′- GGG GTA CCG TGG CAG GAG GCA GTG GGC AG -3′

The cloning sites are underlined: BamHI for the forward primers and KpnI for the reverse primer.

Various trinucleotide and dinucleotide mutant constructs in the context of the −1082 bp reporter were generated using the QuikChange XL mutagenesis kit (Agilent Technologies Canada, Mississauga, ON, Canada), as recommended by the manufacturer, along with the oligonucleotides listed in Table 2 (only the sequence of the sense oligonucleotide is shown) where the mutations are in lowercase.

All the deletion and mutation reporter constructs were confirmed by sequencing (CHUQ Research Centre sequencing platform, Quebec City, QC, Canada). The following expression plasmids were obtained from different research groups: SF1/NR5A1 [49], cJUN [50], CREB [51], C/EBPβ [52], COUP-TFII/NR2F2 [53], and NF-κB p50 [54].

Table 2. Sequence of the oligonucleotides used to generate the various mouse *Insl3* promoter mutant constructs.

M1	5′- CTT GTT TTA AAT GTT GGG tct CGG CTC CTG GCA CAG CGC -3′
M2	5′- GTT TTA AAT GTT GGG GAG Ctt aTC CTG GCA CAG CGC CGC ACC -3′
M3	5′- AAA TGT TGG GGA GCG GCT aag GGC ACA GCG CCG CAC CTG -3′
M4	5′- GTT GGG GAG CGG CTC CTG tac CAG CGC CGC ACC TGG GAG -3′
M5	5′- GGG AGC GGC TCC TGG CAC cta GCC GCA CCT GGG AGA GGA -3′
M6	5′- GCG GCT CCT GGC ACA GCG aat CAC CTG GGA GAG GAC TTC -3′
M7	5′- CTG GCA CAG CGC CGC ACC atG GAG AGG ACT TCA AG -3′
M8	5′- CTG GCA CAG CGC CG ttC CTG GGA GAG GAC TTC AAG -3′

Only the sequence of the sense oligonucleotides is shown. Mutated nucleotides are in lowercase.

4.2. Cells Culture, Transfections, and Reporter Assays

Mouse MA-10 Leydig cells (ATCC, Manassas, VA, USA, Cat# CRL-3050, RRID:CVCL_D789) were grown in DMEM/F12 medium supplemented with 2.438 g/L sodium bicarbonate, 3.57 g/L HEPES, and 15% horse serum on gelatin-coated plates. Penicillin and streptomycin sulphate were added to the cell culture media to a final concentration of 50 mg/L, and cells were kept at 37 °C, 5% CO_2 in a humidified incubator. MA-10 Leydig cells were validated by morphology and by quantifying steroidogenic output, as previously described [55–62]. MA-10 cells were transiently transfected using polyethylenimine hydrochloride (PEI) (Sigma-Aldrich Canada, Oakville, ON, Canada), as previously described [61,63,64], or the calcium phosphate co-precipitation method, as described in [22,35,48,65]. Briefly, MA-10 cells were transfected 24 h after plating at a density of 100,000 cells/well, by using 0.5 µg of *Insl3* promoter construct fused to the Firefly luciferase reporter gene, 20 ng of phRL-TK Renilla luciferase expression vector used as an internal control for transfection efficiency, and pSP64 as carrier DNA up to 1.5 µg/well. For transactivation assays, cells were transfected with 400 ng of the mouse *Insl3* −1082/+5 bp reporter vector along with 100 ng of an empty expression vector (pcDNA3.1 as control), or expression vectors for the various transcription factors (50 ng) individually (completed to 100 ng with the empty pcDNA3.1 expression vector to keep the total amount of expression vector to 100 ng), or the combination of transcription factors (50 ng each). The culture media were renewed 3 h before and 16 h after the transfection. Two days after the transfection, MA-10 Leydig cells were harvested and luciferase activities were measured using the Dual Luciferase Assay System (Promega Corp, Madison, WI, USA) and the Luminoskan Ascent luminometer (Thermo Scientific, Milford, MA, USA). The data reported represent the average of at least four experiments, using different DNA preparations, and each performed in duplicate.

4.3. Preparation of Nuclear Extracts

Nuclear extracts from MA-10 Leydig cells were prepared using the method described by Schreiber [66], with the following modifications: cells were rinsed with PBS-EDTA, harvested and pelleted by centrifugation for 45 s at 9000 RPM at 4 °C. The cells were resuspended in 700 µL of buffer A (HEPES 10 mM, EDTA 0.1 mM, EGTA 0.1 mM, DTT 1 mM, PMSF 0.5 mM, aprotinin 5 pg/mL, pepstatin 5 µg/mL, leupeptin 5 µg/mL) and incubated on ice for 15 min. Next, 50 µL of 10% Igepal was added and mixed vigorously for 10 s, followed by centrifugation for 30 sec at 13,000 RPM at 4 °C to pellet the nuclei. Nuclei were then incubated in 50 µL of buffer B (HEPES 20 mM, NaCl 400 mM, EDTA 1 mM, EGTA 1 mM, DTT 1 mM, PMSF 1 mM, aprotinin 5 µg/mL, pepstatin 5 pg/mL, leupeptin 5 µg/mL) with vigorous shaking for 45 min at 4 °C, followed by centrifugation for 5 min at 13,000 RPM at 4 °C. Protein concentrations were estimated using the standard Bradford assay (Bio-Rad Laboratories, Hercules, CA, USA). Nuclear proteins were stored at −80 °C until needed.

4.4. Electromobility Shift Assays

Electromobility shift assays (EMSAs) were performed as previously described [18,22,43,61,67–71]. Briefly, 5 µg of nuclear extracts from MA-10 Leydig cells were incubated in 20 µL of 4 mM

Tris-HCl (pH 8.0), 24 mM KCl, 0.4 mM EDTA (pH 8.0), 0.4 mM dithiothreitol, 5 mM $MgCl_2$, 100 ng BSA, 10% glycerol, and 500 ng poly(dI-dC) for 1 h on ice. A ^{32}P-labeled 42-bp double-stranded oligonucleotide containing the 35 bp ($-186/-151$ bp) region of the mouse *Insl3* promoter was used as a probe (5′- GTT GGG GAG CGG CTC CTG GCA CAG CGC CGC ACC TGG GAG AGG -3′). Competition experiments were performed using 5× and 25× (molar excess) of unlabeled double-stranded oligonucleotides corresponding to the probe or harboring various mutations (shown in lowercase) in the 35-bp ($-186/-151$ bp) region of the mouse *Insl3* promoter (Table 2). Competitions were also performed using oligonucleotides corresponding to consensus binding sites for different transcription factors (Table 3).

Table 3. Sequence of oligonucleotides corresponding to consensus binding sites for different transcription factors used in EMSA assays.

Transcription Factor	Sense Oligonucleotide	Reference
RBP2	5′- GGG CTC CCG CCC CAC GAA AAG -3′	[72]
GLI	5′- CGT CTT GGG TGG TCC ACG -3′	[73]
IKZF1	5′- TCA GCT TTT GGG AAT GTA TTC CCT GTCA -3′	[74]
SP1	5′- CGG CGC AGG GCG GGG CGG GGC GAG -3′	[75]
KLF6	5′- CCG AGG CCA CAC CCT ACT CTC TGA TAG TTC -3′	[76]
HLTF	5′- TTG ATT GAC ATA TA C CAG GAG ATA GA -3′	[77]
ZEB1	5′- GTG CAC AGT GCA AAG GTG GGG CGG CAG -3′	[78]
HAND1	5′- CAA CCA CAA TGG CGT CGT CTG GCA TTT TT -3′	[79]

Only the sequence of sense oligonucleotide is shown.

4.5. Sequence Analysis

To identify binding sites for potential transcription factors, the 35-bp region of the mouse *Insl3* promoter was analyzed using the bioinformatic tools TFbind (https://tfbind. hgc.jp/, last accessed 10 October 2022) [80] and PROMO (PROMO version 3.0.2 using version 8.3 of TRANSFAC, last accessed 10 October 2022, http://alggen.lsi.upc.es/cgi-bin/ promo_v3/promo/promoinit.cgi?dirDB=TF_8.3) [81,82]. Multiple sequence alignment was performed using the CLUSTAL Omega multiple sequence alignment tool (version 1.2.4, last accessed 12 October 2022, https://www.ebi.ac.uk/Tools/msa/clustalo/) [83].

4.6. Statistical Analysis

Statistical analyses were carried out using Statistics Calculators (Statistics Kingdom, Melbourne, Australia, November 2017, https://www.statskingdom.com/kruskal-wallis-calculator.html, accessed on 7 October 2022). To identify significant differences between multiple groups, statistical analyses were carried out using a nonparametric Kruskal–Wallis one-way ANOVA on ranks followed by a Mann–Whitney U test to detect differences between pairs. For all statistical analyses, $p < 0.05$ was considered significant.

Author Contributions: X.C.G. performed the majority of the experiments along with K.J.P. and N.M.R. X.C.G., K.J.P. and J.J.T. analyzed and interpreted the data. J.J.T. conceived the study, coordinated and supervised the project, and wrote the manuscript. All authors have read and agreed to the published version of the manuscript.

Funding: This research was funded by a grant from the Canadian Institutes of Health Research (CIHR) (funding reference number MOP-81387) to JJT.

Institutional Review Board Statement: Not applicable.

Informed Consent Statement: Not applicable.

Data Availability Statement: All data generated or analyzed during this study are included in this article.

Acknowledgments: We would like to thank Dany Chalbos (cJUN), Marc Montminy (CREB), Ming-Jer Tsai (COUP-TFII), Keith Parker (SF1), Steven McKnight (C/EBPβ), and Richard Pope (NF-κB p50) for generously providing plasmids used in this study.

Conflicts of Interest: The authors declare no conflict of interest.

References

1. Forest, M.G. Role of androgens in fetal and pubertal development. *Horm. Res.* **1983**, *18*, 69–83. [CrossRef]
2. Nef, S.; Parada, L.F. Cryptorchidism in mice mutant for Insl3. *Nat. Genet.* **1999**, *22*, 295–299. [CrossRef] [PubMed]
3. Zimmermann, S.; Steding, G.; Emmen, J.M.; Brinkmann, A.O.; Nayernia, K.; Holstein, A.F.; Engel, W.; Adham, I.M. Targeted disruption of the Insl3 gene causes bilateral cryptorchidism. *Mol. Endocrinol.* **1999**, *13*, 681–691. [CrossRef] [PubMed]
4. Ferlin, A.; Pepe, A.; Gianesello, L.; Garolla, A.; Feng, S.; Giannini, S.; Zaccolo, M.; Facciolli, A.; Morello, R.; Agoulnik, A.I.; et al. Mutations in the insulin-like factor 3 receptor are associated with osteoporosis. *J. Bone Miner. Res.* **2008**, *23*, 683–693. [CrossRef]
5. Kawamura, K.; Kumagai, J.; Sudo, S.; Chun, S.Y.; Pisarska, M.; Morita, H.; Toppari, J.; Fu, P.; Wade, J.D.; Bathgate, R.A.; et al. Paracrine regulation of mammalian oocyte maturation and male germ cell survival. *Proc. Natl. Acad. Sci. USA* **2004**, *101*, 7323–7328. [CrossRef]
6. Minagawa, I.; Murata, Y.; Terada, K.; Shibata, M.; Park, E.Y.; Sasada, H.; Kohsaka, T. Evidence for the role of INSL3 on sperm production in boars by passive immunisation. *Andrologia* **2018**, *50*, e13010. [CrossRef]
7. Ferlin, A.; De Toni, L.; Agoulnik, A.I.; Lunardon, G.; Armani, A.; Bortolanza, S.; Blaauw, B.; Sandri, M.; Foresta, C. Protective role of testicular hormone INSL3 from atrophy and weakness in skeletal muscle. *Front. Endocrinol.* **2018**, *9*, 562. [CrossRef]
8. Sedaghat, K.; Shen, P.J.; Finkelstein, D.I.; Henderson, J.M.; Gundlach, A.L. Leucine-rich repeat-containing G-protein-coupled receptor 8 in the rat brain: Enrichment in thalamic neurons and their efferent projections. *Neuroscience* **2008**, *156*, 319–333. [CrossRef]
9. Hampel, U.; Klonisch, T.; Sel, S.; Schulze, U.; Garreis, F.; Seitmann, H.; Zouboulis, C.C.; Paulsen, F.P. Insulin-like factor 3 promotes wound healing at the ocular surface. *Endocrinology* **2013**, *154*, 2034–2045. [CrossRef]
10. Foresta, C.; Bettella, A.; Vinanzi, C.; Dabrilli, P.; Meriggiola, M.C.; Garolla, A.; Ferlin, A. A novel circulating hormone of testis origin in humans. *J. Clin. Endocrinol. Metab.* **2004**, *89*, 5952–5958. [CrossRef]
11. Ivell, R.; Wade, J.D.; Anand-Ivell, R. INSL3 as a biomarker of Leydig cell functionality. *Biol. Reprod.* **2013**, *88*, 147. [CrossRef] [PubMed]
12. Koskimies, P.; Spiess, A.N.; Lahti, P.; Huhtaniemi, I.; Ivell, R. The mouse relaxin-like factor gene and its promoter are located within the 3′ region of the JAK3 genomic sequence. *FEBS Lett.* **1997**, *419*, 186–190. [CrossRef]
13. Sadeghian, H.; Anand-Ivell, R.; Balvers, M.; Relan, V.; Ivell, R. Constitutive regulation of the Insl3 gene in rat Leydig cells. *Mol. Cell Endocrinol.* **2005**, *241*, 10–20. [CrossRef] [PubMed]
14. Spiess, A.N.; Balvers, M.; Tena-Sempere, M.; Huhtaniemi, I.; Parry, L.; Ivell, R. Structure and expression of the rat relaxin-like factor (RLF) gene. *Mol. Reprod. Dev.* **1999**, *54*, 319–325. [CrossRef]
15. Zimmermann, S.; Schottler, P.; Engel, W.; Adham, I.M. Mouse Leydig insulin-like (Ley I-L) gene: Structure and expression during testis and ovary development. *Mol. Reprod. Dev.* **1997**, *47*, 30–38. [CrossRef]
16. Burkhardt, E.; Adham, I.M.; Brosig, B.; Gastmann, A.; Mattei, M.G.; Engel, W. Structural organization of the porcine and human genes coding for a Leydig cell-specific insulin-like peptide (LEY I-L) and chromosomal localization of the human gene (INSL3). *Genomics* **1994**, *20*, 13–19. [CrossRef]
17. Truong, A.; Bogatcheva, N.V.; Schelling, C.; Dolf, G.; Agoulnik, A.I. Isolation and expression analysis of the canine insulin-like factor 3 gene. *Biol. Reprod.* **2003**, *69*, 1658–1664. [CrossRef]
18. Tremblay, M.A.; Mendoza-Villarroel, R.E.; Robert, N.M.; Bergeron, F.; Tremblay, J.J. KLF6 cooperates with NUR77 and SF1 to activate the human INSL3 promoter in mouse MA-10 leydig cells. *J. Mol. Endocrinol.* **2016**, *56*, 163–173. [CrossRef]
19. Koskimies, P.; Levallet, J.; Sipila, P.; Huhtaniemi, I.; Poutanen, M. Murine relaxin-like factor promoter: Functional characterization and regulation by transcription factors steroidogenic factor 1 and DAX-1. *Endocrinology* **2002**, *143*, 909–919. [CrossRef]
20. Zimmermann, S.; Schwarzler, A.; Buth, S.; Engel, W.; Adham, I.M. Transcription of the Leydig insulin-like gene is mediated by steroidogenic factor-1. *Mol. Endocrinol.* **1998**, *12*, 706–713. [CrossRef]
21. Tremblay, J.J.; Robert, N.M. Role of nuclear receptors in *INSL3* gene transcription in Leydig cells. *Ann. N. Y. Acad. Sci.* **2005**, *1061*, 183–189. [CrossRef] [PubMed]
22. Robert, N.M.; Martin, L.J.; Tremblay, J.J. The orphan nuclear receptor NR4A1 regulates *Insulin-like 3* gene transcription in Leydig cells. *Biol. Reprod.* **2006**, *74*, 322–330. [CrossRef] [PubMed]
23. Laguë, E.; Tremblay, J.J. Antagonistic effects of testosterone and the endocrine disruptor mono-(2-ethylhexyl) phthalate on *INSL3* transcription in Leydig cells. *Endocrinology* **2008**, *149*, 4688–4694. [CrossRef]
24. Tremblay, J.J.; Robert, N.M.; Laguë, E. Nuclear receptors, testosterone and post-translational modifications in human *INSL3* promoter activity in testicular Leydig cells. *Ann. N. Y. Acad. Sci.* **2009**, *1160*, 205–212. [CrossRef]
25. Mendoza-Villarroel, R.E.; Di-Luoffo, M.; Camire, E.; Giner, X.C.; Brousseau, C.; Tremblay, J.J. The *Insl3* gene is a direct target for the orphan nuclear receptor, COUP-TFII, in Leydig cells. *J. Mol. Endocrinol.* **2014**, *53*, 43–55. [CrossRef]
26. Di-Luoffo, M.; Pierre, K.J.; Robert, N.M.; Girard, M.J.; Tremblay, J.J. The nuclear receptors SF1 and COUP-TFII cooperate on the Insl3 promoter in Leydig cells. *Reproduction* **2022**, *164*, 31–40. [CrossRef] [PubMed]
27. Strong, M.E.; Burd, M.A.; Peterson, D.G. Evaluation of the MA-10 cell line as a model of insl3 regulation and Leydig cell function. *Anim. Reprod. Sci.* **2019**, *208*, 106116. [CrossRef]

28. Nguyen, H.T.; Najih, M.; Martin, L.J. The AP-1 family of transcription factors are important regulators of gene expression within Leydig cells. *Endocrine* **2021**, *74*, 498–507. [CrossRef]

29. Esteban-Lopez, M.; Agoulnik, A.I. Diverse functions of insulin-like 3 peptide. *J. Endocrinol.* **2020**, *247*, R1–R12. [CrossRef]

30. Ascoli, M. Characterization of several clonal lines of cultured Leydig tumor cells: Gonadotropin receptors and steroidogenic responses. *Endocrinology* **1981**, *108*, 88–95. [CrossRef] [PubMed]

31. Trbovich, A.M.; Martinelle, N.; O'Neill, F.H.; Pearson, E.J.; Donahoe, P.K.; Sluss, P.M.; Teixeira, J. Steroidogenic activities in MA-10 Leydig cells are differentially altered by cAMP and Mullerian inhibiting substance. *J. Steroid Biochem. Mol. Biol.* **2004**, *92*, 199–208. [CrossRef]

32. Ascoli, M. Immortalized Leydig cell lines as models for studying Leydig cell physiology. In *The Leydig Cell in Health and Disease*, 2nd ed.; Payne, A.H.H., Hardy, M.P., Eds.; Humana Press: Totowa, NJ, USA, 2007; pp. 373–381. [CrossRef]

33. Clark, B.J.; Wells, J.; King, S.R.; Stocco, D.M. The purification, cloning, and expression of a novel luteinizing hormone-induced mitochondrial protein in MA-10 mouse Leydig tumor cells. Characterization of the steroidogenic acute regulatory protein (StAR). *J. Biol. Chem.* **1994**, *269*, 28314–28322. [CrossRef]

34. King, S.R.; Ronen-Fuhrmann, T.; Timberg, R.; Clark, B.J.; Orly, J.; Stocco, D.M. Steroid production after in vitro transcription, translation, and mitochondrial processing of protein products of complementary deoxyribonucleic acid for steroidogenic acute regulatory protein. *Endocrinology* **1995**, *136*, 5165–5176. [CrossRef] [PubMed]

35. Martin, L.J.; Boucher, N.; Brousseau, C.; Tremblay, J.J. The orphan nuclear receptor NUR77 regulates hormone-induced StAR transcription in Leydig cells through a cooperation with CaMKI. *Mol. Endocrinol.* **2008**, *22*, 2021–2037. [CrossRef]

36. Engeli, R.T.; Furstenberger, C.; Kratschmar, D.V.; Odermatt, A. Currently available murine Leydig cell lines can be applied to study early steps of steroidogenesis but not testosterone synthesis. *Heliyon* **2018**, *4*, e00527. [CrossRef] [PubMed]

37. Balvers, M.; Spiess, A.N.; Domagalski, R.; Hunt, N.; Kilic, E.; Mukhopadhyay, A.K.; Hanks, E.; Charlton, H.M.; Ivell, R. Relaxin-like factor expression as a marker of differentiation in the mouse testis and ovary. *Endocrinology* **1998**, *139*, 2960–2970. [CrossRef] [PubMed]

38. Li, Y.; Kobayashi, K.; Murayama, K.; Kawahara, K.; Shima, Y.; Suzuki, A.; Tani, K.; Takahashi, A. FEAT enhances INSL3 expression in testicular Leydig cells. *Genes Cells* **2018**, *23*, 952–962. [CrossRef] [PubMed]

39. Laguë, E.; Tremblay, J.J. Estradiol represses *Insulin-like 3* expression and promoter activity in MA-10 Leydig cells. *Toxicology* **2009**, *258*, 101–105. [CrossRef]

40. Clem, B.F.; Hudson, E.A.; Clark, B.J. Cyclic adenosine 3′,5′-monophosphate (cAMP) enhances cAMP-responsive element binding (CREB) protein phosphorylation and phospho-CREB interaction with the mouse steroidogenic acute regulatory protein gene promoter. *Endocrinology* **2005**, *146*, 1348–1356. [CrossRef]

41. Manna, P.R.; Dyson, M.T.; Eubank, D.W.; Clark, B.J.; Lalli, E.; Sassone-Corsi, P.; Zeleznik, A.J.; Stocco, D.M. Regulation of steroidogenesis and the steroidogenic acute regulatory protein by a member of the cAMP response-element binding protein family. *Mol. Endocrinol.* **2002**, *16*, 184–199. [CrossRef]

42. Manna, P.R.; Eubank, D.W.; Lalli, E.; Sassone-Corsi, P.; Stocco, D.M. Transcriptional regulation of the mouse steroidogenic acute regulatory protein gene by the cAMP response-element binding protein and steroidogenic factor 1. *J. Mol. Endocrinol.* **2003**, *30*, 381–397. [CrossRef] [PubMed]

43. El-Asmar, B.; Giner, X.C.; Tremblay, J.J. Transcriptional cooperation between NF-κB and C/EBPβ regulates *Nur77* transcription in Leydig cells. *J. Mol. Endocrinol.* **2009**, *142*, 131–138.

44. Nalbant, D.; Williams, S.C.; Stocco, D.M.; Khan, S.A. Luteinizing hormone-dependent gene regulation in Leydig cells may be mediated by CCAAT/enhancer-binding protein-beta. *Endocrinology* **1998**, *139*, 272–279. [CrossRef]

45. Reinhart, A.J.; Williams, S.C.; Clark, B.J.; Stocco, D.M. SF-1 (steroidogenic factor-1) and C/EBP beta (CCAAT/enhancer binding protein-beta) cooperate to regulate the murine StAR (steroidogenic acute regulatory) promoter. *Mol. Endocrinol.* **1999**, *13*, 729–741.

46. Tsai-Morris, C.H.; Geng, Y.; Xie, X.Z.; Buczko, E.; Dufau, M.L. Transcriptional protein binding domains governing basal expression of the rat luteinizing hormone receptor gene. *J. Biol. Chem.* **1994**, *269*, 15868–15875. [CrossRef]

47. Tremblay, J.J.; Viger, R.S. Transcription factor GATA-4 enhances Müllerian inhibiting substance gene transcription through a direct interaction with the nuclear receptor SF-1. *Mol. Endocrinol.* **1999**, *13*, 1388–1401. [CrossRef]

48. Tremblay, J.J.; Viger, R.S. GATA factors differentially activate multiple gonadal promoters through conserved GATA regulatory elements. *Endocrinology* **2001**, *142*, 977–986. [CrossRef]

49. Lala, D.S.; Rice, D.A.; Parker, K.L. Steroidogenic factor I, a key regulator of steroidogenic enzyme expression, is the mouse homolog of fushi tarazu-factor I. *Mol. Endocrinol.* **1992**, *6*, 1249–1258. [CrossRef] [PubMed]

50. Teyssier, C.; Belguise, K.; Galtier, F.; Chalbos, D. Characterization of the physical interaction between estrogen receptor alpha and JUN proteins. *J. Biol. Chem.* **2001**, *276*, 36361–36369. [CrossRef]

51. Mayr, B.; Montminy, M. Transcriptional regulation by the phosphorylation-dependent factor CREB. *Nat. Rev. Mol. Cell Biol.* **2001**, *2*, 599–609. [CrossRef] [PubMed]

52. Cao, Z.; Umek, R.M.; McKnight, S.L. Regulated expression of three C/EBP isoforms during adipose conversion of 3T3-L1 cells. *Genes Dev.* **1991**, *5*, 1538–1552. [CrossRef] [PubMed]

53. Pereira, F.A.; Qiu, Y.; Zhou, G.; Tsai, M.J.; Tsai, S.Y. The orphan nuclear receptor COUP-TFII is required for angiogenesis and heart development. *Genes Dev* **1999**, *13*, 1037–1049. [CrossRef] [PubMed]

54. Liu, H.; Sidiropoulos, P.; Song, G.; Pagliari, L.J.; Birrer, M.J.; Stein, B.; Anrather, J.; Pope, R.M. TNF-alpha gene expression in macrophages: Regulation by NF-kappa B is independent of c-Jun or C/EBP beta. *J. Immunol.* **2000**, *164*, 4277–4285. [CrossRef]
55. Abdou, H.S.; Bergeron, F.; Tremblay, J.J. A cell-autonomous molecular cascade initiated by AMP-activated protein kinase represses steroidogenesis. *Mol. Cell Biol.* **2014**, *34*, 4257–4271. [CrossRef] [PubMed]
56. Abdou, H.S.; Villeneuve, G.; Tremblay, J.J. The calcium signaling pathway regulates leydig cell steroidogenesis through a transcriptional cascade involving the nuclear receptor NR4A1 and the steroidogenic acute regulatory protein. *Endocrinology* **2013**, *154*, 511–520. [CrossRef] [PubMed]
57. Daems, C.; Di-Luoffo, M.; Paradis, E.; Tremblay, J.J. MEF2 cooperates with forskolin/cAMP and GATA4 to regulate Star gene expression in mouse MA-10 Leydig cells. *Endocrinology* **2015**, *156*, 2693–2703. [CrossRef]
58. Enangue Njembele, A.N.; Bailey, J.L.; Tremblay, J.J. In vitro exposure of Leydig cells to an environmentally relevant mixture of organochlorines represses early steps of steroidogenesis. *Biol. Reprod.* **2014**, *90*, 118. [CrossRef]
59. Enangue Njembele, A.N.; Demmouche, Z.B.; Bailey, J.L.; Tremblay, J.J. Mechanism of action of an environmentally relevant organochlorine mixture in repressing steroid hormone biosynthesis in Leydig cells. *Int. J. Mol. Sci.* **2022**, *23*, 3997. [CrossRef]
60. Enangue Njembele, A.N.; Tremblay, J.J. Mechanisms of MEHP Inhibitory Action and Analysis of Potential Replacement Plasticizers on Leydig Cell Steroidogenesis. *Int. J. Mol. Sci.* **2021**, *22*, 11456. [CrossRef]
61. Hebert-Mercier, P.O.; Bergeron, F.; Robert, N.M.; Mehanovic, S.; Pierre, K.J.; Mendoza-Villarroel, R.E.; de Mattos, K.; Brousseau, C.; Tremblay, J.J. Growth hormone-induced STAT5B regulates Star gene expression through a cooperation with cJUN in mouse MA-10 Leydig cells. *Endocrinology* **2022**, *163*, 1–13. [CrossRef]
62. Mendoza-Villarroel, R.E.; Robert, N.M.; Martin, L.J.; Brousseau, C.; Tremblay, J.J. The nuclear receptor NR2F2 activates *Star* expression and steroidogenesis in mouse MA-10 and MLTC-1 Leydig cells. *Biol. Reprod.* **2014**, *91*, 26. [CrossRef] [PubMed]
63. Pierre, K.J.; Tremblay, J.J. Differential response of transcription factors to activated kinases in steroidogenic and non-steroidogenic cells. *Int. J. Mol. Sci.* **2022**, *23*, 13153. [CrossRef] [PubMed]
64. Mehanovic, S.; Mendoza-Villarroel, R.E.; Viger, R.S.; Tremblay, J.J. The nuclear receptor COUP-TFII regulates *Amhr2* gene transcription via a GC-rich promoter element in mouse Leydig cells. *J. Endocr. Soc.* **2019**, *3*, 2236–2257. [CrossRef] [PubMed]
65. Martin, L.J.; Boucher, N.; El-Asmar, B.; Tremblay, J.J. cAMP-induced expression of the orphan nuclear receptor Nur77 in testicular Leydig cells involves a CaMKI pathway. *J. Androl.* **2009**, *30*, 134–145. [CrossRef]
66. Schreiber, E.; Matthias, P.; Muller, M.M.; Schaffner, W. Rapid detection of octamer binding proteins with 'mini-extracts', prepared from a small number of cells. *Nucleic Acids Res.* **1989**, *17*, 6419. [CrossRef]
67. Bergeron, F.; Bagu, E.T.; Tremblay, J.J. Transcription of platelet-derived growth factor receptor alpha in Leydig cells involves specificity protein 1 and 3. *J. Mol. Endocrinol.* **2011**, *46*, 125–138. [CrossRef]
68. Daems, C.; Martin, L.J.; Brousseau, C.; Tremblay, J.J. MEF2 is restricted to the male gonad and regulates expression of the orphan nuclear receptor NR4A1. *Mol. Endocrinol.* **2014**, *28*, 886–898. [CrossRef]
69. Di-Luoffo, M.; Daems, C.; Bergeron, F.; Tremblay, J.J. Novel targets for the transcription factors MEF2 in MA-10 Leydig cells. *Biol. Reprod.* **2015**, *93*, 9. [CrossRef]
70. Garon, G.; Bergeron, F.; Brousseau, C.; Robert, N.M.; Tremblay, J.J. FOXA3 is expressed in multiple cell lineages in the mouse testis and regulates *Pdgfra* expression in Leydig cells. *Endocrinology* **2017**, *158*, 1886–1897. [CrossRef]
71. Martin, L.J.; Tremblay, J.J. The human 3β-hydroxysteroid dehydrogenase/Δ^5-Δ^4 isomerase type 2 promoter is a novel target for the immediate early orphan nuclear receptor NUR77 in steroidogenic cells. *Endocrinology* **2005**, *146*, 861–869. [CrossRef]
72. Tu, S.; Teng, Y.C.; Yuan, C.; Wu, Y.T.; Chan, M.Y.; Cheng, A.N.; Lin, P.H.; Juan, L.J.; Tsai, M.D. The ARID domain of the H3K4 demethylase RBP2 binds to a DNA CCGCCC motif. *Nat. Struct. Mol. Biol.* **2008**, *15*, 419–421. [CrossRef]
73. Persengiev, S.P.; Kondova, I.I.; Millette, C.F.; Kilpatrick, D.L. GLI family members are differentially expressed during the mitotic phase of spermatogenesis. *Oncogene* **1997**, *14*, 2259–2264. [CrossRef] [PubMed]
74. Ström, L.; Lundgren, M.; Severinson, E. Binding of Ikaros to germline Ig heavy chain gamma1 and epsilon promoters. *Mol. Immunol.* **2003**, *39*, 771–782. [CrossRef] [PubMed]
75. Giatzakis, C.; Papadopoulos, V. Differential utilization of the promoter of peripheral-type benzodiazepine receptor by steroidogenic versus nonsteroidogenic cell lines and the role of Sp1 and Sp3 in the regulation of basal activity. *Endocrinology* **2004**, *145*, 1113–1123. [CrossRef] [PubMed]
76. Yasuda, K.; Hirayoshi, K.; Hirata, H.; Kubota, H.; Hosokawa, N.; Nagata, K. The Kruppel-like factor ZF9 and proteins in the SP1 family regulate the expression of HSP47, a collagen-specific molecular chaperone. *J. Biol. Chem.* **2002**, *277*, 44613–44622. [CrossRef] [PubMed]
77. Hewetson, A.; Hendrix, E.C.; Mansharamani, M.; Lee, V.H.; Chilton, B.S. Identification of the RUSH consensus-binding site by cyclic amplification and selection of targets: Demonstration that RUSH mediates the ability of prolactin to augment progesterone-dependent gene expression. *Mol. Endocrinol.* **2002**, *16*, 2101–2112. [CrossRef]
78. Jethanandani, P.; Kramer, R.H. Alpha7 integrin expression is negatively regulated by deltaEF1 during skeletal myogenesis. *J. Biol. Chem.* **2005**, *280*, 36037–36046. [CrossRef]
79. Hill, A.A.; Riley, P.R. Differential regulation of Hand1 homodimer and Hand1-E12 heterodimer activity by the cofactor FHL2. *Mol. Cell. Biol.* **2004**, *24*, 9835–9847. [CrossRef]
80. Tsunoda, T.; Takagi, T. Estimating transcription factor bindability on DNA. *Bioinformatics* **1999**, *15*, 622–630. [CrossRef]

81. Farre, D.; Roset, R.; Huerta, M.; Adsuara, J.E.; Rosello, L.; Alba, M.M.; Messeguer, X. Identification of patterns in biological sequences at the ALGGEN server: PROMO and MALGEN. *Nucleic Acids Res.* **2003**, *31*, 3651–3653. [CrossRef]
82. Messeguer, X.; Escudero, R.; Farre, D.; Nunez, O.; Martinez, J.; Alba, M.M. PROMO: Detection of known transcription regulatory elements using species-tailored searches. *Bioinformatics* **2002**, *18*, 333–334. [CrossRef] [PubMed]
83. Madeira, F.; Pearce, M.; Tivey, A.R.N.; Basutkar, P.; Lee, J.; Edbali, O.; Madhusoodanan, N.; Kolesnikov, A.; Lopez, R. Search and sequence analysis tools services from EMBL-EBI in 2022. *Nucleic Acids Res.* **2022**, *50*, W276–W279. [CrossRef] [PubMed]

International Journal of
Molecular Sciences

Article

Identification of the Role of TGR5 in the Regulation of Leydig Cell Homeostasis

Hélène Holota [†], Angélique De Haze [†], Emmanuelle Martinot, Melusine Monrose, Jean-Paul Saru, Françoise Caira, Claude Beaudoin * and David H. Volle *

INSERM U1103, CNRS UMR-6293, Université Clermont Auvergne, GReD Institute, Team-Volle, F-63001 Clermont-Ferrand, France
* Correspondence: claude.beaudoin@uca.fr (C.B.); david.volle@inserm.fr (D.H.V.);
  Tel.: +33-473407415 (C.B. & D.H.V.)
† These authors contributed equally to this work.

Abstract: Understanding the regulation of the testicular endocrine function leading to testosterone production is a major objective as the alteration of endocrine function is associated with the development of many diseases such as infertility. In the last decades, it has been demonstrated that several endogenous molecules regulate the steroidogenic pathway. Among them, bile acids have recently emerged as local regulators of testicular physiology and particularly endocrine function. Bile acids act through the nuclear receptor FXRα (Farnesoid-X-receptor alpha; NR1H4) and the G-protein-coupled bile acid receptor (GPBAR-1; TGR5). While FXRα has been demonstrated to regulate testosterone synthesis within Leydig cells, no data are available regarding TGR5. Here, we investigated the potential role of TGR5 within Leydig cells using cell culture approaches combined with pharmacological exposure to the TGR5 agonist INT-777. The data show that activation of TGR5 results in a decrease in testosterone levels. TGR5 acts through the PKA pathway to regulate steroidogenesis. In addition, our data show that TGR5 activation leads to an increase in cholesterol ester levels. This suggests that altered lipid homeostasis may be a mechanism explaining the TGR5-induced decrease in testosterone levels. In conclusion, the present work highlights the impact of the TGR5 signaling pathway on testosterone production and reinforces the links between bile acid signaling pathways and the testicular endocrine function. The testicular bile acid pathways need to be further explored to increase our knowledge of pathologies associated with impaired testicular endocrine function, such as fertility disorders.

Keywords: bile acids; TGR5; testosterone; cholesterol esters

Citation: Holota, H.; De Haze, A.; Martinot, E.; Monrose, M.; Saru, J.-P.; Caira, F.; Beaudoin, C.; Volle, D.H. Identification of the Role of TGR5 in the Regulation of Leydig Cell Homeostasis. *Int. J. Mol. Sci.* **2022**, 23, 15398. https://doi.org/10.3390/ijms232315398

Academic Editor: Jacques J. Tremblay

Received: 30 September 2022
Accepted: 1 December 2022
Published: 6 December 2022

Publisher's Note: MDPI stays neutral with regard to jurisdictional claims in published maps and institutional affiliations.

1. Introduction

One of the main roles of the testis is to produce sex steroids, corresponding to the endocrine function. These hormones play a major role in the control of testicular physiology but also in the regulation of many other physiological functions. Steroid synthesis is initiated from cholesterol. The latter is taken over by the transporters StAR (steroidogenic acute regulatory protein) or PBR (peripheral benzodiazepine receptor or translocator protein, TSPO); this allows the conversion of cholesterol to pregnenolone within the mitochondria [1,2]. Then, steroidogenesis proceeds through a cascade of enzymatic steps involving CYP11A1 (cytochrome P450, family 11, subfamily A, polypeptide 1), HSD3β (3β-hydroxysteroid dehydrogenase), and CYP17A1 (cytochrome P450, family 17, subfamily A, polypeptide 1) [3]. These different successive steps lead to the production of testosterone.

Bile acid homeostasis and dependent signaling pathways have emerged in recent years as key modulators of testicular physiology, with a major impact on the Leydig cell function [4]. Indeed, the nuclear receptor Farnesoid-X-receptor-α (FXRα; NR1H4) has been defined as a regulator of testicular endocrine function. FXRα activation leads to the repression of testicular steroidogenesis [5]. This negative impact of FXRα on steroid synthesis is mainly due to the regulation of *Shp* (*small heterodimer partner; NR0B2*) [6] and *Dax-1* (dosage-sensitive sex reversal, adrenal hypoplasia critical region, on chromosome X, gene-1; NR0B1) expressions [7], two well-known negative regulators of steroidogenesis. This is in turn is associated with germ cell loss and altered male fertility [7].

In addition to FXRα, bile acids are ligands for the G-protein-coupled membrane bile acid receptor 1 (GPBAR1; TGR5). While the role of TGR5 in the testis has been explored in relation to the regulation of germ cell homeostasis [8–10], no study has thus far analyzed its potential impact on endocrine Leydig cells. The present study, using in vitro approaches, deciphers novel roles for TGR5 to control Leydig cell homeostasis.

2. Results

2.1. TGR5 Was Expressed in Mouse Leydig Cells

To define whether TGR5 is expressed in Leydig cells, we carried out several approaches. First, using RT-PCR analysis, *Tgr5* mRNA was detected in the interstitial compartment of the mouse testis (Figure 1A). Then, we performed primary Leydig cell culture experiments. The data show that *Tgr5* mRNA was expressed in adult mouse Leydig cells (Figure 1A). These data were supported by the detection of *Tgr5* mRNA in the murine immortalized Leydig cell line mLTC1 (murine Leydig tumor cell line 1) (Figure 1A). Combined, these data demonstrated that *Tgr5* mRNA was expressed in mouse Leydig cells.

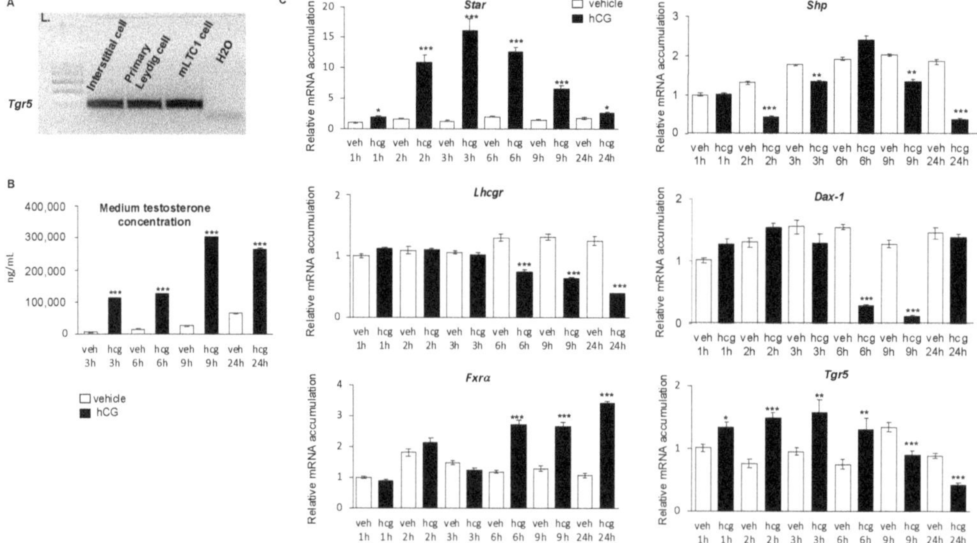

Figure 1. *Tgr5* was expressed in mouse testis. (**A**) Detection of *Tgr5 mRNA* in mouse testicular interstitial compartment, in mouse primary Leydig cells, and in the mLTC1 cell line determined by RT-PCR analysis. (**B**) Testosterone concentrations determined in the culture medium of mLTC1 cells treated with vehicle (NaCl 0.9%) or hCG (25 nM) for 3, 6, 9, or 24 h. (**C**) Relative *Star, Shp, Lhcgr, Dax-1, Fxrα*, and *Tgr5* mRNA accumulations normalized to *β-actin* in mLTC1 cells treated with vehicle (NaCl 0.9%) or hCG (25 nM) for 1, 2, 3, 6, 9, or 24 h. In all panels, n = 9 to 12. Data are expressed as the means ± SEM. Statistical analysis: *, $p < 0.05$; **, $p < 0.01$; ***, $p < 0.001$ versus vehicle-treated cells for the same timing.

2.2. TGR5 mRNA Expression Was Is Regulated by the LH/CG Signaling Pathway in Mouse Leydig Cells

Leydig cell homeostasis is regulated by the hypothalamus–pituitary axis through LH/CG (Luteinizing hormone/choriogonadotropin) signaling [11]. The present data show that mLTC1 cells responded to hCG with the increase in testosterone level as soon as 3 h after treatment (Figure 1B). Regarding gene expression regulations, hCG treatment led, as expected [12], to an upregulation of *Star* mRNA and a downregulation of *Lhcgr* mRNA accumulations (Figure 1C). The present results also confirm that hCG treatment led to lower mRNA accumulations of *Dax-1* and *Shp* (Figure 1C), two nuclear receptors known to repress steroidogenesis. These two receptors have been demonstrated to be target genes of the LH/CG pathway [7,13] and of FXRα within the Leydig cells [6,7]. Moreover, consistently with previous reports [12], hCG treatment led to an increase in *Fxrα* mRNA accumulation (Figure 1C). The mRNA accumulation of *Tgr5* was modulated by hCG. Indeed, hCG treatment led to an increase in *Tgr5* mRNA accumulation in short-term experiments from 1 h up to 6 h after exposure. Then, at 9 h and up to 24 h after hCG treatment, *Tgr5* mRNA levels were lowered compared to vehicle (NaCl 0.9%)-treated cells (Figure 1C). Interestingly, the mRNA expression of *Tgr5* was affected by hCG in a faster manner (starting at 1 h) than the one of *Fxrα*, which was affected only after 6 h. As bile acids have been demonstrated to alter testosterone metabolism, the present data suggest that TGR5 could play a role in the Leydig cells, which led us to explore the roles of TGR5 within Leydig cells.

2.3. TGR5 Activation Regulated Testosterone Production in mLTC1 Cells

To define the role of TGR5 in mouse Leydig cells, we explored its impact on endocrine function by measuring the production of testosterone using the mLTC1 cell line. As TGR5 modulates the protein kinase (PKA) pathway [10,14,15], which is a fast second cellular messenger, we decided to analyze the effect of TGR5 activation in short-time experiments. Treatment of mLTC1 cells with the TGR5 agonist INT-777 for 3 h resulted in a significant decrease in intracellular testosterone levels compared to vehicle-treated cells, while this effect was not observed after 6 h (Figure 2A). A similar decrease in testosterone levels was observed between the culture media of vehicle-treated cells and those of INT-777-treated cells 3 h after treatment (Figure 2B). To circumvent the possibility that immortalized cell lines have a reduced ability to produce testosterone compared to normal Leydig cells, we also measured progesterone, an intermediate in testosterone synthesis. INT-777 treatment was also associated with lower levels of progesterone compared to vehicle-treated cells (Figure 2B). These data support the impact of TGR5 on steroid synthesis.

No impact on the mRNA accumulations of steroidogenic genes such as *Star*, *Cyp11a1*, *Hsd3b1*, or *Cyp17a1* was observed (Figure 2C). In addition, no impact of INT-777 was observed on the mRNA accumulations of *Mrp4* and *Mrp3* (Figure 2D), transporters known to modulate testosterone homeostasis in Leydig cells [16,17]. Note that the other transporter *Mrp2* was not detected in mLTC1 (data not shown). These results suggest that there was no impact of INT-777 on testosterone export.

TGR5 is known to act through the downstream PKA pathway [10,15,18]. To ensure the involvement of the PKA signaling pathway in response to INT-777, experiments using H89, a selective and potent PKA inhibitor, were performed. As expected, H89 led to a lower level of CREB-phosphorylation (Figure S1). The present data highlight the major role of PKA in the mechanism explaining how TGR5 regulates testicular steroidogenesis in Leydig cells. Indeed, the decrease in testosterone levels observed after INT-777 exposure was not detected in cells pretreated with H89 for 2 h, as revealed by intracellular and medium levels of testosterone (Figure 2E). Consistently with the negative impact of INT-777 on steroid synthesis in mLTC1 cells, INT-777 treatment also led to a decrease in CREB-phosphorylation compared to vehicle-treated cells (Figure 2F).

TGR5 shares the same downstream pathways as LH/CG, namely, PKA, which regulates the expression of steroidogenic genes. According to the present results, TGR5 and LH/CG modulate PKA-CREB signaling in an opposite way to mLTC1 cells. We wondered whether activation of TGR5 could modulate the sensitivity of the Leydig cells to the LH/CG pathway. In our hands, no impact on *Lhcgr* mRNA accumulation was noticed following TGR5 activation (Figure 3A). Consistently, no impact of the pre-treatment to INT-777 for

24 h was observed on hCG-induced testosterone levels in mLTC1 cells (Figure 3B). Combined, these data suggest that TGR5 must be a local regulator of testosterone synthesis that does not modulate the central regulation by LH/CG.

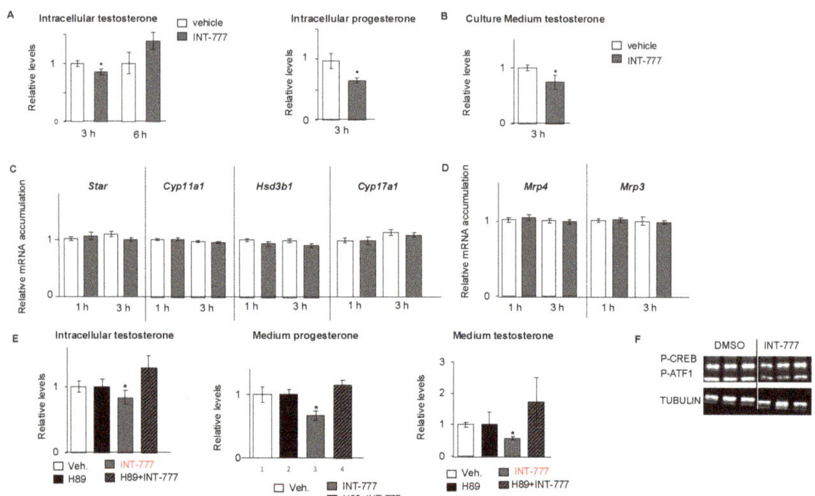

Figure 2. TGR5 regulated testosterone metabolism in a mouse Leydig cell line. (**A**) Relative intracellular levels of testosterone and progesterone normalized to protein concentrations in mLTC1 cells treated for 3 h or 6 h with vehicle or INT-777. (**B**) Relative levels of testosterone in culture medium, normalized to protein concentration, in mLTC1 cells treated for 3 h with vehicle or 25 mM INT-777. (**C**) Relative *Star, Cyp11a1, Hsd3b1*, and *Cyp17a1* mRNA accumulations normalized to *β-actin* in mLTC1 cells treated for 1 or 3 h with vehicle or 25 mM INT-777. (**D**) Relative *Mrp4* and *Mrp3* mRNA accumulations normalized to *β-actin* in mLTC1 cells treated for 1 or 3 h with vehicle or INT-777. (**E**) Relative intracellular levels of testosterone, normalized to protein concentration, in mLTC1 cells pre-treated with H89 for 2 h and then for 3 h with vehicle or INT-777. (**F**) Representative Western blot of P-CREB and TUBULIN in mLTC1 cells treated with vehicle or INT-777. For quantification, vehicle treated cells were arbitrarily set at 1. In all panels, n = 9 to 12. Data are expressed as the means $\pm$ SEM. Statistical analysis: *, $p < 0.05$ compared to vehicle treated cells.

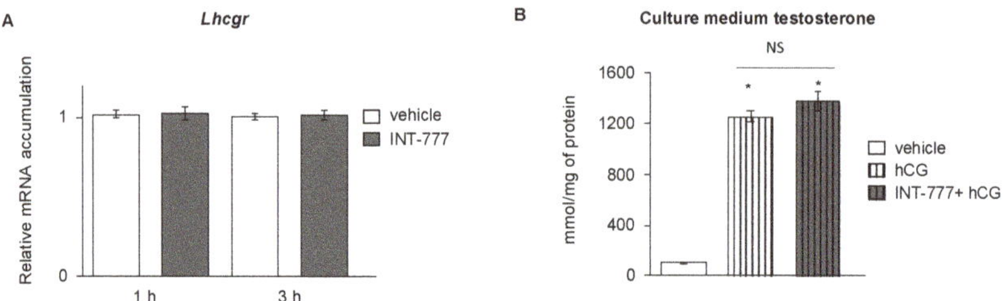

Figure 3. TGR5 did not regulate sensitivity of mLTC1 cells to hCG. (**A**) Relative *Lhcgr* mRNA accumulation normalized to β-actin in mLTC1 cells treated with vehicle or INT-777 for 1 h or 3 h (**B**) Relative intracellular levels of testosterone, normalized to protein concentration, in mLTC1 cells pre-treated with vehicle (DMSO) or INT-777 for 24 h and then for 2 h with vehicle (NaCl 0.9%) or hCG. n = 9 to 12. Data are expressed as the means $\pm$ SEM. Statistical analysis: * $p < 0.05$ compared to vehicle treated cells. NS: non-significant.

2.4. TGR5 Did Not Regulate Glucose Homeostasis in Mouse Leydig Cells

Glucose-6-phophatase (G6pase) and phosphoenol-pyruvate-kinase (Pepck), two enzymes involved in glucose homeostasis, have been shown to regulate the endocrine function of the Leydig cells [19]. In recent years, it has been reported that TGR5 controls glucose metabolism [20]. We wondered whether the impact of TGR5 on testosterone levels could be through an effect on metabolic pathways within the Leydig cells. We thus analyzed whether TGR5 activation could lead to abnormal glucose metabolism in murine Leydig cells. The present data show that activation of TGR5 by INT-777 did not modulate glucose levels in mLTC1 cells (Figure 4). These data suggest that TGR5 might not act through the modulation of glucose homeostasis to alter steroid production.

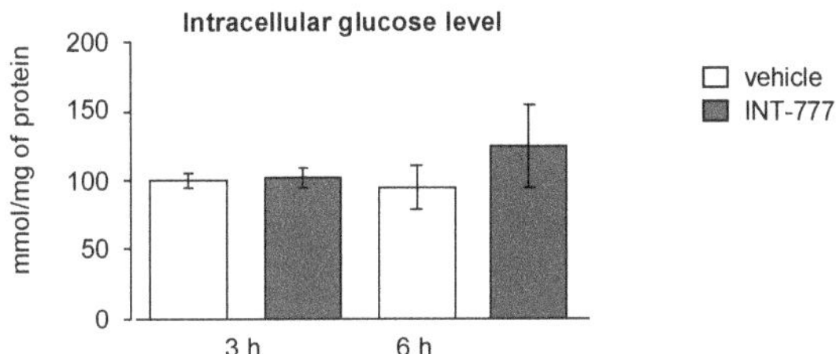

Figure 4. TGR5 did not regulate glucose metabolism in the mouse Leydig cell line. Relative intracellular levels of glucose in mLTC1 cells normalized to protein concentration 3 h after treatment with vehicle (DMSO) or INT-777. n = 9 to 12. Data are expressed as the means ± SEM.

2.5. TGR5 Regulated Lipid Homeostasis in mLTC1 Mouse Leydig Cells

Some regulatory links have been reported between lipid metabolism and TGR5 [21,22]. Among lipids, cholesterol and triglycerides have major impacts on the endocrine capacities of Leydig cells [23,24]. Alteration of lipid metabolism could be visualized in steroidogenic cells using Nile-Red-O staining. We performed staining on mLTC1 treated with vehicle or INT-777 (Figure 5), and it was difficult to define clear data from such analysis as Nile-Red-O staining reveals neutral lipids, triglycerides, and cholesterol esters stored in lipid droplets. No difference was observed on the number of lipid droplets (Figure 5). However, a slight increase in the size of lipid droplets was detected in cells treated with INT-777 compared to vehicle-treated cells (Figure 5). However, non-common variation in the concentrations of either triglycerides or cholesterol esters could lead to misinterpretations.

We then decided to analyze TG levels. The present data show that the activation of TGR5 by the INT-777 had a significant negative impact on the intracellular triglyceride levels (Figure 6A). However, the deregulation of the metabolites was not associated with the alteration of the mRNA accumulations of genes involved in TG synthesis such as *Acc (acetyl-CoA carboxylase)*, *Fas (fatty acid synthase)*, or *Srebp1c (sterol-regulatory-element-binding protein 1)* or other genes such as *Dgat1 (diacylglycerol O-acyltransferase)* and *Dgat2*, for example (Figure 6B).

As no alteration was observed at the mRNA levels, we decided to analyze important actors at the protein level. Indeed, it was demonstrated that the phosphorylation of ACC is key to control its activity (Figure 6C). The data show that the levels of ACC and P-ACC were decreased following INT-777 exposure (Figure 6C).

The effect of INT-777 on TG levels was abolished by pre-treatment with H89 (Figure 6D). These data highlight the complex association between the TGR5 signaling pathway and triglyceride homeostasis. This is consistent with previous data showing the impact of activation of TGR5 signaling in the regulation of hepatic triglyceride homeostasis, which contributes to protection against non-alcoholic fatty liver disease (NAFLD), although the molecular mechanisms have not been elucidated [25].

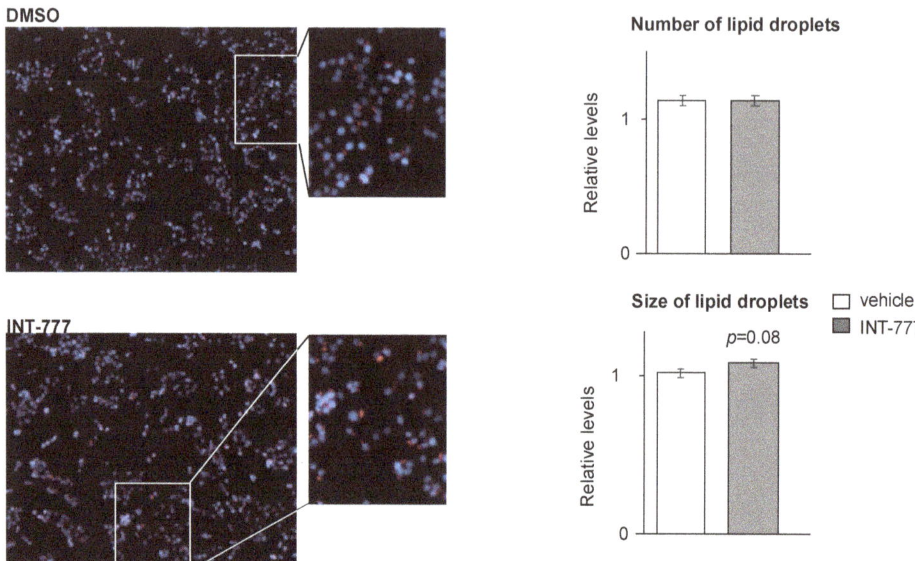

Figure 5. TGR5 modulated lipid metabolism in a mouse Leydig cell line. Representative images mLTC1 cells 3 h after treatment with vehicle (DMSO) or INT-777 and stained with Nile-Red-O. Quantifications of the number and size of lipid droplets in mLTC1 cells 3 h after treatment with vehicle (DMSO) or INT-777 and stained with Nile-Red-O. In all panels, n = 9 to 12. Data are expressed as the means ± SEM.

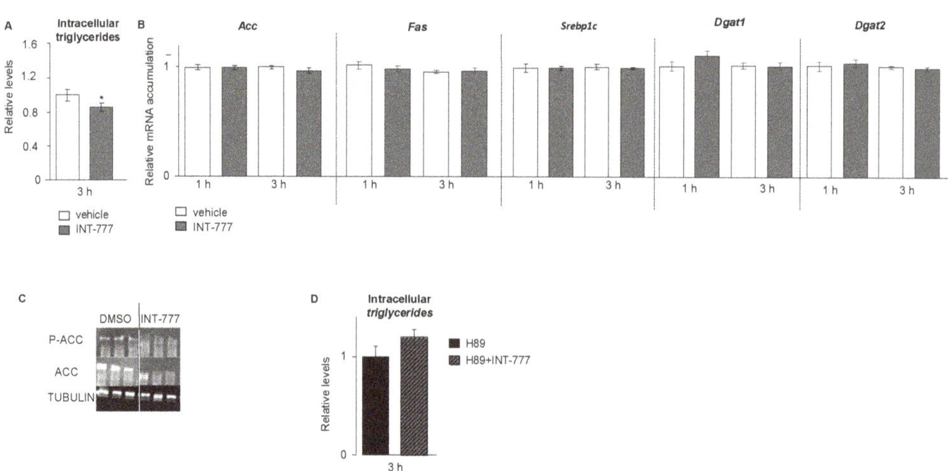

Figure 6. TGR5 modulated TG metabolism in a mouse Leydig cell line. (**A**) Relative intracellular levels of cholesterol in mLTC1 cells normalized to protein concentration 3 h after treatment with vehicle or INT-777. (**B**) Relative mRNA accumulations of *Acc*, Fas, *Srepb1c, Dgat1,* and *Dgat2* normalized to β-actin in a mouse Leydig mLTC1 cell line treated with vehicle or hCG for 3 h. (**C**) Representative Western blots of P-ACC, ACC, and TUBULIN in mLTC1 cells treated with vehicle (DMSO) or INT-777 for 3 h. (**D**) Relative intracellular levels of cholesterol in mLTC1 cells normalized to protein concentration in cells pre-treated 2 h with H89 and then 3 h with vehicle or INT-777. In all panels, n = 9 to 12. Data are expressed as the means ± SEM. Statistical analysis: * $p < 0.05$.

Even though there was a decrease in TG levels, a slight increase in lipid droplet size was observed in INT-777-treated cells (Figure 5), suggesting that there might be an accumulation of other lipid droplet components, such as cholesterol esters. We thus decided to analyze cholesterol levels, which is the initial substrate for steroidogenesis. The data show that the activation of TGR5 by the INT-777 led to an increase in total cellular cholesterol content (Figure 7A). This was not associated with the modulations of the mRNA accumulations of genes involved in their synthesis such as *HmgcoA-reductase or HmgcoA-synthase* (Figure 7B). The altered size of lipid droplets in INT-777-treated cells suggested that there may be an alteration in cholesterol storage. We analyzed the levels of cholesterol esters. The data show that this increase in cholesterol level corresponded to an increase in cholesterol ester levels (Figure 7A). These data suggest that there might be an increased storage or an alteration of the hydrolysis of cholesteryl esters stored in lipid droplets, which could explain both high cholesterol contain and lower testosterone production. For that, we analyzed by RT-qPCR the mRNA accumulation of the *hormone-sensitive lipase* (*Hsl*) that is responsible for neutral cholesteryl ester hydrolase activity. No modulation of the mRNA accumulation of *Hsl* was observed in response to INT-777 (Figure 7B). The effect of the INT-777 on cholesterol levels was abolished by pre-treatment with H89 (Figure 7C).

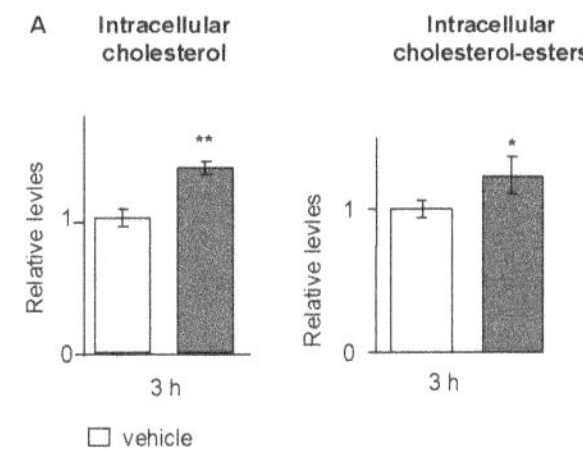
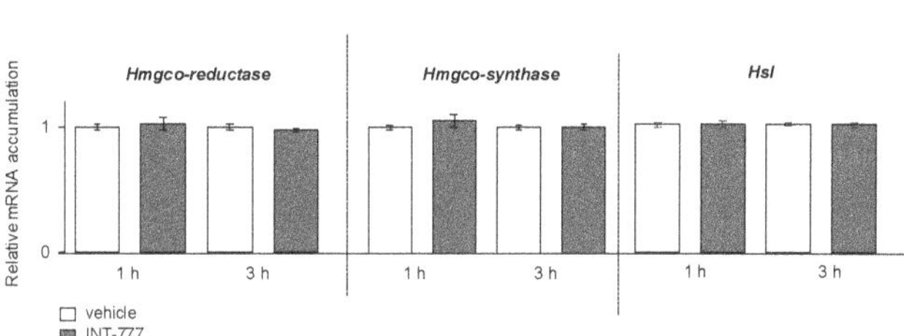
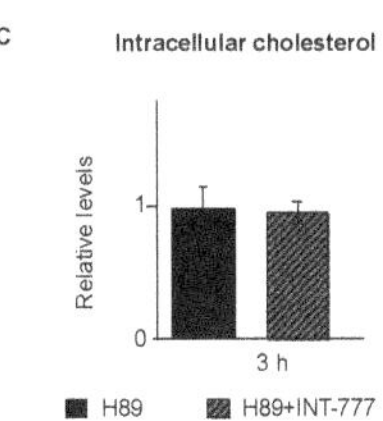

Figure 7. TGR5 regulated triglyceride homeostasis in a mouse Leydig cell line. (**A**) Relative intracellular levels of cholesterol and cholesterol esters in mLTC1 cells normalized to protein concentration 3 h after treatment with vehicle or INT-777. (**B**) Relative mRNA accumulations of *HmgcoA-reductase, HmgcoA-synthase,* and *Hsl* normalized to *β-actin* in a mouse Leydig mLTC1 cell line treated with vehicle or hCG for 3 h. (**C**) Relative intracellular levels of cholesterol in mLTC1 cells normalized to protein concentration in cells pre-treated 2 h with H89 and then 3 h with vehicle or INT-777. In all panels, n = 9 to 12. Data are expressed as the means ± SEM. Statistical analysis: * $p < 0.05$, ** $p < 0.01$.

These data suggest that altered lipid homeostasis induced by TGR5 in Leydig cells could play a role in the lower capacity of these cells to produce testosterone, probably through the alteration of the availability of the substrate for steroidogenesis. However, the exact molecular mechanisms still need to be deciphered.

2.6. TGR5 Regulated Testosterone Production in Mouse Primary Leydig Cells

Since mLTC-1 cells are derived from tumor, and the fact that lipid homeostasis, and particularly cholesterol, is regulated differently between tumoral and normal cells, we decided to assess whether the impacts of INT-777 were reproducible on mouse primary Leydig cells.

Surprisingly no effect of INT-777 on TG was observed in mouse Leydig primary cell cultures (Figure 8A). In contrast, the data show that the impact of INT-777 on cholesterol ester levels was confirmed using primary Leydig cell cultures (Figure 8B). In addition, the experiments performed on mouse Leydig primary cells validated the negative impact of INT-777 treatment on steroid production, as revealed by the measurements of progesterone and testosterone levels in cells and medium (Figure 8C).

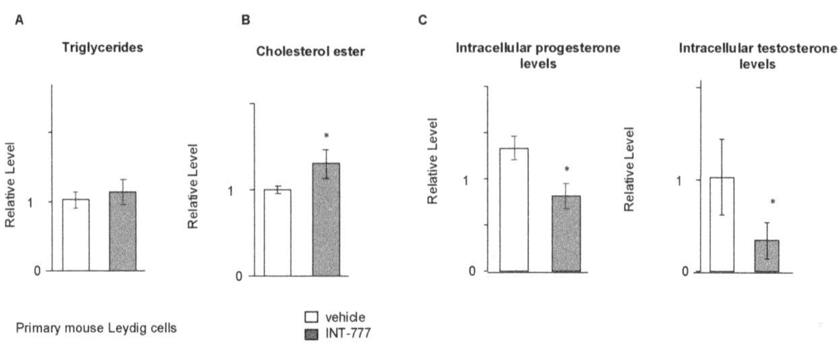

Figure 8. TGR5 signaling was effective in mouse primary Leydig cells. (**A**) Relative intracellular levels of triglycerides in mouse primary Leydig cells normalized to protein concentration 3 h after treatment with vehicle or INT-777. (**B**) Relative intracellular levels of cholesterol esters in mouse primary Leydig cells normalized to protein concentration 3 h after treatment with vehicle or INT-777. (**C**) Relative intracellular levels of progesterone and testosterone in mouse primary Leydig cells normalized to protein concentration 3 h after treatment with vehicle or INT-777. In all panels, n = 3 to 6. Data are expressed as the means $\pm$ SEM. Statistical analysis: * $p < 0.05$.

3. Discussion

The role of the bile acid receptor TGR5 in the testis has been previously demonstrated in germ cell lineage [8–10]. However, its roles on the testicular endocrine cells have not been studied so far. The present work demonstrates the expression of *Tgr5* mRNA in mouse Leydig cells. This led us to study what could be the cellular function regulated by TGR5 in Leydig cells.

We provide evidence for the negative impact of TGR5 activation using INT-777 on the synthesis of testosterone. This effect did not seem to rely on the regulation at the mRNA level of the expression of genes involved in steroidogenesis. However, the steroidogenic pathway has been shown to be regulated by post-translational modifications of key proteins, as demonstrated for STAR [26]. Indeed, it appeared that the phosphorylation of STAR intervenes to support its maximum activity for the transport of cholesterol into the mitochondria and thus to initiate the synthesis of steroids [26]. Thus, the impact of the TGR5 signaling pathway on post-transcriptional and/or post-translational effect must be analyzed to decipher the mechanisms involved before assuming that the effect is through mechanisms independent of gene expression regulation.

Even though it has not been shown before, this involvement of TGR5 in the regulation of steroidogenesis is conceivable. Indeed, it is well known that TGR5 mainly acts through the activation of the adenylate cyclase, leading to the activation of the PKA pathway and by the end to the phosphorylation of CREB [27]. This PKA–CREB pathway has been demonstrated for decades to control the production of testosterone in response to the LH/CG signal [28,29]. Surprisingly, if in the context of LH/CG stimulation, the mobilization of the PKA pathways leads to higher testosterone production, it was unexpected that TGR5 activation could led to lower testosterone levels through PKA, as suggested by the present experiments using H89. The present data suggest that in the mouse Leydig cells, TGR5 activation leads to a decrease in CREB phosphorylation. This is quite unexpected as in other cell types, TGR5 activation was mainly associated with an increase in CREB phosphorylation [10,22]. Further studies are needed to better decipher the molecular mechanisms explaining why in Leydig cells TGR5 acts in an opposite way on PKA-CREB pathway.

In addition, even if the LH/CG and TGR5 pathways share common downstream pathways, the present results suggest that there is no interference by co-treatment with INT-777 and LH/CG. It thus appears that TGR5 pathway only regulates the basal production of testosterone and not the LH/CG induced production. However, it remains to be deciphered as to what could be the mechanisms by which the TGR5-PKA pathway leads to a lower testosterone level.

In recent years, TGR5 has been demonstrated to be involved in the control of several physiological functions [22]. TGR5 controls glucose and lipid metabolisms, and these two metabolisms are known to be involved in the homeostasis of steroid synthesis [19,23]. However, the present data suggest that TGR5 activation has no impacts on glucose levels in Leydig cells. In contrast, within Leydig cells, TGR5 activation led to an alteration of lipid droplets (cholesterol esters and triglycerides), which are a reservoir of substrate for the synthesis of steroids [17]. Using the mLTC1 cell line and mouse primary Leydig cells, it appears that activation of TGR5 by INT-777 leads to an increase in cellular cholesterol content, mainly cholesterol ester levels. This alteration of the lipid metabolism induced by INT-777 could participate in the drop in testosterone levels in the Leydig cells.

The present data highlight the impact of TGR5 signaling pathway on testosterone production. These data must be of importance and require further studies as endocrine homeostasis is involved in the control of many physiological processes, and its alteration is associated with the development of diseases. In addition, the finding that TGR5 activation modulates the steroid production has to be put in line with the actual questioning of the impacts of environmental endocrine disruptors in the programming of health and diseases [30–32].

In the present study, the concentration of INT-777 used (25 µM) is consistent with other published works using this compound [15]. The issue of bile acid concentrations within the testicles has been addressed in the context of hepatic pathologies. It has thus been clearly defined that the testicular concentrations of bile acids observed in this context of hepatic pathologies are compatible with an activation of TGR5. Thus, it will be necessary to better define under which conditions (physiological or pathological) the levels of testicular bile acids will be sufficient to activate the TGR5 in the Leydig cells leading to lower testosterone levels.

Another remaining question is to decipher the endogenous testicular ligand of TGR5. Indeed, if bile acids have been demonstrated to reach the testis from blood [9] and that testis could even produce some bile acids [33], we could not exclude that other endogenous molecules could act as testicular TGR5 agonists. Indeed, some steroids have been demonstrated to activate TGR5, such as pregnandione [34]. This clearly supports the idea of a crosstalk between TGR5 signaling pathway and testicular steroidogenesis.

In the present work, we point out that the G-protein-coupled bile acid receptor TGR5 is expressed in the Leydig cells, where its expression is controlled by the main regulator of the Leydig cell homeostasis, the LH/CG signaling pathway. The regulation of *Tgr5* mRNA accumulation by hCG is dynamic as it is first induced in a short-term experiment and then repressed. This is quite interesting as a similar expression pattern was demonstrated for the mRNA expression of the nuclear bile acid nuclear receptor *Fxrα* [12] for the regulation of the testicular endocrine function [5]. Indeed, in recent years, bile acid homeostasis has been associated with the modulation of the testosterone production by Leydig cells [4]. This is consistent with the fact that in cholestasis disease, testosterone levels are lower than in normal physiological conditions. However, until now, only the nuclear bile acid receptor FXRα was demonstrated to be involved [5,12]. This is consistent with the fact that FXRα pathways repress the expression of steroidogenic genes such as *Star* [6,7]. Moreover, it has been demonstrated that FXRα pathways also regulate the sensitivity of Leydig cells to the

LH/CG, connecting the central regulation of steroidogenesis to local tight control [6,7]. The present data highlight the role of TGR5 in the regulation of the testicular endocrine function. Interestingly, the kinetic of the regulation could be of importance as TGR5 seems to act rapidly on steroidogenesis (3 h, shown herein), whereas for FXRα, the effect was seen on Leydig cells in a longer time frame (12 to 24 h) [6]. This could reflect the differences in the mechanisms of action between a G-protein-coupled receptor versus nuclear receptor. Overall, these data sustain that bile acids either through TGR5 and FXRα exert a negative effect on testicular androgen synthesis. It is also interesting to note that the LH/CG pathway leads first to an upregulation of the expression of both *Tgr5* and *Fxra*. It is conceivable that the upregulation of their expression levels in response to LH/CG is a way to initiate a kind of negative feedback. However, it is interesting to note that LH/CG modulates the expression of *Tgr5* more rapidly than that of *Fxra*. We could thus hypothesize that TGR5 could be "in the first line" to ensure the feedback of testosterone synthesis, and then a second wave arrives through FXRα signaling, which takes longer via the regulation of *Shp* and *Dax-1* expressions.

In addition to the established role of FXRα in the Leydig cells, the identification of TGR5 signaling as a regulator of testicular steroidogenesis, with a similar negative effect on steroid synthesis, opens new perspectives to better understand the complex impacts of bile acids on Leydig cell homeostasis and on the testicular endocrine function.

Our data reinforce the links between the bile acid signaling pathways through TGR5 and/or FXRα and the control of the endocrine function of the Leydig cells. Combined with previous data, the present results define TGR5 and FXRα as critical actors that need to be more deeply explored to increase our knowledge of pathologies associated with altered testicular endocrine function such as fertility disorders.

4. Materials and Methods

4.1. Animals

C57Bl/6J mice were purchased from Charles River Laboratories (L'Arbresle, France). Mice were acclimated at least 2 weeks before experiments. Mice were housed in temperature-controlled rooms with 12 h light/dark cycles. Mice had ad libitum access to food and water. The refinement is based on the housing and monitoring of the animals as well as the development of protocols that consider animal welfare. This has been achieved by enriching the cages (cardboard tunnel and mouse houses). This study was conducted in accordance with current regulations and standards approved by Institut National de la Santé et de la Recherche Médicale Animal Care Committee and by the animal care committee.

4.2. Interstitial Cell Enrichment

For the generation of data on enriched interstitial cells compared to tubular preparations or whole testis samples, the methodology used relies on mechanical and enzymatic processes for dissociating the fractions as previously described [12]. The interstitial enriched cells and tubular compartments were generated as described in a previous study. Briefly, testes from 90-day-old male mice were decapsulated and incubated for 20 min at 33 °C in Dulbecco's modified Eagle's medium (DMEM)/Ham's F12 (1:1), transferrin (5 µg/mL), insulin (4 µg/mL), and vitamin E (0.2 µg/mL) medium containing collagenase (0.8 mg/mL) (Life Technologies, Invitrogen, Cergy-Pontoise, France). Extracts were collected by centrifugation for 10 min at $200 \times g$, and the pellet was resuspended in fresh medium.

4.3. Cell Line Approaches

mLTC1 cells were used as previously described [35]. Cells were plated in 12-well plates. Twenty-four hours after plating, cells were put in serum-free medium overnight. Then, cells were treated with vehicles (DMSO 1/1000 (for H89) or NaCl0.9% 1/1000 (for hCG)), INT777 (25 µM; Sigma-Aldrich, St. Louis, MO, USA), or hCG (25 nM, Sigma-Aldrich,

St. Louis, MO, USA). Following this, cells were harvested at different time points, and messenger RNA (mRNA) or protein extractions were performed.

4.4. Real-Time RT-qPCR

RNA from cell samples were isolated using RNAzol. cDNA was synthesized from total RNA with the MMLV and random hexamer primers (Promega, Charbonnières-les-Bains, France). The real-time PCR measurement of individual cDNA was performed using SYBR green dye (Master mix Plus for SYBR Assay, *Takara Bio*) to measure duplex DNA formation with the Eppendorf–Realplex system. For each experiment, standard curves were generated with pools of cDNA from cells with different genotypes and/or treatments. The results were analyzed using the $\Delta\Delta$ct method. Primer sequences are reported in Table 1.

Table 1. Sequences of primers used in this study.

Gene Name	Forward	Reverse
Actin	TCATCACTATTGGCAACGAGC	AGTTTCATGGATGCCACAGG
Lhcgr	AGCTAATGCCTTTGACAACC	GATGGACTCATTATTCATCC
Star	TGTCAAGGAGATCAAGGTCCT	CGATAGGACCTGGTTGATGAT
Cyp11a1	CTGCCTCCAGACTTCTTTCG	TTCTTGAAGGGCAGCTTGTT
Hsd3b1	ATGGTCTGCCTGGGAATGAC	ACTGCAGGAGGTCAGAGCT
Cyp17a1	CCTGATACGAAGCACTTCTCG	CCAGGACCCCAAGTGTGTTCT
Shp	CGATCCTCTTCAACCCAGATG	AGGGCTCCAAGACTTCACACA
Dax-1	GTCCAGGCCATCAAGAGTTTC	CAGCTTTGCACAGAGCATCTC
Acc	GCCTTTCACATGAGATCCAGC	CTGCAATACCATTGTTGGCGA
Fas	AAGCGGTCTGGAAAGCTGAA	AGGCTGGGTTGATACCTCCA
Srebp1c	GAACAGACACTGGCCGAGAT	GAGGCCAGAGAAGCAGAAGAG
HmgcoA-reductase	ACAGAAACTCCACGTGACGA	TTCAGCAGTGCTTTCTCCGT
HmgcoA-Synthase	GCAAAAAGATCCGTGCCCAG	GTCATTCAGGAACATCCGAGC
Mrp3	AGTCTTCGGGAGTGCTCATCA	AGGATTTGTGTCAAGATTCTCCG
Mrp4	TTAGATGGGCCTCTGGTTCT	GCCCACAATTCCAACCTTT
Dgat1	TGGCTGCATTTCAGATTGAG	GCTGGGAAGCAGATGATTGT
Dgat2	CTTCCTGGTGCTAGGAGTGGC	GCTGGATGGGAAAGTAGTCTCGG
Hsl	CATATCCGCTCTCCAGTTGACC	CCTATCTTCTCCATCGACTACTCC

4.5. Nile-Red-O Staining

Cells were plated in 6-well plates. Twenty-four hours after plating, cells were put in serum-free medium overnight. Then, cells were treated for with vehicle (DMSO 1/1000) or INT777 (25 µM; Sigma-Aldrich, St. Louis, MO, USA). Three hours after treatment, cells were fixed with 4% PFA fixation (10 min room temperature agitation). Cells were washed 3 times for 5 min with PBS-Tween 0.01% at room temperature. Then, saturation was performed for 1 h at room temperature with PBS-Tween 0.01% + BSA. Following this, cells were incubated in PBS-Tween 0.01% with Nile-Red-O (1 ng/mL) for 30 min. Four washes were performed with PBS. Cells were then counterstained with Hoechst medium (1 mg/mL) and then mounted on PBS/glycerol (50/50).

4.6. Glucose, Cholesterol, Cholesterol Ester, and Triglyceride Measurements

Cells were plated in 6-well plates. Twenty-four hours after plating, cells were put in serum-free medium overnight. Then, cells were treated for with vehicle (DMSO 1/1000 (for H89) or NaCl0.9% 1/1000 (for hCG)), INT777 (25 µM; Sigma-Aldrich, St. Louis, MO, USA), or hCG (25 nM, Sigma-Aldrich, St. Louis, MO, USA). Following this, cells were harvested in

PBS1X. After this, measurements conducted using colorimetric assays were performed on cells or medium following the recommendations of the manufacturer (Glucose RTU, 61269, Biomerieux SA, France). Triglyceride measurements were performed using a colorimetric assay kit (Triglycerides FS*, DiaSys Diagnostic Systems GmbH, Holzheim, Germany).

4.7. Testosterone Measurements

Cells were plated in 6-well plates. Twenty-four hours after plating, cells were placed in serum-free medium overnight. Then, cells were treated with vehicles (DMSO 1/1000 (for H89) or NaCl0.9% 1/1000 (for hCG)), INT777 (25 μM; Sigma-Aldrich, St. Louis, MO, USA), or hCG (25 nM, Sigma-Aldrich, St. Louis, MO, USA). Following this, cells were harvested at different time points. Testosterone was measured in the cell extract of MLTC1 cells as previously described using a commercial kit (AR-K032-H5, ARBOR ASSAYS, INC), which was used for the assays.

4.8. Statistical Analyses

All numerical data are represented as mean $\pm$ SEM. Significant difference was set at $p < 0.05$. Differences between groups were determined by t-test.

Supplementary Materials: The following supporting information can be downloaded at: https://www.mdpi.com/article/10.3390/ijms232315398/s1. Figure S1: Representative western blots of P-CREB and TUBULIN in mLTC1 cells treated with vehicle (DMSO) or H89 for 2 h.

Author Contributions: Conceptualization, H.H., A.D.H. and D.H.V.; data curation, H.H. and A.D.H.; formal analysis, D.H.V.; funding acquisition, D.H.V.; investigation, H.H., A.D.H., E.M., M.M. and J.-P.S.; methodology, H.H., A.D.H., E.M. and M.M.; writing—original draft, F.C., C.B. and D.H.V. All authors have read and agreed to the published version of the manuscript.

Funding: The study was funded by Inserm; CNRS; Université Clermont Auvergne; *Ligue contre le cancer* (comité Puy de Dôme). Laura Thirouard and Manon Garcia received support from the *Fondation Recherche Médicale* (FDT202001010780 and R17089CC). Volle's team received support by the Cancer-Inserm (C20010CS) and the French government IDEX-ISITE initiative 16-IDEX-0001 (CAP 20-25).

Institutional Review Board Statement: Not applicable.

Data Availability Statement: Data are available upon request to corresponding author.

Acknowledgments: We thank Sandrine Plantade, Khirredine Ouchen, and Philippe Mazuel for their help at the animal facility.

Conflicts of Interest: The authors declare no conflict of interest.

References

1. Papadopoulos, V.; Baraldi, M.; Guilarte, T.R.; Knudsen, T.B.; Lacapère, J.-J.; Lindemann, P.; Norenberg, M.D.; Nutt, D.; Weizman, A.; Zhang, M.-R.; et al. Translocator Protein (18 kDa): New Nomenclature for the Peripheral-Type Benzodiazepine Receptor Based on Its Structure and Molecular Function. *Trends Pharmacol. Sci.* **2006**, *27*, 402–409. [CrossRef] [PubMed]
2. Stocco, D.M. The Role of the StAR Protein in Steroidogenesis: Challenges for the Future. *J. Endocrinol.* **2000**, *164*, 247–253. [CrossRef]
3. Midzak, A.; Papadopoulos, V. Steroidogenesis: The Classics and Beyond. *Steroids* **2015**, *103*, 1–2. [CrossRef] [PubMed]
4. Sèdes, L.; Martinot, E.; Baptissart, M.; Baron, S.; Caira, F.; Beaudoin, C.; Volle, D.H. Bile Acids and Male Fertility: From Mouse to Human? *Mol. Asp. Med.* **2017**, *56*, 101–109. [CrossRef] [PubMed]
5. Holota, H.; Thiouard, L.; Monrose, M.; Garcia, M.; De Haze, A.; Saru, J.; Caira, F.; Beaudoin, C.; Volle, D. FXRα Modulates Leydig Cell Endocrine Function in Mouse. *Mol. Cell. Endocrinol.* **2020**, *518*, 303–315. [CrossRef] [PubMed]
6. Volle, D.H.; Duggavathi, R.; Magnier, B.C.; Houten, S.M.; Cummins, C.L.; Lobaccaro, J.-M.A.; Verhoeven, G.; Schoonjans, K.; Auwerx, J. The Small Heterodimer Partner Is a Gonadal Gatekeeper of Sexual Maturation in Male Mice. *Genes Dev.* **2007**, *21*, 303–315. [CrossRef] [PubMed]
7. Baptissart, M.; Martinot, E.; Vega, A.; Sédes, L.; Rouaisnel, B.; de Haze, A.; Baron, S.; Schoonjans, K.; Caira, F.; Volle, D.H. Bile Acid-FXRα Pathways Regulate Male Sexual Maturation in Mice. *Oncotarget* **2016**, *7*, 19468–19482. [CrossRef]

8. Baptissart, M.; Sèdes, L.; Holota, H.; Thirouard, L.; Martinot, E.; de Haze, A.; Rouaisnel, B.; Caira, F.; Beaudoin, C.; Volle, D.H. Multigenerational Impacts of Bile Exposure Are Mediated by TGR5 Signaling Pathways. *Sci. Rep.* **2018**, *8*, 16875. [CrossRef]

9. Baptissart, M.; Vega, A.; Martinot, E.; Pommier, A.J.; Houten, S.M.; Marceau, G.; de Haze, A.; Baron, S.; Schoonjans, K.; Lobaccaro, J.-M.A.; et al. Bile Acids Alter Male Fertility through G-Protein-Coupled Bile Acid Receptor 1 Signaling Pathways in Mice. *Hepatology* **2014**, *60*, 1054–1065. [CrossRef]

10. Thirouard, L.; Holota, H.; Monrose, M.; Garcia, M.; de Haze, A.; Damon-Soubeyrand, C.; Renaud, Y.; Saru, J.-P.; Perino, A.; Schoonjans, K.; et al. Identification of a Crosstalk among TGR5, GLIS2, and TP53 Signaling Pathways in the Control of Undifferentiated Germ Cell Homeostasis and Chemoresistance. *Adv. Sci.* **2022**, *9*, e2200626. [CrossRef]

11. Kaprara, A.; Huhtaniemi, I.T. The Hypothalamus-Pituitary-Gonad Axis: Tales of Mice and Men. *Metab. Clin. Exp.* **2018**, *86*, 3–17. [CrossRef] [PubMed]

12. Holota, H.; Thirouard, L.; Garcia, M.; Monrose, M.; de Haze, A.; Saru, J.-P.; Caira, F.; Beaudoin, C.; Volle, D.H. Fxralpha Gene Is a Target Gene of HCG Signaling Pathway and Represses HCG Induced Steroidogenesis. *J. Steroid Biochem. Mol. Biol.* **2019**, *194*, 105460. [CrossRef] [PubMed]

13. Vega, A.; Martinot, E.; Baptissart, M.; De Haze, A.; Saru, J.; Baron, S.; Caira, F.; Schoonjans, K.; Lobaccaro, J.; Volle, D. Identification of the Link between the Hypothalamo-Pituitary Axis and the Testicular Orphan Nuclear Receptor NR0B2 in Male Mice. *Endocrinology* **2014**. submitted. [CrossRef] [PubMed]

14. Wu, S.; Romero-Ramírez, L.; Mey, J. Taurolithocholic Acid but Not Tauroursodeoxycholic Acid Rescues Phagocytosis Activity of Bone Marrow-derived Macrophages under Inflammatory Stress. *J. Cell Physiol.* **2022**, *237*, 1455–1470. [CrossRef] [PubMed]

15. Perino, A.; Velázquez-Villegas, L.A.; Bresciani, N.; Sun, Y.; Huang, Q.; Fénelon, V.S.; Castellanos-Jankiewicz, A.; Zizzari, P.; Bruschetta, G.; Jin, S.; et al. Central Anorexigenic Actions of Bile Acids Are Mediated by TGR5. *Nat. Metab.* **2021**, *3*, 595–603. [CrossRef]

16. Li, C.Y.; Basit, A.; Gupta, A.; Gáborik, Z.; Kis, E.; Prasad, B. Major Glucuronide Metabolites of Testosterone Are Primarily Transported by MRP2 and MRP3 in Human Liver, Intestine and Kidney. *J. Steroid Biochem. Mol. Biol.* **2019**, *191*, 105350. [CrossRef]

17. Morgan, J.A.; Cheepala, S.B.; Wang, Y.; Neale, G.; Adachi, M.; Nachagari, D.; Leggas, M.; Zhao, W.; Boyd, K.; Venkataramanan, R.; et al. Deregulated Hepatic Metabolism Exacerbates Impaired Testosterone Production in Mrp4-Deficient Mice. *J. Biol. Chem.* **2012**, *287*, 14456–14466. [CrossRef]

18. Masyuk, T.V.; Masyuk, A.I.; Lorenzo Pisarello, M.; Howard, B.N.; Huang, B.Q.; Lee, P.-Y.; Fung, X.; Sergienko, E.; Ardecky, R.J.; Chung, T.D.Y.; et al. TGR5 Contributes to Hepatic Cystogenesis in Rodents with Polycystic Liver Diseases through Cyclic Adenosine Monophosphate/Gαs Signaling. *Hepatology* **2017**, *66*, 1197–1218. [CrossRef]

19. Ahn, S.W.; Gang, G.-T.; Tadi, S.; Nedumaran, B.; Kim, Y.D.; Park, J.H.; Kweon, G.R.; Koo, S.-H.; Lee, K.; Ahn, R.-S.; et al. Phosphoenolpyruvate Carboxykinase and Glucose-6-Phosphatase Are Required for Steroidogenesis in Testicular Leydig Cells. *J. Biol. Chem.* **2012**, *287*, 41875–41887. [CrossRef]

20. Thomas, C.; Gioiello, A.; Noriega, L.; Strehle, A.; Oury, J.; Rizzo, G.; Macchiarulo, A.; Yamamoto, H.; Mataki, C.; Pruzanski, M.; et al. TGR5-Mediated Bile Acid Sensing Controls Glucose Homeostasis. *Cell Metab.* **2009**, *10*, 167–177. [CrossRef]

21. Holter, M.M.; Chirikjian, M.K.; Govani, V.N.; Cummings, B.P. TGR5 Signaling in Hepatic Metabolic Health. *Nutrients* **2020**, *12*, 2598. [CrossRef] [PubMed]

22. Pols, T.W.H.; Noriega, L.G.; Nomura, M.; Auwerx, J.; Schoonjans, K. The Bile Acid Membrane Receptor TGR5: A Valuable Metabolic Target. *Dig. Dis.* **2011**, *29*, 37–44. [CrossRef] [PubMed]

23. Shen, W.-J.; Azhar, S.; Kraemer, F.B. Lipid Droplets and Steroidogenic Cells. *Exp. Cell Res.* **2016**, *340*, 209–214. [CrossRef] [PubMed]

24. Sedes, L.; Thirouard, L.; Maqdasy, S.; Garcia, M.; Caira, F.; Lobaccaro, J.-M.A.; Beaudoin, C.; Volle, H. David Cholesterol: A Gatekeeper of Male Fertility? *Front. Endocrinol.* **2018**, *9*, 369. [CrossRef]

25. Donepudi, A.C.; Boehme, S.; Li, F.; Chiang, J.Y.L. G Protein-Coupled Bile Acid Receptor Plays a Key Role in Bile Acid Metabolism and Fasting-Induced Hepatic Steatosis. *Hepatology* **2017**, *65*, 813–827. [CrossRef]

26. Strauss, J.F.; Kallen, C.B.; Christenson, L.K.; Watari, H.; Devoto, L.; Arakane, F.; Kiriakidou, M.; Sugawara, T. The Steroidogenic Acute Regulatory Protein (StAR): A Window into the Complexities of Intracellular Cholesterol Trafficking. *Recent Prog. Horm. Res.* **1999**, *54*, 369–394; discussion 394–395.

27. Martinot, E.; Sèdes, L.; Baptissart, M.; Lobaccaro, J.-M.; Caira, F.; Beaudoin, C.; Volle, D.H. Bile Acids and Their Receptors. *Mol. Asp. Med.* **2017**, *56*, 2–9. [CrossRef]

28. Manna, P.R.; Dyson, M.T.; Eubank, D.W.; Clark, B.J.; Lalli, E.; Sassone-Corsi, P.; Zeleznik, A.J.; Stocco, D.M. Regulation of Steroidogenesis and the Steroidogenic Acute Regulatory Protein by a Member of the CAMP Response-Element Binding Protein Family. *Mol. Endocrinol.* **2002**, *16*, 184–199. [CrossRef] [PubMed]

29. King, S.R.; LaVoie, H.A. Gonadal Transactivation of STARD1, CYP11A1 and HSD3B. *Front. Biosci. (Landmark Ed.)* **2012**, *17*, 824–846. [CrossRef]

30. Bateman, M.E.; Strong, A.L.; McLachlan, J.A.; Burow, M.E.; Bunnell, B.A. The Effects of Endocrine Disruptors on Adipogenesis and Osteogenesis in Mesenchymal Stem Cells: A Review. *Front. Endocrinol.* **2016**, *7*, 171. [CrossRef]

31. De Coster, S.; van Larebeke, N. Endocrine-Disrupting Chemicals: Associated Disorders and Mechanisms of Action. *J. Environ. Public Health* **2012**, *2012*, 713696. [CrossRef] [PubMed]

32. Haverinen, E.; Fernandez, M.F.; Mustieles, V.; Tolonen, H. Metabolic Syndrome and Endocrine Disrupting Chemicals: An Overview of Exposure and Health Effects. *Int. J. Environ. Res. Public Health* **2021**, *18*, 13047. [CrossRef] [PubMed]

33. Martinot, E.; Baptissart, M.; Véga, A.; Sèdes, L.; Rouaisnel, B.; Vaz, F.; Saru, J.-P.; de Haze, A.; Baron, S.; Caira, F.; et al. Bile Acid Homeostasis Controls CAR Signaling Pathways in Mouse Testis through FXRalpha. *Sci. Rep.* **2017**, *7*, 42182. [CrossRef]
34. Kawamata, Y.; Fujii, R.; Hosoya, M.; Harada, M.; Yoshida, H.; Miwa, M.; Fukusumi, S.; Habata, Y.; Itoh, T.; Shintani, Y.; et al. A G Protein-Coupled Receptor Responsive to Bile Acids. *J. Biol. Chem.* **2003**, *278*, 9435–9440. [CrossRef] [PubMed]
35. Nguyen, T.M.D.; Filliatreau, L.; Klett, D.; Hai, N.V.; Duong, N.T.; Combarnous, Y. Effect of Soluble Adenylyl Cyclase (ADCY10) Inhibitors on the LH-Stimulated CAMP Synthesis in Mltc-1 Leydig Cell Line. *Int. J. Mol. Sci.* **2021**, *22*, 4641. [CrossRef]

International Journal of
Molecular Sciences

Article

$Lhb^{-/-}$ $Lhr^{-/-}$ Double Mutant Mice Phenocopy $Lhb^{-/-}$ or $Lhr^{-/-}$ Single Mutants and Display Defects in Leydig Cells and Steroidogenesis

Zhenghui Liu [1], Mark Larsen [1], Zhenmin Lei [2], C. V. Rao [3] and T. Rajendra Kumar [1,4,*]

[1] Division of Reproductive Sciences, Department of Obstetrics and Gynecology, University of Colorado
     Anschutz Medical Campus, Aurora, CO 80045, USA
[2] Department of Obstetrics and Gynecology, University of Louisville School of Medicine,
     Louisville, KY 40202, USA
[3] Department of Cell Biology, Molecular and Human Genetics, Obstetrics and Gynecology, Herbert Wertheim
     College of Medicine, Florida International University, Miami, FL 33199, USA
[4] Division of Reproductive Endocrinology, Department of Obstetrics and Gynecology, University of Colorado
     Anschutz Medical Campus, Aurora, CO 80045, USA
*   Correspondence: raj.kumar@cuanschutz.edu

Citation: Liu, Z.; Larsen, M.; Lei, Z.; Rao, C.V.; Kumar, T.R. $Lhb^{-/-}$ $Lhr^{-/-}$ Double Mutant Mice Phenocopy $Lhb^{-/-}$ or $Lhr^{-/-}$ Single Mutants and Display Defects in Leydig Cells and Steroidogenesis. *Int. J. Mol. Sci.* **2022**, 23, 15725. https://doi.org/10.3390/ijms232415725

Academic Editor: Jacques J. Tremblay

Received: 16 November 2022
Accepted: 9 December 2022
Published: 11 December 2022

Publisher's Note: MDPI stays neutral with regard to jurisdictional claims in published maps and institutional affiliations.

Abstract: In the mouse, two distinct populations of Leydig cells arise during testis development. Fetal Leydig cells arise from a stem cell population and produce T required for masculinization. It is debated whether they persist in the adult testis. A second adult Leydig stem cell population gives rise to progenitor-immature-mature adult type Leydig cells that produce T in response to LH to maintain spermatogenesis. In testis of adult null male mice lacking either only LH ($Lhb^{-/-}$) or LHR ($Lhr^{-/-}$), mature Leydig cells are absent but fetal Leydig cells persist. Thus, it is not clear whether other ligands signal via LHRs in *Lhb* null mice or LH signals via other receptors in the absence of LHR in *Lhr* null mice. Moreover, it is not clear whether truncated LHR isoforms generated from the same *Lhr* gene promoter encode functionally relevant LH receptors. To determine the in vivo roles of LH-LHR signaling pathway in the Leydig cell lineage, we generated double null mutant mice lacking both LH Ligand and all forms of LHR. Phenotypic analysis indicated testis morpho-histological characteristics are identical among double null and single mutants which all showed poorly developed interstitium with a reduction in Leydig cell number and absence of late stage spermatids. Gene expression analyses confirmed that the majority of the T biosynthesis pathway enzyme-encoding mRNAs expressed in Leydig cells were all suppressed. Expression of thrombospondin-2, a fetal Leydig cell marker gene was upregulated in single and double null mutants indicating that fetal Leydig cells originate and develop independent of LH-LHR signaling pathway in vivo. Serum and intratesticular T levels were similarly suppressed in single and double mutants. Consequently, expression of AR-regulated genes in Sertoli and germ cells were similarly affected in single and double mutants without any evidence of any additive effect in the combined absence of both LH and LHR. Our studies unequivocally provide genetic evidence that in the mouse testis, fetal Leydig cells do not require LH-LHR signaling pathway and a one-to-one LH ligand-LHR signaling pathway exists in vivo to regulate adult Leydig cell lineage and spermatogenesis.

Keywords: LH; Leydig cell; LH-receptor; testosterone; spermatogenesis; thrombospondin-2

1. Introduction

Luteinizing hormone (LH) is a pituitary-derived heterodimeric glycoprotein synthesized in gonadotropes. It consists of an α- subunit that is non-covalently associated with the hormone-specific β-subunit [1–3]. In the male, LH binds and signals via G-protein coupled receptors (GPCRs) known as LH receptors (LHRs) expressed on Leydig cells within the testis interstitium [1–3]. LHR-mediated intracellular signaling cascades regulate a battery of enzymatic steps resulting in testosterone (T) biosynthesis in Leydig cells [1–3]. T diffuses into testis tubules and binds androgen receptors (ARs) expressed in Sertoli cells and regulates spermatogenesis and male fertility [4,5].

During mouse testis development, two distinct populations of Leydig cells arise within the testis interstitium [6–13]. The fetal Leydig stem cells give rise to fetal Leydig cells which produce testosterone required for masculinization. Although originally thought these fetal Leydig cells do not persist, the idea that they do exist in the adult testis was confirmed by elegant lineage tracing studies [14–17]. Subsequently, a distinct adult Leydig stem cell population in the adult testis gives rise to highly proliferative Leydig progenitor cells from which immature Leydig cells are produced. Immature Leydig cells have limited proliferating capacity and terminally differentiate into mature adult Leydig cells. Mature Leydig cells produce testosterone in response to LH to maintain spermatogenesis in the adult [6–9,12]. The distinct actions of LH- and LHR-mediated signaling in fetal versus adult Leydig cells remain unclear.

Genetically engineered mouse models confirm LH action is essential for spermatogenesis. *Lhb* null mice (and hence LH-deficient) are hypogonadal and infertile [18]. These null mice demonstrate defects in adult Leydig cell development, profound suppression of serum and intratesticular T and spermiogenesis defect [18,19]. However, *Lhb*$^{+/-}$ heterozygous mice are fertile and do not show any overt testis phenotypes. Similarly, two different and independent null mutations were engineered at the *Lhr* locus and *Lhr* null mice were generated [20,21]. In both cases, *Lhr* null mutants exhibit hypogonadism and infertility and defects in Leydig cells and steroidogenesis, similar to that observed in *Lhb* null mice [20,21]. *Lhr*$^{+/-}$ heterozygous mice are fertile and do not show any overt testis phenotypes, similar to *Lhb*$^{+/-}$ heterozygous mice.

It is interesting to note that in either the LH ligand (*Lhb*) or receptor (*Lhr*) null mice, fetal Leydig cells persist compared to those in control male mice [12,18,20,21]. This raises the possibility that ligands other than LH may signal through LHRs or LH may signal via receptors other than LHR to maintain fetal Leydig cells. Here, we genetically intercrossed heterozygous *Lhb*$^{+/-}$ and *Lhr*$^{+/-}$ mice and generated *Lhb Lhr* double null mutants. We report testis phenotypes and quantitative expression data on key steroidogenesis enzymes and other testis cell marker genes in these double null mutants. Our studies reveal that a one-to-one LH-LHR signaling exists in vivo in mouse testis and genetically confirm that *Lhb Lhr* double mutant mice do not exhibit any additional testis phenotypes compared to individual null mutants lacking either *Lhb* or *Lhr*. Furthermore, our studies unequivocally confirm that fetal Leydig cells arise independent of LH-LHR-mediated signaling during mouse testis development.

2. Results

2.1. Generation of Lhb$^{-/-}$ Lhr$^{-/-}$ Double Mutant Mice

In the mouse *Lhb* is localized to chromosome 7 and *Lhr* is localized to chromosome 17. To generate double mutant mice lacking both *Lhb* and *Lhr*, we set up a two-step breeding scheme. We first obtained *Lhb*$^{+/-}$ *Lhr*$^{+/-}$ double heterozygous mice and subsequently intercrossed these *Lhb*$^{+/-}$ *Lhr*$^{+/-}$ double heterozygous mice (Supplementary Figure S1). This scheme successfully resulted in generation of *Lhb*$^{-/-}$ *Lhr*$^{-/-}$ double mutant male mice in the expected 1:32 frequency. *Lhb*$^{-/-}$ *Lhr*$^{-/-}$ double mutant male mice were viable and developed normally.

2.2. Lhb$^{-/-}$ Lhr$^{-/-}$ Double Mutant Male Mice Display Testis Phenotypes Similar to Single Null Mice Lacking Only Lhb or Lhr

Testis histology of single mutant adult male mice lacking either LH ligand (*Lhb*$^{-/-}$) or LHR (*Lhr*$^{-/-}$) demonstrate severely reduced interstitium between tubules and spermatogenesis arrest, when compared to testis histology in control WT mice [18,20,21] and as shown in Figure 1. To determine how the combined absence of LH ligand and LHR would affect testis development, we performed histological analysis on testis obtained from *Lhb*$^{-/-}$

$Lhr^{-/-}$ double mutant male mice at age 56d and compared to that in age-matched controls (*Ctrl*). Grossly, we did not find any histological differences among PAS-hematoxylin-stained testis sections obtained from $Lhb^{-/-}$ or $Lhr^{-/-}$ single or $Lhb^{-/-}$ $Lhr^{-/-}$ double mutants (Figure 1A–C). The interstitium in double mutant testis tubules and lumen were similarly insignificant and spermatogenesis was arrested at the round spermatid stage and elongated spermatids were absent (Figure 1D). Consistent with the testis histology, testis weight (Figure 1E) and tubule diameter (Figure 1F) were significantly suppressed compared to *Ctrl* group ($p < 0.05$, One way ANOVA, *Ctrl* vs. $Lhb^{-/-}$ or $Lhr^{-/-}$ or $Lhb^{-/-}$ $Lhr^{-/-}$; n = 3 mice for testis weights and n = 350 tubules for tubule diameter measurements from multiple sections obtained from 3 mice) but were comparable between $Lhb^{-/-}$ or $Lhr^{-/-}$ single and $Lhb^{-/-}$ $Lhr^{-/-}$ double mutants ($p > 0.05$, One way ANOVA, $Lhb^{-/-}$ vs. $Lhr^{-/-}$ or $Lhb^{-/-}$ $Lhr^{-/-}$; $Lhr^{-/-}$ vs. $Lhb^{-/-}$ $Lhr^{-/-}$. Expression of 3-beta-hydroxysteroid dehydrogenase-1 (HSD3B1) is found in both fetal and adult Leydig cells [22–24]. We performed immunofluorescence analysis (Figure 2A–D) on testis sections and counted HSD3B1$^+$ Leydig cells within interstitial spaces. There was ~58–64% reduction in HSD3B1$^+$ Leydig cells in testis of single or double mutants lacking only LH or only LHR or both LH and LHR compared to control mice (Figure 2E). Thus, testis histo-morphological phenotypes were similar with no additional apparent abnormalities in $Lhb^{-/-}$ $Lhr^{-/-}$ mice when compared to those in mice lacking only *Lhb* or *Lhr*.

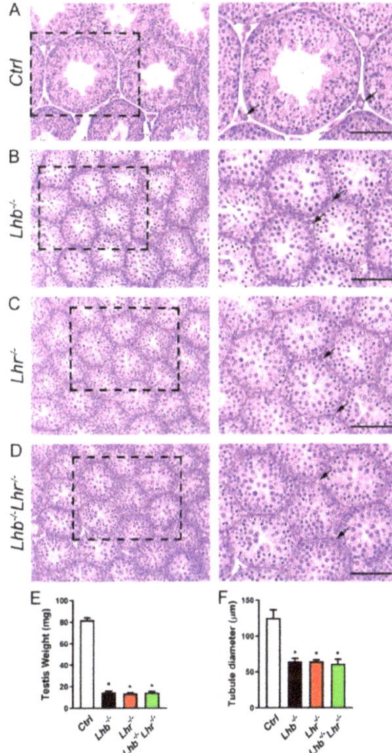

Figure 1. Histo-morphological analysis of testis from single and double null mutant males. PAS-hematoxylin-stained testis sections (**A–D**) indicate that compared to the presence of well-defined interstitium (black arrow in high magnification image of panel (**A**)) and all stages of spermatogenesis including late stage spermatids in testis of a control (*Ctrl*) mouse (**A**), the interstitium is poorly developed in single ($Lhb^{-/-}$, panel (**B**) and $Lhr^{-/-}$, panel (**C**)) and double mutant ($Lhb^{-/-}$ $Lhr^{-/-}$, panel (**D**)) mouse testis sections (black arrows in high magnification images of panels (**B–D**)). Testis

weights (**E**) and tubule dimeter (**F**) are significantly reduced in single and double mutant mice compared to controls (* $p < 0.05$, *Ctrl* vs. *Lhb*$^{-/-}$ or *Lhr*$^{-/-}$ or *Lhb*$^{-/-}$ *Lhr*$^{-/-}$, One-way ANOVA) but are similar among single and double mutants ($p > 0.05$, *Lhb*$^{-/-}$ vs. *Lhr*$^{-/-}$ or *Lhb*$^{-/-}$ *Lhr*$^{-/-}$; *Lhr*$^{-/-}$ vs. *Lhb*$^{-/-}$ *Lhr*$^{-/-}$, One-way ANOVA). Testis obtained from adult male mice at 56d of age were used for histo-morphological analysis. Black squares in panels (**A–D**) are the regions magnified and represented on the right panels. In panel (**E**): * $p < 0.05$, one-way ANOVA, n = 3. In panel (**F**): * $p < 0.05$, one-way ANOVA, n = 350 tubules from multiple testis sections obtained from 3 mice. Scale bar in panels (**A–D**) represents 50 μm.

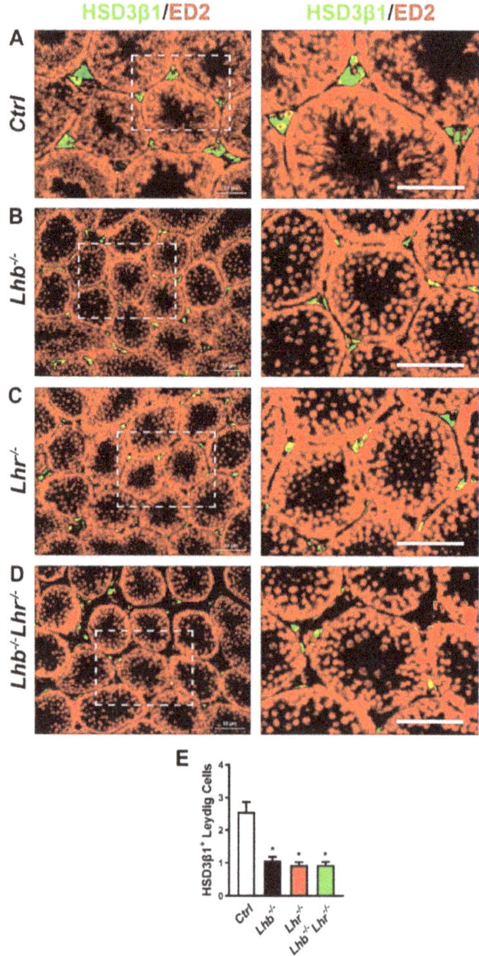

Figure 2. Immunofluorescence analysis of Leydig cells in testis sections. HSD3β1$^+$ Leydig cells are visualized in interstitial spaces in testis sections obtained from control mice (*Ctrl*; (**A**)), mice lacking either only LH (**B**), or only LHR (**C**) or both LH and LHR (**D**). Regions indicated in white squares are shown as higher magnification images in right side panels in each case. Nuclei are stained in red. White bar represents 50 μm. HSD3β1$^+$ Leydig cells in multiple interstitial spaces are quantified (**E**). Loss of LH or LHR or both results in significant suppression (58–64%) of HSD3β1$^+$ Leydig cells compared to the control group (* $p < 0.05$, single or double mutant vs. *Ctrl*; at least 100 interstitial spaces per genotype) but no differences were noted in HSD3β1$^+$ Leydig cells between single or double mutant sections ($p > 0.05$, single or double mutant vs. *Ctrl*; at least 100 interstitial spaces per genotype).

2.3. Testicular Gene Expression Changes and T Levels in Lhb$^{-/-}$ Lhr$^{-/-}$ Double Mutant Are Similar to Those in Mice Lacking either Lhb or Lhr

Our previous work identified that loss of LH ligand or LHR results in profound changes in gene expression in multiple testis cell types including Leydig, Sertoli and germ cells in male mice [18,20,21,25,26]. To test the effect of combined loss of both the LH ligand and LHR on testis gene expression, we analyzed testis cell type-specific key marker genes by Taqman qPCR analysis (as shown below and Supplementary Figure S2). As predicted, expression of mature adult type Leydig cell marker genes was suppressed in the absence of either LH or LHR in testes of *Lhb* or *Lhr* single null mutants. Loss of both LH and LHR in *Lhb$^{-/-}$ Lhr$^{-/-}$* double mutants did not show any synergistic effect or additional changes in the expression of these marker genes (Figure 3A–G,I–L). One exception noted was that of *Srd5a1* which did not show any change in the absence of LH or LHR or in the combined absence of both LH ligand and receptor (Figure 3H). Similarly, *Hsd17b1* showed a trend towards suppression (Figure 3I). *Lifr* is normally expressed in multiple testis cell types including adult progenitor Leydig cells, Sertoli and peritubular myoid cells [27,28]. Its expression was also similarly suppressed in testis of double mutants as well as single mutants compared to that in controls (Figure 3M). Of the two *Hsd3b* genes, expression of the mature Leydig cell-specific (*Hsd3b6*) but not the fetal onset form (*Hsd3b1*) was profoundly suppressed (Figure 3D,L). Similarly, expression of *Insl3*, an adult mature Leydig cell-specific marker gene was significantly suppressed in both single and double mutants (Figure 3K).

Most importantly, expression of *Thbs2*, which is the fetal Leydig cell-specific marker was significantly upregulated in testis of single and double mutants compared to that in *Ctrl* mice (Figure 3N). These data indicate that while most of the adult mature Leydig cell marker genes which encode key T biosynthesis enzymes, were similarly suppressed in the absence of either LH or LHR or both, fetal Leydig cell-specific marker gene (*Thbs2*) was upregulated in the absence of LH or LHR or combined absence of both. Thus, fetal Leydig cells accumulate in testis in the absence of LH or LHR or in the combined absence of both LH and LHR.

In response to LH stimulation, Leydig cell-derived T diffuses into tubules and binds androgen receptor (AR) in Sertoli cells to regulate spermatogenesis. The clear suppression of most of the T biosynthesis enzyme-encoding mRNAs suggests that T levels must also be suppressed. To further determine, if loss of both LH and LHR results in any additive effect, we quantified both serum (Figure 4A) and intratesticular T (ITT) levels (Figure 4B) by a specific ELISA. Loss of LH or LHR resulted in significant suppression of both serum (Mean $\pm$ SEM values are *Ctrl* = 0.5 $\pm$ 0.15 ng/mL; *Lhb$^{-/-}$* = 0.13 $\pm$ 0.03 ng/mL; *Lhr$^{-/-}$* = 0.17 $\pm$ 0.07 ng/mL, $p < 0.05$ *Ctrl* vs. single mutant, one-way ANOVA, n = 4–6) and ITT (Mean $\pm$ SEM values are *Ctrl* = 4.58 $\pm$ 0.04 ng/mg; *Lhb$^{-/-}$* = 0.04 $\pm$ 0.003 ng/mg; *Lhr$^{-/-}$* = 0.04 $\pm$ 0.003 ng/mg, $p < 0.05$ *Ctrl* vs. single mutant, one-way ANOVA, n = 4–6) compared to those in controls as reported earlier [18,20,21], but did not get further suppressed in the combined absence of both LH and LHR in double null mutants (Figure 4A,B). Serum T (0.15 $\pm$ 0.04 ng/mL) and ITT (0.03 $\pm$ 0.006 ng/mg) values in *Lhb$^{-/-}$ Lhr$^{-/-}$* mice were not significantly different compared to those in single mutants ($p > 0.05$ *Lhb$^{-/-}$ Lhr$^{-/-}$* vs. *Lhb$^{-/-}$* or *Lhr$^{-/-}$*, one-way ANOVA, n = 4–6).

Next, we assessed expression of Sertoli cell-specific marker genes in testis of *Lhb* or *Lhr* single null and *Lhb Lhr* double null mutants. The expression of *Fshr* and *Ar* was under negative regulation of T [29–32]. Accordingly, testicular expression of *Fshr* and *Ar* was upregulated in the absence of only LH or LHR or combined absence of both LH and LHR (Figure 5A,B). The expression of other androgen-responsive Sertoli cell targets including *Nr5a1*, *Rhox5* and *Clu* was suppressed similarly in single and double null mutants (Figure 5C–E). Thus, absence of LH signaling pathway results in defects in Sertoli cell gene expression secondary to loss of androgen action.

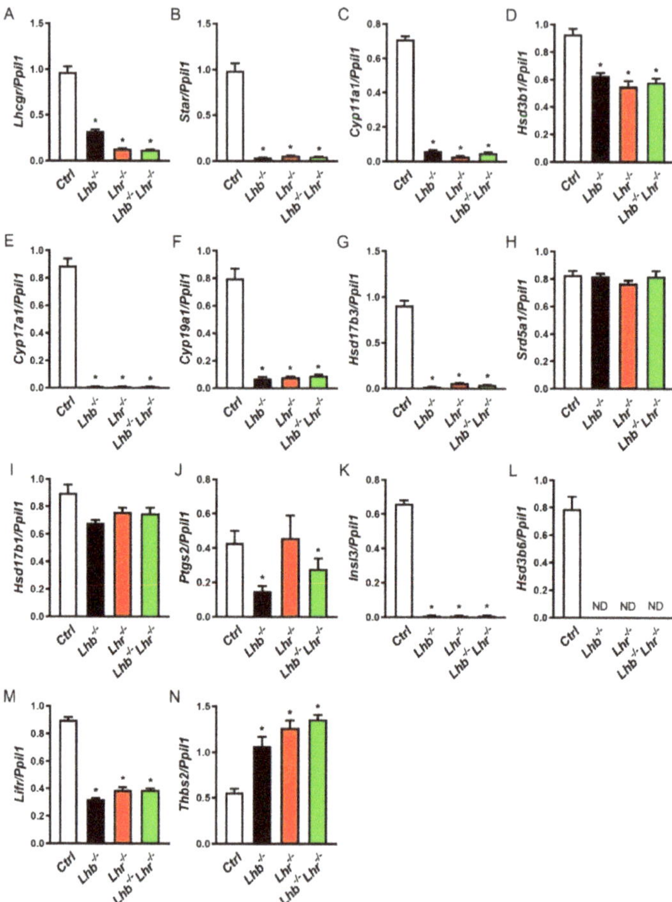

Figure 3. Taqman qPCR analysis of Leydig cell marker genes. Expression of Leydig cell marker genes was quantified in testis samples obtained from adult male mice (n = 3–4) at 56 d of age. Expression of almost all of the selected mature Leydig cell genes was suppressed in the absence of only LH or LHR or both compared to that in *Ctrl* group but is similar among single and double mutants. *Srd5a1* expression was unaffected in the absence of LH-LHR signaling (**H**). Expression of *Hsd17b1* showed a trend towards suppression (**I**). Adult Leydig cell progenitor marker gene *Lifr* was suppressed in single or double mutants compared to that in *Ctrl* (**M**). Expression of *Thbs2* was significantly upregulated in single and double mutants compared to *Ctrl* (**N**) indicating that fetal Leydig cells develop and accumulate in the absence of either only LH or LHR or both. (**A–N**) panels: * $p < 0.05$, One-way ANOVA, *Ctrl* vs. $Lhb^{-/-}$ or $Lhr^{-/-}$ or $Lhb^{-/-}$ $Lhr^{-/-}$; $p > 0.05$, One-way ANOVA, $Lhb^{-/-}$ vs. $Lhr^{-/-}$ or $Lhb^{-/-}$ $Lhr^{-/-}$; $Lhr^{-/-}$ vs. $Lhb^{-/-}$ $Lhr^{-/-}$, in all panels except panel J, where a discordant expression was noted. For all qPCR assays, expression of *Ppil1* was used as internal control and cDNA samples in triplicate were used from testis obtained from 3 mice.

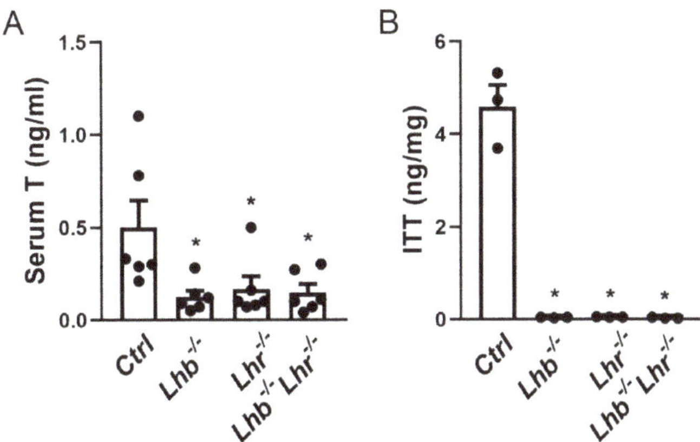

Figure 4. Serum and ITT assays in single and double mutants. Serum (**A**) and intratesticular T (ITT) (**B**) was quantified by an ELISA. Compared to *Ctrl*, both serum (*Ctrl* = 0.5 ± 0.15 ng/mL; *Lhb*$^{-/-}$ = 0.13 ± 0.03 ng/mL; *Lhr*$^{-/-}$ = 0.17 ± 0.07 ng/mL; *Lhb*$^{-/-}$ *Lhr*$^{-/-}$ = 0.15 ± 0.04 ng/mL, * $p < 0.05$, One-way ANOVA, *Ctrl* vs. *Lhb*$^{-/-}$, *Lhr*$^{-/-}$ or *Lhb*$^{-/-}$ *Lhr*$^{-/-}$, n = 4–6 mice) and ITT (*Ctrl* = 4.58 ± 0.04 ng/mg; *Lhb*$^{-/-}$ = 0.04 ± 0.003 ng/mg; *Lhr*$^{-/-}$ = 0.04 ± 0.003 ng/mg; *Lhb*$^{-/-}$ *Lhr*$^{-/-}$ = 0.03 ± 0.006 ng/mg, * $p < 0.05$, One-way ANOVA, *Ctrl* vs. *Lhb*$^{-/-}$, *Lhr*$^{-/-}$ or *Lhb*$^{-/-}$ *Lhr*$^{-/-}$, n = 4–6 mice) are significantly suppressed in single or double mutants. No additive effect of loss both LH and LHR are noted. The serum T and ITT values are not significantly different among single or double mutants ($p > 0.05$, One-way ANOVA, *Lhb*$^{-/-}$ vs. *Lhr*$^{-/-}$ or *Lhb*$^{-/-}$ *Lhr*$^{-/-}$, *Lhr*$^{-/-}$ vs. *Lhb*$^{-/-}$ *Lhr*$^{-/-}$, n = 4–6 mice).

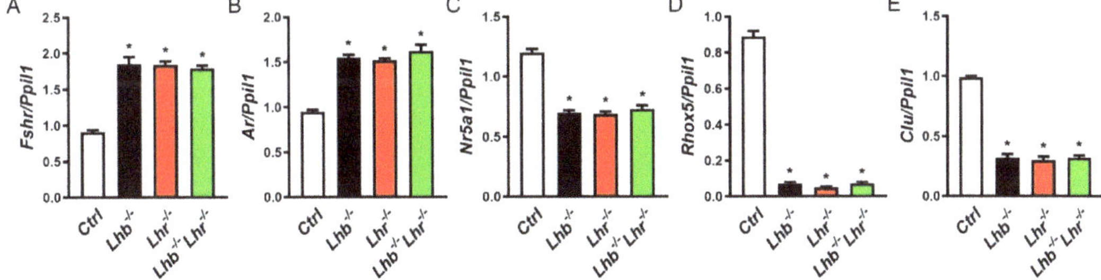

Figure 5. Taqman qPCR analysis of Sertoli cell marker genes. T-repressed genes namely, *Fshr* (**A**) and *Ar* (**B**) are upregulated in the absence of LH or LHR or both. In contrast, those that are positively regulated by T, such as *Nr5a1*, *Rhox5* and *Clu* are similarly suppressed in the absence of LH or LHR or both (**C–E**). * $p < 0.05$, One-way ANOVA, *Ctrl* vs. *Lhb*$^{-/-}$ or *Lhr*$^{-/-}$ or *Lhb*$^{-/-}$ *Lhr*$^{-/-}$; $p > 0.05$, One-way ANOVA, *Lhb*$^{-/-}$ vs. *Lhr*$^{-/-}$ or *Lhb*$^{-/-}$ *Lhr*$^{-/-}$; *Lhr*$^{-/-}$ vs. *Lhb*$^{-/-}$ *Lhr*$^{-/-}$. For all qPCR assays, expression of *Ppil1* was used as internal control and cDNA samples in triplicate were used from testis obtained from 3 mice.

Finally, we assessed expression of germ cell marker genes by qPCR assays. As predicted from the histological analysis of testis cell types, early stage germ cell markers may not be affected in the absence of LH or LHR or both. Indeed, expression of *Plzf* (stem cell progenitor marker), *Stra8* and *Kit* (spermatocyte markers) was not suppressed in the single or double mutants (Figure 6A–C). Expression of other meiotic / post-meiotic (*Tex14*, *Sycp 2* and *Sycyp 3*) and spermatid (*Acrv1*) markers were significantly suppressed in testis of single and double mutants compared to those in controls (Figure 6D,F–H). Thus, similar to gene expression changes in Leydig and Sertoli cells, combined loss of both LH and LHR did not also result in any additive effects in expression of germ cell marker genes in testis of double mutants.

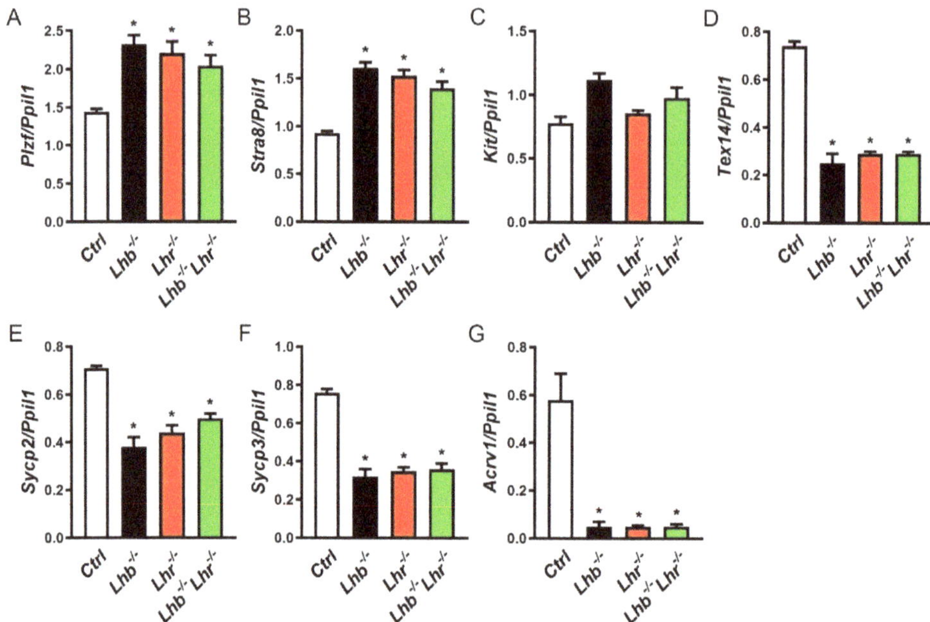

Figure 6. Taqman qPCR analysis of germ cell marker genes. Early germ cell markers such as those expressed in male germline stem cell progenitor (*Plzf*, (**A**)) and spermatocyte (*Stra8*, (**B**); *Kit*, (**C**)) are upregulated or not affected in the absence of LH or LHR or both. Genes that play critical roles in meiosis (*Tex14*, *Sycp2*, *Sycp3*, (**D–F**)) and spermatids (*Acrv1*, (**G**)) are significantly suppressed in single and double mutants. No additive effect of loss of both LH and LHR are noted on expression of any of the genes tested. * $p < 0.05$, One-way ANOVA, *Ctrl* vs. $Lhb^{-/-}$ or $Lhr^{-/-}$ or $Lhb^{-/-}$ $Lhr^{-/-}$; $p > 0.05$, One-way ANOVA, $Lhb^{-/-}$ vs. $Lhr^{-/-}$ or $Lhb^{-/-}$ $Lhr^{-/-}$; $Lhr^{-/-}$ vs. $Lhb^{-/-}$ $Lhr^{-/-}$. For all qPCR assays, expression of *Ppil1* was used as internal control and cDNA samples in triplicate were used from testis obtained from 3 mice.

3. Discussion

Earlier studies by us and others established that single mutants lacking genes encoding LH ligand and LHR phenocopy each other [18,20,21]. Using these well-characterized genetic models, here we have generated double mutant mice lacking both the LH Ligand and LHR. This genetic approach allowed us to directly test in vivo LH ligand-independent actions mediated via LHRs and whether LH acts via receptors other than LHR. Our histomorphological analysis of testis and testicular gene expression analysis in single and double null mutants clearly indicate that combined absence of LH-LHR signaling pathway does not result in any additional phenotypes not observed in single mutants lacking only LH ligand or LHR. Our in vivo genetic approach confirms that it is unlikely LH signals through non-LHRs or other ligands signal through LHR to regulate testis development, spermatogenesis and male fertility. Our *Lhb Lhr* double null mutant mice represent the first genetic model in which signaling via the LH ligand-LH receptor pair is completely absent with only FSH action uniquely present in the male. Additionally, both serum T and ITT are profoundly suppressed in double mutants, similar to those observed in single mutants.

As evident from testicular histology, single and double mutants are indistinguishable. In all three genotypes ($Lhb^{-/-}$, $Lhr^{-/-}$ and $Lhb^{-/-}$ $Lhr^{-/-}$), the interstitium is poorly developed with no evidence of adult mature Leydig cells within the interstitium and the lumen is reduced in tubules. Further ultrastructural studies will be needed to confirm the exact morphological identity of the Leydig cell lineage in the absence of only LH or LHR or both. During mouse testis development, two distinct populations of Leydig cells arise. Fetal Leydig cell lineage is present in the absence of only LH or only LHR [18,20,21]. These earlier studies suggested a possibility that non-LH-LHR mediated signaling events may regulate fetal Leydig cells because in single mutants lacking LH ligand, LHRs are present and may mediate actions of ligands other than LH. Similarly, in mutants lacking LHRs, LH ligand is present and may bind receptors other than LHR. The continued presence of fetal Leydig cells as indicated by significantly upregulated *Thbs2* gene expression in double null mutants lacking both LH and LHR and similarly observed in single mutants, provides unequivocal direct genetic evidence that LH-LHR signaling is not required for fetal Leydig cell development in vivo. Our observations on fetal Leydig cells are consistent with those by Shima et al. who used lineage tracing methods and adult *Ar* knockout mice [16] and as summarized [13,24]. It is possible that other ligands such as oxytocin and ACTH and their corresponding signaling pathways may play a role in mouse fetal Leydig cell specification in the absence of LH/LHR [10,11,33–38]. Alternatively, other neuroendocrine or locally produced peptides within the testis may also regulate Leydig cells [36,38].

Testis cell type-specific marker gene expression analysis (Leydig cell, Sertoli cell and germ cell-specific markers) revealed identical changes in double null mutants compared to those in single mutants further reinforcing that no additive effects are observed in the combined absence of both LH and LHR. Because we noted both up- and down regulation of genes, these reflect true gene expression changes but not as a result of cell death (Figures 3, 5 and 6 and Supplementary Figure S2). Interestingly, expression of *Ptgs2* that encodes prostaglandin synthase-2 is discordant between single and double mutants. Similarly, expression of *Srd5a1* is not altered in single and double mutants compared to that in controls. The exact mechanism for regulating expression of these two enzyme-encoding genes in T biosynthesis pathway remains to be identified. All other mature Leydig cell-, Sertoli- and germ cell- marker genes were identically regulated in the absence of only LH or LHR or both and secondary to absence of T action. Our future RNASeq studies may allow us to identify large-scale gene expression changes and gene networks differentially regulated in testis of single and double mutants.

In summary, genetic analysis of double mutant male mice lacking both LH ligand and LHR reveals identical testis phenotypes observed in individual mutants lacking only LH or only LHR. Significantly upregulated expression of *Thbs2* in testis of single mutants or double mutants lacking both LH and LHR confirms that LH-LHR signaling is not essential for fetal Leydig cell development. Thus, our genetic model provides a novel source of factors that regulate fetal Leydig cells and may provide novel insights into fetal Leydig cell biology and early events in masculinization during testis development. Loss of LH-LHR pathway results in profound suppression of T and pharmacological rescue of single mutants lacking only LH or only LHR by T has already been achieved [18,20,39]. Therefore, our double mutants may prove useful to further understand LH/LHR-independent actions of T in Leydig cell development and spermatogenesis.

4. Materials and Methods

4.1. Mice

All experimental procedures were carried out on adult male mice (8–10 weeks) and are in accordance with NIH guidelines and approved by the Institutional Animal Care and Use Committee at the University of Colorado Anschutz Medical Campus. *Lhb* null mutant mice were previously generated as described [18]. *Lhr* null mutant mice were generated as previously described [20]. To generate *Lhb Lhr* double null mutants, we initially crossed *Lhb*$^{+/-}$ mice with *Lhr*$^{+/-}$ mice and obtained *Lhb*$^{+/-}$ *Lhr*$^{+/-}$ double heterozygous mice. In the second step, we intercrossed these *Lhb*$^{+/-}$ *Lhr*$^{+/-}$ double heterozygous mice and obtained *Lhb*$^{-/-}$ *Lhr*$^{-/-}$ double mutant mice. Double mutant male mice were used for all present studies and maintained on C57/BL6/129SvEv/129SVJ hybrid genetic background. Mice were housed in rooms equipped with controlled conditions of temperature and humidity and maintained on a 12 h light: dark light cycle with autoclaved standard rodent chow and water supplied ad libitum. For genotyping, tail DNA samples prepared by Millipore (Millipore-Sigma, St. Louis, MO, USA) columns were used in PCR reactions using *Lhb* and *Lhr* allele-specific primer pairs. These primers distinguish the wild-type (WT) and mutant alleles in each case by the size of the amplified DNA fragments separated on ethidium bromide-stained agarose gels as described [20,40].

4.2. Histological and Immunofluorescence Analysis

Testes were harvested from adult mice (n = 3) under isoflurane anesthesia, weighed and one testis was immediately fixed in Bouin's reagent solution (Millipore-Sigma, St. Louis, MO, USA) overnight with constant shaking at room temperature and changed into 70% ethanol. The paraffin-embedded sections were cut at 6 μm thickness, deparaffinized and rehydrated by serial immersion in xylene, followed by graded series of alcohol and stained with periodic acid-Schiff's reagent (PAS) and hematoxylin as described [18,41–43]. The images of stained testis sections were digitally captured using a Leica microscope and used for tubule diameter calculations as described [43,44]. Immunofluorrescence was performed on testes sections using a rabbit antibody (1:2000) against HSD3B1 (gift from Dr. Buck Hales) and counterstained with Ethidium Homodimer-2 (E3599, Invitrogen, Carlsbad, CA, USA) for visualization of nuclei in cells as described [42,45]. A goat anti-rabbit IgG-Alexa flour-488 conjugated second antibody (Invitrogen, Carlsbad, CA, USA) was used at a dilution of 1:200. Antibody-stained sections were mounted with ProLong Diamond Antifade Mountant (P36970, Life Technologies, Carlsbad, CA, USA), and observed under an epifluorescence microscope (Leica).

4.3. RNA Isolation and Taqman qPCR Assays

Total RNA was extracted from testes using RNeasy Mini Kit (74106, QIAGEN, Germantown, MD, USA). One μg of RNA was reverse transcribed into cDNA using SuperScript™—III Reverse Transcriptase (18080-093, Invitrogen, Carlsbad, CA, USA). PCR was performed in 10 μL of reaction volume containing 2 μL of cDNA which was diluted at 1:40, 0.05 μM each of Primer/probe combos, and 5 μL of 2× PrimeTime Gene Expression Master Mix (1055772, Integrated DNA Technologies, Coralville, IA, USA) using QuantStudio 6 Flex Real-Time PCR machine. The relative standard curve method was used for gene expression quantification as described [42,43]. For each primer, a series of dilution of standard cDNA at 1:5, 1:10 and 1:50 were assigned the quantity as 2000, 1000 and 200. Relative mRNA levels of target genes normalized to *Ppil1* expression were obtained and the ratios were presented. Predesigned mouse qPCR Primer/probe combos were purchased from Integrated DNA Technologies, Inc., Coralville, IA, USA. For qPCR assays, triplicate cDNA samples were used from testis obtained from at least 3 mice.

4.4. Testosterone Assays

Mice were exsanguinated under isoflurane anesthesia and serum was separated in a table top centrifuge at room temperature and stored frozen at −80 °C until further use. Frozen testis samples were processed as described [46] and the resulting supernatant aliquots were used for intratesticular T measurements. Total protein in testicular extracts was quantified by BCA protein assay kit (BioRad, Hercules, CA, USA) using bovine serum albumin as standard. Serum (ng/mL) and intratesticular T (ng/mg) levels were measured in 25 μL sample aliquots by an ELISA kit (TE187S-100, Calbiotech, Inc., El Cajon, CA, USA) according to manufacturer's instructions. Samples were prepared from tissues collected from 3–6 mice. The assay sensitivity is 0.1 ng/mL and intra-assay and inter-assay % CVs are 6.4 and 9.7, respectively.

4.5. Statistical Analysis

Data are presented as the mean $\pm$ standard error of the mean (SEM). Statistical significance was determined using ANOVA followed by Turkey's post hoc test. $p < 0.05$ was considered statistically significant. Statistical analyses were performed using PRISM software (version 9.4.1, GraphPad Software, Inc., San Diego, CA, USA).

Supplementary Materials: The following supporting information can be downloaded at: https://www.mdpi.com/article/10.3390/ijms232415725/s1.

Author Contributions: Conceptualization and Research design, T.R.K.; Data generation and analysis, Z.L. (Zhenghui Liu), M.L. and T.R.K.; resources, Z.L. (Zhenmin Lei) and C.V.R.; original draft and manuscript writing, T.R.K.; supervision and project administration, T.R.K.; funding acquisition, T.R.K. All authors have read and agreed to the published version of the manuscript.

Funding: T.R.K. received funding in part for this research project by The Makowski Family Endowment (CUE 63470698) and Gonadotropin Research Fund (CU 0222933) established at the University of Colorado Anschutz Medical Campus, Aurora, CO, USA.

Institutional Review Board Statement: The animal study protocol was approved by the University of Colorado Anschutz Medical Campus (protocol number 0161, T.R.K.), Aurora, CO, USA and approved on 11 March 2022. All experiments were in accordance with the National Institute of Health animal research policies.

Informed Consent Statement: Not applicable.

Acknowledgments: We thank Carolyn Sadler for help with mouse genotyping and Buck Hales for generously providing HSD3B1 antibody used in immunofluorescence experiments.

Conflicts of Interest: The authors declare no conflict of interest.

Abbreviations

AR = Androgen receptor; LH = Luteinizing hormone; LHR = LH-receptor; T = Testosterone

References

1. Bousfield, G.R.; Jia, L.; Ward, D.N. Gonadotropins: Chemistry and biosynthesis. In *Knobil's Physiology of Reproduction*, 3rd ed.; Neill, J.D., Ed.; Academic Press: Amsterdam, The Netherlands, 2006; Volume I, pp. 1581–1634.
2. Narayan, P.; Ulloa-Aguirre, A.; Dias, J.A. *Yen & Jaffe's Reproductive Endocrinology*, 8th ed.; Strauss, J.F., III, Barbieri, R.L., Eds.; Elsevier: Philadelphia, PA, USA, 2019; pp. 25–57.
3. Ulloa-Aguirre, A.; Dias James, A.; Bousfield, G.R. Gonadotropins. In *Endocrinology of the Testis and Male Reproduction*; Simoni, M., Huhtaniemi, I., Eds.; Springer International, AG: Berlin/Heidelberg, Germany, 2017; pp. 1–52.
4. O'Hara, L.; Smith, L.B. Androgen receptor roles in spermatogenesis and infertility. *Best Pract. Res. Clin. Endocrinol. Metab.* **2015**, *29*, 595–605. [CrossRef] [PubMed]
5. O'Shaughnessy, P.J. Hormonal control of germ cell development and spermatogenesis. *Semin. Cell Dev. Biol.* **2014**, *29*, 55–65. [CrossRef] [PubMed]
6. Chen, H.; Ge, R.S.; Zirkin, B.R. Leydig cells: From stem cells to aging. *Mol. Cell Endocrinol.* **2009**, *306*, 9–16. [CrossRef] [PubMed]
7. Chen, H.; Wang, Y.; Ge, R.; Zirkin, B.R. Leydig cell stem cells: Identification, proliferation and differentiation. *Mol. Cell Endocrinol.* **2017**, *445*, 65–73. [CrossRef] [PubMed]
8. Chen, P.; Zirkin, B.R.; Chen, H. Stem Leydig Cells in the Adult Testis: Characterization, Regulation and Potential Applications. *Endocr. Rev.* **2020**, *41*, 22–32. [CrossRef]
9. Inoue, M.; Baba, T.; Morohashi, K.I. Recent progress in understanding the mechanisms of Leydig cell differentiation. *Mol. Cell Endocrinol.* **2018**, *468*, 39–46. [CrossRef]
10. O'Shaughnessy, P.J.; Baker, P.; Sohnius, U.; Haavisto, A.M.; Charlton, H.M.; Huhtaniemi, I. Fetal development of Leydig cell activity in the mouse is independent of pituitary gonadotroph function. *Endocrinology* **1998**, *139*, 1141–1146. [CrossRef]
11. O'Shaughnessy, P.J.; Baker, P.J.; Johnston, H. The foetal Leydig cell—Differentiation, function and regulation. *Int. J. Androl.* **2006**, *29*, 90–95; discussion 105–108. [CrossRef]
12. Teerds, K.J.; Huhtaniemi, I.T. Morphological and functional maturation of Leydig cells: From rodent models to primates. *Hum. Reprod Update* **2015**, *21*, 310–328. [CrossRef]
13. Wen, Q.; Cheng, C.Y.; Liu, Y.X. Development, function and fate of fetal Leydig cells. *Semin. Cell Dev. Biol.* **2016**, *59*, 89–98. [CrossRef]

14. Barsoum, I.B.; Yao, H.H. Fetal Leydig cells: Progenitor cell maintenance and differentiation. *J. Androl.* **2010**, *31*, 11–15. [CrossRef]
15. Inoue, M.; Shima, Y.; Miyabayashi, K.; Tokunaga, K.; Sato, T.; Baba, T.; Ohkawa, Y.; Akiyama, H.; Suyama, M.; Morohashi, K. Isolation and Characterization of Fetal Leydig Progenitor Cells of Male Mice. *Endocrinology* **2016**, *157*, 1222–1233. [CrossRef] [PubMed]
16. Shima, Y.; Matsuzaki, S.; Miyabayashi, K.; Otake, H.; Baba, T.; Kato, S.; Huhtaniemi, I.; Morohashi, K. Fetal Leydig Cells Persist as an Androgen-Independent Subpopulation in the Postnatal Testis. *Mol. Endocrinol.* **2015**, *29*, 1581–1593. [CrossRef] [PubMed]
17. Shima, Y.; Morohashi, K.I. Leydig progenitor cells in fetal testis. *Mol. Cell Endocrinol.* **2017**, *445*, 55–64. [CrossRef] [PubMed]
18. Ma, X.; Dong, Y.; Matzuk, M.M.; Kumar, T.R. Targeted disruption of luteinizing hormone beta-subunit leads to hypogonadism, defects in gonadal steroidogenesis, and infertility. *Proc. Natl. Acad. Sci. USA* **2004**, *101*, 17294–17299. [CrossRef]
19. Kumar, T.R. Functional analysis of LHbeta knockout mice. *Mol. Cell Endocrinol.* **2007**, *269*, 81–84. [CrossRef] [PubMed]
20. Lei, Z.M.; Mishra, S.; Zou, W.; Xu, B.; Foltz, M.; Li, X.; Rao, C.V. Targeted disruption of luteinizing hormone/human chorionic gonadotropin receptor gene. *Mol. Endocrinol.* **2001**, *15*, 184–200. [CrossRef]
21. Zhang, F.P.; Poutanen, M.; Wilbertz, J.; Huhtaniemi, I. Normal prenatal but arrested postnatal sexual development of luteinizing hormone receptor knockout (LuRKO) mice. *Mol. Endocrinol.* **2001**, *15*, 172–183. [CrossRef]
22. O'Shaughnessy, P.J.; Willerton, L.; Baker, P.J. Changes in Leydig cell gene expression during development in the mouse. *Biol. Reprod.* **2002**, *66*, 966–975. [CrossRef]
23. Yokoyama, C.; Chigi, Y.; Baba, T.; Ohshitanai, A.; Harada, Y.; Takahashi, F.; Morohashi, K.I. Three populations of adult Leydig cells in mouse testes revealed by a novel mouse HSD3B1-specific rat monoclonal antibody. *Biochem. Biophys. Res. Commun.* **2019**, *511*, 916–920. [CrossRef]
24. Zirkin, B.R.; Papadopoulos, V. Leydig cells: Formation, function, and regulation. *Biol. Reprod* **2018**, *99*, 101–111. [CrossRef] [PubMed]
25. Rao, C.V.; Lei, Z.M. Consequences of targeted inactivation of LH receptors. *Mol. Cell Endocrinol.* **2002**, *187*, 57–67. [CrossRef] [PubMed]
26. Zhang, F.P.; Pakarainen, T.; Zhu, F.; Poutanen, M.; Huhtaniemi, I. Molecular characterization of postnatal development of testicular steroidogenesis in luteinizing hormone receptor knockout mice. *Endocrinology* **2004**, *145*, 1453–1463. [CrossRef] [PubMed]
27. Curley, M.; Darbey, A.; O'Donnell, L.; Kilcoyne, K.R.; Wilson, K.; Mungall, W.; Rebourcet, D.; Guo, J.; Mitchell, R.T.; Smith, L.B. Leukemia inhibitory factor-receptor signalling negatively regulates gonadotrophin-stimulated testosterone production in mouse Leydig Cells. *Mol. Cell Endocrinol.* **2022**, *544*, 111556. [CrossRef]
28. Curley, M.; Milne, L.; Smith, S.; Atanassova, N.; Rebourcet, D.; Darbey, A.; Hadoke, P.W.F.; Wells, S.; Smith, L.B. Leukemia Inhibitory Factor-Receptor is Dispensable for Prenatal Testis Development but is Required in Sertoli cells for Normal Spermatogenesis in Mice. *Sci. Rep.* **2018**, *8*, 11532. [CrossRef]
29. Dong, L.; Jelinsky, S.A.; Finger, J.N.; Johnston, D.S.; Kopf, G.S.; Sottas, C.M.; Hardy, M.P.; Ge, R.S. Gene expression during development of fetal and adult Leydig cells. *Ann. N. Y. Acad Sci.* **2007**, *1120*, 16–35. [CrossRef]
30. Heckert, L.L.; Griswold, M.D. The expression of the follicle-stimulating hormone receptor in spermatogenesis. *Recent Prog. Horm. Res.* **2002**, *57*, 129–148. [CrossRef]
31. Shan, L.X.; Bardin, C.W.; Hardy, M.P. Immunohistochemical analysis of androgen effects on androgen receptor expression in developing Leydig and Sertoli cells. *Endocrinology* **1997**, *138*, 1259–1266. [CrossRef]
32. Shan, L.X.; Zhu, L.J.; Bardin, C.W.; Hardy, M.P. Quantitative analysis of androgen receptor messenger ribonucleic acid in developing Leydig cells and Sertoli cells by in situ hybridization. *Endocrinology* **1995**, *136*, 3856–3862. [CrossRef]
33. Bardin, C.W.; Chen, C.L.; Morris, P.L.; Gerendai, I.; Boitani, C.; Liotta, A.S.; Margioris, A.; Krieger, D.T. Proopiomelanocortin-derived peptides in testis, ovary, and tissues of reproduction. *Recent Prog. Horm. Res.* **1987**, *43*, 1–28.
34. Martinez-Arguelles, D.B.; Campioli, E.; Culty, M.; Zirkin, B.R.; Papadopoulos, V. Fetal origin of endocrine dysfunction in the adult: The phthalate model. *J. Steroid Biochem. Mol. Biol.* **2013**, *137*, 5–17. [CrossRef] [PubMed]
35. Nicholson, H.D.; Pickering, B.T. Oxytocin, a male intragonadal hormone. *Regul. Pept.* **1993**, *45*, 253–256. [CrossRef] [PubMed]
36. O'Shaughnessy, P.J.; Baker, P.J.; Johnston, H. Neuroendocrine regulation of Leydig cell development. *Ann. N. Y. Acad. Sci.* **2005**, *1061*, 109–119. [CrossRef] [PubMed]
37. O'Shaughnessy, P.J.; Fleming, L.M.; Jackson, G.; Hochgeschwender, U.; Reed, P.; Baker, P.J. Adrenocorticotropic hormone directly stimulates testosterone production by the fetal and neonatal mouse testis. *Endocrinology* **2003**, *144*, 3279–3284. [CrossRef]
38. Sharpe, R.M.; Maddocks, S.; Kerr, J.B. Cell-cell interactions in the control of spermatogenesis as studied using Leydig cell destruction and testosterone replacement. *Am. J. Anat.* **1990**, *188*, 3–20. [CrossRef]
39. Yuan, F.P.; Lin, D.X.; Rao, C.V.; Lei, Z.M. Cryptorchidism in LhrKO animals and the effect of testosterone-replacement therapy. *Hum. Reprod.* **2006**, *21*, 936–942. [CrossRef]
40. Nagaraja, A.K.; Agno, J.E.; Kumar, T.R.; Matzuk, M.M. Luteinizing hormone promotes gonadal tumorigenesis in inhibin-deficient mice. *Mol. Cell Endocrinol.* **2008**, *294*, 19–28. [CrossRef]
41. Kumar, T.R.; Wang, Y.; Lu, N.; Matzuk, M.M. Follicle stimulating hormone is required for ovarian follicle maturation but not male fertility. *Nat. Genet.* **1997**, *15*, 201–204. [CrossRef]
42. Liu, Z.; Wang, H.; Larsen, M.; Gunewardana, S.; Cendali, F.I.; Reisz, J.A.; Akiyama, H.; Behringer, R.R.; Ma, Q.; Hammoud, S.S.; et al. The solute carrier family 7 member 11 (SLC7A11) is regulated by LH/androgen and required for cystine/glutathione homeostasis in mouse Sertoli cells. *Mol. Cell Endocrinol.* **2022**, *549*, 111641. [CrossRef]

43. Wang, H.; Larson, M.; Jablonka-Shariff, A.; Pearl, C.A.; Miller, W.L.; Conn, P.M.; Boime, I.; Kumar, T.R. Redirecting intracellular trafficking and the secretion pattern of FSH dramatically enhances ovarian function in mice. *Proc. Natl. Acad. Sci. USA* **2014**, *111*, 5735–5740. [CrossRef]
44. Kumar, T.R.; Low, M.J.; Matzuk, M.M. Genetic rescue of follicle-stimulating hormone beta-deficient mice. *Endocrinology* **1998**, *139*, 3289–3295. [CrossRef] [PubMed]
45. Oury, F.; Ferron, M.; Huizhen, W.; Confavreux, C.; Xu, L.; Lacombe, J.; Srinivas, P.; Chamouni, A.; Lugani, F.; Lejeune, H.; et al. Osteocalcin regulates murine and human fertility through a pancreas-bone-testis axis. *J. Clin. Investig.* **2013**, *123*, 2421–2433. [CrossRef] [PubMed]
46. Kumar, T.R.; Varani, S.; Wreford, N.G.; Telfer, N.M.; de Kretser, D.M.; Matzuk, M.M. Male reproductive phenotypes in double mutant mice lacking both FSHbeta and activin receptor IIA. *Endocrinology* **2001**, *142*, 3512–3518. [CrossRef] [PubMed]

International Journal of
Molecular Sciences

Article

Differential Response of Transcription Factors to Activated Kinases in Steroidogenic and Non-Steroidogenic Cells [†]

Kenley Joule Pierre [1] and Jacques J. Tremblay [1,2,*]

[1] Reproduction, Mother and Child Health, Room T3-67, CHU de Québec—Université Laval Research Centre, Québec, QC G1V 4G2, Canada
[2] Centre for Research in Reproduction, Development and Intergenerational Health, Department of Obstetrics, Gynecology and Reproduction, Faculty of Medicine, Université Laval, Québec, QC G1V 0A6, Canada
* Correspondence: jacques-j.tremblay@crchudequebec.ulaval.ca; Tel.: +1-418-525-4444 (ext. 46254)
[†] This research was conducted as part of the requirements for a PhD degree.

Abstract: Hormone-induced Leydig cell steroidogenesis requires rapid changes in gene expression in response to various hormones, cytokines, and growth factors. These proteins act by binding to their receptors on the surface of Leydig cells leading to activation of multiple intracellular signaling cascades, downstream of which are several kinases, including protein kinase A (PKA), Ca^{2+}/calmodulin-dependent protein kinase I (CAMKI), and extracellular signal-regulated protein kinase 1 and 2 (ERK1/2). These kinases participate in hormone-induced steroidogenesis by phosphorylating numerous proteins including transcription factors leading to increased steroidogenic gene expression. How these various kinases and transcription factors come together to appropriately induce steroidogenic gene expression in response to specific stimuli remains poorly understood. In the present work, we compared the effect of PKA, CAMKI and ERK1/2 on the transactivation potential of 15 transcription factors belonging to 5 distinct families on the activity of the *Star* gene promoter. We not only validated known cooperation between kinases and transcription factors, but we also identified novel cooperations that have not yet been before reported. Some transcription factors were found to respond to all three kinases, whereas others were only activated by one specific kinase. Differential responses were also observed within a family of transcription factors. The diverse response to kinases provides flexibility to ensure proper genomic response of steroidogenic cells to different stimuli.

Keywords: Leydig cells; Steroidogenesis; signaling cascade; Nuclear receptors; bZIP factors; MEF2 factors; GATA4; CAMKI; PKA; ERK1/2

Citation: Pierre, K.J.; Tremblay, J.J. Differential Response of Transcription Factors to Activated Kinases in Steroidogenic and Non-Steroidogenic Cells. *Int. J. Mol. Sci.* **2022**, *23*, 13153. https://doi.org/10.3390/ijms232113153

Academic Editor: Jerome F. Strauss, III

Received: 2 October 2022
Accepted: 28 October 2022
Published: 29 October 2022

Publisher's Note: MDPI stays neutral with regard to jurisdictional claims in published maps and institutional affiliations.

1. Introduction

Leydig cells are located in the interstitial space between the seminiferous tubules in the mammalian testis where they produce two hormones, androgens (mainly testosterone) and insulin-like 3 (INSL3). During fetal life, these hormones are essential for masculinization on the male embryo while in postnatal life they are responsible for the development of internal and external male characteristics that occur at puberty, for the initiation and maintenance of spermatogenesis, and for bone metabolism. Testosterone synthesis requires several transporters and enzymes. The process begins with the shuttling of cholesterol, the substrate for all steroid hormones, from the outer to the inner mitochondrial membrane to deliver it to the CYP11A1 enzyme which makes pregnenolone. Pregnenolone then diffuses out of the mitochondria and reaches the smooth endoplasmic reticulum where it is transformed into testosterone by the sequential action of CYP17A1, HSD3B1, and HSD17B3 (reviewed in [1–3]). The transport of cholesterol, which constitutes the rate-limiting step in steroidogenesis, involves a large protein complex [4] which includes the steroidogenic acute regulatory (STAR) protein [5]. The importance of the STAR protein in Leydig cell steroidogenesis is supported by the existence of naturally occurring mutations in the human

Star gene responsible for lipoid congenital adrenal hyperplasia and by inactivation of the *Star* gene in the mouse where males display female external genitalia, consistent with impaired testosterone production (reviewed in [6]).

Steroidogenesis in Leydig cells is regulated by several growth factors, cytokines, and hormones, the main one being the pituitary luteinizing hormone (LH). Binding of LH to its G protein-coupled receptor on the surface of Leydig cells activates adenylate cyclase leading to cAMP synthesis, which in turns activates several signaling pathways (reviewed in [1,7]). Similarly, growth factors and cytokines bind to their respective receptor leading to activation of intracellular signaling cascades (reviewed in [7–11]). Downstream of these pathways are several kinases, including protein kinase A (PKA), Ca^{2+}/calmodulin-dependent protein kinase I (CAMKI), and extracellular signal-regulated protein kinase 1 and 2 (ERK1/2) (reviewed in [1]). These kinases participate in hormone-induced steroidogenesis by phosphorylating numerous proteins that include several transcription factors leading to increased steroidogenic gene expression (reviewed in [1,12]).

Because of the vital role of the STAR protein in steroidogenesis and because *Star* gene expression is strongly and rapidly induced in response to LH/cAMP stimulation (reviewed in [13]), the transcriptional regulation of the *Star* gene has been actively studied to identify *Star* promoter regulatory elements and their corresponding transcription factors that bind to these elements. Multiple transcription factors have been proposed to increase *Star* transcription (reviewed in [1,14,15]). These include members of the nuclear receptor family (SF1/NR5A1, LRH1/NR5A1, NUR77/NR4A1, COUP-TFII/NR2F2), bZIP family (cJUN, cFOS, CREB, CREM, C/EBPβ), GATA family (GATA4, GATA6), MADS box family (MEF2A, MEF2D), and the STAT domain family (STAT5B) (reviewed in [1,12,16–21]). While some of these transcription factors are de novo synthesized in response to LH/cAMP, such as NUR77 [22–24], most are activated by phosphorylation by one of three main kinases (PKA, CAMKI, ERK1/2).

Despite the significant progress that have been made in identifying the signaling pathways and transcription factors regulating hormone-induced steroidogenesis in Leydig cells, important questions remain. For instance, in the majority of the studies performed so far, the effect of a given kinase on transcription factor activation and *Star* promoter activity has been limited to studying individual transcription factors. However, PKA, CAMKI, and ERK1/2 are all activated in response to LH/cAMP and therefore can target and activate several transcription factors simultaneously. In the present study, we have determined the ability of PKA, CAMKI and ERK1/2 to activate and functionally cooperate with 15 different transcription factors belonging to 5 distinct families. This allowed us to describe distinct transcription factor/kinase cooperation profiles that lead to increased *Star* gene transcription.

2. Results

To determine whether the three main kinases (PKA, CAMKI, and ERK1/2), previously shown to activate *Star* transcription, can increase the transactivation potential of various transcription factors belonging to different families, we performed transient transfections in both a steroidogenic (MA-10 Leydig) and a non-steroidogenic (CV-1 fibroblast) cell line. Cells were cotransfected with a *Star* reporter plasmid along with expression vectors for the various transcription factors with or without expression vectors for the PKA catalytic subunit α, constitutively active CAMKI, and constitutively active MEK1 which phosphorylates and activates ERK1/2.

2.1. Cell- and Kinase-Specific Cooperations with bZIP Family Members

We first tested three members of the bZIP family (cJUN, CREB, and C/EBPβ) alone. As shown in Figure 1, CREB and cJUN activated the *Star* promoter in both MA-10 Leydig and CV-1 fibroblast cells, with the activation by cJUN being stronger in MA-10 cells (~10-fold vs. ~4-fold in CV-1 cells). In contrast, CREB-mediated activation was stronger in CV-1 cells (~4-fold). There was a tendency towards an activation by C/EBPβ but it did not reach statistical significance (Figure 1). We next repeated these experiments but in the presence of a kinase. All three kinases (CAMKI CA, PKA Cα, MEK1 CA) increased *Star* promoter activity (up to 7-fold in MA-10 cells and up to 3.8-fold in CV-1 cells) on their own. The activation by CAMKI CA and PKA Cα was much stronger in MA-10 Leydig cells than in CV-1 fibroblast cells, while the reverse was true for MEK1 CA. This indicated that the kinases activate transcription factors already present in the cells, mainly in MA-10 cells for CAMKI CA and PKA Cα, and in CV-1 cells for MEK1 CA. When assessed for functional cooperation, CAMKI CA potently enhanced cJUN- (up to 25-fold) and CREB- (up to 10-fold) dependent activation of the *Star* promoter in both MA-10 Leydig and CV-1 fibroblast cells (Figure 1). PKA Cα also cooperated with cJUN and with CREB but only in CV-1 cells (Figure 1). Finally, a cooperation between MEK1 CA and cJUN (up to 15-fold) and C/EBPβ (up to 10-fold) was observed solely in CV-1 fibroblasts (Figure 1). Together, these results indicated that although the transactivation by all three bZIP family members is enhanced in the presence of a kinase, the resulting cooperation is kinase- and cell type-specific, with cJUN and CREB activity being mainly enhanced by CAMKI CA and PKA Cα, and MEK1 CA (ERK1/2) with C/EBPβ and cJUN.

2.2. Some Nuclear Receptors Cooperate with All Three Kinases while Others Are Stimulated by a Single Kinase

The nuclear receptor family is composed of 48 members, several of which are expressed in Leydig cells where they regulate expression of numerous genes, including *Star* (reviewed in [16]). As previously demonstrated [23,25], members of the NR2F (COUP-TFI, COUP-TFII) and NR4A families activated the *Star* promoter both in MA-10 and CV-1 cells (Figure 2). We next determined whether cooperation, between the kinases and seven nuclear receptors belonging to three distinct families, occurs on the *Star* promoter. As shown in Figure 2, the activation of the *Star* promoter by COUP-TFI and COUP-TFII (NR2F family members) was significantly enhanced by CAMKI CA and PKA Cα in both MA-10 and CV-1 cells. On the other hand, all three kinases activated at least one of the NR4A family members (CAMKI CA with NURR1 and NOR1; PKA Cα with NOR1, and MEK1 CA with NURR1 and NOR1) in MA-10 Leydig and/or CV-1 fibroblast cells. Finally, the two members of the NR5A family, SF1 and LRH1, were only stimulated by MEK1 CA and exclusively in CV-1 fibroblast cells (Figure 2). Together these results establish that stimulation of nuclear receptor activity is highly kinase-dependent, which is determined by the family the nuclear receptor belongs to.

2.3. MEF2 Factors Cooperate with PKA Cα and MEK1 CA in CV-1 Cells

Of the four MEF2 family members, three are expressed in Leydig cells (MEF2A, 2C and 2D) where they are known to regulate the expression of several genes [26–31]. To determine whether the three kinases could enhance MEF2 transactivation potential, transient transfections were performed in MA-10 Leydig and CV-1 fibroblast cells. As shown in Figure 3, cooperation was observed between PKA Cα and MEF2C and MEF2D as well as between MEK1 CA and MEF2A and MEF2D. These cooperations were observed exclusively in CV-1 cells, which is consistent with the fact that MA-10 Leydig cells already contain high levels of MEF2 proteins [28].

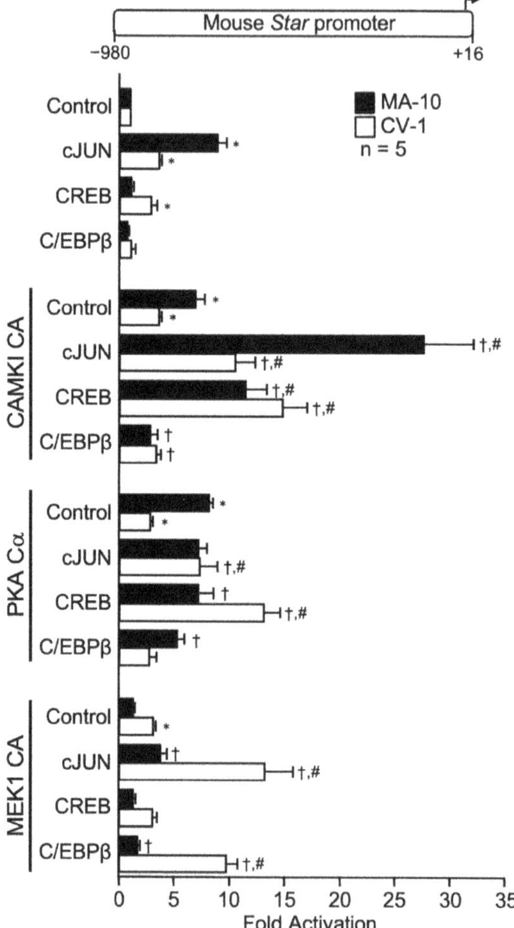

Figure 1. Transcriptional cooperation between bZIP family members and the kinases CAMKI constitutively active (CAMKI CA), PKA catalytic subunit alpha (PKA Cα), and MEK1 constitutively active (MEK1 CA activates ERK1/2) on the mouse *Star* promoter. MA-10 Leydig (black bars) and CV-1 fibroblast (white bars) cells were cotransfected with either an empty expression vector as a control or expression vectors for the different bZIP factors (cJUN, CREB, C/EBPaddress information is correct) and kinases (CAMKI CA, PKA Cα, MEK1 CA) individually or in combination as indicated, along with a −980 to +16 bp mouse *Star* reporter. Results are shown as Fold Activation over control ± SEM. An asterisk (*) represents a statistically significant difference in activation by the transcription factor or the kinase compared to control (empty expression vector, value set at 1, $p < 0.05$). A dagger (†) represents a statistically significant difference in activation between the transcription factor + the kinase compared to the corresponding transcription factor alone ($p < 0.05$). A hashtag (#) represents a statistically significant difference in activation between the transcription factor + the kinase compared to the kinase alone ($p < 0.05$). The number (n) of replicates is indicated.

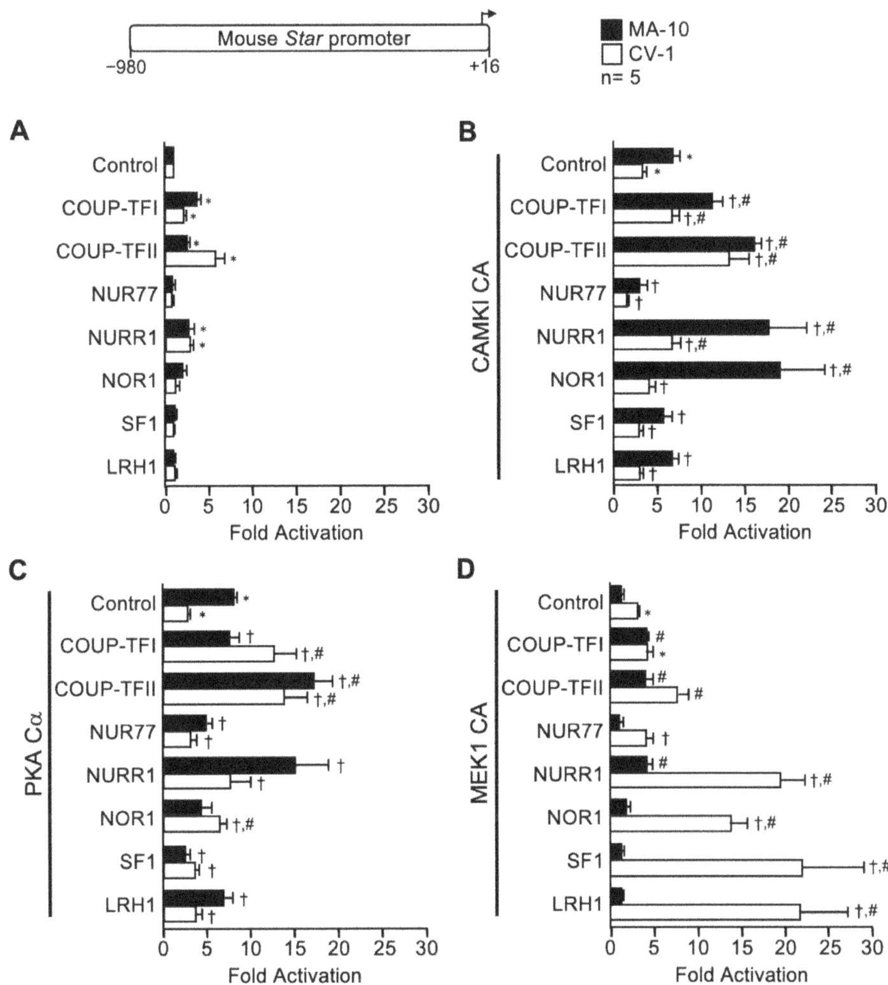

Figure 2. Specific cooperation between select members of the nuclear receptor family and the kinase CAMKI constitutively active (CAMKI CA), PKA catalytic subunit alpha (PKA Cα), and MEK1 constitutively active (MEK1 CA activates ERK1/2) on the mouse *Star* promoter. MA-10 Leydig (black bars) and CV-1 fibroblast (white bars) cells were cotransfected with a −980 to +16 bp mouse *Star* reporter along with an empty expression vector as a control or expression vectors for the different nuclear receptors (COUP-TFI, COUP-TFII, NUR77, NURR1, NOR1, SF1, LRH1) used individually (**A**) or in combination with CAMKI CA (**B**), PKA Cα (**C**), and MEK1 CA (**D**). Results are shown as Fold Activation over control ± SEM. An asterisk (*) represents a statistically significant difference in activation by the transcription factor or the kinase compared to the control empty expression vector (whose value was set at 1, $p < 0.05$). A dagger (†) represents a statistically significant difference in activation between the transcription factor + the kinase compared to the corresponding transcription factor alone ($p < 0.05$). A hashtag (#) represents a statistically significant difference in activation between the transcription factor + the kinase compared to the kinase alone ($p < 0.05$). The number of replicates is indicated.

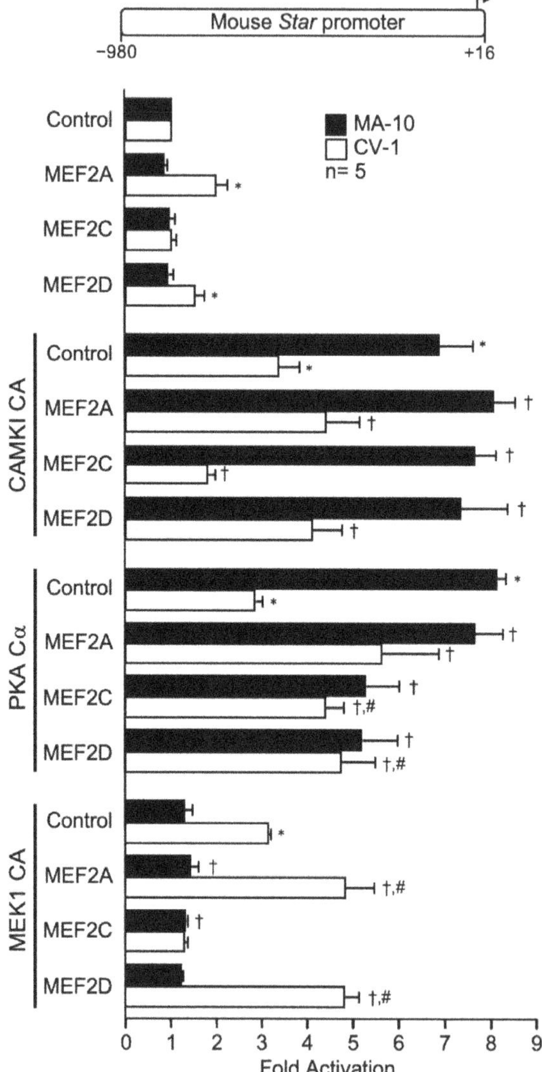

Figure 3. Cooperation between members of MADS box family of transcription factors and the kinases PKA catalytic subunit alpha (PKA Cα) and MEK1 constitutively active (MEK1 CA activates ERK1/2) on the mouse *Star* promoter. MA-10 Leydig (black bars) and CV-1 fibroblast (white bars) cells were cotransfected with an empty expression vector as a control or expression vectors for various MADS box family members (MEF2A, MEF2C, MEF2D) and kinases (CAMKI CA, PKA Cα, MEK1 CA) individually or in combination as indicated, along with a −980 to +16 bp mouse *Star* reporter. Results are shown as Fold Activation over control ± SEM. An asterisk (*) represents a statistically significant difference in activation by the transcription factor or the kinase compared to control (empty expression vector, value set at 1, $p < 0.05$). A dagger (†) represents a statistically significant difference in activation between the transcription factor + the kinase compared to the corresponding transcription factor alone ($p < 0.05$). A hashtag (#) represents a statistically significant difference in activation between the transcription factor + the kinase compared to the kinase alone ($p < 0.05$). The number (n) of replicates is indicated.

2.4. GATA4 Activity Is Strongly Enhanced in the Presence of CAMKI CA, PKA Cα, and MEK1 CA

The GATA4 transcription factor is a known activator of *Star* promoter activity in Leydig cells [27,32–37]. Consistent with this, GATA4 was found to activate the *Star* promoter both in MA-10 Leydig (3-fold) and CV-1 fibroblast (5-fold) cells (Figure 4). Combination of GATA4 with CAMKI CA (up to 13-fold), and PKA Cα (up to 16-fold), enhanced GATA4 activation both in MA-10 and CV-1 cells (Figure 4). A MEK1 CA/GATA4 cooperation was only observed in MA-10 Leydig cells (Figure 4). These results suggest that GATA4 is a common target downstream of diverse intracellular signaling pathways and kinases.

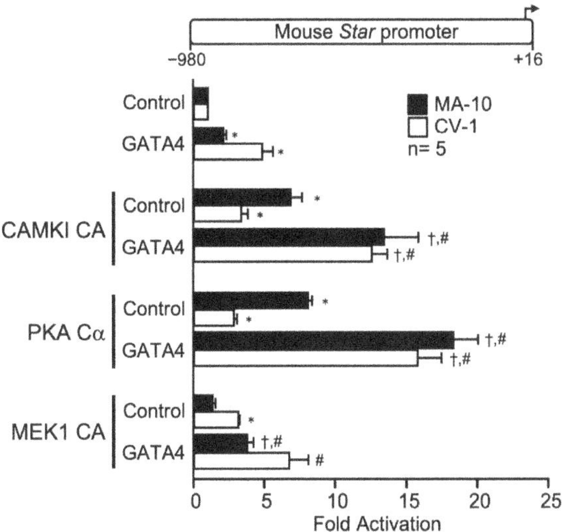

Figure 4. The zinc finger transcription factor GATA4 cooperates with CAMKI constitutively active (CAMKI CA), PKA catalytic subunit alpha (PKA Cα), and MEK1 constitutively active (MEK1 CA activates ERK1/2) on the mouse *Star* promoter. MA-10 Leydig (black bars) and CV-1 fibroblast (white bars) cells were cotransfected with a control empty expression vector or expression vectors for GATA4 and the different kinases (CAMKI CA, PKA Cα, MEK1 CA) individually or in combination as indicated, along with a −980 to +16 bp mouse *Star* reporter. Results are shown as Fold Activation over control ± SEM. An asterisk (*) represents a statistically significant difference in activation by the transcription factor or the kinase compared to control (empty expression vector, value set at 1, $p < 0.05$). A dagger (†) represents a statistically significant difference in activation between the transcription factor + the kinase compared to the corresponding transcription factor alone ($p < 0.05$). A hashtag (#) represents a statistically significant difference in activation between the transcription factor + the kinase compared to the kinase alone ($p < 0.05$). The number (n) of replicates is indicated.

2.5. STAT5B Cooperates with All Kinases

STAT5B is a transcription factor activated in response to growth hormone (reviewed in [38]). Recently, STAT5B was found to mediate GH-induced *Star* gene expression in Leydig cells [39]. In agreement with this, a constitutively active form of STAT5B (STAT5B CA) activated the *Star* promoter by ~3-fold in MA-10 Leydig and CV-1 fibroblast cells (Figure 5). Combination of STAT5B CA with any kinase (CAMKI CA, PKA Cα, MEK1 CA) resulted in a synergistic activation of the *Star* promoter reaching nearly 20-fold in both MA-10 Leydig and CV-1 fibroblast cells (Figure 5). This indicates that STAT5B is a versatile transcription factor that can be stimulated by various kinases.

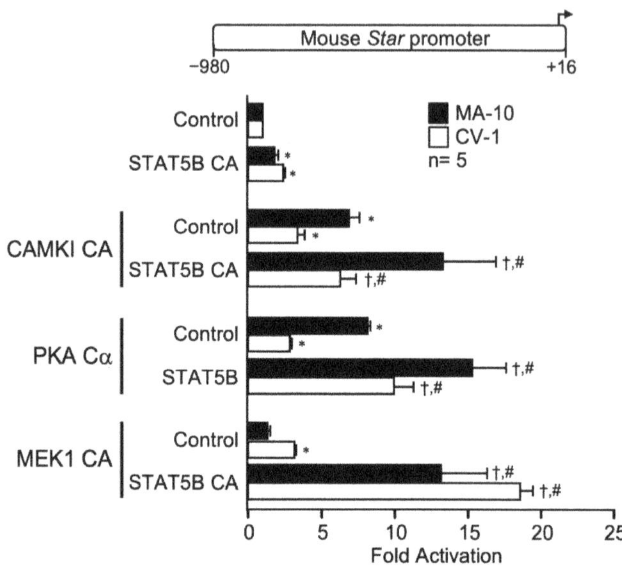

Figure 5. Cooperation between STAT5B and CAMKI constitutively active (CAMKI CA), PKA catalytic subunit alpha (PKA Cα), and MEK1 constitutively active (MEK1 CA activates ERK1/2) on the mouse *Star* promoter. MA-10 Leydig (black bars) and CV-1 fibroblast (white bars) cells were cotransfected with an empty expression vector as a control or expression vectors for constitutively active STAT5B (STAT5B CA) and the different kinases (CAMKI CA, PKA Cα, MEK1 CA) individually or in combination as indicated, along with a −980 to +16 bp mouse *Star* reporter. Results are shown as Fold Activation over control ± SEM. An asterisk (*) represents a statistically significant difference in activation by the transcription factor or the kinase compared to control (empty expression vector, value set at 1, $p < 0.05$). A dagger (†) represents a statistically significant difference in activation between the transcription factor + the kinase compared to the corresponding transcription factor alone ($p < 0.05$). A hashtag (#) represents a statistically significant difference in activation between the transcription factor + the kinase compared to the kinase alone ($p < 0.05$). The number (n) of replicates is indicated.

3. Discussion

Hormone-induced Leydig cell steroidogenesis is a strictly regulated process that requires changes in gene expression and protein phosphorylation. Expression of the *Star* gene, which is induced upon stimulation of steroidogenesis has been the subject of intense studies since its discovery nearly 30 years ago [5]. This has led to the identification of several transcription factors that act via regulatory motifs clustered within the proximal *Star* promoter (reviewed in [1,40,41]). Some of these transcription factors are activated by various kinases, including PKA, CAMKI and ERK1/2, that are themselves activated in response to different hormones known to increase *Star* expression and steroidogenesis in Leydig cells (reviewed in [1]). In other systems, some of these transcription factors are known to be activated downstream of different signaling pathways and kinases. How these kinases and transcription factors are integrated to appropriately induce *Star* transcription in response to specific stimuli remains poorly understood. The main objective of this work was to compare the effect of three key kinases on the transactivation potential of 15 transcription factors belonging to 5 distinct families on *Star* promoter activity. As described below, this allowed us to validate existing cooperations as well as to identify new ones. As summarized in Figure 6, some transcription factors were activated by multiple kinases indicating that they may act downstream of multiple signaling cascades. Furthermore, the fact that all the transcription factors and kinases were assessed simultaneously allowed for a direct comparison of the significance (or strength) of the transcription factor/kinase cooperation (summarized in Table 1).

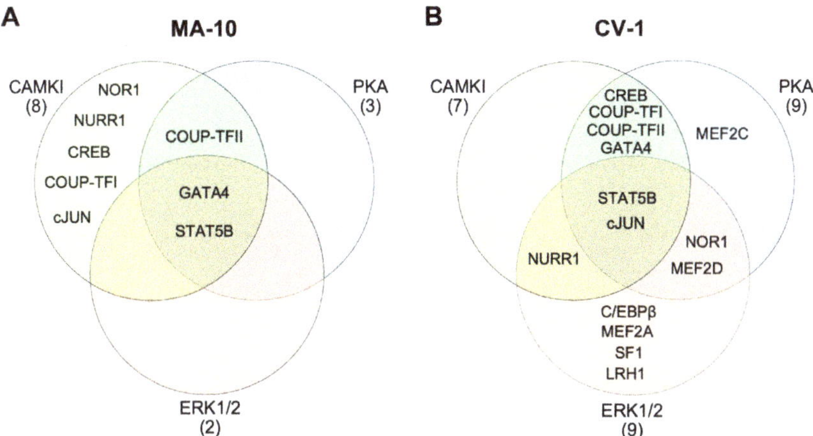

Figure 6. Cooperation of transcription factors and kinases on the *Star* promoter in steroidogenic (MA-10) vs. non-steroidogenic (CV-1) cells. The three-set Venn diagram highlights the transcription factors that cooperate with CAMKI CA (green circles), PKA Cα (blue circles), and MEK1 CA (activating ERK1/2, orange circles) in MA-10 Leydig (**A**) and in CV-1 fibroblast (**B**) cells.

Table 1. Transactivation potential of transcription factors by different kinases.

		Control		CAMKI CA		PKA Cα		MEK1 CA	
		MA-10	**CV-1**	**MA-10**	**CV-1**	**MA-10**	**CV-1**	**MA-10**	**CV-1**
	Control	1.0	1.0	6.9	3.4	8.1	2.8	1.3	3.1
bZIP	cJUN	8.8	3.5	27.5	10.4	7.2	7.2	3.8	13.2
	CREB	1.0	2.9	11.4	14.7	7.1	13.1	1.3	3.1
	C/EBPβ	0.7	1.1	2.8	3.4	5.1	2.8	1.7	9.7
Nuclear receptors	COUP-TFI	3.7	2.1	11.4	6.7	7.6	12.8	4.1	4.2
	COUP-TFII	2.5	5.8	16.0	13.4	17.2	13.9	4.0	7.6
	NUR77	0.8	0.7	3.1	1.6	5.0	3.1	1.0	4.1
	NURR1	2.6	2.8	17.9	6.7	15.1	7.7	4.1	19.3
	NOR1	2.0	1.1	19.2	4.1	4.3	6.5	1.8	13.7
	SF1	1.0	0.9	5.8	2.9	2.5	3.6	1.3	21.8
	LRH1	0.9	1.1	6.7	3.2	6.9	3.7	1.3	21.5
MADS box	MEF2A	0.8	2.0	8.0	4.4	7.6	5.6	1.4	4.8
	MEF2C	1.0	1.0	7.6	1.8	5.3	4.4	1.3	1.3
	MEF2D	0.9	1.5	7.3	4.1	5.2	4.7	1.2	4.8
GATA	GATA4	2.1	4.9	13.5	12.6	18.4	15.9	3.8	6.8
STAT	STAT5B	1.8	2.4	13.3	6.3	15.3	10.0	13.1	18.6

Fold activations are compared to the control (empty expression vector) for which the value was set to 1.

Experiments were performed in two cell lines, the MA-10 Leydig cell line and the CV-1 fibroblast cell line. MA-10 Leydig cells endogenously express the various transcription factors and kinases tested as well as the *Star* gene. Therefore, expression of a kinase often leads to a significant activation since it activates transcription factors already present in the cells. The transcription factor could be the factor of interest or another transcription factor. On the other hand, CV-1 cells are considered heterologous cells since they do not express the *Star* gene and all transcription factors normally found in a Leydig cell. Using a heterologous cell line therefore allows to detect promoter activation by a transcription factor or a cooperation between factors that would otherwise be undetectable in a homologous cell line like MA-10 Leydig cells. In agreement with this, the activity of some transcription factors was only stimulated by a kinase in CV-1 cells (Figure 6).

3.1. New Cooperations between Transcription Factors and Kinases

The roles of PKA and MEK1 (activating ERK1/2) have been well characterized and they are known to stimulate various transcription factors in different cell types. However, their effects on most of the transcription factors in the activation of the *Star* promoter remained to be characterized.

3.1.1. PKA-Induced Transcription Factor Activity

Of the 15 transcription factors tested, 3 were activated by PKA Cα in MA-10 Leydig cells while in CV-1 fibroblast cells, the same 3 plus an additional 6 (9 in total) were activated (blue circles in Figure 6). PKA is known to stimulate cJUN activity, but this was shown using a synthetic reporter [42]. We found that cJUN-dependent activation of the *Star* promoter is significantly induced by PKA Cα in CV-1 cells. Similarly, CREB is a classic target of PKA [43] and therefore the cooperation between CREB and PKA Cα on the *Star* promoter was not unexpected. We also found that PKA Cα stimulated the activity of the orphan nuclear receptors COUP-TFI and COUP-TFII on the *Star* promoter in both MA-10 and CV-1 cells, which was unknown for the *Star* gene. A previous study in a different system did report that PKA could cooperate with COUP-TFI on the *Vitronectin* promoter [44]. Similarly, the stimulatory effect of PKA Cα on NR4A family members (NUR77, NURR1, NOR1) has only been observed on the *Pomc* promoter in pituitary corticotrope cells [45]. Here, we found that PKA Cα enhanced NOR1-dependent activation of the *Star* promoter. Of the different MEF2 family members, we found that PKA Cα increased the transactivation potential of MEF2C and MEF2D on the *Star* promoter. A previous study in the heart found that PKA represses MEF2A activity [46], indicating the existence of cell type-specific responses. The transcription factor GATA4 has been described as a direct target for PKA, which increases its transactivation potential on the *Star* promoter [47]. Our current study reproduces this observation, thus validating the appropriateness of our experimental system. We also observed a potent stimulation of STAT5B-dependent activation of the *Star* promoter by PKA Cα both in MA-10 Leydig and CV-1 fibroblast cells. STAT5B is known to be phosphorylated by members of the JAK kinase family in response to growth hormone (reviewed in [48]) in many cell types, including in Leydig cells [49]. Whether PKA directly phosphorylates STAT5B or whether it acts on STAT5B-interacting partner remains to be established.

3.1.2. MEK1-Induced Transcription Factor Activity

We found MEK1 CA to enhance the transactivation potential of 2 transcription factors in MA-10 Leydig cells and 9 in CV-1 fibroblast cells (orange circles in Figure 6). The ERK1/2 kinases activated by MEK1 are known to phosphorylate and activate GATA4-dependent transactivation in the heart [50] as well in the mouse testis where phosphorylation of GATA4 Ser105 is required for testosterone production [51]. However, a direct effect of MEK1 CA on GATA4-dependent activation of the *Star* promoter as we observed had never been reported. This suggests that in addition to being stimulated by PKA, GATA4 activity is also activated by ERK1/2 on the *Star* promoter. Similar to GATA4, the transactivation potential of STAT5B on the *Star* promoter was also enhanced in the presence of MEK1 CA. A cooperation between STAT5B and ERK1/2 has never been reported. We also found that MEK1 CA enhances the activity of the NR5A nuclear receptors SF1 and LRH1. Although ERK1/2-mediated SF1 and LRH1 phosphorylation can stimulate their transactivation potential [52,53] in HeLa and JEG-3 cells on either the *Cyp7a1* or a synthetic promoter, this potentiation had not been reported on the *Star* promoter. With respect to the NR4A family of nuclear receptors, ERK1/2 was previously reported to phosphorylate and stimulate NUR77 leading to enhanced activity of an artificial promoter in AtT-20 corticotrope cells [54]. Although we did not observe any effect of MEK1 CA (ERK1/2) on NUR77-dependent activation of the *Star* promoter, MEK1 CA did nonetheless significantly enhance the activity of the other two NR4A family members NURR1 and NOR1. This suggests the existence of a cell- and promoter-dependent response of transcription factors to different kinases. The bZIP factors cJUN and C/EBPβ were also found to cooperate with MEK1 CA on the *Star* promoter. This was unexpected since activated ERK1/2 is believed to inactivate cJUN, while other MAPK members such as JNKs phosphorylate and activate cJUN (reviewed in [55]). In 3T3-L1 preadipocytes, C/EBPβ is phosphorylated in a MEK1-dependent manner, stimulating its transactivation potential [56,57], which is similar to what we observed on the *Star* promoter.

3.1.3. CAMKI-Induced Transcription Factor Activity

CAMKI is the most recently identified kinase in Leydig cells and consequently the least studied in these cells. The transactivation potential of 8 transcription factors in MA-10 cells and 7 in CV-1 cells was enhanced by CAMKI CA (green circles in Figure 6). Similar to what we observed in the present study, the transactivation potential of all members of the NR4A family of nuclear receptors (NUR77, NURR1 and NOR1) and of SF1 (NR5A1) was previously shown to be increased in the presence of CAMKI CA on the *Star* promoter [23]. The nuclear receptors COUP-TFI and COUP-TFII were both activated in the presence of CAMKI CA, as revealed by the synergistic activation of the *Star* promoter in both MA-10 and CV-1 cells. The two COUP-TF factors have not been reported to be stimulated by CAMKI, although a study showed that the activity of COUP-TFI is potentiated by the related CAMKIV in neuronal cells [58]. Our study also revealed that CAMKI CA enhances the activity of cJUN on the *Star* promoter both in MA-10 and in CV-1 cells, similar what was recently reported on the *Cx43* promoter in MA-10 Leydig cells [59]. We observed a strong stimulation of the activity of CREB by CAMKI CA on the *Star* promoter. CREB is known to be phosphorylated in different cell types by CAMKI and CAMKIV leading to an increase in its transactivation potential [60,61]. We found that CAMKI CA significantly enhanced the transactivation potential of GATA4 and STAT5B on the *Star* promoter in both MA-10 and CV-1 cells. Both transcription factors were not previously known to cooperate with CAMKI.

3.2. Versatility in Transcription Factor Response to Different Kinases

An interesting observation in our findings is the fact that the transactivation potential of most transcription factors was stimulated by more than one kinase. This is clearly apparent in the Venn diagrams presented in Figure 6. Some transcription factors were found to respond to all three kinases such as GATA4 and STAT5B in MA-10 Leydig cells, and cJUN and STAT5B in CV-1 cells. This flexibility in how a given transcription factor responds to different kinases suggests that the transcription factor can likely mediate the effects of different stimuli thus ensuring proper genomic response.

Another form of versatility exists within a family of transcription factors where different members respond to different kinases. For instance, in the nuclear receptor family, NR4A members (NUR77, NURR1, NOR1) responded to all three kinases (PKA Cα, CAMKI CA, MEK1 CA), NRF2 members (COUP-TFI, COUP-TFII) were stimulated by two kinases (PKA Cα, CAMKI), and NR5A members (SF1, LRH1) were only activated by MEK1 CA. Since most of these nuclear receptors can bind to the same response element in a promoter, this differential response to a kinase might provide the specificity needed to ensure the proper nuclear receptor is activated downstream of a signaling cascade leading to increased gene expression.

In conclusion, our current work has identified several transcription factors whose transactivation potential is stimulated by different kinases. Some of these transcription factors were previously reported to be directly phosphorylated by the kinase. However, for several others identified in our current work, it remains to be determined whether they are a direct target of the kinase. It is possible that the kinase might instead phosphorylate another protein that can then interact with the transcription factor leading to a cooperation between the two transcription factors. In this case, kinase-mediated phosphorylation of a factor might render it more receptive to interactions and cooperations with other transcription factors. This could be tested by determining whether the different kinases further enhance known cooperations between two transcription factors. Additional work is needed to answer these questions.

4. Materials and Methods

4.1. Plasmids

The mouse *Star* luciferase promoter construct (−980/+16 bp) was previously described [34]. The MEF2A and MEF2D [27,28,30] and GATA4 [62] expression plasmids were previously described. The mouse MEF2C expression plasmid was generated by amplifying the complete coding sequence by PCR (forward primer 5′-CCC AAG CTT ATG GGG AGA AAA AAG ATT CAG ATT-3′, reverse primer 5′-GCT CTA GAT CAT GTT GCC CAT CCT TCA-3′) and subcloning the resulting PCR product into the HindIII and XbaI cloning sites of the pcDNA3.1 expression vector (Invitrogen Canada, Burlington, ON, Canada). The following expression plasmids were sourced from different research groups: rat NUR77/NR4A1, NURRI/NURR1, NOR1/NR4A3 [63], mouse SF1/NR5A1 [64], human LRH1/NR5A2 [65], cJUN [66], CREB and PKA catalytic subunit α [67], constitutively active MEK1 [68], C/EBPβ [69], mouse COUP-TFI/NR2F1 and COUP-TFII/NR2F2 [70], constitutively active STAT5B [71], and constitutively active CAMKI [72].

4.2. Cells Culture, Transfections, and Reporter Assays

Mouse MA-10 Leydig cells (ATCC, Manassas, VA, USA, Cat# CRL-3050, RRID:CVCL_D789) were grown in DMEM/F12 medium supplemented with 2.438 g/L sodium bicarbonate, 3.57 g/L HEPES, and 15% horse serum on gelatin-coated plates. African green monkey kidney fibroblast CV-1 cells (ATCC, Cat# CRL-6305, RRID:CVCL_0229) were grown in DMEM medium supplemented with 3.7 g/L HEPES, and 10% newborn calf serum. Penicillin and streptomycin sulphate were added to the cell culture media to a final concentration of 50 mg/L, and all cell lines were kept at 37 °C, 5% CO_2 in a humidified incubator. All cell lines were validated by morphology and Leydig cell lines by quantifying steroidogenic output (progesterone for MA-10) as previously described [25,27,39,73–77]. MA-10 (60,000 cells per well) and CV-1 (25,000 cells per well) were transiently transfected using polyethylenimine hydrochloride (PEI) (Sigma-Aldrich Canada, Oakville, ON, Canada) as previously described [39,78] or the calcium phosphate co-precipitation method as described in [23,24]. Briefly, the cells were seeded in 24-well plates and cotransfected with 400 ng of the mouse *Star* −980/+16 bp reporter vector along with 100 ng of an empty expression vector (pcDNA3.1 as control), or expression vectors for the various transcription factors (50 ng) or kinases (30 ng) individually (completed to 100 ng with the empty pcDNA3.1 expression vector to keep the total amount of expression vector to 100 ng), or the combination of a transcription factor (50 ng) plus a kinase (30 ng) and empty pcDNA3.1 (20 ng). For the calcium phosphate co-precipitation method, 1 μg of SP64 inert plasmid was also added as carrier. Sixteen hours post transfection, the media was replaced, and the cells were grown for additional 32 h. Cells were then lysed, the lysates were collected, and the luciferase measurements was performed using a Tecan Spark 10M multimode plate reader (Tecan, Morrisville, NC, USA) as previously described [39,78]. The number of experiments, each performed in triplicate, is indicated in each figure. All the cDNAs are cloned in an expression plasmid containing a strong promoter (CMV), which leads to high expression levels. The quantity of expression vector needed to obtain specific and optimal promoter activation was determined by performing a dose–response curve as described in [25]. Western blots were routinely performed to ensure overexpression was achieved, especially when using a new DNA plasmid preparation.

4.3. Statistical Analysis

Comparisons between two groups were performed using an unpaired Student *t*-test (GraphPad Prism, GraphPad Software, San Diego, CA, USA, version 9.4.1 (458)). For all statistical analyses, $p < 0.05$ was considered significant.

Author Contributions: K.J.P. performed all the experiments. K.J.P. and J.J.T. analyzed and interpreted the data. J.J.T. conceived the study, coordinated and supervised the project, and wrote the manuscript. All authors have read and agreed to the published version of the manuscript.

Funding: This research was funded by a grant from the Canadian Institutes of Health Research (CIHR) (funding reference number MOP-81387) to J.J.T.

Institutional Review Board Statement: Not applicable.

Informed Consent Statement: Not applicable.

Data Availability Statement: All data generated or analyzed during this study are included in this article.

Acknowledgments: We would like to thank Toshio Kitamura (constitutively active STAT5B), Dany Chalbos (cJUN), Jacques Drouin (NUR77, NURR1, NOR1), Luc Bélanger (LRH1), Marc Montminy (CREB, PKA catalytic subunit α), Ming-Jer Tsai (COUP-TFI, COUP-TFII), Robert Viger (GATA4), Keith Parker (SF1), Steven McKnight (C/EBPβ), Natalie Ahn (constitutively active MEK1), and Thomas Soderling (constitutively active CAMKI) for generously providing plasmids used in this study. We are thankful to Luc Martin for his assistance with the MEF2C plasmid preparation.

Conflicts of Interest: The authors declare no conflict of interest.

References

1. Tremblay, J.J. Molecular regulation of steroidogenesis in endocrine Leydig cells. *Steroids* **2015**, *103*, 3–10. [CrossRef] [PubMed]
2. Zirkin, B.R.; Papadopoulos, V. Leydig cells: Formation, function, and regulation. *Biol. Reprod.* **2018**, *99*, 101–111. [CrossRef] [PubMed]
3. Selvaraj, V.; Stocco, D.M.; Clark, B.J. Current knowledge on the acute regulation of steroidogenesis. *Biol. Reprod.* **2018**, *99*, 13–26. [CrossRef] [PubMed]
4. Midzak, A.; Rone, M.; Aghazadeh, Y.; Culty, M.; Papadopoulos, V. Mitochondrial protein import and the genesis of steroidogenic mitochondria. *Mol. Cell. Endocrinol.* **2011**, *336*, 70–79. [CrossRef] [PubMed]
5. Clark, B.J.; Wells, J.; King, S.R.; Stocco, D.M. The purification, cloning, and expression of a novel luteinizing hormone-induced mitochondrial protein in MA-10 mouse Leydig tumor cells. Characterization of the steroidogenic acute regulatory protein (StAR). *J. Biol. Chem.* **1994**, *269*, 28314–28322. [CrossRef]
6. Stocco, D.M. Clinical disorders associated with abnormal cholesterol transport: Mutations in the steroidogenic acute regulatory protein. *Mol. Cell. Endocrinol.* **2002**, *191*, 19–25. [CrossRef]
7. Haider, S.G. Cell biology of Leydig cells in the testis. *Int. Rev. Cytol.* **2004**, *233*, 181–241. [CrossRef]
8. Ipsa, E.; Cruzat, V.F.; Kagize, J.N.; Yovich, J.L.; Keane, K.N. Growth Hormone and Insulin-Like Growth Factor Action in Reproductive Tissues. *Front. Endocrinol.* **2019**, *10*, 777. [CrossRef]
9. Syriou, V.; Papanikolaou, D.; Kozyraki, A.; Goulis, D.G. Cytokines and male infertility. *Eur. Cytokine Netw.* **2018**, *29*, 73–82. [CrossRef]
10. Roumaud, P.; Martin, L.J. Roles of leptin, adiponectin and resistin in the transcriptional regulation of steroidogenic genes contributing to decreased Leydig cells function in obesity. *Horm. Mol. Biol. Clin. Investig.* **2015**, *24*, 25–45. [CrossRef]
11. Bornstein, S.R.; Rutkowski, H.; Vrezas, I. Cytokines and steroidogenesis. *Mol. Cell. Endocrinol.* **2004**, *215*, 135–141. [CrossRef] [PubMed]
12. de Mattos, K.; Viger, R.S.; Tremblay, J.J. Transcription factors in the regulation of Leydig cell gene expression and function. *Front. Endocrinol.* **2022**, *13*, 881309. [CrossRef] [PubMed]
13. Manna, P.R.; Stetson, C.L.; Slominski, A.T.; Pruitt, K. Role of the steroidogenic acute regulatory protein in health and disease. *Endocrine* **2016**, *51*, 7–21. [CrossRef] [PubMed]
14. Stocco, D.M.; Wang, X.; Jo, Y.; Manna, P.R. Multiple signaling pathways regulating steroidogenesis and steroidogenic acute regulatory protein expression: More complicated than we thought. *Mol. Endocrinol.* **2005**, *19*, 2647–2659. [CrossRef]
15. Tugaeva, K.V.; Sluchanko, N.N. Steroidogenic Acute Regulatory Protein: Structure, Functioning, and Regulation. *Biochemistry* **2019**, *84*, S233–S253. [CrossRef]
16. Martin, L.J.; Tremblay, J.J. Nuclear receptors in Leydig cell gene expression and function. *Biol. Reprod.* **2010**, *83*, 3–14. [CrossRef]
17. Tremblay, J.J. Transcription factors as regulators of gene expression during Leydig cell differentiation and function. In *The Leydig Cell in Health and Disease*; Payne, A.H., Hardy, M.P., Eds.; Humana Press: Totowa, NJ, USA, 2007; Volume 2, pp. 333–343.

18. Nguyen, H.T.; Najih, M.; Martin, L.J. The AP-1 family of transcription factors are important regulators of gene expression within Leydig cells. *Endocrine* **2021**, *74*, 498–507. [CrossRef]

19. Tremblay, J.J.; Viger, R.S. Novel roles for GATA transcription factors in the regulation of steroidogenesis. *J. Steroid Biochem. Mol. Biol.* **2003**, *85*, 291–298. [CrossRef]

20. Viger, R.S.; Taniguchi, H.; Robert, N.M.; Tremblay, J.J. Role of the GATA family of transcription factors in andrology. *J. Androl.* **2004**, *25*, 441–452. [CrossRef]

21. Viger, R.S.; Guittot, S.M.; Anttonen, M.; Wilson, D.B.; Heikinheimo, M. Role of the GATA family of transcription factors in endocrine development, function, and disease. *Mol. Endocrinol.* **2008**, *22*, 781–798. [CrossRef]

22. Song, K.H.; Park, J.I.; Lee, M.O.; Soh, J.; Lee, K.; Choi, H.S. LH induces orphan nuclear receptor Nur77 gene expression in testicular Leydig cells. *Endocrinology* **2001**, *142*, 5116–5123. [CrossRef] [PubMed]

23. Martin, L.J.; Boucher, N.; Brousseau, C.; Tremblay, J.J. The orphan nuclear receptor NUR77 regulates hormone-induced StAR transcription in Leydig cells through a cooperation with CaMKI. *Mol. Endocrinol.* **2008**, *22*, 2021–2037. [CrossRef] [PubMed]

24. Martin, L.J.; Boucher, N.; El-Asmar, B.; Tremblay, J.J. cAMP-induced expression of the orphan nuclear receptor Nur77 in testicular Leydig cells involves a CaMKI pathway. *J. Androl.* **2009**, *30*, 134–145. [CrossRef] [PubMed]

25. Mendoza-Villarroel, R.E.; Robert, N.M.; Martin, L.J.; Brousseau, C.; Tremblay, J.J. The nuclear receptor NR2F2 activates *Star* expression and steroidogenesis in mouse MA-10 and MLTC-1 Leydig cells. *Biol. Reprod.* **2014**, *91*, 26. [CrossRef]

26. Abdou, H.S.; Robert, N.M.; Tremblay, J.J. Calcium-dependent Nr4a1 expression in mouse Leydig cells requires distinct AP1/CRE and MEF2 elements. *J. Mol. Endocrinol.* **2016**, *56*, 151–161. [CrossRef]

27. Daems, C.; Di-Luoffo, M.; Paradis, E.; Tremblay, J.J. MEF2 cooperates with forskolin/cAMP and GATA4 to regulate Star gene expression in mouse MA-10 Leydig cells. *Endocrinology* **2015**, *156*, 2693–2703. [CrossRef]

28. Daems, C.; Martin, L.J.; Brousseau, C.; Tremblay, J.J. MEF2 is restricted to the male gonad and regulates expression of the orphan nuclear receptor NR4A1. *Mol. Endocrinol.* **2014**, *28*, 886–898. [CrossRef]

29. Di-Luoffo, M.; Brousseau, C.; Bergeron, F.; Tremblay, J.J. The transcription factor MEF2 is a novel regulator of Gsta gene class in mouse MA-10 Leydig cells. *Endocrinology* **2015**, *156*, 4695–4706. [CrossRef]

30. Di-Luoffo, M.; Brousseau, C.; Tremblay, J.J. MEF2 and NR2F2 cooperate to regulate Akr1c14 gene expression in mouse MA-10 Leydig cells. *Andrology* **2016**, *4*, 335–344. [CrossRef]

31. Di-Luoffo, M.; Daems, C.; Bergeron, F.; Tremblay, J.J. Novel targets for the transcription factors MEF2 in MA-10 Leydig cells. *Biol. Reprod.* **2015**, *93*, 9. [CrossRef]

32. Martin, L.J.; Bergeron, F.; Viger, R.S.; Tremblay, J.J. Functional cooperation between GATA factors and cJUN on the Star promoter in MA-10 Leydig cells. *J. Androl.* **2012**, *33*, 81–87. [CrossRef] [PubMed]

33. Martin, L.J.; Tremblay, J.J. The nuclear receptors NUR77 and SF1 play additive roles with c-JUN through distinct elements on the mouse *Star* promoter. *J. Mol. Endocrinol.* **2009**, *42*, 119–129. [CrossRef]

34. Tremblay, J.J.; Viger, R.S. GATA factors differentially activate multiple gonadal promoters through conserved GATA regulatory elements. *Endocrinology* **2001**, *142*, 977–986. [CrossRef] [PubMed]

35. Hiroi, H.; Christenson, L.K.; Chang, L.; Sammel, M.D.; Berger, S.L.; Strauss, J.F., III. Temporal and spatial changes in transcription factor binding and histone modifications at the steroidogenic acute regulatory protein (StAR) locus associated with StAR transcription. *Mol. Endocrinol.* **2004**, *18*, 791–806. [CrossRef] [PubMed]

36. Nishida, H.; Miyagawa, S.; Vieux-Rochas, M.; Morini, M.; Ogino, Y.; Suzuki, K.; Nakagata, N.; Choi, H.S.; Levi, G.; Yamada, G. Positive regulation of steroidogenic acute regulatory protein gene expression through the interaction between Dlx and GATA-4 for testicular steroidogenesis. *Endocrinology* **2008**, *149*, 2090–2097. [CrossRef]

37. Wooton-Kee, C.R.; Clark, B.J. Steroidogenic factor-1 influences protein-deoxyribonucleic acid interactions within the cyclic adenosine 3,5-monophosphate-responsive regions of the murine steroidogenic acute regulatory protein gene. *Endocrinology* **2000**, *141*, 1345–1355. [CrossRef]

38. Rotwein, P. Regulation of gene expression by growth hormone. *Mol. Cell. Endocrinol.* **2020**, *507*, 110788. [CrossRef]

39. Hebert-Mercier, P.O.; Bergeron, F.; Robert, N.M.; Mehanovic, S.; Pierre, K.J.; Mendoza-Villarroel, R.E.; de Mattos, K.; Brousseau, C.; Tremblay, J.J. Growth hormone-induced STAT5B regulates Star gene expression through a cooperation with cJUN in mouse MA-10 Leydig cells. *Endocrinology* **2022**, *163*, bqab267. [CrossRef]

40. Reinhart, A.J.; Williams, S.C.; Stocco, D.M. Transcriptional regulation of the StAR gene. *Mol. Cell. Endocrinol.* **1999**, *151*, 161–169. [CrossRef]

41. Manna, P.R.; Wang, X.J.; Stocco, D.M. Involvement of multiple transcription factors in the regulation of steroidogenic acute regulatory protein gene expression. *Steroids* **2003**, *68*, 1125–1134. [CrossRef]

42. de Groot, R.P.; Sassone-Corsi, P. Activation of Jun/AP-1 by protein kinase A. *Oncogene* **1992**, *7*, 2281–2286. [PubMed]

43. Gonzalez, G.A.; Montminy, M.R. Cyclic AMP stimulates somatostatin gene transcription by phosphorylation of CREB at serine 133. *Cell* **1989**, *59*, 675–680. [CrossRef]

44. Gay, F.; Barath, P.; Desbois-Le Peron, C.; Metivier, R.; Le Guevel, R.; Birse, D.; Salbert, G. Multiple phosphorylation events control chicken ovalbumin upstream promoter transcription factor I orphan nuclear receptor activity. *Mol. Endocrinol.* **2002**, *16*, 1332–1351. [CrossRef] [PubMed]

45. Maira, M.; Martens, C.; Batsche, E.; Gauthier, Y.; Drouin, J. Dimer-specific potentiation of NGFI-B (Nur77) transcriptional activity by the protein kinase A pathway and AF-1-dependent coactivator recruitment. *Mol. Cell. Biol.* **2003**, *23*, 763–776. [CrossRef] [PubMed]

46. Du, M.; Perry, R.L.; Nowacki, N.B.; Gordon, J.W.; Salma, J.; Zhao, J.; Aziz, A.; Chan, J.; Siu, K.W.; McDermott, J.C. Protein kinase A represses skeletal myogenesis by targeting myocyte enhancer factor 2D. *Mol. Cell. Biol.* **2008**, *28*, 2952–2970. [CrossRef]

47. Tremblay, J.J.; Viger, R.S. Transcription factor GATA-4 is activated by phosphorylation of serine 261 via the cAMP/PKA pathway in gonadal cells. *J. Biol. Chem.* **2003**, *278*, 22128–22135. [CrossRef]

48. Grimley, P.M.; Dong, F.; Rui, H. Stat5a and Stat5b: Fraternal twins of signal transduction and transcriptional activation. *Cytokine Growth Factor Rev.* **1999**, *10*, 131–157. [CrossRef]

49. Kanzaki, M.; Morris, P.L. Lactogenic hormone-inducible phosphorylation and gamma-activated site-binding activities of Stat5b in primary rat Leydig cells and MA-10 mouse Leydig tumor cells. *Endocrinology* **1998**, *139*, 1872–1882. [CrossRef]

50. van Berlo, J.H.; Elrod, J.W.; Aronow, B.J.; Pu, W.T.; Molkentin, J.D. Serine 105 phosphorylation of transcription factor GATA4 is necessary for stress-induced cardiac hypertrophy in vivo. *Proc. Natl. Acad. Sci. USA* **2011**, *108*, 12331–12336. [CrossRef]

51. Bergeron, F.; Boulende Sab, A.; Bouchard, M.F.; Taniguchi, H.; Souchkova, O.; Brousseau, C.; Tremblay, J.J.; Pilon, N.; Viger, R.S. Phosphorylation of GATA4 serine 105 but not serine 261 is required for testosterone production in the male mouse. *Andrology* **2019**, *7*, 357–372. [CrossRef]

52. Lee, Y.K.; Choi, Y.H.; Chua, S.; Park, Y.J.; Moore, D.D. Phosphorylation of the hinge domain of the nuclear hormone receptor LRH-1 stimulates transactivation. *J. Biol. Chem.* **2006**, *281*, 7850–7855. [CrossRef] [PubMed]

53. Hammer, G.D.; Krylova, I.; Zhang, Y.; Darimont, B.D.; Simpson, K.; Weigel, N.L.; Ingraham, H.A. Phosphorylation of the nuclear receptor SF-1 modulates cofactor recruitment: Integration of hormone signaling in reproduction and stress. *Mol. Cell* **1999**, *3*, 521–526. [CrossRef]

54. Kovalovsky, D.; Refojo, D.; Liberman, A.C.; Hochbaum, D.; Pereda, M.P.; Coso, O.A.; Stalla, G.K.; Holsboer, F.; Arzt, E. Activation and induction of NUR77/NURR1 in corticotrophs by CRH/cAMP: Involvement of calcium, protein kinase A, and MAPK pathways. *Mol. Endocrinol.* **2002**, *16*, 1638–1651. [CrossRef]

55. Meng, Q.; Xia, Y. c-Jun, at the crossroad of the signaling network. *Protein Cell* **2011**, *2*, 889–898. [CrossRef] [PubMed]

56. Park, B.H.; Qiang, L.; Farmer, S.R. Phosphorylation of C/EBPbeta at a consensus extracellular signal-regulated kinase/glycogen synthase kinase 3 site is required for the induction of adiponectin gene expression during the differentiation of mouse fibroblasts into adipocytes. *Mol. Cell. Biol.* **2004**, *24*, 8671–8680. [CrossRef] [PubMed]

57. Piwien-Pilipuk, G.; MacDougald, O.; Schwartz, J. Dual regulation of phosphorylation and dephosphorylation of C/EBPbeta modulate its transcriptional activation and DNA binding in response to growth hormone. *J. Biol. Chem.* **2002**, *277*, 44557–44565. [CrossRef]

58. Kane, C.D.; Means, A.R. Activation of orphan receptor-mediated transcription by Ca(2+)/calmodulin-dependent protein kinase IV. *EMBO J.* **2000**, *19*, 691–701. [CrossRef]

59. Najih, M.; Nguyen, H.T.; Martin, L.J. Involvement of calmodulin-dependent protein kinase I in the regulation of the expression of connexin 43 in MA-10 tumor Leydig cells. *Mol. Cell. Biochem.* **2022**. [CrossRef]

60. Senga, Y.; Ishida, A.; Shigeri, Y.; Kameshita, I.; Sueyoshi, N. The phosphatase-resistant isoform of CaMKI, Ca²⁺/Calmodulin-Dependent Protein Kinase I delta (CaMKIdelta), remains in its "primed" form without Ca²⁺ stimulation. *Biochemistry* **2015**, *54*, 3617–3630. [CrossRef]

61. Sun, P.; Lou, L.; Maurer, R.A. Regulation of activating transcription factor-1 and the cAMP response element-binding protein by Ca2+/calmodulin-dependent protein kinases type I, II, and IV. *J. Biol. Chem.* **1996**, *271*, 3066–3073. [CrossRef]

62. Tremblay, J.J.; Viger, R.S. Nuclear receptor Dax1 represses the transcriptional cooperation between GATA-4 and SF-1 in Sertoli cells. *Biol. Reprod.* **2001**, *64*, 1191–1199. [CrossRef] [PubMed]

63. Philips, A.; Lesage, S.; Gingras, R.; Maira, M.H.; Gauthier, Y.; Hugo, P.; Drouin, J. Novel dimeric Nur77 signaling mechanism in endocrine and lymphoid cells. *Mol. Cell. Biol.* **1997**, *17*, 5946–5951. [CrossRef] [PubMed]

64. Lala, D.S.; Rice, D.A.; Parker, K.L. Steroidogenic factor I, a key regulator of steroidogenic enzyme expression, is the mouse homolog of fushi tarazu-factor I. *Mol. Endocrinol.* **1992**, *6*, 1249–1258. [CrossRef]

65. Galarneau, L.; Pare, J.F.; Allard, D.; Hamel, D.; Levesque, L.; Tugwood, J.D.; Green, S.; Belanger, L. The alpha1-fetoprotein locus is activated by a nuclear receptor of the Drosophila FTZ-F1 family. *Mol. Cell. Biol.* **1996**, *16*, 3853–3865. [CrossRef] [PubMed]

66. Teyssier, C.; Belguise, K.; Galtier, F.; Chalbos, D. Characterization of the physical interaction between estrogen receptor alpha and JUN proteins. *J. Biol. Chem.* **2001**, *276*, 36361–36369. [CrossRef] [PubMed]

67. Mayr, B.; Montminy, M. Transcriptional regulation by the phosphorylation-dependent factor CREB. *Nat. Rev. Mol. Cell Biol.* **2001**, *2*, 599–609. [CrossRef]

68. Mansour, S.J.; Matten, W.T.; Hermann, A.S.; Candia, J.M.; Rong, S.; Fukasawa, K.; Vande Woude, G.F.; Ahn, N.G. Transformation of mammalian cells by constitutively active MAP kinase kinase. *Science* **1994**, *265*, 966–970. [CrossRef]

69. Cao, Z.; Umek, R.M.; McKnight, S.L. Regulated expression of three C/EBP isoforms during adipose conversion of 3T3-L1 cells. *Genes Dev.* **1991**, *5*, 1538–1552. [CrossRef]

70. Pereira, F.A.; Qiu, Y.; Zhou, G.; Tsai, M.J.; Tsai, S.Y. The orphan nuclear receptor COUP-TFII is required for angiogenesis and heart development. *Genes Dev.* **1999**, *13*, 1037–1049. [CrossRef]

71. Ariyoshi, K.; Nosaka, T.; Yamada, K.; Onishi, M.; Oka, Y.; Miyajima, A.; Kitamura, T. Constitutive activation of STAT5 by a point mutation in the SH2 domain. *J. Biol. Chem.* **2000**, *275*, 24407–24413. [CrossRef]

72. Wayman, G.A.; Kaech, S.; Grant, W.F.; Davare, M.; Impey, S.; Tokumitsu, H.; Nozaki, N.; Banker, G.; Soderling, T.R. Regulation of axonal extension and growth cone motility by calmodulin-dependent protein kinase I. *J. Neurosci.* **2004**, *24*, 3786–3794. [CrossRef] [PubMed]

73. Abdou, H.S.; Bergeron, F.; Tremblay, J.J. A cell-autonomous molecular cascade initiated by AMP-activated protein kinase represses steroidogenesis. *Mol. Cell. Biol.* **2014**, *34*, 4257–4271. [CrossRef] [PubMed]

74. Abdou, H.S.; Villeneuve, G.; Tremblay, J.J. The calcium signaling pathway regulates leydig cell steroidogenesis through a transcriptional cascade involving the nuclear receptor NR4A1 and the steroidogenic acute regulatory protein. *Endocrinology* **2013**, *154*, 511–520. [CrossRef]

75. Enangue Njembele, A.N.; Bailey, J.L.; Tremblay, J.J. In vitro exposure of Leydig cells to an environmentally relevant mixture of organochlorines represses early steps of steroidogenesis. *Biol. Reprod.* **2014**, *90*, 118. [CrossRef] [PubMed]

76. Enangue Njembele, A.N.; Demmouche, Z.B.; Bailey, J.L.; Tremblay, J.J. Mechanism of action of an environmentally relevant organochlorine mixture in repressing steroid hormone biosynthesis in Leydig cells. *Int. J. Mol. Sci.* **2022**, *23*, 3997. [CrossRef]

77. Enangue Njembele, A.N.; Tremblay, J.J. Mechanisms of MEHP Inhibitory Action and Analysis of Potential Replacement Plasticizers on Leydig Cell Steroidogenesis. *Int. J. Mol. Sci.* **2021**, *22*, 11456. [CrossRef]

78. Mehanovic, S.; Mendoza-Villarroel, R.E.; Viger, R.S.; Tremblay, J.J. The nuclear receptor COUP-TFII regulates *Amhr2* gene transcription via a GC-rich promoter element in mouse Leydig cells. *J. Endocr. Soc.* **2019**, *3*, 2236–2257. [CrossRef]

International Journal of
Molecular Sciences

Article

Targeted Disruption of Lats1 and Lats2 in Mice Impairs Testis Development and Alters Somatic Cell Fate

Nour Abou Nader, Amélie Ménard, Adrien Levasseur, Guillaume St-Jean , Derek Boerboom, Gustavo Zamberlam and Alexandre Boyer *

Centre de Recherche en Reproduction et Fertilité, Faculté de Médecine Vétérinaire, Université de Montréal, Saint-Hyacinthe, QC J2S 7C6, Canada
* Correspondence: alexandre.boyer.1@umontreal.ca; Tel.: +1-450-773-8521 (ext. 8345)

Abstract: Hippo signaling plays an essential role in the development of numerous tissues. Although it was previously shown that the transcriptional effectors of Hippo signaling Yes-associated protein (YAP) and transcriptional coactivator with PDZ-binding motif (TAZ) can fine-tune the regulation of sex differentiation genes in the testes, the role of Hippo signaling in testis development remains largely unknown. To further explore the role of Hippo signaling in the testes, we conditionally deleted the key Hippo kinases large tumor suppressor homolog kinases 1 and -2 (*Lats1* and *Lats2*, two kinases that antagonize YAP and TAZ transcriptional co-regulatory activity) in the somatic cells of the testes using an *Nr5a1*-cre strain (*Lats1*$^{flox/flox}$;*Lats2*$^{flox/flox}$;*Nr5a1*-cre). We report here that early stages of testis somatic cell differentiation were not affected in this model but progressive testis cord dysgenesis was observed starting at gestational day e14.5. Testis cord dysgenesis was further associated with the loss of polarity of the Sertoli cells and the loss of SOX9 expression but not WT1. In parallel with testis cord dysgenesis, a loss of steroidogenic gene expression associated with the appearance of myofibroblast-like cells in the interstitial space was also observed in mutant animals. Furthermore, the loss of YAP phosphorylation, the accumulation of nuclear TAZ (and YAP) in both the Sertoli and interstitial cell populations, and an increase in their transcriptional co-regulatory activity in the testes suggest that the observed phenotype could be attributed at least in part to YAP and TAZ. Taken together, our results suggest that Hippo signaling is required to maintain proper differentiation of testis somatic cells.

Keywords: Hippo signaling; Lats1/2; fetal Leydig cells; Sertoli cells; transgenic mouse model

Citation: Abou Nader, N.; Ménard, A.; Levasseur, A.; St-Jean, G.; Boerboom, D.; Zamberlam, G.; Boyer, A. Targeted Disruption of Lats1 and Lats2 in Mice Impairs Testis Development and Alters Somatic Cell Fate. *Int. J. Mol. Sci.* **2022**, *23*, 13585. https://doi.org/10.3390/ijms232113585

Academic Editor: Jacques J. Tremblay

Received: 3 October 2022
Accepted: 2 November 2022
Published: 5 November 2022

Publisher's Note: MDPI stays neutral with regard to jurisdictional claims in published maps and institutional affiliations.

1. Introduction

In mice, testes are first derived from the intermediate mesoderm and arise from the proliferation of coelomic epithelial cells. Thickening of the coelomic epithelia initially form the adrenogonadal primordium (AGP) at embryonic day 9.5 (e9.5), a structure from which the adrenal cortex also develops [1,2]. Following primordial germ cell migration (around e10.5), the AGP separates into two distinct tissues as the adrenal progenitor cells migrate dorsomedially at e11.0 [3]. Shortly after, the sex determining region of chromosome Y (SRY) is first expressed in a subpopulation of somatic cells of the gonadal primordium (GP) [4,5] which induces their differentiation into Sertoli cells via a positive feedback loop between SRY (sex-determining region Y)-box 9 (SOX9) and fibroblast growth factor 9 (FGF9) [6–8]. After Sertoli cell differentiation, another subpopulation of somatic cells differentiates into fetal Leydig cells (FLCs) following the secretion of desert hedgehog (DHH) by the differentiating Sertoli cells and subsequent activation of Hedgehog signaling [9–11]. Tracing experiments either alone or in combination with single-cell RNAseq suggest that both Sertoli cells and FLCs originate from a common progenitor cell population [12–14]. Numerous factors, including Wilm's tumor 1 (WT1) [15–18], nuclear receptor subfamily 5, group A, member 1 (NR5A1) [19], GATA binding protein 4 (GATA4) [20–22], insulin/insulin-like growth factor [23,24], Sine-oculis-related homeobox-1 and 4 (SIX1/4) [25], or Chromobox2 (Cbx2) [26,27], are involved in early stages of testicular somatic cell differentiation but additional factors and pathways are likely involved.

Hippo is an evolutionarily conserved signaling pathway with well-established roles in cell fate determination and proliferation during embryonic development (reviewed in [28,29]). It consists of a kinase cascade that regulates two functionally redundant transcriptional co-activators, Yes-associated protein (YAP) and transcriptional coactivator with PDZ-binding motif (TAZ). In response to various extracellular signals, the mammalian STE20-like protein kinases 1 and -2 (MST1, MST2) are activated and phosphorylate the large tumor suppressor homolog kinases 1 and -2 (LATS1, LATS2), which in turn phosphorylate and inactivate YAP and TAZ [30–33]. When the cascade is inactivated, YAP and TAZ accumulate in the nucleus and interact principally with members of the TEA domain (TEAD) family of transcription factors to regulate the transcription of genes involved in cell growth, apoptosis, and proliferation [29,34].

Very little is known about the role of Hippo signaling in the testis, and most studies thus far have focused on its involvement in postnatal testicular physiology. Notably, it was shown that YAP is negatively regulated by FSH/PKA signaling in Sertoli cells [35,36] and that the number of YAP-positive Leydig cells increases in the mouse testis following chronic heat stress [37]. It was also shown that the loss of *Yap* and *Taz* decreases the expression of male sex-determining genes and increases the expression of female sex-determining genes in an immature Sertoli cell culture model [38], while the loss of *Lats1* and *Lats2* in granulosa cells have the opposite effect [39]. Finally, it was also demonstrated that the loss of Kindlin-2 (also known as fermitin family homolog 2; *Fermt2*) in Sertoli cells inhibits their proliferation and impairs cell–cell junction and blood–testis barrier maintenance by enhancing LATS1 interaction with YAP [40]. However, to this day, no studies have directly evaluated the role of LATS1 and LATS2, the core kinases of Hippo signaling, in testicular development.

In this study, we generated a mouse model in which *Lats1* and *Lats2* were conditionally deleted in testicular somatic cells to characterize the role of the Hippo signaling pathway in the development of the testes.

2. Results

2.1. Testicular Dysgenesis Is Observed following Concomitant Deletion of Lats1 and Lats2 in Testis Somatic Cells

To investigate the role of Hippo signaling in testicular development, we genetically disrupted the core Hippo pathway kinases *Lats1*, *Lats2*, or both in testicular somatic cells. To do this, mice bearing floxed alleles for *Lats1* and/or *Lats2* were crossed with a *Nr5a1*-cre [41] strain that drives the expression of Cre recombinase in the precursors of both Sertoli cells and FLCs of the developing testes. *Lats1*$^{\text{flox/flox}}$;*Nr5a1*-cre and *Lats2*$^{\text{flox/flox}}$;*Nr5a1*-cre mice developed normally, were healthy, had normal lifespans, and were fertile. Testes from *Lats1*$^{\text{flox/flox}}$; *Nr5a1*-cre and *Lats2*$^{\text{flox/flox}}$; *Nr5a1*-cre mice appeared normal at the gross and histologic levels and were therefore not further studied. Conversely, double cKO (*Lats1*$^{\text{flox/flox}}$;*Lats2*$^{\text{flox/flox}}$;*Nr5a1*-cre) mice died between 2 and 3 weeks of age of adrenal failure due to Cre expression in the adrenal cortex [42]. Gross morphological assessment showed that the testes from 2-week-old *Lats1*$^{\text{flox/flox}}$;*Lats2*$^{\text{flox/flox}}$;*Nr5a1*-cre mice were smaller than the testes from the control *Lats1*$^{\text{flox/flox}}$;*Lats2*$^{\text{flox/flox}}$ animals and had a distinctive, lobular appearance (Figure 1A).

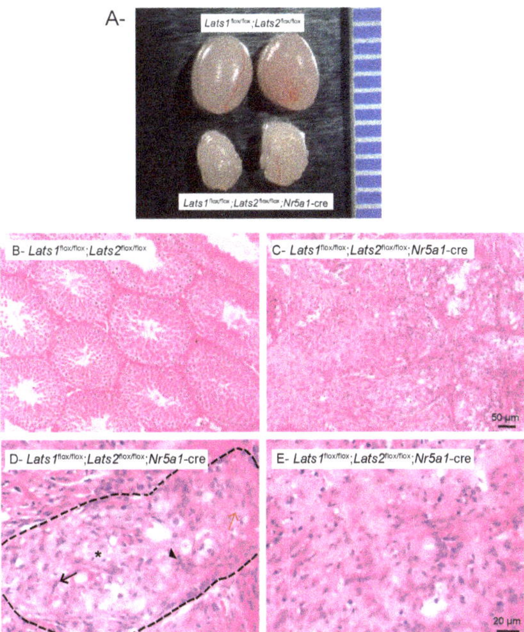

Figure 1. Testes of *Lats1*$^{\text{flox/flox}}$;*Lats2*$^{\text{flox/flox}}$;*Nr5a1*-cre mice are larger than controls and are histomorphologically abnormal. (**A**) Photographs of testes from 2-week-old *Lats1*$^{\text{flox/flox}}$;*Lats2*$^{\text{flox/flox}}$ (control) and *Lats1*$^{\text{flox/flox}}$;*Lats2*$^{\text{flox/flox}}$;*Nr5a1*-cre mice. Ruler graduations are in millimeters. (**B,C**) Photomicrographs comparing the testes of 2-week-old (**B**) *Lats1*$^{\text{flox/flox}}$;*Lats2*$^{\text{flox/flox}}$ and (**C**) *Lats1*$^{\text{flox/flox}}$;*Lats2*$^{\text{flox/flox}}$;*Nr5a1*-cre mice. (**D,E**) Photomicrographs illustrating the remaining seminiferous tubules (**D**) and interstitium (**E**) in *Lats1*$^{\text{flox/flox}}$;*Lats2*$^{\text{flox/flox}}$;*Nr5a1*-cre mice at higher magnification. Arrowhead = cells with a large nucleus and pale abundant cytoplasm. Red arrow = cell with an elongated large nucleus with an abundant cytoplasm. Black arrow = spindle-shaped cells. Asterisk = pyknotic cell. Dashes = delimitation of the seminiferous tubules. Scale bar in C is valid for B and scale bar in E is valid for D. Hematoxylin and eosin stain.

Histological analyses revealed striking abnormalities in the testes of *Lats1*$^{\text{flox/flox}}$;*Lats2*$^{\text{flox/flox}}$;*Nr5a1*-cre mice. Seminiferous tubules were rarely observed and were very large, with disorganized cells and no lumen (Figure 1B–E). Several different types of cells were present within these tubules, with some cells having elongated large nuclei with an abundant cytoplasm, others having round large nuclei and abundant pale cytoplasm, others having a spindle shape, and others being pyknotic (Figure 1D). Furthermore, the tubules were surrounded by spindle-shaped interstitial cells reminiscent of fibroblasts or myofibroblasts (Figure 1E).

2.2. Testicular Dysgenesis Is First Observed at e14.5 in Lats1$^{\text{flox/flox}}$;Lats2$^{\text{flox/flox}}$;Nr5a1-cre Mice

To analyze the onset and evolution of the phenotype observed in the testes of *Lats1*$^{\text{flox/flox}}$;*Lats2*$^{\text{flox/flox}}$;*Nr5a1*-cre mice, embryos were collected from embryonic days e12.5 onward, and histopathologic examination of the developing testes was performed. These analyses showed that testes from e12.5 *Lats1*$^{\text{flox/flox}}$;*Lats2*$^{\text{flox/flox}}$;*Nr5a1*-cre mice were phenotypically indistinguishable from age-matched controls (Figure 2) ($n = 6$). However, starting at e14.5, dysgenesis of a few testis cords became apparent in 30% of the mutant animals ($n = 10$) (Figure 2, dashed lines), while no abnormal phenotype was apparent in the remaining animals (Figure S1A). By e16.5, dysgenesis was apparent in a larger portion of the testis cords of all mutant animals ($n = 4$) and an accumulation of spindle-shaped cells was apparent in some regions of the interstitium (Figure 2, red arrow). At e17.5, the phenotypic changes observed in the testes of *Lats1*^{flox/flox};*Lats2*^{flox/flox};*Nr5a1*-cre mice were exacerbated and dysgenesis was observed in all of the testis cords, with the Sertoli cells being disorganized and having a large prominent cytoplasm and the remaining germ cells being mostly apoptotic (Figure S2A, arrowhead) and pushed over to one side of the cords ($n = 10$) (Figure 2, arrow, Figure S2B). Testis cords were also larger than their control counterparts and the interstitium was mostly occupied by spindle-shaped cells (Figure 2). Phenotypic changes observed in the testes of *Lats1*^{flox/flox};*Lats2*^{flox/flox};*Nr5a1*-cre mice were even more pervasive and pronounced in testes of 1dpp mutant animals (Figure 2), with germ cells being absent in the majority of the tubules (Figure 2, Figure S2B), the presence of some spindle-shaped cells of indeterminate origin located inside the testis cords (Figure 2, arrowhead), and the interstitium being completely occupied by spindle-shaped cells.

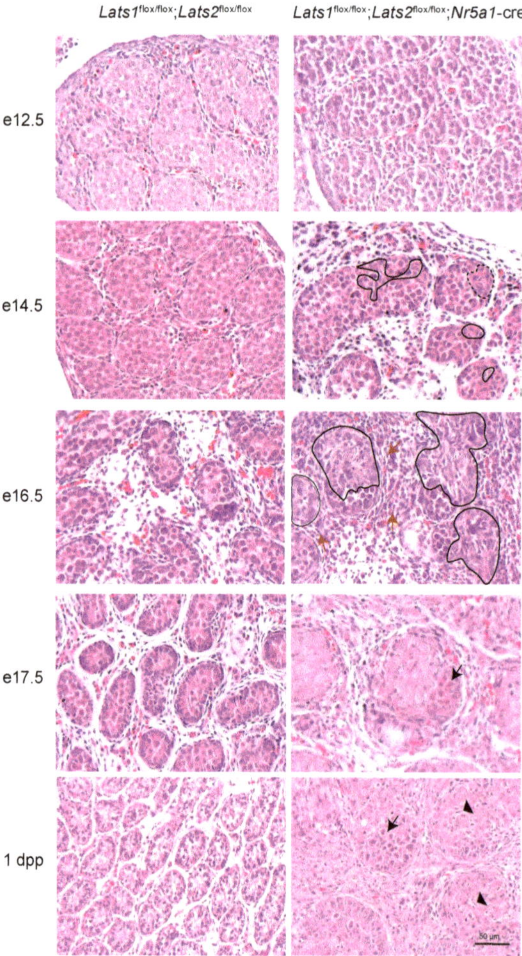

Figure 2. Progressive testicular dysgenesis in the *Lats1*^{flox/flox};*Lats2*^{flox/flox};*Nr5a1*-cre mice. Photomicrographs comparing testis histology of *Lats1*^{flox/flox};*Lats2*^{flox/flox};*Nr5a1*-cre with that of *Lats1*^{flox/flox};*Lats2*^{flox/flox} controls during development. Dashed lines = testis cords dysgenesis, black arrow = germ cells, red arrow = interstitial spindle-shaped cells, arrowhead = spindle-shaped cells in the testis cords. Scale bar (lower right) is valid for all images. Hematoxylin and eosin stain.

2.3. Hippo Signaling Is Inactivated in Late Stages of Testicular Differentiation

Although the *Nr5a1* promoter is active from the early stages of adrenal–gonadal primordium formation at around e9.0–e10 in mice [43], the phenotypic changes observed in the testes of *Lats1*$^{flox/flox}$;*Lats2*$^{flox/flox}$;*Nr5a1*-cre mice were not apparent until e14.5. This is later than expected but is consistent with the timing of phenotypic changes and inactivation of *Lats1* and *Lats2* observed in the developing adrenal glands of mutant animals, which is also first observed at e14.5 [42]. To determine if this was also the case in the testes, RT-qPCR analyses were performed on developing testes at embryonic days e14.5, e17.5 and 1dpp. The results showed a small but significant reduction (33%) of testis *Lats1* mRNA levels (but no reduction of *Last2* mRNA levels) in e14.5 mutant animals (Figure 3A); 56% and 42% decreases in testis *Lats1* and *Lats2* mRNA levels in e17.5 mutant animals (Figure 3B); and 60% and 58% decreases in testis *Lats1* and *Lats2* mRNA levels in 1dpp mutant animals (Figure 3C).

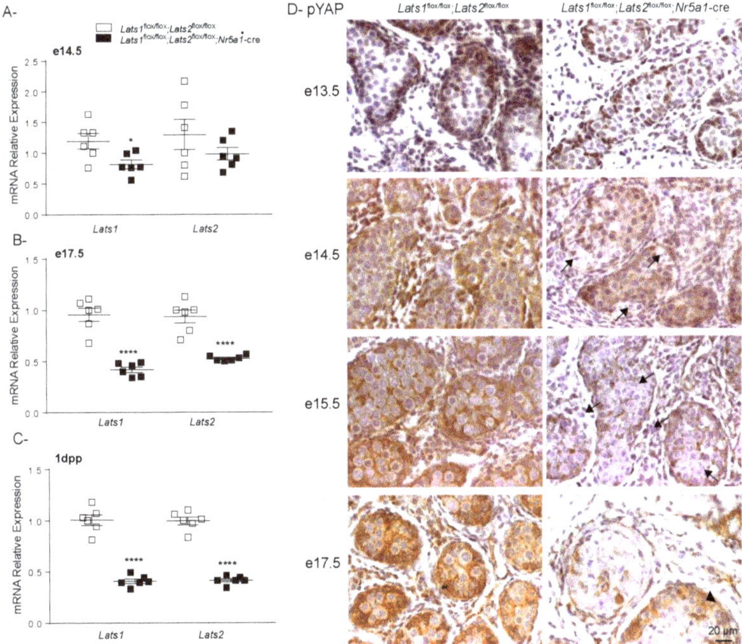

Figure 3. Efficiency of *Lats1* and *Lats2* knockdown in the testes of *Lats1*$^{flox/flox}$;*Lats2*$^{flox/flox}$;*Nr5a1*-cre mice. (**A–C**) RT-qPCR analysis of *Lats1* and *Lats2* mRNA levels in the testes of (**A**) e14.5, (**B**) e17.5, and (**C**) 1 dpp mice of the indicated genotypes (*n* = 6 animals/genotype). All data were normalized to the housekeeping gene *Rpl19* and are expressed as means (columns) ± SEM (error bars). Asterisks = significantly different from control (* *p* < 0.05; **** *p* < 0.0001). (**D**) Immunohistochemical analysis of phospho-YAP expression in the testes of mice of the indicated genotypes. Scale bar (lower right) is valid for all images. Arrowhead = cells with inefficient recombination. Arrow = cells with efficient recombination.

We were unable to obtain quality immunohistochemistry for LATS1/2 to complement our RT-qPCR *Lats1/2* expression data. Loss of *Lats1* and *Lats2* was therefore assessed indirectly by evaluating the phosphorylation of their substrate YAP. At e13.5, phospho-YAP was readily detected in the nucleus of the Sertoli cells of both control and mutant testes (Figure 3D). Weak expression of phospho-YAP could also be detected in some interstitial cells of both control and mutant testes. These results suggest that Hippo signaling was not inactivated in mutant animals at this time point. At e14.5, phospho-YAP expression was readily detected in the Sertoli cells and interstitial cells of the control testes (Figure 3D) and some mutant animals (Figure S1B). However, a decrease in the expression of phospho-YAP could be observed in a few Sertoli cells of some mutant animals (Figure 3D, Arrow). Furthermore, phospho-YAP expression remained weak and mostly cytoplasmic in the interstitium of these animals (Figure 3D). At e15.5 and e17.5, phospho-YAP expression remained elevated in both the interstitial cells and the Sertoli cells of control testes (Figure 3D). However, in the testes of the mutant animals, phospho-YAP was no longer detected in the majority of the interstitial cells (Figure 3D). Phospho-YAP expression was also lost in the majority (but not all) of the Sertoli cells, suggesting that recombination had not yet occurred in all Sertoli cells (Figure 3D, arrowhead). The pattern of expression of phospho-YAP is in accordance with the loss of *Lats1* and *Lats2* in the mutant animals and confirmed that inactivation of Hippo signaling occurs in both the FLCs and Sertoli cells. Variation between phospho-YAP expression in e14.5 mutant animals combined with the RT-qPCR data also suggests that recombination is initiated around that time point in mutant animals.

Since the inactivation of Hippo signaling is also normally associated with an increase in the nuclear localization of both YAP and TAZ, the expression of YAP and TAZ and their classic downstream targets were further evaluated at e17.5 by immunohistochemistry (using antibodies that mark both their phosphorylated and unphosphorylated forms). At this time point, YAP expression was detected in the cytoplasm and nucleus of most interstitial cells and in the Sertoli cells of control animals (Figure 4A,C). In the testes of mutant animals, expression of YAP was observed in the nucleus of most interstitial cells (Figure 4B,D). Nuclear, but generally weak, expression of YAP was detected in the Sertoli cells located in the middle of the tubules (Figure 4B,E) while strong YAP expression was observed in the Sertoli cells located at the periphery of some tubules (Figure 4B, arrow). However, this last population also expressed phospho-YAP (Figure 3D), suggesting that they correspond to Sertoli cells in which recombination did not occur. Weak TAZ expression was detected in the cytoplasm of the majority of Sertoli cells and some interstitial cells, as well as in the nucleus of rare interstitial cells in the control animals (Figure 4F,H). However, nuclear expression of TAZ was detected in a larger proportion of interstitial cells in the testes of *Lats1*^{flox/flox};*Lats2*^{flox/flox};*Nr5a1*-cre mice (Figure 4G,I). Furthermore, TAZ expression was detected in both the cytoplasm and nucleus of the recombined Sertoli cells (Figure 4G,J), suggesting that the increase of TAZ activity might play a greater role in the observed phenotype.

To further determine if the transcriptional co-regulatory activity of YAP and TAZ was increased in *Lats1*^{flox/flox};*Lats2*^{flox/flox};*Nr5a1*-cre mice, the mRNA levels of the well-established YAP/TAZ target genes ankyrin repeat domain 1 (*Ankrd1*), connective tissue growth factor (*Ctgf*), and cysteine-rich and angiogenic inducer 61 (*Cyr61*) were quantified by RT-qPCR. A 40-fold increase in the mRNA levels of *Ankrd1* and an 8-fold increase in the mRNA levels of *Ctgf* and *Cyr61* were observed in the testes of e17.5 *Lats1*^{flox/flox};*Lats2*^{flox/flox};*Nr5a1*-cre mice (Figure 4K), confirming that the expression of known targets of YAP and TAZ is increased in the testes of mutant animals. Taken together, these data suggest that transcriptional regulatory activity of YAP and/or TAZ is increased in the somatic cells of *Lats1*^{flox/flox};*Lats2*^{flox/flox};*Nr5a1*-cre mice, and most likely play a role in the observed phenotypic changes.

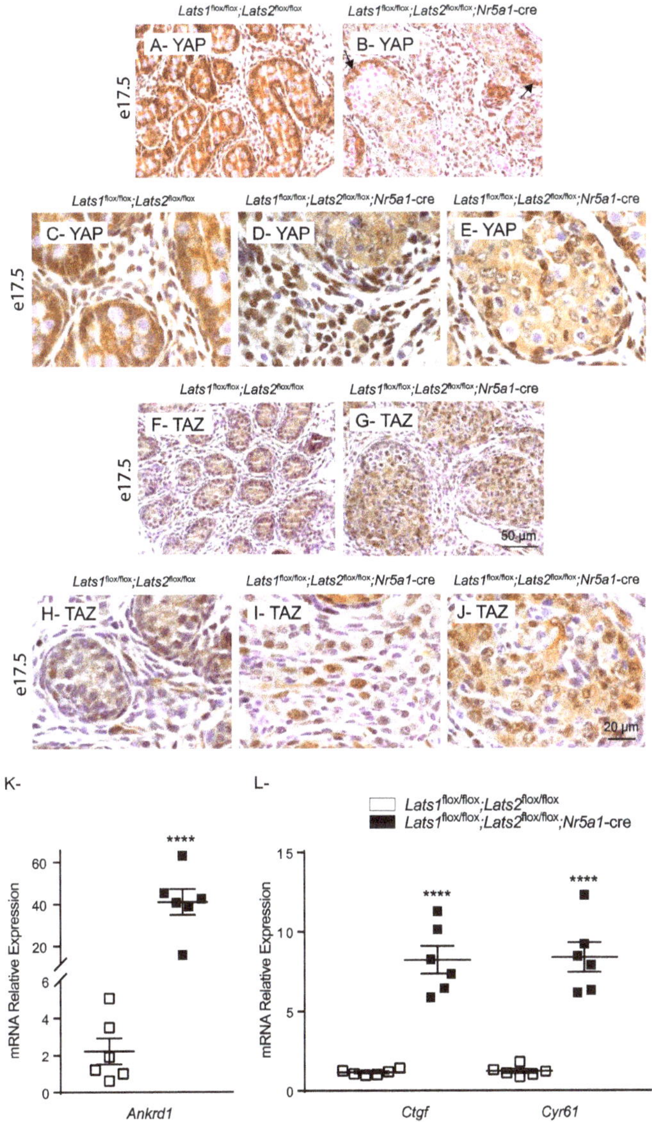

Figure 4. YAP/TAZ activity increases in the testes of $Lats1^{flox/flox};Lats2^{flox/flox};Nr5a1^{cre/+}$ mice. (**A–E**) Immunohistochemical analysis of YAP expression in testes of e17.5 mice of the indicated genotypes. (**F–J**) Immunohistochemical analysis of TAZ expression in testes of e17.5 mice of the indicated genotypes. Arrow = YAP expression in non-recombined Sertoli cells. Scale bar in (**G**) is valid for (**A,B,F**). Scale bar in (**J**) is valid for (**C–E,H,I**). (**K,L**) RT-qPCR analysis of YAP/TAZ downstream targets mRNA levels in the testes of e17.5 mice of the indicated genotypes (n = 6 animals/genotype). All data were normalized to the housekeeping gene *Rpl19* and are expressed as means (columns) ± SEM (error bars). Asterisks = significantly different from control (**** $p < 0.0001$).

2.4. Disruption of Lats1 and Lats2 Alters the Epithelization and the Differentiation of the Sertoli Cells during Testis Development

YAP and TAZ activation have been known to play key roles in processes such as epithelial-to-mesenchymal (EMT) transition and fibrosis, two cellular processes that share many molecular characteristics [44–46]. Misexpression of mesenchymal-associated pathways could explain both the loss polarity of Sertoli cells and the appearance of fibroblast/myofibroblast-like cells in the interstitial tissue. To determine if normal epithelization of the Sertoli cells was affected in the testes of mutant animals, markers of Sertoli cell junctions were then evaluated. Expression of GAP junction protein alpha-1 (GJA1), (an important component of Sertoli/Sertoli interactions [47,48]) expression was first analyzed. At e14.5, expression of GJA1 was marginal in both the control and mutant animals, with most of the expression detected in the interstitial tissue and only weak expression observed in the testis cords (Figure 5A). A marked increase in the expression of GJA1 was observed in the testis cords of control and mutant animals at e16.5. However, GJA1 expression was more important at the periphery of the testis cords in control animals, a pattern of expression that is not observed in mutant animals (Figure 5A). At e17.5, GJA1 expression was mainly observed at the periphery of testis cords (at the presumed Sertoli cell junctions), while GJA1 expression was lost in the majority of the testis cords of the mutant animals (Figure 5A). Furthermore, the mRNA levels for occludin (*Ocln*), a major contributor to Sertoli cell tight junctions, and for par-6 family cell polarity regulator beta (*Pard6b*), a gene important for apicobasal polarity, also decreased in the testes of mutant animals (Figure 5B) suggesting that epithelization of the Sertoli cells was affected in *Lats1*$^{\text{flox/flox}}$;*Lats2*$^{\text{flox/flox}}$;*Nr5a1*-cre mice.

To further determine if the loss of *Lats1* and *Lats2* in Sertoli cells affects their cellular identity, immunohistochemistry for WT1 and SOX9 was performed. In control animals, WT1+ and SOX9+ positive cells were mostly located at the periphery of the testis cords from e14.5 to 1dpp, though rare positive cells were observed in the middle of the cords at e14.5 (Figure 6A,B). In the testes of mutant animals, an increasing portion of WT1+ was found in the middle of the testis cords between e14.5 and 1dpp (Figure 6A, arrow), again suggesting that these Sertoli cells had lost their polarity. Interestingly, even if localization of SOX9+ cells were initially similar to WT1+ cells in *Lats1*$^{\text{flox/flox}}$;*Lats2*$^{\text{flox/flox}}$;*Nr5a1*-cre, only rare SOX9+ cells could be detected in the seminiferous tubules of 1dpp mice (Figure 6B). To confirm the results obtained by immunohistochemistry and to further characterize the Sertoli cells in *Lats1*$^{\text{flox/flox}}$;*Lats2*$^{\text{flox/flox}}$;*Nr5a1*-cre mice, the expression of the Sertoli cell markers cytochrome P450, 26, retinoic acid B1 (*Cyp26b1*), desert hedgehog (*Dhh*), doublesex and mab-3-related transcription factor 1 (*Dmrt1*), *Sox9*, and *Wt1* were evaluated by RT-qPCR at 1dpp. Interestingly, the mRNA levels of all Sertoli cell markers decreased significantly except for the mRNA levels of *Wt1*, which were maintained in the testes of mutant animals (Figure 6C). To determine if Sertoli cells could differentiate into granulosa cells, RT-qPCR were also performed for forkhead box L2 (*Foxl2*) and wingless-type MMTV integration site family, member 4 (*Wnt4*). However, expression levels of both markers did not increase in the testes of mutant animals (Figure S3). Taken together, the epithelization defect and the expression of Sertoli cell markers suggest that the loss of *Lats1/2* affects the identity of the Sertoli cells but that these cells do not transdifferentiate into granulosa cells.

A- GJA1

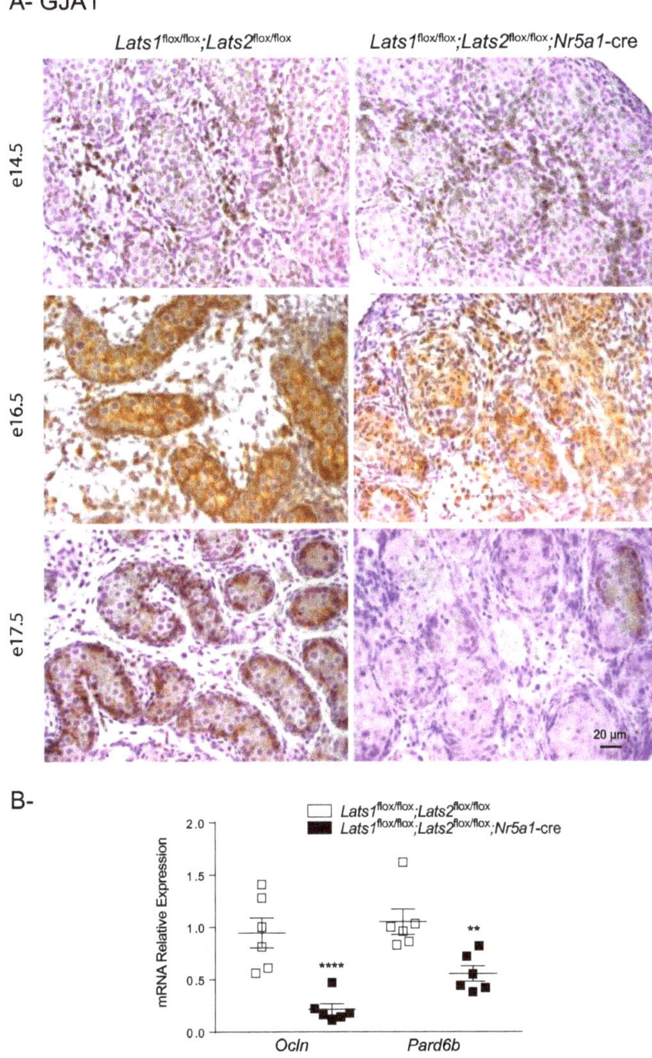

Figure 5. Loss of *Lats1* and *Lats2* affects the polarity of the Sertoli cells. (**A**) Immunohistochemical analysis of GJA1 expression in testes of mice of the indicated genotypes. Scale bar (lower right) is valid for all images. (**B**) RT-qPCR analysis of *Ocln* and *Pard6b* mRNA levels in the testes of 1 dpp mice of the indicated genotypes ($n = 6$ animals/genotype). All data were normalized to the housekeeping gene *Rpl19* and are expressed as means (columns) $\pm$ SEM (error bars). Asterisks = significantly different from control (** $p < 0.01$; **** $p < 0.0001$).

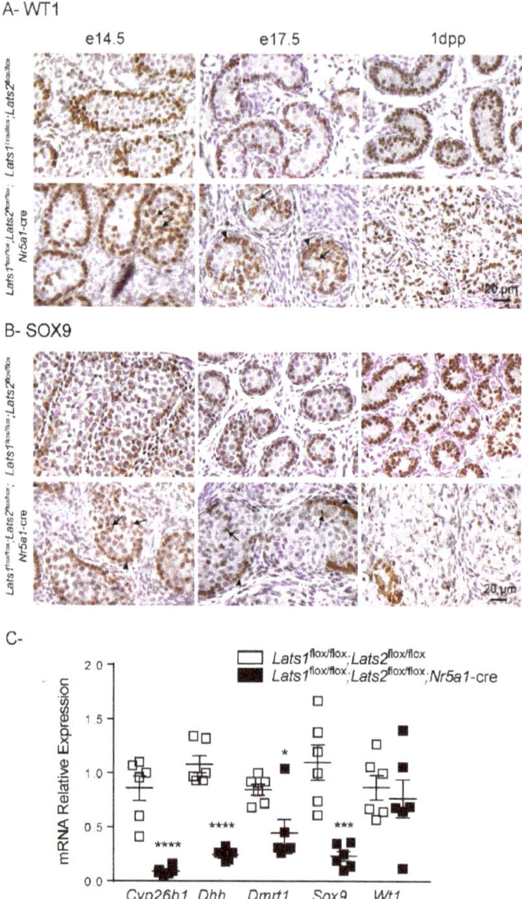

Figure 6. Loss of *Lats1* and *Lats2* affects expression of key Sertoli cell markers. (**A**) Immunohistochemical analysis of WT1 expression in testes of mice of the indicated genotypes. (**B**) Immunohistochemical analysis of SOX9 expression in testes of mice of the indicated genotypes. Arrow = immunopositive cells that lost their polarity. Arrowhead = immunopositive cells with normal polarity. Scale bar (lower right) is valid for all images. (**C**) RT-qPCR analysis of Sertoli cell markers in testes of 1dpp mice of the indicated genotypes (n = 6 animals/genotype). All data were normalized to the housekeeping gene *Rpl19* and are expressed as means (columns) ± SEM (error bars). Asterisks = significantly different from control (* $p < 0.05$; *** $p < 0.001$; **** $p < 0.0001$).

2.5. Disruption of Lats1 and Lats2 Causes the Loss of Steroidogenic Markers and Leads to Fibrosis of the Testis Interstitium

To characterize the consequences of *Lats1* and *Lats2* deletion in FLCs, CYP17A1 expression was evaluated by immunohistochemistry. At e14.5, the expression of CYP17A1 was indistinguishable between control and mutant animals (Figure 7A). However, at e17.5, the number of CYP17A1+ cells was considerably reduced in the testes of mutant animals compared to their control counterparts (Figure 7A).

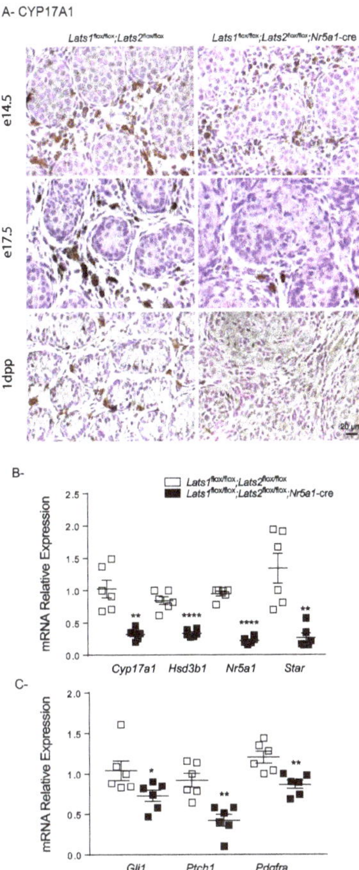

Figure 7. Loss of *Lats1* and *Lats2* affects FLCs. (**A**) Immunohistochemical analysis of CYP17A1 expression in testes of mice of the indicated genotypes. Scale bar (lower right) is valid for all images. (**B,C**) RT-qPCR analysis of steroidogenesis (**B**) and progenitor FLC markers (**C**) in testes of 1dpp mice of the indicated genotypes ($n = 6$ animals/genotype). All data were normalized to the housekeeping gene *Rpl19* and are expressed as means (columns) $\pm$ SEM (error bars). Asterisks = significantly different from control (* $p < 0.05$; ** $p < 0.01$; **** $p < 0.0001$).

By 1dpp, CYP17A1+ cells were completely absent in the testes of mutant animals (Figure 7A). To confirm the loss of steroidogenic cells, *Cyp17a1*, hydroxy-delta-5-steroid dehydrogenase, 3 beta- and steroid-delta-isomerase 1 (*Hsd3b1*), nuclear receptor subfamily 5, group A, member 1 (*Nr5a1*) and steroidogenic acute regulatory protein (*Star*) mRNA levels were evaluated. As expected, a marked loss of every steroidogenic marker was observed in testes of *Lats1*$^{\text{flox/flox}}$;*Lats2*$^{\text{flox/flox}}$;*Nr5a1*-cre mice (Figure 7B). Together, these results suggest that like Sertoli cells, FLC differentiation initially occurs but that FLC subsequently lose their steroidogenic capacity.

To further characterize the FLC population, expression of platelet-derived growth factor receptor, alpha polypeptide (*Pdgfra*), and patched homolog 1 (*Ptch1*), which are initially expressed in both interstitial cells and non-steroidogenic FLCs progenitors [49–52], as well as the effector of Hedgehog signaling *Gli1*, were also evaluated by RT-qPCR at 1dpp. Again, a significant decrease in the mRNA levels of all three markers were observed in testes of *Lats1*$^{\text{flox/flox}}$;*Lats2*$^{\text{flox/flox}}$;*Nr5a1*-cre mice (Figure 7C), arguing against the possibility that FLCs dedifferentiate into non-steroidogenic FLCs progenitors.

The progressive loss of steroidogenesis and the adoption of a myofibroblast-like spindle shape by the testis interstitial cells was reminiscent of similar changes that occur in the adrenal gland of *Lats1*$^{flox/flox}$;*Lats2*$^{flox/flox}$;*Nr5a1*-cre mice [42]. In the latter, adrenocortical cells (which share a common embryonic origin with FLCs) overexpress the mesenchymal cell and myofibroblast marker vimentin (VIM), followed by an increase in the expression of the myocyte and myofibroblast marker α-SMA [42]. To determine if this was also the case in the testes, VIM and α-SMA expression were evaluated in the testes of control and mutant mice. A marked increase in the number of VIM+ cells was detected in the interstitial tissue of the mutant testes compared to controls (Figure 8A,B) at e17.5. Interestingly, the expression pattern of VIM was also modified in the Sertoli cells (Figure 8A,B). An increase in α-SMA expression was detected in the majority of the interstitial cells of mutant animals (Figure 8C,D) but only at 1dpp. This increase of expression was also associated with an increase in the mRNA levels of actin, alpha 2, smooth muscle, aorta (*Acta2*, the gene coding for α-SMA), and of the myofibroblast markers caldesmon 1 (*Cald1*) and secreted phosphoprotein 1 (*Spp1*) (Figure 8E), as well as with the eventual accumulation of collagen fibers in the interstitial tissue of older animals (Figure 8F, G). Together these results suggest that the transdifferentiation of the FLCs cells could be at least in part responsible for the interstitial fibrosis.

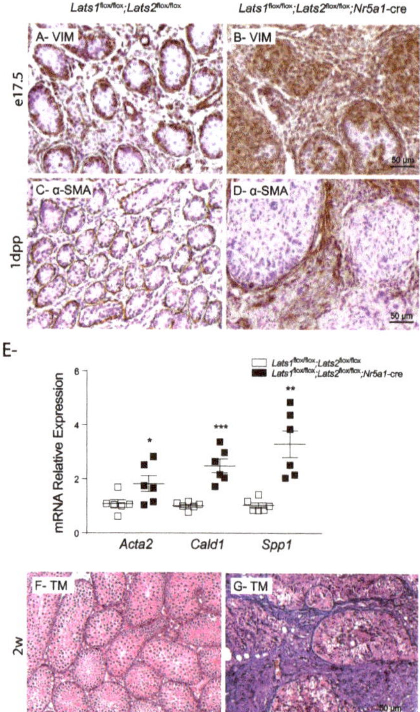

Figure 8. Loss of *Lats1* and *Lats2* leads to fibrosis in the testis interstitial tissue. (**A,B**) Immunohistochemical analysis of VIM expression in testes of e17.5 mice of the indicated genotypes. (**C,D**) Immunohistochemical analysis of α-SMA expression in testes of 1dpp mice of the indicated genotypes. (**E**) RT-qPCR analysis of myofibroblast markers in testes of 1dpp mice of the indicated genotypes (n = 6 animals/genotype). All data were normalized to the housekeeping gene *Rpl19* and are expressed as means (columns) ± SEM (error bars). Asterisks = significantly different from control (* $p < 0.05$; ** $p < 0.01$; *** $p < 0.001$). (**F,G**) Masson's trichrome staining of 2-week-old mice of the indicated phenotype. Scale bars in the right panels are valid for the left panels.

3. Discussion

In recent years, Hippo has been identified as one of the most important signaling pathways involved in tissue development; however, no study has evaluated its function in the development of the testes. We report herein that the concomitant inactivation of the two core kinases of the Hippo pathway, *Lats1* and *Lats2*, alters the fate of the testicular somatic cells.

Even though NR5A1 is normally expressed early during gonadal development (e9.5–10.0) in a progenitor cell population that differentiate into both Sertoli cells and FLCs [13], the loss of *Lats1/2* observed in the *Lats1*$^{flox/flox}$;*Lats2*$^{flox/flox}$;*Nr5a1*-cre model was only apparent in late stages of testis development, as suggested by the loss of *Lats1* and *Lats2* mRNA levels and concomitant loss of YAP phosphorylation. This result could be explained in one of two ways. First, it is possible that recombination of the *Lats1* and *Lats2* floxed alleles is not efficient in testis somatic cells. Second, it is possible that the *Nr5a1*-cre model used in the present study does not sufficiently express Cre in the early stages of testis development to allow proper recombination. In that regard, it is important to note that the *Nr5a1*-cre model used for this study was the one developed by the Lowell group [41], which has never been fully characterized in the early stages of testis development and which has been used more sparingly than the model developed by the Parker group [53]. No matter the reason, our results suggest that efficient recombination only occurs when Sertoli cells and FLCs are already committed to their respective lineage, and it can be concluded that the phenotypic changes observed in the mutants herein are most likely due to the transdifferentiation of the Sertoli cells and FLCs rather than to the altered differentiation of their common progenitors.

Inactivation of *Lats1/2* seems to impair the development of both Sertoli cells and FLCs cells in the *Lats1*$^{flox/flox}$;*Lats2*$^{flox/flox}$;*Nr5a1*-cre animals. In Sertoli cells, the redistribution/increase of VIM expression, the absence of increase in junction protein, alpha 1, 43 KDA (GJA1) expression, the decrease in the expression of *Ocln* and *Pard6b*, and the presence of WT1+ and SOX9+ cells at the center of the tubules together suggest that Sertoli cells lose their epithelial nature and potentially gain characteristics of mesenchymal cells in mutant mice. Epithelial–mesenchymal transition (EMT) is a common event following the downregulation of Hippo signaling and the increase in YAP/TAZ activity [39,54–59] and has notably been reported to occur in ovarian granulosa cells [39], which share a common precursor with Sertoli cells [4]. Interestingly, it was shown that YAP and WT1 act in synergy to loosen cell–cell contacts and trigger EMT in the epithelial MDCK cell line [59], suggesting that YAP or TAZ could act with WT1 in a similar manner in Sertoli cells. It was also previously observed that Sertoli cells initially have mesenchyme-like behaviors and morphology before migrating to the periphery of the testis cords and gaining characteristics of epithelial cells [60]. The phenotype observed in the *Lats1*$^{flox/flox}$;*Lats2*$^{flox/flox}$;*Nr5a1*-cre model could therefore suggest that the Sertoli cells regain characteristics of mesenchymal cells and dedifferentiate to a more primitive stage of development. The fact that WT1 expression (whose expression precedes the differentiation of the Sertoli cell expression [16,61,62]) is maintained in the majority of the presumptive Sertoli cells while SOX9 and *Dhh* are lost also argues in favor of the dedifferentiation of the Sertoli cells. Further experiments are required to delineate the role of Hippo signaling in the maintenance of the identity of the Sertoli cells.

Loss of SOX9 expression in the presumptive immature Sertoli cells of the *Lats1*$^{flox/flox}$;*Lats2*$^{flox/flox}$;*Nr5a1*-cre animals was surprising as it was previously shown that loss of *Yap* and *Taz* leads to the downregulation of SOX9 expression in primary cultures of immature Sertoli cells [38] and that loss of *Lats1* and *Lats2* in postnatal granulosa cells leads to an increase in SOX9 expression and the appearance of Sertoli-like cells in the ovary [39]. This finding suggests that interactions between Hippo signaling and *Sox9* expression are more complex than expected. Since YAP and TAZ can bind several transcription factors, it is possible that the presence or absence of other transcription factors during different stages of Sertoli cell development dictate the regulation of *Sox9* by Hippo signaling. Another possibility is that very tight regulation of the transcriptional activity of YAP and TAZ is necessary to direct proper levels of *Sox9* expression and to determine Sertoli cell fate. For instance, basal levels of YAP and TAZ expression may be necessary for proper expression of SOX9 in the developing Sertoli cells, while overexpression of YAP and TAZ could lead to its inactivation. Indeed, a phenomenon similar to this has been reported to occur in the developing kidney, where the deletion of *Yap* in the cap mesenchyme prevents nephron formation [63], whereas the overexpression of YAP/TAZ leads to the differentiation of nephron progenitors into myofibroblasts [64].

Aside from the Sertoli cells, the interstitial tissues were also affected in the *Lats1*$^{\text{flox/flox}}$; *Lats2*$^{\text{flox/flox}}$;*Nr5a1*-cre animals. Loss of steroidogenic FLCs and the concomitant appearance of a spindle-shaped cell population leading to the development of interstitial fibrosis was observed in the testis of mutant animals at e16.5. Because steroidogenic cells progressively transdifferentiate into myofibroblast-like cells in the adrenal cortex of *Lats1*$^{\text{flox/flox}}$;*Lats2*$^{\text{flox/flox}}$;*Nr5a1*-cre animals [42], we hypothesized that FLCs (which share a common origin with the steroidogenic cells of the adrenal cortex) also transdifferentiate into myofibroblast-like cells in these animals. Interestingly, it was previously shown that disruption of *Nr5a1* in FLCs leads to massive interstitial fibrosis caused by the synthesis of extracellular matrix by the *Nr5a1*-disrupted FLCs [65], suggesting that loss of *Nr5a1* in our model could be responsible for the observed fibrosis. Though it seems likely that some myofibroblast-like cells were of Leydig cell origin, we cannot rule out the possibility that others originated from peritubular myoid cells or other stromal cell types. Lineage tracing experiments would be needed to determine their origin unequivocally. Hippo signaling inactivation and YAP/TAZ transcriptional activation have also been shown to play a major roles in the transdifferentiation of cells into myofibroblasts and fibrosis in non-endocrine tissues such as the lung [66] and the heart [67] as well as in tissues derived from the intermediate mesoderm such as the kidney [68] and the Mullerian duct [69]. Among the known downstream targets of YAP/TAZ, CTGF could play an important role in the transdifferentiation of FLCs into myofibroblast-like cells. Indeed, CTGF has been previously described as a key driver of fibrosis [70–73] and myofibroblast formation [74–76]. Furthermore, it was shown that TAZ was dramatically enriched on the promoter of *Ctgf* following inactivation of *Lats1/2* in mouse Mullerian ducts, leading to *Ctgf* overexpression and myofibroblast transdifferentiation [69].

Even though we attributed the phenotypic changes observed in the Sertoli cells and FLCs of *Lats1*$^{\text{flox/flox}}$;*Lats2*$^{\text{flox/flox}}$;*Nr5a1*-cre mice to the direct loss of *Lats1/2* in these cell types, there is also a possibility that the loss of *Lats1/2* in Sertoli cells affected the FLCs and vice versa. For example, loss of DHH secretion by Sertoli cells in *Lats1*$^{\text{flox/flox}}$;*Lats2*$^{\text{flox/flox}}$;*Nr5a1*-cre animals could affect FLC differentiation [11]. Conversely, loss of Activin A secretion by FLCs could indirectly affect Sertoli cells and testis cord formation [77]. Activin A increases SOX9 nuclear localization in the esophageal adenocarcinoma cell line FLO-1 [78], suggesting that Activin A could also regulate SOX9 in the Sertoli cells. Furthermore, transdifferentiation of the FLC population into myofibroblast-like cells could mechanically interfere with testis cord development. In agreement with this hypothesis, it was shown that inactivation of *Lats1/2* in the Müllerian duct mesenchyme cells prior to the degradation of the Müllerian ducts in males leads to their differentiation into myofibroblasts, which in turn interfere with the development and the coiling of the adjacent epididymides [69]. Transgenic models that target *Lats1/2* specifically in the Sertoli cells and the FLCs are needed to distinguish between the direct and indirect effects in each cell population. Lastly, though an increase in YAP/TAZ transcriptional activity in the testes of *Lats1*$^{\text{flox/flox}}$;*Lats2*$^{\text{flox/flox}}$;*Nr5a1*-cre animals is the presumptive driving force behind the observed phenotypic changes, it is important to mention that LATS1/2 can occasionally act independently of the canonical Hippo pathway [79,80]. The generation of a *Lats1*$^{\text{flox/flox}}$;*Lats2*$^{\text{flox/flox}}$;*Yap*^{floxflox};*Taz*^{floxflox};*Nr5a1*-cre mouse model is needed to determine if YAP/TAZ stabilization is solely responsible for the effects of *Lats1/2* loss.

In summary, we report here a previously unheard role of the Hippo pathway in the development of the testis, as loss of *Lats1/2* results in testis cord dysgenesis and interstitial fibrosis. Further studies are required to define the mechanism of action of Hippo signaling throughout testis development in both FLC and Sertoli cell differentiation.

4. Material and Methods

4.1. Ethics

Animal procedures were approved by the Comité d'Éthique de l'Utilisation des Animaux of the Université de Montréal (protocol numbers Rech-1739 and Rech-1909 respectively approved in 2015 and 2017) and conformed to the guidelines of the Canadian Council on Animal Care.

4.2. Transgenic Mouse Strains

Nr5a1-cre mice (FVB-Tg-*Nr5a1*$^{Cre7Lowl/J}$) were obtained from the Jackson Laboratory and maintained by crossing Cre-positive males with wild-type females (C57BL/6J). *Lats1*$^{flox/flox}$ (*Lats1*$^{tm1.1JFm/RjoJ}$) and *Lats2*$^{flox/flox}$ (*Lats2*$^{tm1.1JFm/RjoJ}$) mice were obtained from Dr. Randy L. Johnson (M.D. Anderson Cancer Center, Houston, TX, USA). Mice were selectively bred over several generations to obtain the *Lats1*$^{flox/flox}$;*Nr5a1*-cre, *Lats2*$^{flox/flox}$;*Nr5a1*-cre and *Lats1*$^{flox/flox}$;*Lats2*$^{flox/flox}$;*Nr5a1*-cre genotypes. Genotype analyses were done on tail biopsies by PCR as previously described for Cre [41] and *Lats1/2* [81].

4.3. Tissue Collection

All embryos or testes were collected from e12.5 to e17.5, 1dpp (day post-partum) and 2-week-old *Lats1*$^{flox/flox}$;*Lats2*$^{flox/flox}$;*Nr5a1*-cre male mice and *Lats1*$^{flox/flox}$;*Lats2*$^{flox/flox}$ control littermates and were fixed in 4% paraformaldehyde for 4 h (whole embryos, 1 dpp testes) or Bouin's fixative overnight (testes from 2-week-old mice) and embedded in paraffin for histopathologic analyses or immunohistochemistry (IHC). Some testes from e14.5, e17.5, and 1dpp animals were flash frozen followed by homogenization for quantitative RT-PCR (RT-qPCR).

4.4. Histopathology and Immunohistochemistry

Histopathology analyses were performed on paraffin embedded, 5 µm thick tissues stained with hematoxylin and eosin (H&E) or Masson's trichrome. Immunohistochemistry analyses were performed on paraffin-embedded, 5 µm thick tissue sections using VectaStain Elite avidin–biotin complex method kits (Vector Laboratories, Newark, CA, USA) or mouse on mouse (M.O.M.) elite peroxidase kit (Vector Laboratories) as directed by the manufacturers. Sections were probed with antibodies against cCASP3, CYP17A1, DDX4, GJA1, α-SMA, SOX9, TAZ, VIM, WT1, YAP, or phospho-YAP. Staining was done using the 3,3′-diaminobenzidine peroxidase substrate kit (Vector Laboratories) or with a chromogen (Agilent) and counterstained with hematoxylin before mounting. Negative controls consisted of slides for which the primary antibody was omitted. Antibodies used are listed in Supplemental Table S1.

4.5. Reverse-Transcription-Quantitative PCR

Total RNA from testes of e14.5, e17.5, and 1 dpp animals was extracted using the Total RNA Mini Kit (FroggaBio, Concord, ON, Canada) according to the manufacturer's protocol. Total RNA was reverse-transcribed using 100 ng of RNA and the SuperScriptVilo™ cDNA synthesis kit (Thermo Fisher scientific, Waltham, Ma, U.S.). Real-time PCR reactions were run on a CFX96 Touch instrument (Bio-Rad, Hercules, CA, USA), using Supergreen Advanced qPCR MasterMix (Wisent, St-Bruno, Qc, Canada). Each PCR reaction consisted of 7.5 µL of Power SYBR Green PCR Master Mix, 2.3 µL of water, 4 µL of cDNA sample and 0.6 µL (400 nmol) of gene-specific primers. PCR reactions run without cDNA (water blank) served as negative controls. A common thermal cycling program (3 min at 95 °C, m of 15 s at 95 °C, 30 s at 60 °C and 30 s at 72°C) was used to amplify each transcript. To quantify relative gene expression, the Ct of genes of interest was compared with that of *Rpl19*, according to the ratio R = [E$^{Ct\ Rpl19}$/E$^{Ct\ target}$] where E is the amplification efficiency for each primer pair. *Rpl19* Ct values did not change significantly between tissues, and *Rpl19* was therefore deemed suitable as an internal reference gene. The specific primer sequences used are listed in Supplemental Table S2.

4.6. Statistical Analyses

All statistical analyses were performed with Prism software version 6.0d (GraphPad Software Inc., CA, USA, RRID: SCR_002798). All the datasets were subjected to the F test to determine the equality of variances. Student's t-test was used for all comparisons between genotypes. Means were considered significantly different when p value was < 0.05. All data are presented as means $\pm$ SEM.

Supplementary Materials: The following supporting information can be downloaded at: https://www.mdpi.com/article/10.3390/ijms232113585/s1.

Author Contributions: N.A.N. performed most of the experiments described in this manuscript, analyzed the data, and wrote the first draft of the manuscript. A.M., A.L., and G.S.-J. helped to perform some experiments. G.Z. and D.B. provided supervision to trainees and manuscript revision. A.B. conceived and designed the experiments, supervised trainees, and wrote the final draft of the manuscript. All authors have read and agreed to the published version of the manuscript.

Funding: This work was supported by Discovery Grants and an accelerator supplement (RGPIN-2020-05230 and RGPAS-2020-00026 to AB) a Discovery grant (RGPIN-2018-06470 to GZ) from the Natural Sciences and Engineering Research Council of Canada (NSERC). NAN is a recipient of a PhD scholarship from the Fonds de Recherche Québec—Nature et technologies (FRQNT). AM is a recipient of a graduate scholarship from NSERC.

Institutional Review Board Statement: Animal procedures were approved by the Comité d'Éthique de l'Utilisation des Animaux of the Université de Montréal (protocol numbers Rech-1739 and Rech-1909) and conformed to the guidelines of the Canadian Council on Animal Care.

Informed Consent Statement: Not applicable.

Data Availability Statement: All relevant data are within the paper and its supporting information files.

Acknowledgments: The authors would like to thank Manon Salvas, Marie LeGad-LeRoy, and Nadia Ménard for technical assistance and Randy L. Johnson (M.D. Anderson Cancer Center, Houston Tx) for generously providing the *Lats1/2* floxed mice.

Conflicts of Interest: The authors have no conflicts of interest to declare.

References

1. Hatano, O.; Takakusu, A.; Nomura, M.; Morohashi, K. Identical origin of adrenal cortex and gonad revealed by expression profiles of Ad4BP/SF-1. *Genes Cells* **1996**, *1*, 663–671. [CrossRef] [PubMed]
2. Bandiera, R.; Sacco, S.; Vidal, V.P.; Chaboissier, M.C.; Schedl, A. Steroidogenic organ development and homeostasis: A WT1-centric view. *Mol. Cell. Endocrinol.* **2015**, *408*, 145–155. [CrossRef] [PubMed]
3. Val, P.; Martinez-Barbera, J.P.; Swain, A. Adrenal development is initiated by Cited2 and Wt1 through modulation of Sf-1 dosage. *Development* **2007**, *134*, 2349–2358. [CrossRef] [PubMed]
4. Albrecht, K.H.; Eicher, E.M. Evidence that Sry is expressed in pre-Sertoli cells and Sertoli and granulosa cells have a common precursor. *Dev. Biol.* **2001**, *240*, 92–107. [CrossRef]
5. Schmahl, J.; Eicher, E.M.; Washburn, L.L.; Capel, B. Sry induces cell proliferation in the mouse gonad. *Development* **2000**, *127*, 65–73. [CrossRef]
6. Chaboissier, M.C.; Kobayashi, A.; Vidal, V.I.; Lützkendorf, S.; van de Kant, H.J.; Wegner, M.; de Rooij, D.G.; Behringer, R.R.; Schedl, A. Functional analysis of Sox8 and Sox9 during sex determination in the mouse. *Development* **2004**, *131*, 1891–1901. [CrossRef]
7. Kim, Y.; Kobayashi, A.; Sekido, R.; DiNapoli, L.; Brennan, J.; Chaboissier, M.C.; Poulat, F.; Behringer, R.R.; Lovell-Badge, R.; Capel, B. Fgf9 and Wnt4 act as antagonistic signals to regulate mammalian sex determination. *PLoS Biol.* **2006**, *4*, e187. [CrossRef]
8. Schmahl, J.; Kim, Y.; Colvin, J.S.; Ornitz, D.M.; Capel, B. Fgf9 induces proliferation and nuclear localization of FGFR2 in Sertoli precursors during male sex determination. *Development* **2004**, *131*, 3627–3636. [CrossRef]
9. Clark, A.M.; Garland, K.K.; Russell, L.D. Desert hedgehog (Dhh) gene is required in the mouse testis for formation of adult-type Leydig cells and normal development of peritubular cells and seminiferous tubules. *Biol. Reprod.* **2000**, *63*, 1825–1838. [CrossRef]

10. Pierucci-Alves, F.; Clark, A.M.; Russell, L.D. A developmental study of the Desert hedgehog-null mouse testis. *Biol. Reprod.* **2001**, *65*, 1392–1402. [CrossRef]

11. Yao, H.H.; Whoriskey, W.; Capel, B. Desert Hedgehog/Patched 1 signaling specifies fetal Leydig cell fate in testis organogenesis. *Genes Dev.* **2002**, *16*, 1433–1440. [CrossRef] [PubMed]

12. Liu, C.; Rodriguez, K.; Yao, H.H. Mapping lineage progression of somatic progenitor cells in the mouse fetal testis. *Development* **2016**, *143*, 3700–3710. [CrossRef] [PubMed]

13. Stévant, I.; Neirijnck, Y.; Borel, C.; Escoffier, J.; Smith, L.B.; Antonarakis, S.E.; Dermitzakis, E.T.; Nef, S. Deciphering Cell Lineage Specification during Male Sex Determination with Single-Cell RNA Sequencing. *Cell Rep.* **2018**, *22*, 1589–1599. [CrossRef]

14. Stévant, I.; Kühne, F.; Greenfield, A.; Chaboissier, M.C.; Dermitzakis, E.T.; Nef, S. Dissecting Cell Lineage Specification and Sex Fate Determination in Gonadal Somatic Cells Using Single-Cell Transcriptomics. *Cell Rep.* **2019**, *26*, 3272–3283.e3273. [CrossRef] [PubMed]

15. Shimamura, R.; Fraizer, G.C.; Trapman, J.; Lau Yf, C.; Saunders, G.F. The Wilms' tumor gene WT1 can regulate genes involved in sex determination and differentiation: SRY, Müllerian-inhibiting substance, and the androgen receptor. *Clin. Cancer Res.* **1997**, *3*, 2571–2580.

16. Bradford, S.T.; Wilhelm, D.; Bandiera, R.; Vidal, V.; Schedl, A.; Koopman, P. A cell-autonomous role for WT1 in regulating Sry in vivo. *Hum. Mol. Genet.* **2009**, *18*, 3429–3438. [CrossRef] [PubMed]

17. Wen, Q.; Zheng, Q.S.; Li, X.X.; Hu, Z.Y.; Gao, F.; Cheng, C.Y.; Liu, Y.X. Wt1 dictates the fate of fetal and adult Leydig cells during development in the mouse testis. *Am. J. Physiol. Endocrinol. Metab.* **2014**, *307*, E1131–E1143. [CrossRef]

18. Zhang, L.; Chen, M.; Wen, Q.; Li, Y.; Wang, Y.; Wang, Y.; Qin, Y.; Cui, X.; Yang, L.; Huff, V.; et al. Reprogramming of Sertoli cells to fetal-like Leydig cells by Wt1 ablation. *Proc. Natl. Acad. Sci. USA* **2015**, *112*, 4003–4008. [CrossRef]

19. Luo, X.; Ikeda, Y.; Parker, K.L. A cell-specific nuclear receptor is essential for adrenal and gonadal development and sexual differentiation. *Cell* **1994**, *77*, 481–490. [CrossRef]

20. Tevosian, S.G.; Albrecht, K.H.; Crispino, J.D.; Fujiwara, Y.; Eicher, E.M.; Orkin, S.H. Gonadal differentiation, sex determination and normal Sry expression in mice require direct interaction between transcription partners GATA4 and FOG2. *Development* **2002**, *129*, 4627–4634. [CrossRef]

21. Hu, Y.C.; Okumura, L.M.; Page, D.C. Gata4 is required for formation of the genital ridge in mice. *PLoS Genet.* **2013**, *9*, e1003629. [CrossRef] [PubMed]

22. Viger, R.S.; de Mattos, K.; Tremblay, J.J. Insights Into the Roles of GATA Factors in Mammalian Testis Development and the Control of Fetal Testis Gene Expression. *Front. Endocrinol.* **2022**, *13*, 902198. [CrossRef] [PubMed]

23. Nef, S.; Verma-Kurvari, S.; Merenmies, J.; Vassalli, J.D.; Efstratiadis, A.; Accili, D.; Parada, L.F. Testis determination requires insulin receptor family function in mice. *Nature* **2003**, *426*, 291–295. [CrossRef] [PubMed]

24. Pitetti, J.L.; Calvel, P.; Romero, Y.; Conne, B.; Truong, V.; Papaioannou, M.D.; Schaad, O.; Docquier, M.; Herrera, P.L.; Wilhelm, D.; et al. Insulin and IGF1 receptors are essential for XX and XY gonadal differentiation and adrenal development in mice. *PLoS Genet.* **2013**, *9*, e1003160. [CrossRef]

25. Fujimoto, Y.; Tanaka, S.S.; Yamaguchi, Y.L.; Kobayashi, H.; Kuroki, S.; Tachibana, M.; Shinomura, M.; Kanai, Y.; Morohashi, K.; Kawakami, K.; et al. Homeoproteins Six1 and Six4 regulate male sex determination and mouse gonadal development. *Dev. Cell* **2013**, *26*, 416–430. [CrossRef]

26. Garcia-Moreno, S.A.; Lin, Y.T.; Futtner, C.R.; Salamone, I.M.; Capel, B.; Maatouk, D.M. CBX2 is required to stabilize the testis pathway by repressing Wnt signaling. *PLoS Genet.* **2019**, *15*, e1007895. [CrossRef]

27. Katoh-Fukui, Y.; Miyabayashi, K.; Komatsu, T.; Owaki, A.; Baba, T.; Shima, Y.; Kidokoro, T.; Kanai, Y.; Schedl, A.; Wilhelm, D.; et al. Cbx2, a polycomb group gene, is required for Sry gene expression in mice. *Endocrinology* **2012**, *153*, 913–924. [CrossRef]

28. Maugeri-Sacca, M.; De Maria, R. The Hippo pathway in normal development and cancer. *Pharmacol. Ther.* **2018**, *186*, 60–72. [CrossRef]

29. Piccolo, S.; Dupont, S.; Cordenonsi, M. The biology of YAP/TAZ: Hippo signaling and beyond. *Physiol. Rev.* **2014**, *94*, 1287–1312. [CrossRef]

30. Hong, A.W.; Meng, Z.; Yuan, H.X.; Plouffe, S.W.; Moon, S.; Kim, W.; Jho, E.H.; Guan, K.L. Osmotic stress-induced phosphorylation by NLK at Ser128 activates YAP. *EMBO Rep.* **2017**, *18*, 72–86. [CrossRef]

31. Moon, S.; Kim, W.; Kim, S.; Kim, Y.; Song, Y.; Bilousov, O.; Kim, J.; Lee, T.; Cha, B.; Kim, M.; et al. Phosphorylation by NLK inhibits YAP-14-3-3-interactions and induces its nuclear localization. *EMBO Rep.* **2017**, *18*, 61–71. [CrossRef] [PubMed]

32. Wehling, L.; Keegan, L.; Fernández-Palanca, P.; Hassan, R.; Ghallab, A.; Schmitt, J.; Schirmacher, P.; Kummer, U.; Hengstler, J.G.; Sahle, S.; et al. Spatial modeling reveals nuclear phosphorylation and subcellular shuttling of YAP upon drug-induced liver injury. *bioRxiv* **2022**. [CrossRef]

33. Li, W.; Cooper, J.; Zhou, L.; Yang, C.; Erdjument-Bromage, H.; Zagzag, D.; Snuderl, M.; Ladanyi, M.; Hanemann, C.O.; Zhou, P.; et al. Merlin/NF2 loss-driven tumorigenesis linked to CRL4(DCAF1)-mediated inhibition of the hippo pathway kinases Lats1 and 2 in the nucleus. *Cancer Cell* **2014**, *26*, 48–60. [CrossRef] [PubMed]

34. Varelas, X. The Hippo pathway effectors TAZ and YAP in development, homeostasis and disease. *Development* **2014**, *141*, 1614–1626. [CrossRef]

35. Sen Sharma, S.; Majumdar, S.S. Transcriptional co-activator YAP regulates cAMP signaling in Sertoli cells. *Mol. Cell. Endocrinol.* **2017**, *450*, 64–73. [CrossRef]

36. Sen Sharma, S.; Vats, A.; Majumdar, S. Regulation of Hippo pathway components by FSH in testis. *Reprod. Biol.* **2019**, *19*, 61–66. [CrossRef]

37. Badr, G.; Abdel-Tawab, H.S.; Ramadan, N.K.; Ahmed, S.F.; Mahmoud, M.H. Protective effects of camel whey protein against scrotal heat-mediated damage and infertility in the mouse testis through YAP/Nrf2 and PPAR-gamma signaling pathways. *Mol. Reprod. Dev.* **2018**, *85*, 505–518. [CrossRef]

38. Levasseur, A.; Paquet, M.; Boerboom, D.; Boyer, A. Yes-associated protein and WW-containing transcription regulator 1 regulate the expression of sex-determining genes in Sertoli cells, but their inactivation does not cause sex reversal. *Biol. Reprod.* **2017**, *97*, 162–175. [CrossRef]

39. Tsoi, M.; Morin, M.; Rico, C.; Johnson, R.L.; Paquet, M.; Gévry, N.; Boerboom, D. Lats1 and Lats2 are required for ovarian granulosa cell fate maintenance. *FASEB J.* **2019**, *33*, 10819–10832. [CrossRef]

40. Chi, X.; Luo, W.; Song, J.; Li, B.; Su, T.; Yu, M.; Wang, T.; Wang, Z.; Liu, C.; Li, Z.; et al. Kindlin-2 in Sertoli cells is essential for testis development and male fertility in mice. *Cell Death Dis.* **2021**, *12*, 604. [CrossRef]

41. Dhillon, H.; Zigman, J.M.; Ye, C.; Lee, C.E.; McGovern, R.A.; Tang, V.; Kenny, C.D.; Christiansen, L.M.; White, R.D.; Edelstein, E.A.; et al. Leptin directly activates SF1 neurons in the VMH, and this action by leptin is required for normal body-weight homeostasis. *Neuron* **2006**, *49*, 191–203. [CrossRef] [PubMed]

42. Ménard, A.; Abou Nader, N.; Levasseur, A.; St-Jean, G.; Le Roy, M.L.G.; Boerboom, D.; Benoit-Biancamano, M.O.; Boyer, A. Targeted Disruption of Lats1 and Lats2 in Mice Impairs Adrenal Cortex Development and Alters Adrenocortical Cell Fate. *Endocrinology* **2020**, *161*, bqaa052. [CrossRef] [PubMed]

43. Ikeda, Y.; Shen, W.H.; Ingraham, H.A.; Parker, K.L. Developmental expression of mouse steroidogenic factor-1, an essential regulator of the steroid hydroxylases. *Mol. Endocrinol.* **1994**, *8*, 654–662. [CrossRef] [PubMed]

44. Kalluri, R.; Neilson, E.G. Epithelial-mesenchymal transition and its implications for fibrosis. *J. Clin. Investig.* **2003**, *112*, 1776–1784. [CrossRef]

45. Marconi, G.D.; Fonticoli, L.; Rajan, T.S.; Pierdomenico, S.D.; Trubiani, O.; Pizzicannella, J.; Diomede, F. Epithelial-Mesenchymal Transition (EMT): The Type-2 EMT in Wound Healing, Tissue Regeneration and Organ Fibrosis. *Cells* **2021**, *10*, 1587. [CrossRef] [PubMed]

46. Nieto, M.A.; Huang, R.Y.; Jackson, R.A.; Thiery, J.P. EMT: 2016. *Cell* **2016**, *166*, 21–45. [CrossRef]

47. Carette, D.; Weider, K.; Gilleron, J.; Giese, S.; Dompierre, J.; Bergmann, M.; Brehm, R.; Denizot, J.P.; Segretain, D.; Pointis, G. Major involvement of connexin 43 in seminiferous epithelial junction dynamics and male fertility. *Dev. Biol.* **2010**, *346*, 54–67. [CrossRef]

48. Sridharan, S.; Simon, L.; Meling, D.D.; Cyr, D.G.; Gutstein, D.E.; Fishman, G.I.; Guillou, F.; Cooke, P.S. Proliferation of adult sertoli cells following conditional knockout of the Gap junctional protein GJA1 (connexin 43) in mice. *Biol. Reprod.* **2007**, *76*, 804–812. [CrossRef]

49. Inoue, M.; Shima, Y.; Miyabayashi, K.; Tokunaga, K.; Sato, T.; Baba, T.; Ohkawa, Y.; Akiyama, H.; Suyama, M.; Morohashi, K. Isolation and Characterization of Fetal Leydig Progenitor Cells of Male Mice. *Endocrinology* **2016**, *157*, 1222–1233. [CrossRef]

50. McClelland, K.S.; Bell, K.; Larney, C.; Harley, V.R.; Sinclair, A.H.; Oshlack, A.; Koopman, P.; Bowles, J. Purification and Transcriptomic Analysis of Mouse Fetal Leydig Cells Reveals Candidate Genes for Specification of Gonadal Steroidogenic Cells. *Biol. Reprod.* **2015**, *92*, 145. [CrossRef]

51. Brennan, J.; Tilmann, C.; Capel, B. Pdgfr-alpha mediates testis cord organization and fetal Leydig cell development in the XY gonad. *Genes Dev.* **2003**, *17*, 800–810. [CrossRef]

52. Karl, J.; Capel, B. Sertoli cells of the mouse testis originate from the coelomic epithelium. *Dev. Biol.* **1998**, *203*, 323–333. [CrossRef]

53. Bingham, N.C.; Verma-Kurvari, S.; Parada, L.F.; Parker, K.L. Development of a steroidogenic factor 1/Cre transgenic mouse line. *Genesis* **2006**, *44*, 419–424. [CrossRef]

54. Xie, D.; Cui, J.; Xia, T.; Jia, Z.; Wang, L.; Wei, W.; Zhu, A.; Gao, Y.; Xie, K.; Quan, M. Hippo transducer TAZ promotes epithelial mesenchymal transition and supports pancreatic cancer progression. *Oncotarget* **2015**, *6*, 35949–35963. [CrossRef] [PubMed]

55. Li, S.; Zhang, X.; Zhang, R.; Liang, Z.; Liao, W.; Du, Z.; Gao, C.; Liu, F.; Fan, Y.; Hong, H. Hippo pathway contributes to cisplatin resistant-induced EMT in nasopharyngeal carcinoma cells. *Cell Cycle* **2017**, *16*, 1601–1610. [CrossRef] [PubMed]

56. Lei, Q.Y.; Zhang, H.; Zhao, B.; Zha, Z.Y.; Bai, F.; Pei, X.H.; Zhao, S.; Xiong, Y.; Guan, K.L. TAZ promotes cell proliferation and epithelial-mesenchymal transition and is inhibited by the hippo pathway. *Mol. Cell. Biol.* **2008**, *28*, 2426–2436. [CrossRef]

57. Shao, D.D.; Xue, W.; Krall, E.B.; Bhutkar, A.; Piccioni, F.; Wang, X.; Schinzel, A.C.; Sood, S.; Rosenbluh, J.; Kim, J.W.; et al. KRAS and YAP1 converge to regulate EMT and tumor survival. *Cell* **2014**, *158*, 171–184. [CrossRef] [PubMed]

58. Liu, Y.; He, K.; Hu, Y.; Guo, X.; Wang, D.; Shi, W.; Li, J.; Song, J. YAP modulates TGF-beta1-induced simultaneous apoptosis and EMT through upregulation of the EGF receptor. *Sci. Rep.* **2017**, *7*, 45523. [CrossRef] [PubMed]

59. Park, J.; Kim, D.H.; Shah, S.R.; Kim, H.N.; Kshitiz; Kim, P.; Quinones-Hinojosa, A.; Levchenko, A. Switch-like enhancement of epithelial-mesenchymal transition by YAP through feedback regulation of WT1 and Rho-family GTPases. *Nat. Commun.* **2019**, *10*, 2797. [CrossRef] [PubMed]

60. Nel-Themaat, L.; Jang, C.W.; Stewart, M.D.; Akiyama, H.; Viger, R.S.; Behringer, R.R. Sertoli cell behaviors in developing testis cords and postnatal seminiferous tubules of the mouse. *Biol. Reprod.* **2011**, *84*, 342–350. [CrossRef]

61. Armstrong, J.F.; Pritchard-Jones, K.; Bickmore, W.A.; Hastie, N.D.; Bard, J.B. The expression of the Wilms' tumour gene, WT1, in the developing mammalian embryo. *Mech. Dev.* **1993**, *40*, 85–97. [CrossRef]

62. Kreidberg, J.A.; Sariola, H.; Loring, J.M.; Maeda, M.; Pelletier, J.; Housman, D.; Jaenisch, R. WT-1 is required for early kidney development. *Cell* **1993**, *74*, 679–691. [CrossRef]
63. Reginensi, A.; Scott, R.P.; Gregorieff, A.; Bagherie-Lachidan, M.; Chung, C.; Lim, D.S.; Pawson, T.; Wrana, J.; McNeill, H. Yap- and Cdc42-dependent nephrogenesis and morphogenesis during mouse kidney development. *PLoS Genet.* **2013**, *9*, e1003380. [CrossRef] [PubMed]
64. McNeill, H.; Reginensi, A. Lats1/2 Regulate Yap/Taz to Control Nephron Progenitor Epithelialization and Inhibit Myofibroblast Formation. *J. Am. Soc. Nephrol.* **2017**, *28*, 852–861. [CrossRef] [PubMed]
65. Shima, Y.; Miyabayashi, K.; Sato, T.; Suyama, M.; Ohkawa, Y.; Doi, M.; Okamura, H.; Suzuki, K. Fetal Leydig cells dedifferentiate and serve as adult Leydig stem cells. *Development* **2018**, *145*, dev169136. [CrossRef] [PubMed]
66. Liu, F.; Lagares, D.; Choi, K.M.; Stopfer, L.; Marinković, A.; Vrbanac, V.; Probst, C.K.; Hiemer, S.E.; Sisson, T.H.; Horowitz, J.C.; et al. Mechanosignaling through YAP and TAZ drives fibroblast activation and fibrosis. *Am. J. Physiol. Lung Cell. Mol. Physiol.* **2015**, *308*, L344–L357. [CrossRef]
67. Byun, J.; Del Re, D.P.; Zhai, P.; Ikeda, S.; Shirakabe, A.; Mizushima, W.; Miyamoto, S.; Brown, J.H.; Sadoshima, J. Yes-associated protein (YAP) mediates adaptive cardiac hypertrophy in response to pressure overload. *J. Biol. Chem.* **2019**, *294*, 3603–3617. [CrossRef]
68. Rinschen, M.M.; Grahammer, F.; Hoppe, A.-K.; Kohli, P.; Hagmann, H.; Kretz, O.; Bertsch, S.; Höhne, M.; Göbel, H.; Bartram, M.P.; et al. YAP-mediated mechanotransduction determines the podocyte's response to damage. *Sci. Signal.* **2017**, *10*, eaaf8165. [CrossRef]
69. St-Jean, G.; Tsoi, M.; Abedini, A.; Levasseur, A.; Rico, C.; Morin, M.; Djordjevic, B.; Miinalainen, I.; Kaarteenaho, R.; Paquet, M.; et al. Lats1 and Lats2 are required for the maintenance of multipotency in the Mullerian duct mesenchyme. *Development* **2019**, *146*, dev180430. [CrossRef]
70. Sonnylal, S.; Shi-Wen, X.; Leoni, P.; Naff, K.; Van Pelt, C.S.; Nakamura, H.; Leask, A.; Abraham, D.; Bou-Gharios, G.; de Crombrugghe, B. Selective expression of connective tissue growth factor in fibroblasts in vivo promotes systemic tissue fibrosis. *Arthritis Rheum.* **2010**, *62*, 1523–1532. [CrossRef]
71. Lipson, K.E.; Wong, C.; Teng, Y.; Spong, S. CTGF is a central mediator of tissue remodeling and fibrosis and its inhibition can reverse the process of fibrosis. *Fibrogenesis Tissue Repair* **2012**, *5*, S24. [CrossRef] [PubMed]
72. Sakai, N.; Nakamura, M.; Lipson, K.E.; Miyake, T.; Kamikawa, Y.; Sagara, A.; Shinozaki, Y.; Kitajima, S.; Toyama, T.; Hara, A.; et al. Inhibition of CTGF ameliorates peritoneal fibrosis through suppression of fibroblast and myofibroblast accumulation and angiogenesis. *Sci. Rep.* **2017**, *7*, 5392. [CrossRef] [PubMed]
73. Yang, Z.; Sun, Z.; Liu, H.; Ren, Y.; Shao, D.; Zhang, W.; Lin, J.; Wolfram, J.; Wang, F.; Nie, S. Connective tissue growth factor stimulates the proliferation, migration and differentiation of lung fibroblasts during paraquat-induced pulmonary fibrosis. *Mol. Med. Rep.* **2015**, *12*, 1091–1097. [CrossRef] [PubMed]
74. Sonnylal, S.; Xu, S.; Jones, H.; Tam, A.; Sreeram, V.R.; Ponticos, M.; Norman, J.; Agrawal, P.; Abraham, D.; de Crombrugghe, B. Connective tissue growth factor causes EMT-like cell fate changes in vivo and in vitro. *J. Cell Sci.* **2013**, *126*, 2164–2175. [CrossRef] [PubMed]
75. Garrett, Q.; Khaw, P.T.; Blalock, T.D.; Schultz, G.S.; Grotendorst, G.R.; Daniels, J.T. Involvement of CTGF in TGF-beta1-stimulation of myofibroblast differentiation and collagen matrix contraction in the presence of mechanical stress. *Investig. Ophthalmol. Vis. Sci.* **2004**, *45*, 1109–1116. [CrossRef] [PubMed]
76. Grotendorst, G.R.; Rahmanie, H.; Duncan, M.R. Combinatorial signaling pathways determine fibroblast proliferation and myofibroblast differentiation. *FASEB J.* **2004**, *18*, 469–479. [CrossRef]
77. Archambeault, D.R.; Yao, H.H. Activin A, a product of fetal Leydig cells, is a unique paracrine regulator of Sertoli cell proliferation and fetal testis cord expansion. *Proc. Natl. Acad. Sci. USA* **2010**, *107*, 10526–10531. [CrossRef]
78. Taylor, C.; Loomans, H.A.; Le Bras, G.F.; Koumangoye, R.B.; Romero-Morales, A.I.; Quast, L.L.; Zaika, A.I.; El-Rifai, W.; Andl, T.; Andl, C.D. Activin a signaling regulates cell invasion and proliferation in esophageal adenocarcinoma. *Oncotarget* **2015**, *6*, 34228–34244. [CrossRef]
79. Torigata, K.; Daisuke, O.; Mukai, S.; Hatanaka, A.; Ohka, F.; Motooka, D.; Nakamura, S.; Ohkawa, Y.; Yabuta, N.; Kondo, Y.; et al. LATS2 Positively Regulates Polycomb Repressive Complex 2. *PLoS ONE* **2016**, *11*, e0158562. [CrossRef]
80. Britschgi, A.; Duss, S.; Kim, S.; Couto, J.P.; Brinkhaus, H.; Koren, S.; De Silva, D.; Mertz, K.D.; Kaup, D.; Varga, Z.; et al. The Hippo kinases LATS1 and 2 control human breast cell fate via crosstalk with ERα. *Nature* **2017**, *541*, 541–545. [CrossRef]
81. Heallen, T.; Zhang, M.; Wang, J.; Bonilla-Claudio, M.; Klysik, E.; Johnson, R.L.; Martin, J.F. Hippo pathway inhibits Wnt signaling to restrain cardiomyocyte proliferation and heart size. *Science* **2011**, *332*, 458–461. [CrossRef] [PubMed]

International Journal of
Molecular Sciences

Article

Impact of Fetal Exposure to Endocrine Disrupting Chemical Mixtures on FOXA3 Gene and Protein Expression in Adult Rat Testes

Casandra Walker [1], Annie Boisvert [2,3], Priyanka Malusare [1] and Martine Culty [1,2,3,*]

1 Department of Pharmacology and Pharmaceutical Sciences, Alfred E. Mann School of Pharmacy and Pharmaceutical Sciences, University of Southern California, Los Angeles, CA 90089-9121, USA
2 The Research Institute of the McGill University Health Centre, McGill University, Montreal, QC H4A 3J1, Canada
3 Department of Medicine, McGill University, Montreal, QC H4A 3J1, Canada
* Correspondence: culty@usc.edu; Tel.: +1-323-865-1677

Abstract: Perinatal exposure to endocrine disrupting chemicals (EDCs) has been shown to affect male reproductive functions. However, the effects on male reproduction of exposure to EDC mixtures at doses relevant to humans have not been fully characterized. In previous studies, we found that in utero exposure to mixtures of the plasticizer di(2-ethylhexyl) phthalate (DEHP) and the soy-based phytoestrogen genistein (Gen) induced abnormal testis development in rats. In the present study, we investigated the molecular basis of these effects in adult testes from the offspring of pregnant SD rats gavaged with corn oil or Gen + DEHP mixtures at 0.1 or 10 mg/kg/day. Testicular transcriptomes were determined by microarray and RNA-seq analyses. A protein analysis was performed on paraffin and frozen testis sections, mainly by immunofluorescence. The transcription factor forkhead box protein 3 (FOXA3), a key regulator of Leydig cell function, was identified as the most significantly downregulated gene in testes from rats exposed in utero to Gen + DEHP mixtures. FOXA3 protein levels were decreased in testicular interstitium at a dose previously found to reduce testosterone levels, suggesting a primary effect of fetal exposure to Gen + DEHP on adult Leydig cells, rather than on spermatids and Sertoli cells, also expressing FOXA3. Thus, FOXA3 downregulation in adult testes following fetal exposure to Gen + DEHP may contribute to adverse male reproductive outcomes.

Keywords: endocrine disruptors; EDC mixtures; in utero exposures; transcriptome analysis; testicular function; Leydig cells

Citation: Walker, C.; Boisvert, A.; Malusare, P.; Culty, M. Impact of Fetal Exposure to Endocrine Disrupting Chemical Mixtures on FOXA3 Gene and Protein Expression in Adult Rat Testes. *Int. J. Mol. Sci.* **2023**, *24*, 1211. https://doi.org/10.3390/ijms24021211

Academic Editor: Jacques J. Tremblay

Received: 1 November 2022
Revised: 30 December 2022
Accepted: 1 January 2023
Published: 7 January 2023

1. Introduction

Infertility is a global problem in which male factors have been found to account for nearly half of the cases [1,2]. The etiology of male infertility includes defects in sperm quality, low sperm count, ductal obstruction or dysfunction, or hypothalamic–pituitary axis disturbances [3,4]. Researchers have found lower sperm counts and decreased quality of semen in certain geographical regions, suggesting the influence of socioeconomic, nutritional, and/or environmental differences [1,5]. Decreased quality of semen has also been found to coincide with increasing incidence rates in male genital tract abnormalities such as cryptorchidism, a major risk factor for testicular cancer [6,7]. Additionally, infertility and male reproductive pathologies such as hypospadias and testicular cancer are on the rise in the Western world, and an estimated 10% of couples in the United States are classified as infertile [4].

The male reproductive system is one of the main targets of endocrine disrupting chemicals (EDCs), because of the requirement of sex hormones for its development and functioning, and the fact that many EDCs disrupt androgen and estrogen production and/or signaling [8–10]. While androgens produced by fetal testes drive the development of all male reproductive tissues, adult testes are dedicated to androgen and spermatozoa production, as well as regulating non-reproductive tissues. The testis is a complex and highly plastic tissue that comprises germ cells at different stages of development and several types of somatic cells. The main somatic cells are Leydig cells that produce androgens critical for the development and steady-state functions of the testis; Sertoli cells that regulate germ cell development and survival; peritubular myoid cells that contribute to interstitium components and germ cell regulation; and immune cells that maintain testis immune privilege and interact with other cell types.

EDCs are hypothesized to be causative agents of male reproductive disorders. Indeed, many studies, usually with individual EDCs given to pregnant dams at doses exceeding human exposure, have shown the disruptive effects of EDCs on male offspring reproductive functions, as summarized in the Endocrine Society's statement review by Gore and colleagues [11]. EDCs can be natural compounds, such as genistein (Gen), a plant phytoestrogen found in soy products, baby soy formula, and vegetarian diets, or artificial compounds, such as 2-diethylhexyl-phthalate (DEHP), a plasticizer with anti-androgenic properties found in many consumer products. Gen acts as an estrogen-receptor agonist, and this mechanism of action forms the basis of its classification as a "phytoestrogen" [12]. Gen has been reported to be useful in the treatment of some cancers and chronic diseases by increasing apoptosis and differentiation. It has also been shown to alter early testicular germ cell development in rats [13–15] and to delay puberty in male primates [16]. Gen inhibits ATP-utilizing enzymes such as specific tyrosine kinases in vitro. Additionally, it has been found to have antioxidant effects and to inhibit angiogenesis. Of note is that some of these benefits only occur after consumption of a soy-rich diet. Moreover, genistein has been found to have low toxicity.

DEHP is a synthetic chemical used to increase the flexibility of plastics; it has been used for years in consumer goods such as household appliances, packaging, medical tubing, flooring, and other products [17–19]. Over 98% of the United States population has detectable levels of DEHP and its major metabolite MEHP in their urine. MEHP is also found in breastmilk. Studies have shown that there is a positive correlation between fast food consumption and DEHP levels in the body, and have stressed that people can easily get contaminated through common diets [20]. DEHP can reach the systemic circulation through ingestion and absorption by the skin. DEHP is an anti-androgenic compound that decreases Leydig cell production of testosterone in males [21]. It has also been shown to decrease Sertoli cell function and anogenital distance, an androgen-dependent process, in male rodents and humans [22].

Gen and DEHP have both been linked to male reproductive pathologies. However, the effects of EDC mixtures at environmentally relevant doses have not been well characterized. This identifies a need to evaluate the effects of perinatal exposure to Gen + DEHP mixtures at doses relevant to humans, to determine their impact on male reproduction. Our previous studies have found that in utero exposure to mixtures of Gen and DEHP (Gen + DEHP) at a dose mimicking the exposure level of the general population, and a higher dose mimicking that of more susceptible populations (such as hospitalized neonates), resulted in abnormal testicular development in adult (PND120) male rats [23]. Other investigators have reported similar responses to EDCs in germ cells from human and rat fetal/neonatal testes, validating the use of rat models to study the impact of EDCs on early germ cell development [7,24,25]. Since disrupting perinatal germ cells can hamper spermatogenesis and reproduction later in life, we hypothesize that fetal exposures to Gen + DEHP mixtures at doses relevant to humans impact the adult testes by disrupting the developmental program of key testicular cell types and altering their adult functions. Our goal is to identify the functional pathways altered by in utero exposure to Gen + DEHP mixtures, the testicular cell types in which these changes occur, and the mechanisms driving them that could explain the adverse reproductive effects observed [26].

2. Results

The present study used two transcriptomic approaches to establish DEG profiles and to perform a functional pathway analysis. First, we extended the analysis of microarray data collected during a previous study for a dose of 10 mg/kg/day [26]. Then, we performed a whole transcriptome RNA-seq analysis on additional samples to compare the effects of both 0.1 and 10 mg/kg/day treatments and to determine whether they disrupted the same molecular pathways and genes (See Flow-Chart in Materials and Methods).

2.1. Sertoli and Germ Cell Functional Pathways in Adult Offspring Are Altered by In Utero Exposure to Gen + DEHP Mixtures

We imported the list of 1184 differentially expressed genes affected by Gen + DEHP at a 10 mg/kg/day dose into the Database for Annotation, Visualization and Integrated Discovery (DAVID) linked with the Kyoto Encyclopedia of Genes and Genomes (KEGG). The KEGG pathway revealed that many DEGs were related to Sertoli and germ cell pathways (Table 1). Among those, the Hippo signaling pathway is a conserved growth control pathway that plays a role in regulating proliferation of various cell types and has been found to be important for Sertoli cell function [27]. The Wnt pathway has been found to promote spermatogonial stem cell maintenance by suppressing apoptosis via the beta-catenin pathway [28]. Beta-catenin mRNA and protein are predominant in the seminiferous tubules of fetal mice and beta-catenin has been found to be abundant in Sertoli cells. Retinoic acid is critical for spermatogenesis and male fertility [29]. Additionally, retinoids are important for proliferation and differentiation of type A spermatogonia and spermiogenesis. Adherens junctions are found between Sertoli cells and Sertoli and germ cells, ensuring nutrient transfer from Sertoli to germ cells, and proper movement of germ cells from the basement membrane to the lumen. The MAPK pathway has been found to regulate the dynamics of tight junctions and adherens junctions and is involved in the proliferation and meiosis of germ cells [30]. These data suggest that in utero exposure to Gen + DEHP affects cells within the seminiferous tubules of the testis, Sertoli and germ cells.

Table 1. In utero exposure to a 10 mg/kg/day Gen + DEHP mixture affects Sertoli and germ cell pathways in adult offspring. KEGG analysis of differentially expressed genes (DEGs) by microarray analysis reveals canonical pathways related to Sertoli and germ cells in rat testes exposed in utero to 10 mg/kg/day of Gen + DEHP. Statistical cut-offs of 40% fold change with an unadjusted *p*-value of 0.05 were used to obtain gene lists. Three to four pups were used per treatment.

Term	Gene Count	Percentage	*p*-Value
Hippo signaling pathway	16	1.6	0.0041
Retinol metabolism	11	1.1	0.0042
Wnt signaling pathway	14	1.4	0.0120
cGMP-PKG signaling pathway	15	1.5	0.0160
Adherens junction	9	0.9	0.0170
Galactose metabolism	5	0.5	0.0560
Metabolic pathways	70	6.9	0.0740
MAPK signaling pathway	18	1.8	0.0770
Oxytocin signaling pathway	22	1.2	0.0860

2.2. Forkhead Box A3 (Foxa3) Is the Most Downregulated Gene in Adult Testes from Rats Exposed in Utero to Gen + DEHP Mixtures

Next, we examined changes in the expression of genes identified as differentially expressed by conducting a microarray analysis in rats exposed in utero to a Gen + DEHP mixture at 10 mg/kg/day as compared with control rats. Among 1184 DEGs in the Gen + DEHP-exposed rats, the transcription factor forkhead box A3 (*Foxa3*) (also called hepatocyte nuclear factor 3γ) was the most downregulated gene, positioning *Foxa3* as a long-term testicular target gene of fetal exposure to EDC mixtures (Table 2). Additionally, it was the only transcription factor on this list. Interestingly, CYP11A1, the cytochrome P450 metabolizing cholesterol to pregnenolone, the first step in steroid formation, was significantly upregulated. Using ingenuity pathway analysis (IPA), next, we compared the effects of fetal exposure to Gen and DEHP alone to those of the mixture at 10 mg/kg/day, using the search term "transcription". This analysis determined that *Foxa3* was highly differentially expressed only in in utero Gen + DEHP exposed rat testes (Figure 1). Whereas *Foxa3* was significantly downregulated by 62% of control values by Gen + DEHP fetal exposure (-2.60-fold change, $p = 0.0014$), it was only decreased by 30% of controls in rats exposed to DEHP (-1.42-fold change, $p = 0.0271$), and not significantly altered by fetal exposure to Gen (-1.12-fold change, $p = 0.0816$). Among these genes, only a few genes showed significant changes following fetal exposure to DEHP: Tmem249 (-1.31-fold change, $p = 0.007$), Tmem210 (-1.45-fold change, $p = 0.007$), Krt86 (-1.43-fold change, $p = 0.033$), and Pkmyt1 (-1.69-fold change, $p = 0.005$). No genes were altered by fetal exposure to Gen alone. Regarding the 10 most upregulated genes, Cyp2a1 was increased to a lesser extend by fetal exposure to Gen alone (2.99-fold change, $p = 0.047$) than by the mixture, and there was no significant increase in rats exposed in utero to DEHP alone. Cyp11a1 was slightly increased by GEN exposure (1.29-fold change, $p = 0.001$) but not significantly changed with DEHP alone. None of the other genes upregulated by the mixture were altered by Gen or DEHP alone. FOXA3 has previously been reported as the only FOX A family member identified in testes. More specifically, it has been found to be expressed in Leydig, Sertoli, and germ cells. A study by Behr and colleagues, in 2007, found that mice that were homozygous or heterozygous for the FOXA3 null allele exhibited reduced male fertility secondary to increased germ cell apoptosis [31]. These data provide further rationale to exploring the long-term effects of fetal EDC exposure on adult phenotypes, such as FOXA3 downregulation on testicular functions.

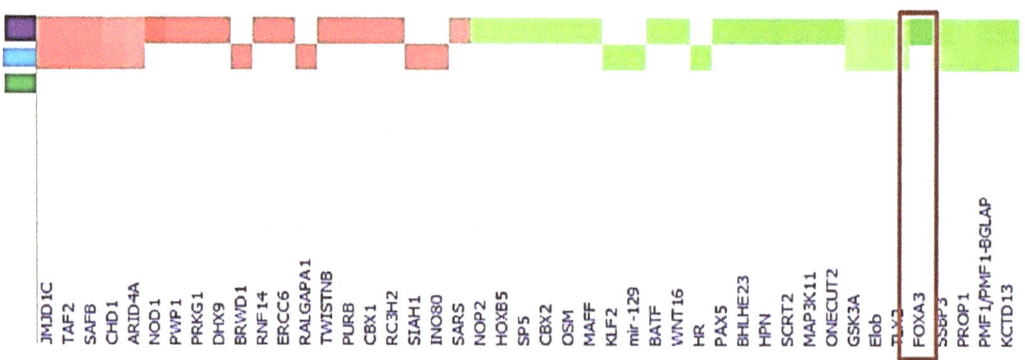

Figure 1. *Foxa3* is significantly decreased in rat testes exposed in utero to Gen + DEHP mixtures, but not by Gen or DEHP alone. Ingenuity pathway analysis (IPA) of the microarray data searching the term "transcription" identified *Foxa3* (red box) as downregulated in testes from adult rats exposed in utero to 10 mg/kg/day Gen + DEHP (p-value ≤ 0.05). Treatments: Gen, green; DEHP, blue; Gen-DEHP, purple. Expression changes: Red, upregulation and green, downregulation.

The whole transcriptome RNA sequencing analysis (RNA-seq) conducted on additional samples revealed that there was a higher number of differentially expressed genes at the lower dose of 0.1 mg/kg/day as compared with the higher dose of 10 mg/kg/day for Gen + DEHP (Figure 2A). The RNA-seq showed that *Foxa3* was decreased, similarly to the microarray data, with overlapping values between GEN + DEHP doses of 0.1 and 10 mg/kg/day (Figure 2B). Similar data were obtained by qPCR analysis, showing significant decreases in *Foxa3* at both doses (Figure 2C). These results suggest non-monotonic effects of fetal exposure to EDC mixtures, and the further need to evaluate them at doses below currently documented NOAELs.

Table 2. FOXA3 is the most downregulated DEG in adult rat testes exposed in utero to a 10 mg/kg/day Gen + DEHP mixture. The KEGG analysis of microarrays shows the 10 most downregulated (green) and 10 most upregulated (red) genes.

Symbol	Entrez Gene Name	Fold Change	*p*-Value
Foxa3	forkhead box A3	−2.609	0.0001
Tmem249	transmembrane protein 249	−2.439	0.0412
Krt86	keratin 86	−2.397	0.0381
Tmem210	transmembrane protein 210	−2.378	0.0420
Ldoc1	LDOC1 regulator of NFKB signaling	−2.279	0.0109
Pkmyt1	protein kinase, membrane associated tyrosine/threonine 1	−2.214	0.0053
Olr1749	olfactory receptor 1749	−2.201	0.0054
Cyp2g1	cytochrome P450, family 2, subfamily g, polypeptide 1	−2.183	0.0332
Krtap10-7	keratin associated protein 10-7	−2.149	0.0060
Or1f1	olfactory receptor family 1 subfamily F member 1	−2.124	0.0010
Gpr34	G protein-coupled receptor 34	2.497	0.0192
Sat1	spermidine/spermine N1-acetyltransferase 1	2.549	0.0319
Scarb2	scavenger receptor class B member 2	2.551	0.0362
St3gal4	ST3 beta-galactoside alpha-2,3-sialyltransferase 4	2.611	0.0393
Xiap	X-linked inhibitor of apoptosis	2.64	0.0357
Abca6	ATP binding cassette subfamily A member 6	2.719	0.0496
Clic2	chloride intracellular channel 2	2.728	0.0448
Cyp11a1	cytochrome P450 family 11 subfamily A member 1	2.865	0.0419
Cyp2a6	cytochrome P450 family 2 subfamily A member 6	3.128	0.0235
Cyp2a1	cytochrome P450, family 2, subfamily a, polypeptide 1	4.044	0.0410

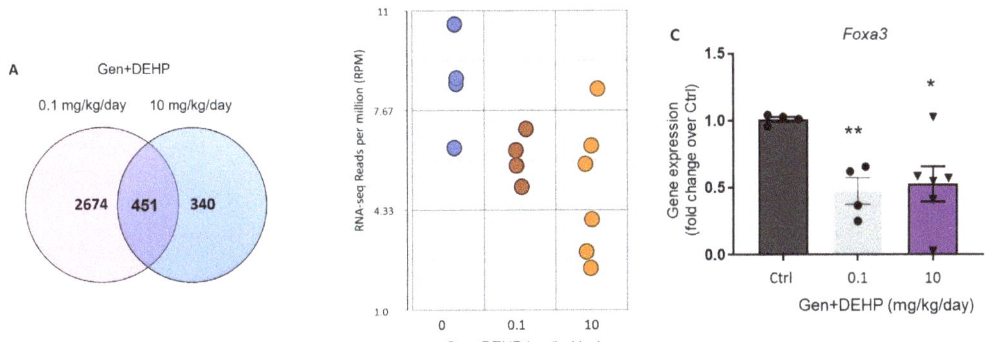

Figure 2. *Foxa3* expression is decreased in rat testes exposed in utero to Gen + DEHP mixtures: (**A**) Venn diagram of DEGs for both exposure doses, from Partek Flow analysis of RNA sequencing data; (**B**) RNA-seq analysis of *Foxa3* expression performed on rat testes exposed to 0.1 (Red dots) or 10 mg/kg/day (Orange dots) of Gen + DEHP mixture as compared with control rats (Blue dots). $N = 4$ rats from independent dams; (**C**) qPCR analysis assessing *Foxa3* mRNA expression in rat testes exposed to Gen + DEHP mixtures. $N = 4$ to 6 rats from different dams. Significance p values: * $p \leq 0.05$; ** $p \leq 0.01$.

To assess global FOXA3 protein levels in testes, we used MALDI imaging mass spectrometry (MALDI IMS) to quantify the amount of FOXA3 protein in frozen adult testis sections from control rats and in utero Gen + DEHP exposed rats as compared with ferredoxin 2, another Leydig cell protein, and to the Sertoli cell marker, androgen-binding protein (ABP). The protein levels of Foxa3 and Ferredoxin 2 were both reduced in testes from rats exposed in utero to 0.1 mg/kg/day Gen + DEHP mixture, as shown by decreases in yellow signals on the sections and the histograms quantifying the normalized signal intensity by surface unit of sections as compared with the control rats (Figure 3A,B). In contrast, the intensity of the Sertoli cell protein Abp was strong in the control and EDC-exposed samples (Figure 3C). These results confirmed Foxa3 as a critical long-term testicular target of fetal exposure to Gen + DEHP, in agreement with the mRNA data.

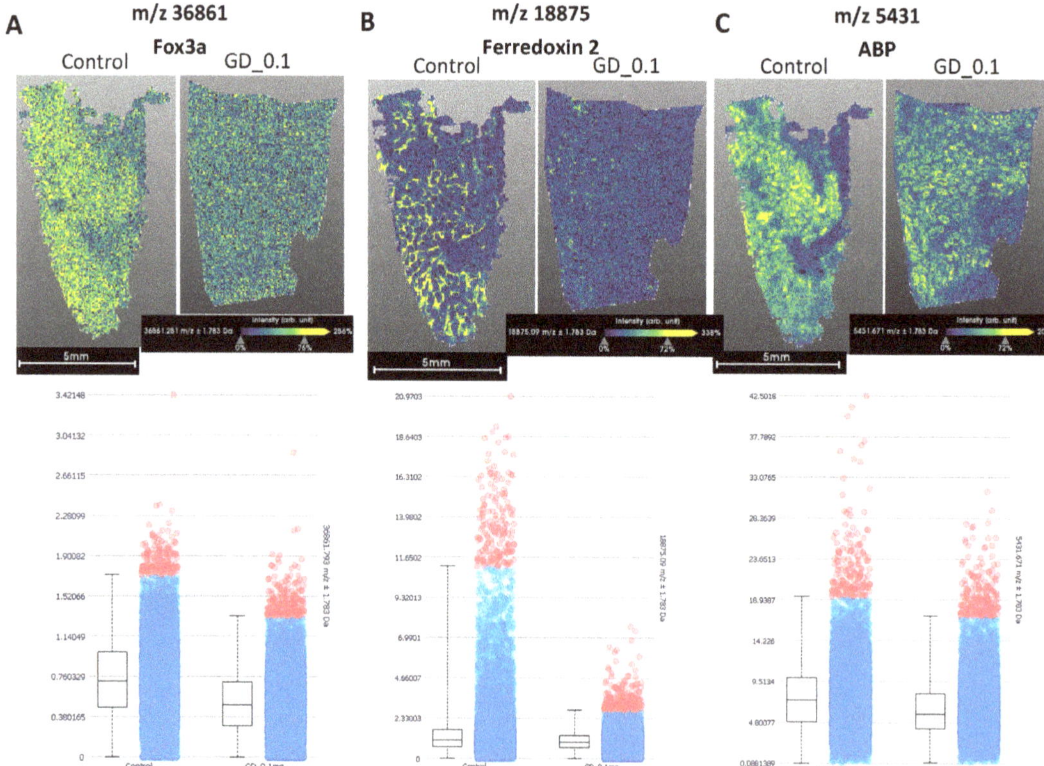

Figure 3. MALDI imaging mass spectrometry reveals that FOXA3 protein is decreased in adult testes after in utero exposure to Gen + DEHP mixture. Proteins were identified according to their mass in frozen sections of PND120 rat testes treated with vehicle (control) and Gen + DEHP mixture (GD) at 0.1 mg/kg/day, analyzed by MALDI IMS: (**A**) Foxa3; (**B**) Ferredoxin 2; (**C**) Androgen-binding protein (ABP). For each protein, the left panel shows the intensity of protein signal, with yellow representing the highest intensity; while the Box plots in the right panel show the quantification of relative signal intensity in testis sections, normalized by surface unit.

2.3. FOXA3-Interacting Genes Are Differentially Expressed in Testes from in Utero Gen + DEHP-Exposed Rats

Our next goal was to identify genes regulated by FOXA3 within the testes. First, this was accomplished by running an in silico search of the IPA database for DEGs predicted to be targets of FOXA3 in the literature. Noticeably, the two doses of Gen + DEHP mixtures identified different putative targets of FOXA3, suggesting that doses with a 100-fold difference did not act on the same molecular mechanisms (Table 3A,B). Interestingly, the lower dose mainly downregulated genes, including the transcription factor Foxa1 and other DNA-interacting proteins (Table 3A), whereas the higher dose induced gene upregulation (Table 3B).

Table 3. In silico search for FOXA3-interacting genes in testes. The IPA database was searched for genes with a relationship to "FOXA3" in adult rat testes exposed in utero to (**A**) 0.1 and (**B**) 10 mg/kg/day Gen + DEHP mixtures using the IPA of RNA-seq data. Data are shown in fold change of control samples. N = 4 per treatment, $p \leq 0.05$, FC = 2; (**C**) TRANSFAC-generated list of FOXA3 target genes in testes.

A

Symbol	Gene Name	Fold Change	*p*-Value
Cox4I2	cytochrome c oxidase subunit 4I2	−4.026	0.000
Zbtb33	zinc finger and BTB domain containing 33	−3.544	0.008
Foxa1	forkhead box A1	−2.333	0.007
Nrob2	nuclear receptor subfamily 0 group B member 2	−2.076	0.004
Khdrbs2	KH RNA binding domain containing, signal transduction associated 2	−2.021	0.028
Lmnb2	lamin B2	2.164	0.001

B

Symbol	Gene Name	Fold Change	*p*-Value
Foxa3	forkhead box A3	−2.609	0.000
Klhl13	kelch like family member 13	1.408	0.036
P4ha1	prolyl 4-hydroxylase subunit alpha 1	1.492	0.044
Nsf	N-ethylmaleimide sensitive factor, vesicle fusing ATPase	1.705	0.050
Cycs	cytochrome c, somatic	1.971	0.026

C

Symbol	Gene Name	Testis Cell Type(s)	Relation to Foxa3
Hoxa10	Homeobox protein A10	Leydig	Target gene
Hsd3β	Hydroxysteroid dehydrogenase 3 β	Leydig	Target gene
Apo(a-I)	Apolipoprotein (A-I)	Leydig	Target genes (*ApoE*) In Foxa2/Foxa3 TF network
Pck-1	Phosphoenolpyruvate carboxy kinase	Leydig (Steroidogenesis)	Target gene In Foxa2/Foxa3 TF network
Hmgb1	High mobility group box 1	Sertoli, germ	Protein-protein interaction
Pparα	Peroxisome proliferator receptor alpha	Sertoli, germ, Leydig	Protein-protein interaction
G6p	Glucose-6-phosphatase	Leydig (steroidogenesis)	In Foxa2/Foxa3 TF network
Nur77	Nuclear receptor subfamily 4 group A member 1	Leydig (Steroidogenesis)	Target gene
Tf	Transferrin	Sertoli	Target gene
Tle3	Transducin like enhancer of split 3	Sertoli	Target gene

Since we used whole testis extracts, it was important to determine the specific cell type(s) where FOXA3 was decreased. We used TRANSFAC, a transcription factor database, to further explore FOXA3-interacting genes. This analysis showed that FOXA3 target genes were located in Leydig, Sertoli, and germ cells, with many of these genes found in Leydig cells (Table 3C). These data suggest that FOXA3 may be significantly decreased in Leydig cells of the testes and that this decrease could affect FOXA3 target genes and disrupt Leydig cell functions, including steroidogenesis. Among the genes listed in Table 3A,B, none of the genes were significantly altered by in utero treatments with 10 mg/kg/day of Gen or DEHP alone, although Cox4i2 and Khdrbs2 showed decreasing trends of −2.78-fold change ($p = 0.083$) and −2.07-fold change ($p = 0.312$), respectively, in rats exposed to DEHP alone. These data further highlighted the unique position of FOXA3 as a long-term target of in utero exposure to Gen + DEHP mixtures (Supplemental Table S1).

2.4. Fetal Exposure to Gen + DEHP Mixtures Decreases the Protein Expression of FOXA3 in Adult Testicular Interstitium

A study by Behr and colleagues reported that mice that were homozygous for the FOXA3 −/−null allele were infertile [31]. Another study found that FOXA3 bound to PDGFRa in Leydig cells, which was necessary for Leydig cell differentiation and embryonic development [32]. To determine in which cell types FOXA3 protein was downregulated, we examined the protein expression profiles of FOXA3 in testes of adult rats exposed in utero to vehicle, 0.1, or 10 mg/kg/day of Gen + DEHP mixtures by IF analysis. FOXA3 gave positive signals in the interstitium of control rat testes (Figure 4). Positive FOXA3 signal was also seen in germ cells within the seminiferous tubules in control samples, in agreement with published data. Although FOXA3-positive interstitial cells were visible in all rat testes, suggesting no change in cell numbers, the signal intensity of FOXA3 in these cells was reduced by fetal exposure to EDC mixtures, indicating that these treatments did not prevent the formation of adult Leydig cells, but rather altered their protein profiles (Figure 4A). FOXA3 levels were also decreased in seminiferous tubules of testes from rats in utero exposed to Gen + DEHP mixtures at both doses, while DAPI nuclear signal showed that germ cells were present in these tubules (Figure 4A). Quantification of FOXA3 relative signal intensity for several images per sample using ImageJ after grey scale conversion showed that the FOXA3 relative signal intensity in interstitial cells was reduced by 27% of control values for rats in utero exposed to the lower dose, and was only reduced by 12% for the higher dose (244.7 ± 9.9 in control rats, 179.0 ± 8.9 for Gen + DEHP 0.1, and 216 ± 5.9 for Gen + DEHP 10). The reduction at the low dose was similar to that measured by MALDI IMS. Inside the seminiferous tubules, FOXA3 relative signal intensity was reduced by 29% and 27% of control values for the lower and higher doses, respectively, (251.6 ± 14.6 for control rats, 178.5 ± 10.6 for Gen + DEHP 0.1, and 184.7 ± 10.8 for Gen + DEHP 10). IF analysis in control samples of FOXA3 and PDGFRa, both expressed in Leydig cells, showed that the two proteins colocalized in the interstitium (Figure 4B) [31]. The colocalization of FOXA3 and PDGFRa, and the appearance of FOXA3-positive cells in the interstitium, suggest that the adult interstitial cells in which FOXA3 expression was reduced by fetal exposure to the EDC mixtures were Leydig cells. Thus, FOXA3 expression in adult testes was altered by fetal exposure to the EDC mixtures both in interstitial cells—most likely Leydig cells—and germ cells, with fetal exposure to the lower dose having stronger effect than the higher dose in interstitial cells, but similar effects in germ cells.

2.5. Gen + DEHP Mixtures Decrease the Expression of Genes/Protein Related to Steroidogenesis

We further used IPA to identify genes in our dataset that were related to the search term "steroidogenesis", the main function of Leydig cells. All genes identified were significantly down- or upregulated, at least two-fold changes, in adult testes by fetal exposure to 0.1 mg/kg/day Gen + DEHP mixture as compared with the control rats (Table 4).

Growth hormone releasing hormone (GHRH) has been reported to be present in interstitial cells and germ cells in rat testes [33]. The same study also found that GHRH secreted by Leydig cells in adult rats could stimulate cAMP formation via induction of luteinizing hormone (LH), and that it could regulate Sertoli cell function. Cytochromes 7B1, 7A1, and 2R1 were also found to be decreased in the study. Cytochrome P450s in the testes are important for metabolizing cholesterol into testosterone [34]. Cytochrome P450 family 7 subfamily B1 has been found to regulate 11 beta-hydroxysteroid dehydrogenase 1 (11b-HSD1) in rat Leydig cells [34]. CYP7A1 and CYP2R1 are also part of the cholesterol synthesis network. Translocator protein TSPO binds cholesterol and transports it to the inner mitochondrial membrane for steroid biosynthesis to occur [35]. The family of apolipoproteins (i.e., APOA1) are also responsible for cholesterol transport [36]. Forkhead box A1 (FOXA1) is a member of the forkhead box A family that has been found to bind to the androgen receptor in the prostate [37]. From these genes, only *Apoa1* was significantly increased following fetal exposure to both doses of Gen + DEHP mixtures. The other genes were significantly altered only at the lower dose. None of these genes were significantly altered by fetal exposure to Gen or DEHP alone at the 10 mg/kg/day by RNA-seq analysis (Supplemental Table S2). These data suggest that steroid biosynthesis is affected in adult Leydig cells of the testes, while a reduction in GHRH could also suggest disturbances within the hypothalamic-pituitary axis, thus, affecting testosterone feedback. Since these are lists of FOXA3 target genes, these data suggest that a reduction in FOXA3 results in a reduction of GHRH, CYP7B1, CYP7A1, CYP2CR1, TSPO, FOXA1, and HSD17B, while the opposite may occur for APOA1.

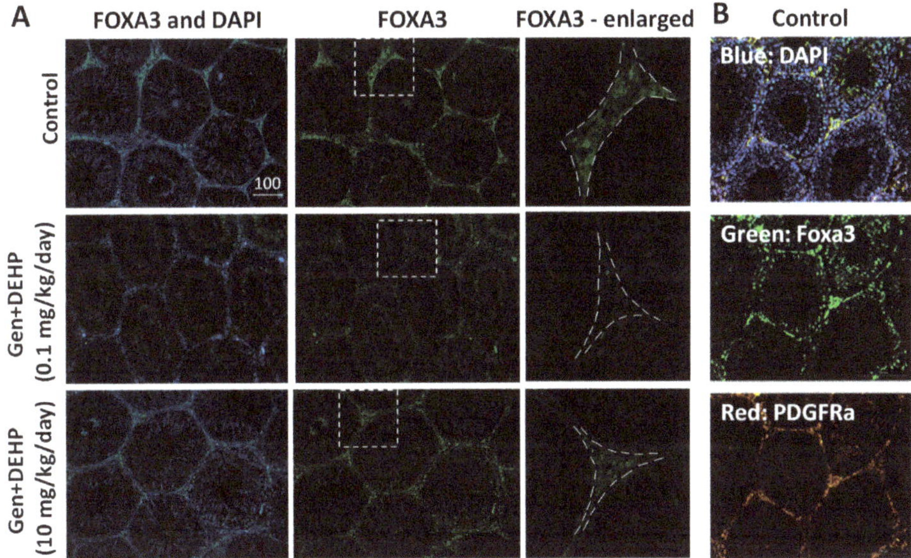

Figure 4. FOXA3 protein expression is decreased in testicular interstitial cells from adult rats exposed in utero to Gen + DEHP mixtures: (**A**) FOXA3 IF signal in testes from control rats and from rats exposed to Gen + DEHP mixture at 0.1 or 10 mg/kg/day. Left panel, merged IF images of FOXA3 (green) and DAPI nuclear staining (blue); middle column, FOXA3 IF alone; right column, magnification of white dotted inserts indicated in middle column; (**B**) IF colocalization of FOXA3 (green) and PDGFRa (red) in control rat testes. Representative photos are shown. Same scale in μm used for left and center columns in panel A, and in all pictures of panel B.

We also examined the mRNA levels of several key genes of the steroidogenic cascade by qPCR in the testes of control rates and Gen + DEHP-treated rats after excluding infertile rats, to identify steroidogenic genes that might be related to less serious adverse phenotypes than infertility in in utero EDC exposed rats. We found that *Tspo*, *Hsd3b*, *Cyp11a1*, and *Cyp17* showed decreasing trends in rats exposed to 0.1 mg/kg/day, with *Hsd3b* decrease reaching close to significance with a *p*-value of 0.0504 at dose 0.1, and a *p*-value of 0.0713 at dose 10 (Figure 5A). Additional studies will be needed to determine whether these alterations are due to FOXA3 mediation or are a result of other disturbances. TSPO is a well characterized protein that is critical for steroid biosynthesis in testes [12]. While the qPCR analysis showed a decreasing trend in *Tspo* for fetal exposure to the dose of 0.1 mg/kg/day, TSPO protein levels showed some decreases in testes of adult rats exposed in utero to both doses of Gen + DEHP mixtures (Figure 5B). Quantification of the relative signal intensity of TSPO in several pictures/samples using ImageJ after grey scale conversion showed that TSPO protein levels were decreased by 14% in interstitial cells of rats exposed to the lower dose of mixture, and only 9% for the higher dose of EDC mixture (224.5 ± 1.5 for control rats, 193.7 ± 3.9 for Gen + DEHP 0.1, and 205.0 ± 4.9 for Gen + DEHP 10), while there was no noticeable change in the tubules, in which spermatogenesis was visible. The small changes in TSPO expression in the treated rats as compared with the significant changes in mRNA-seq analysis performed on different sets of rats likely reflect the variability in the adult reproductive phenotypes observed in in utero-treated rats, where 25% and 38% had abnormal testes, including Sertoli-only tubules, and/or small litters at both doses of Gen + DEHP mixtures, respectively, and 13% were infertile for both doses [26].

Table 4. Fetal exposure to Gen + DEHP mixtures affects steroidogenesis-related genes. Differentially expressed genes related to "steroidogenesis" in adult rat testes exposed to 0.1 mg/kg/day Gen + DEHP mixture generated from ingenuity pathway analysis of RNA-seq data. Data are shown in fold change of control samples. N = 4 per treatment, $p \leq 0.05$, FC = 2.

Symbol	Steroidogenesis-Related Gene Name	Fold Change	p-Value
Ghrh	growth hormone releasing hormone	−2.690	0.008
Foxa1	forkhead box A1	−2.333	0.007
Hsd17b1	hydroxysteroid 17-beta dehydrogenase 1	−2.272	0.032
Cypb1	cytochrome P450 family 7 subfamily B member 1	−2.260	0.020
Cyp7a1	cytochrome P450 family 7 subfamily A member 1	−2.164	0.018
Cyp2r1	cytochrome P450 family 2 subfamily R member 1	−2.066	0.005
Tspo	translocator protein	−2.047	0.012
Apoa1	apolipoprotein A1	2.080	0.002

Taken together, these data suggest that FOXA3 may bind to and affect the transcription of several genes important for steroid biosynthesis, ultimately affecting steroidogenesis in testes. This is consistent with our previous study using the same treatments, in which we found that circulating testosterone levels were significantly reduced in adult rats exposed in utero to 0.1 mg/kg/day of Gen + DEHP mixture [26]. This further suggests that *Tspo* gene expression may be regulated by FOXA3, in a direct or indirect manner, but not in all in utero EDC-exposed rats.

2.6. Identification of FOXA3 Target Genes in Rat Testes by ChIP-Seq Analysis

We used ChIP-seq analysis to identify potential gene targets of FOXA3 in adult control rat testes (Table 5), and then examined the expression levels of these genes in the RNA-seq dataset. Only one gene, *Phlpp1*, showed a significant change, in that case a 20% increase, in the testes of rats exposed to 0.1 mg/kg/day of Gen + DEHP mixture, while *Phlpp1* showed a 68% increasing trend for the 10 mg/kg/day dose (Table 6). *Tmeff2* was another gene presenting increasing trends of 50% and 55% above control levels for 0.1 and 10 mg/kg/day of Gen + DEHP mixtures, respectively. In contrast, *Cxcl13* expression showed a decreasing trend in the testes of rats exposed to 0.1 mg/kg/day of Gen + DEHP mixture. Since a RNA-seq transcriptome analysis is performed on total testis RNA, and FOXA3 is expressed not only in Leydig cells, but also in germ and Sertoli cells, it is possible that adult FOXA3 expression might be differentially affected by fetal exposure to Gen + DEHP mixtures and might have different target genes in these cell types, potentially masking its effect in Leydig cells. Further studies are needed to examine this possibility. None of these genes were significantly altered in adult testes by fetal exposures to Gen or DEHP alone (Supplemental Table S3).

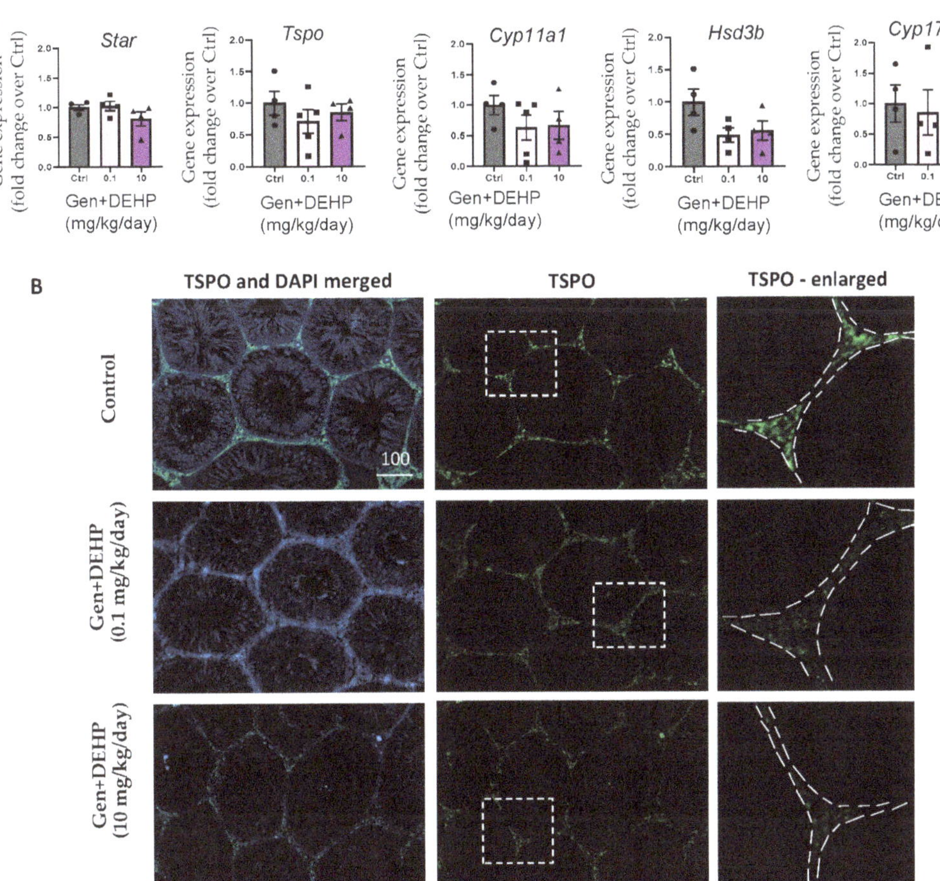

Figure 5. Fetal exposures to Gen + DEHP mixtures affect adult steroidogenesis-related genes: (**A**) qPCR analysis of steroidogenic genes in control (Crtl), Gen + DEHP mixtures at 0.1 and 10 mg/kg/day, excluding infertile rats. N = 3–5 per treatment; (**B**) protein expression of TSPO identified by immunofluorescence (green) showing reduced protein expression in testes from rats exposed in utero to 0.1 and 10 mg/kg/day of Gen + DEHP mixtures as compared with control rats. Representative photos are shown. Same scale in μm used for left and center columns in Figure 5B. Pictures in the right column are enlarged views of the white dotted area in pictures from center column.

Table 5. Annotation table of genes identified by ChIP-seq as binding to FOXA3. In total, 18 protein coding genes and one ribosomal RNA were identified after sequencing the libraries of DNA fragments pulled down using 1.5 or 3 µg of two anti-FOXA3 antibodies. A ChIP-seq analysis was performed on DNA from frozen adult control rat testes. Distance to TSS, nearest promoter ID, and gene name are indicated. Several non-coding sequences were removed from the table.

Antibody	Peak ID	Chromosome	Start	End	Strand	Annotation	Detailed Annotation	Distance to TSS	Nearest Promoter ID	Gene Name	Gene Description	Gene Type
Invitrogen 1.5 ug	001_0788_0 01_peak_1	chrY	1841330	1841557	+	Intergenic	SATI_RN | Satellite | Satellite	−438021	NM_001329902	Usp9y	ubiquitin specific peptidase 9, Y-linked	protein-coding
Invitrogen 3 ug	001_0788_0 02_peak_7	chr20	27473893	27474238	+	Intergenic	Intergenic	78160	NM_147146	Rwdd1	RWD domain containing 1	protein-coding
	001_0788_0 02_peak_5	chr14	15463091	15463463	+	Intergenic	Lx2B | LINE | L1	−205092	NM_001017496	Cxcl13	C-X-C motif chemokine ligand 13	protein-coding
	001_0788_0 02_peak_2	chr1	147551697	147552007	+	intron (intron 9)	RNHAL1 | LINE | L1	−162049	NM_001013904	Cyp2c6v1	cytochrome P450, CYP2C6, variant 1	protein-coding
	001_0788_0 02_peak_8	chr5	91563314	91563577	+	Intergenic	Lx2B | LINE | L1	428461	NR_046238	Rn5-8s	5.8S ribosomal RNA	rRNA
	001_0788_0 02_peak_9	chr7	48411144	48411477	+	Intergenic	Intergenic	−138146	NM_001108745	Ppfia2	PTPRF interacting protein alpha 2	protein-coding
	001_0788_0 02_peak_3	chr1	233681071	233681328	+	Intergenic	Lx2A | LINE | L1	298421	NM_031036	Gnaq	G protein subunit alpha q	protein-coding
	001_0788_0 02_peak_4	chr10	74769530	74769818	+	Intergenic	Intergenic	−45727	NM_001129777	Rad51c	RAD51 paralog C	protein-coding
SantaCruz 1.5 ug	001_0788_0 03_peak_9	chr20	27473924	27474228	+	Intergenic	Intergenic	78149	NM_147146	Rwdd1	RWD domain containing 1	protein-coding
	001_0788_0 03_peak_3	chr15	92078631	92078967	+	Intergenic	ERVB2_2-I_RN | LTR | ERVK	1228621	NM_001107283	Slitrk1	SLIT and NTRK-like family, member 1	protein-coding
	001_0788_0 03_peak_6	chr2	132257139	132257414	+	Intergenic	Intergenic	−1599074	NM_001106429	Pabpc4l	poly(A) binding protein, cytoplasmic 4-like	protein-coding
	001_0788_0 03_peak_13	chr9	114715355	114715651	+	Intergenic	Intergenic	−5890	NM_013017	Rab12	RAB12, member RAS oncogene family	protein-coding

Table 5. *Cont.*

Antibody	Peak ID	Chromosome	Start	End	Strand	Annotation	Detailed Annotation	Distance to TSS	Nearest Promoter ID	Gene Name	Gene Description	Gene Type
SantaCruz 1.5 ug	001_0788_0 03_peak_10	chr6	89503088	89503335	+	Intergenic	Intergenic	−211184	NM_199269	Mdga2	MAM domain containing glycosylphos-phatidylinositol anchor 2	protein-coding
	001_0788_0 03_peak_12	chr9	56664665	56664957	+	Intergenic	Lx2 ǀ LINE ǀ L1	−991107	NM_001108795	Tmeff2	transmembrane protein with EGF-like and two follistatin-like domains 2	protein-coding
	001_0788_0 03_peak_2	chr13	109148367	109148629	+	Intergenic	Intergenic	−307016	NM_001107200	Ptpn14	protein tyrosine phosphatase, non-receptor type 14	protein-coding
	001_0788_0 03_peak_4	chr15	103206833	103207086	+	Intergenic	Intergenic	133280	NM_001107286	Tgds	TDP-glucose 4,6-dehydratase	protein-coding
	001_0788_0 03_peak_7	chr2	173864996	173865243	+	Intergenic	Lx6 ǀ LINE ǀ L1	147557	NM_001009542	Pdcd10	programmed cell death 10	protein-coding
	001_0788_0 03_peak_11	chr8	125968054	125968301	+	Intergenic	RNLTR8C2 ǀ LTR ǀ ERVK	−443652	NM_001025705	Azi2	5-azacytidine induced 2	protein-coding
	001_0788_0 04_peak_5	chr20	27473930	27474176	+	Intergenic	MamRTE1 ǀ LINE ǀ RTE-BovB	78172	NM_147146	Rwdd1	RWD domain containing 1	protein-coding
	001_0788_0 04_peak_3	chr19	58032198	58032458	+	Intron (intron 10)	B4A ǀ SINE ǀ B4	212068	NM_175596	Disc1	DISC1 scaffold protein	protein-coding
SantaCruz 3 ug	001_0788_0 04_peak_2	chr13	26223458	26223722	+	intron (intron 4)	intron (intron 4)	51375	NM_021657	Phlpp1	PH domain and leucine rich repeat protein phosphatase 1	protein-coding
	001_0788_0 04_peak_7	chr8	53458384	53458659	+	Intergenic	Intergenic	47205	NM_153311	Tmprss5	transmembrane serine protease 5	protein-coding
	001_0788_0 04_peak_6	chr20	39622048	39622391	+	Intergenic	RNLTR3d ǀ LTR ǀ ERVK	−614218	NM_001106392	Hs3st5	heparan sulfate-glucosamine 3-sulfotransferase 5	protein-coding

Table 6. mRNA expression changes in genes identified by ChIP-seq as binding to FOXA3. The 18 genes found to bind FOXA3 by ChIP-seq analysis of an adult control testis sample were examined in the RNA-seq data of control rats vs. in utero Gen + DEHP-exposed rat testes.

Gene	GD 0.1 vs. Ctrl		GD 10 vs. Ctrl	
	Fold Change	*p* Value	Fold Change	*p* Value
Cxcl13	−1.36	0.605	−1.04	0.209
Mdga2	−1.14	0.916	1.05	0.277
Gnaq	−1.13	0.178	−1.12	0.951
Pabpc4	−1.08	0.384	1.10	0.106
Rn5-8s	−1.06	0.845	−1.02	0.745
Ppfia2	−1.06	0.435	−1.07	0.920
Rab12	−1.05	0.635	1.04	0.280
Tgds	−1.02	0.473	1.11	0.338
Usp9y	1.02	0.276	1.17	0.234
Ptpn14	1.04	0.422	1.05	0.859
Tmprss5	1.04	0.952	−1.49	0.891
Rad51c	1.05	0.318	−1.08	0.116
Pdcd10	1.07	0.728	−1.02	0.271
Rwdd1	1.08	0.904	1.05	0.720
Azi2	1.15	0.493	−1.04	0.00757
Cyp2c6v1	1.16	0.221	1.11	0.585
Phlpp1	1.20	0.0244	1.68	0.110
Tmeff2	1.50	0.347	1.55	0.249

3. Discussion

DEHP, a phthalate plasticizer used in many commercial products and medical devices, and Gen, a phytoestrogen abundant in baby soy formula and vegetarian diets, are among the hundreds of chemicals with potential EDC activity to which we are exposed in our daily life. Our previous studies have shown that in utero exposure to Gen + DEHP mixtures increased the rates of infertility and abnormal testis development, altered gene expression, and induced inflammatory processes in adult (postnatal day (PND) 120) rats, differently from exposure to Gen or DEHP alone [26]. In the present study, we focused on the effects of fetal exposure to Gen + DEHP mixtures, with the goal of identifying long-term alterations of functional pathways and genes in adult rat testes, to gain insight into the etiology of the observed reproductive phenotypes in adult testes. In this study, we used mixtures of Gen and DEHP given at doses equivalent to levels measured in humans, to increase the chance of identifying long-term target pathways that could be meaningful for humans.

We found that in utero exposure to Gen and DEHP mixtures altered functional pathways related to Sertoli, germ, and Leydig cell development and function, in adult rat testes. More importantly, the transcription factor FOXA3 was downregulated in adult testes exposed to the mixtures as fetuses, but not to the same extent by fetal exposure to Gen or DEHP alone. FOXA3 is a transcription factor that has previously been identified in Leydig, Sertoli, and germ cells and is critical for testicular function [31,32]. In the present study, we showed that *FOXA3* mRNA was decreased in adult rat testes exposed in utero to 0.1 and 10 mg/kg/day of Gen + DEHP mixtures, using microarrays, RNA-seq, and qPCR analyses. Although gene expression changes could reflect changes in the proportion of specific cell types within the tissue rather than true changes of expression in the cells, for example, a large loss of germ cells as seen in the testes of rats exposed as fetuses to Gen + DEHP with Sertoli-only phenotypes [26] could increase the proportion of Leydig cells and the representation of Leydig cell markers in the samples, this does not seem to be the case here, since FOXA3 protein was decreased, not increased, and this was observed in samples from adult rats exposed in utero to Gen + DEHP where spermatogenesis was visible. FOXA3 was originally identified in the liver as one of three members of the forkhead box A family. Forkhead box A family members are winged helix proteins that function as transcriptional regulators by binding to target sites on DNA [38]. Studies by Garon and Behr teams previously reported FOXA3 to be the only FoxA family member identified in the testes [31,32]. Here, we identified another forkhead box A member, *Foxa1*, in adult rat testes, that was downregulated (−2.33-fold change) in the RNA-seq dataset and was found to have a protein–DNA interaction with FOXA3 using a network interaction search in IPA. FOXA1 was reported to bind to the androgen receptor in the prostate. These data suggest that Foxa1 is indeed present in adult rat testes and may be altered by fetal exposure to EDCs, either directly or through interactions with Foxa3. The study identified Foxa3 to be a novel gene downregulated in adult rats by in utero exposure to environmentally relevant doses of Gen + DEHP mixtures.

Taken together, the fact that the EDC exposures took place in fetuses and that FOXA3, a gene known to be expressed in adult Leydig cells, was reduced in the testicular interstitium of adult rats exposed as fetuses to these EDCs, suggests that the developmental programming of adult type Leydig cells was disrupted in utero. Although adult Leydig cells are different from fetal Leydig cells, and differentiate from early postnatal progenitor cells that progress to immature LCs before fully differentiating to adult LCs, they have been proposed to share an undefined early fetal common precursor [39,40]. Thus, one could interpret our data to imply the disruption of a fetal precursor of adult Leydig cells jeopardizing their future differentiation in adult cell type. This is reminiscent of the Developmental Origins of Health and Disease (DOHaD) theory, originally developed based on studies relating food scarcity in parents or grandparents to metabolic syndrome in children, which has been extended to many biological functions, including reproduction [11,41]. This complex phenomenon most likely involves epigenetic remodeling as well as adaptive responses of the fetus to changes in the environment which can be retained in the adult, as well as across generations [42]. However, one cannot rule out that fetal EDC exposure altered the developmental programming of other cell types in the testes, or elements of the H–P–T axis, or even other endocrine tissues interacting with testes, contributing to the effects observed in adult Leydig cells.

The search of gene targets of FOXA3 by ChIP-seq analysis identified three genes pulled down with FOXA3, *Phlpp1*, *Tmeff2*, and *Cxcl13* that have been found in testes and/or androgen-responsive tissues and may warrant further examination. PHLPP1 (PH domain and leucine-rich repeat protein phosphatase 1) has been found to be expressed in human testes, including spermatogonia [43]. TMEFF2 has also been identified in human testes, shown to prevent PDGF-AA-induced cellular proliferation, to be upregulated by androgen in some prostate cancer cells, and to have oncogenic and onco-suppressive actions depending on the tissue/cancer type context [44,45]. CXCL13 is also an androgen-responsive gene and has been shown to be involved in androgen-induced prostate cancer [46].

We also aimed to link the effect of FOXA3 downregulation in adult testes to changes in testicular function. The exact function played by FOXA3 in Leydig, Sertoli, and germ cells is not fully understood [31,32]. We used TRANSFAC, a transcription factor database, to identify FOXA3 target genes and their specific cell types. This approach pinpointed at genes associated with Leydig cells including *Hsd3b*, *Apo (A-I)*, *Pck*, *Nur77*, and *G6P*. A network interaction search on genes in our dataset related to "steroidogenesis" also found that *Tspo*, *Hsd17b1*, *Foxa1*, *Ghrh*, and a number of cytochrome P450s were downregulated. These genes are critical for the proper functioning of steroidogenesis [47,48]. A qPCR analysis further identified decreasing trends in the expression of *Cyp17*, *HSD3b*, *Cyp11a1*, and *Tspo,*which was more pronounced in adult rat testes exposed in utero to the 0.1 mg/kg/day dose. Tspo protein levels were also decreased in the interstitium, but not in seminiferous tubules, of adult rats, suggesting differential cell-specific long-term consequences of fetal EDC exposure in adult testes. TSPO is a translocator protein that plays a role in cholesterol-mediated transport from the outer to the inner mitochondrial membrane, highly expressed in and an important component of Leydig cells, but also expressed at lower levels in pachytene spermatocytes and dividing spermatogonia in adult rat testes [49,50]. These data may explain the reduction in serum testosterone levels that we previously observed in rats in utero exposed to Gen + DEHP mixtures, where the lower dose had more dramatic effects than the higher dose, further suggesting that fetal exposure to these EDC mixtures at doses lower than their individual NOAEL can affect adult steroidogenesis, a key function of the testes [26].

4. Materials and Methods

4.1. Animal Treatments and Tissue Collection

Animal treatments and tissue collection were performed as previously described [26]. Timed pregnant Sprague-Dawley rats were purchased from Charles Rivers Laboratories (Saint-Constant, QC, Canada) and switched to a casein/cornstarch-based, phytoestrogen-free diet (casein diet) AIN-93G (Teklad diet, Envigo, Indianapolis, IN, USA) from 2 days before gavage to weaning, to avoid dietary exposure to genistein. The rats were maintained on a 12L:12D photoperiod with ad libitum access to food and water and handled according to protocols approved by the McGill University Health Centre Animal Care Committee and the Canadian Council on Animal Care. Pregnant rats were treated by gavage from gestational day 14 to parturition with either vehicle (corn oil) alone or containing GEN, DEHP, or GEN + DEHP mixtures at the doses of 0.1 and 10 mg/kg/day, encompassing exposure levels found in the general population to those measured in vegetarian/vegan women and more susceptible populations such as hospitalized neonates exposed to DEHP via medical equipment and fed soy formula (Figure 6) [16,51–55]. Doses were adjusted to changes in dam weights. Offspring were weighed and euthanized at PND120. The testes were collected, weighed, and either fixed in 4% paraformaldehyde or snap frozen for gene and protein expression analyses.

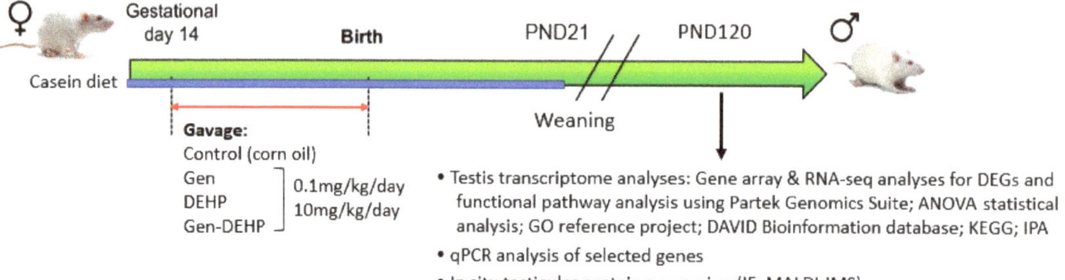

Figure 6. Flow chart of treatments and experiments performed.

4.2. RNA Extraction and Quantitative Real-Time PCR

RNA was extracted from testes using a Nucleospin XS Kit and digested with DNase I (Takara Bio, San Jose, CA, USA). Complementary DNA was synthesized using the transcriptor synthesis kit (Roche Diagnostics, Indianapolis, IN, USA). Quantitative real-time PCR (qPCR) was performed as previously described with a LightCycler 480 using SYBR Green Supermix (BioRad, Hercules, CA, USA) and a Master Mix kit (Roche Diagnostics). Glyceraldehyde-3-phosphate dehydrogenase was used as reference to normalize gene expression. A minimum of 3–8 male offspring from different litters were assessed in triplicate. The comparative Ct method was used to calculate relative gene expression. Primers specific for the genes of interest were designed with the NCBI Primer Design Database.

4.3. Immunofluorescence (IF)

The IF analysis was performed as previously described [49,56]. Briefly, slides were first dewaxed and rehydrated using Citrisol and Trilogy (Cell Marque, Rocklin, CA, USA) solution. Following treatment with Dako Target Antigen Retrieval Solution (DAKO, Troy, MI, USA), the sections were incubated with PBS containing 10% BSA and 10% donkey serum for one hour to block non-specific protein interactions. The sections were then incubated with anti-forkhead box A3 (FOXA3) antibody (SAB2108468, Sigma-Aldrich, Saint Louis, MO, USA) (1:50 dilution), anti-platelet-derived growth factor receptor alpha (Pdgfrα) (sc-398206, Bioss, Woburn, MA USA) (1:50 dilution), or translocator protein (TSPO) (1:400 dilution) antibodies [57] diluted in PBS containing 10% BSA, 0.1% Triton-X, and donkey serum overnight at 4 °C. Then, the slides were incubated with a fluorescent goat anti-rabbit Alexa Fluor 488 (Thermo Fisher, Waltham, MA, USA) diluted in PBS containing 1% BSA for one hour at room temperature. Nuclear staining was performed using nuclear DAPI anti-fade and mounting medium (Vector Labs, Newark, CA, USA), coverslipped, and then imaged. Immunofluorescence signals were viewed using the appropriate filters on a Biotek Cytation 5 slide imager. The IF analysis was conducted on 2–3 independent offspring per treatment, and representative pictures are shown.

4.4. MALDI Imaging Mass Spectrometry

The matrix-assisted laser desorption/ionization with imaging mass spectrometry (MALDI IMS) analysis was performed by the Core Facility of the School of Pharmacy of USC, to generate "intensity maps" showing the in situ relative abundance of FOXA3 using previously described methodology [57]. Briefly, frozen PND120 adult rat testes were cryosectioned at 12 uM thickness at −21 °C and thawed on precooled ITO-coated slides. Then, sections were washed in 70% ethanol for 120 s two times, followed by washing with 100% ethanol for 120 s. The MALDI matrix consisting of sinapic acid at 10 mg/mL in 50% acetonitrile/0.1% formic acid was sprayed on sections. Matrix-coated sections were recrystallized using 50% formic acid at 80 °C for 10 min. Then, sections were imaged using a Rapiflex MALDI IMS system (Bruker, rapifleX system, Billerica, MA, USA) at 100 uM spatial resolution.

4.5. Microarray Analysis

The microarrays were performed as previously described [26]. Briefly, RNA was extracted with a PicoPure RNA isolation kit (Arcturus, San Diego, CA, USA) from the testes of PND120 rat offspring from three different dams per treatment. The gene array analysis was performed on Affymetrix 2.0 ST microarray chips by Genome Quebec, as previously described [56,58]. A statistical analysis was performed using the Partek Genomics Suite software, to identify differentially expressed genes (DEGs) between the Gen and DEHP treatments and control samples using ANOVA. In total, 19,786 protein-coding genes and microRNAs were analyzed. DEGs were identified using an unadjusted p-value of 0.05 as cutoff and applying a fold-change cutoff of at least 40% above or below the control values. The gene lists from Partek were analyzed for functional pathways and networks using the Ingenuity Pathway Analysis (IPA) software (https://www.nihlibrary.nih.gov/resources/tools/ingenuity-pathways-analysis-ipa) and the Database for Association, Visualization and Integrated Discovery (DAVID) software (https://david.ncifcrf.gov/) linked to the Kyoto Encyclopedia of Genes and Genomes (KEGG) database (https://www.genome.jp/kegg/). The 2^log2 values of the signal intensities (originally expressed as log2 values) were calculated, and genes expressed at a relative intensity below 40 (representing 33% of all genes on the arrays) in all conditions were not given priority. Gene and pathway relevance to the study were assessed via PubMed keyword searches.

4.6. Whole Transcriptome RNA Sequencing

Transcriptomic RNA sequencing was performed at the USC Norris Molecular Genomics Core. Total RNA were isolated using a Qiagen (Germatown, MD, USA) All prep Extraction kit following the manufacturer's protocol (Qiagen Cat. No. 80284). Libraries were simultaneously prepared using an Illumina Truseq Stranded mRNA Library Preparation kit (Illumina Cat. No. 20020594; San Diego, CA, USA). Transcriptomic RNAseq libraries were sequenced on Illumina Nextseq500 at 25 million reads per sample at 2×75 read length. Data were trimmed normalized and analyzed using Partek Flow. Gene lists were created to detect differentially expressed genes between

4.7. ChIP-Seq Analysis

The ChIP-seq analysis was performed by Diagenode s.a. (Denville, NJ, USA) to identify potential target genes of FOXA3 in rat testes, using DNA from frozen adult control rat testes, which contains higher levels of FOXA3 protein than testes from rats exposed in utero to Gen + DEHP. Briefly, chromatin was prepared from 175 mg frozen tissue by Diagenode ChIP-seq/ChIP-qPCR Profiling service (Diagenode Cat. No. G02010000) using an iDeal ChIP-seq Kit for Transcription Factors (Diagenode Cat. No. C01010170). Chromatin shearing was performed with a Bioruptor® Pico sonication device (Diagenode Cat. No. B01060001) for 4 cycles. ChIP immunoprecipitation was performed on 8 µg of chromatin using 1.5 µg and 3 µg of two anti-FOXA3 antibodies (Invitrogen PA1-813, Lot VD296799, Santa Cruz sc-74424 X, Lot A2320). Some chromatin was kept as input. ChIP efficiency was assessed by qPCR analyses, and rat primers for H3K4me3/Pol2 were used as internal ChIP controls prior to library preparation. Libraries were prepared from input and ChIP's DNA using MicroPlex Library Preparation from Diagenode with 24 UDI for MicroPlex v3. Optimal library amplification was assessed and the libraries were purified, quantified, and fragment sizes determined. Libraries were analyzed using Illumina sequencing (NextSeq 2000 P2 200 cycles), paired-end reads, 2×50 bp, 30 million raw reads per mark on average. Quality check, alignment to reference genome, identification of enriched regions, and annotation of ChIP-Seq peaks with genomic regions (introns, exons, promoters, 1-to-5 kb upstream-TSS, and intergenic regions) were performed.

4.8. Statistical Analysis

The statistical analysis was performed using one-way ANOVA with post hoc Dunnett's test or unpaired two-tailed Student's *t*-test for qPCR data analysis, using the statistical analysis functions in GraphPad Prism 7.04 program (GraphPad Inc. San Diego, CA, USA). Because the two EDCs used were not expected to have similar effects, an unpaired two-tailed Student's *t*-test was used to determine the statistical significance between each control-EDC pair for qPCR analysis. The gene array analysis was performed on three independent N (one offspring/dam) per treatment condition, using the ANOVA application from the bioinformatics Partek platform. Fertility was assessed using 8 to 9 offspring from different litters per treatment condition [26]. For the qPCR analysis, the results are presented as mean ± SEM of fold changes relative to vehicle control. Experimental points were performed in triplicate for each sample, from 3 to 8 rats from different dams per treatment condition. Asterisks indicate a significant change relative to control, with p-values ≤ 0.05 considered to be statistically significant.

5. Conclusions

In this study, we identified functional pathways related to Leydig, Sertoli, and germ cells altered in adult rat testes following in utero exposure to mixtures of genistein and DEHP at doses encountered by humans. FOXA3, a transcription factor critical for Leydig cell function, was downregulated, as well as several genes in its interactome and steroidogenic genes. The protein expression of FOXA3 was reduced in testicular interstitium by in utero exposure to genistein + DEHP mixtures. The results of this study suggest that in utero exposure to low-dose mixtures of Gen and DEHP can disrupt the developmental program of key testicular cell types, including Leydig cells, and implies that FOXA3 is a pivotal target of the long-term adverse effects of fetal exposure to EDC mixtures on male reproduction. Our findings also suggest that FOXA3 could be used as sentinel gene to screen for potential long-term effects of fetal exposure to EDC mixtures with potential adverse male reproductive effects, in conjunction with the determination of cell-specific markers and/or morphometric analyses, to assess the possibility of effects due to changes in cellularity rather than cell-specific gene/protein alterations.

Supplementary Materials: The following supporting information can be downloaded at: https://www.mdpi.com/article/10.3390/ijms24021211/s1.

Author Contributions: Conceptualization, C.W. and M.C.; methodology, C.W., A.B., P.M. and M.C.; formal analysis, C.W. and M.C.; investigation, C.W., A.B. and P.M.; resources, M.C.; writing—original draft preparation, C.W.; writing—review and editing, M.C.; supervision, M.C.; funding acquisition, M.C. All authors have read and agreed to the published version of the manuscript.

Funding: This research was funded in part by funds from a grant from the Canadian Institutes of Health Research (CIHR) (Operating grant # MOP-133456) and funds from the USC Alfred E. School of Pharmacy and Pharmaceutical Sciences to M.C. W.C. was supported by a USC Alfred E. School of Pharmacy and Pharmaceutical Sciences Dean's award.

Institutional Review Board Statement: The animal study protocol was approved by the Institutional Review Board, the McGill University Health Centre Animal Care Committee, and the Canadian Council on Animal Care.

Informed Consent Statement: Not applicable.

Data Availability Statement: The data supporting reported results can be found in the manuscript.

Acknowledgments: We thank the USC Libraries Bioinformatics Services for assisting with data analysis of the microarray data, for performing the RNA-seq analysis and for their assistance in analyzing the data. The bioinformatics software and computing resources used in the analysis were funded by the USC Office of Research and the USC Libraries. We thank Alireza Abdolvahabi from the Mass Spectrometry Core at USC Alfred E. School of Pharmacy and Pharmaceutical Sciences for performing the MALDI IMS analysis of the samples and assisting in the interpretation of the data. We thank the histology core of the USC Alfred E. School of Pharmacy and Pharmaceutical Sciences for preparing paraffin blocks and sections for protein analysis, and frozen sections for MALDI IMS analysis.

Conflicts of Interest: The authors declare no conflict of interest.

References

1. Barratt, C.L.R.; Björndahl, L.; De Jonge, C.J.; Lamb, D.J.; Osorio Martini, F.; McLachlan, R.; Oates, R.D.; Van Der Poel, S.; St John, B.; Sigman, M.; et al. The diagnosis of male infertility: An analysis of the evidence to support the development of global WHO guidance—Challenges and future research opportunities. *Hum. Reprod. Update* **2017**, *23*, 660–680. [CrossRef] [PubMed]
2. Krausz, C.; Riera-Escamilla, A. Genetics of male infertility. *Nat. Rev. Urol.* **2018**, *15*, 369–384. [CrossRef] [PubMed]
3. Sharma, A.; Minhas, S.; Dhillo, W.S.; Jayasena, C.N. Male infertility due to testicular disorders. *J. Clin. Endocrinol. Metab.* **2020**, *106*, e442–e459. [CrossRef] [PubMed]
4. Sharma, A.; Mollier, J.; Brocklesby, R.W.K.; Caves, C.; Jayasena, C.; Minhas, S. Endocrine-disrupting chemicals and male reproductive health. *Reprod. Med. Biol.* **2020**, *19*, 243–253. [CrossRef] [PubMed]
5. Levine, H.; Jorgensen, N.; Martino-Andrade, A.; Mendiola, J.; Weksler-Derri, D.; Jolles, M.; Pinotti, R.; Swan, S.H. Temporal trends in sperm count: A systematic review and meta-regression analysis of samples collected globally in the 20th and 21st centuries. *Hum. Reprod. Update* **2022**. [CrossRef]
6. Sharpe, R.M.; Skakkebaek, N.E. Testicular dysgenesis syndrome: Mechanistic insights and potential new downstream effects. *Fertil. Steril.* **2008**, *89*, e33–e38. [CrossRef]
7. Heger, N.E.; Hall, S.J.; Sandrof, M.A.; McDonnell, E.V.; Hensley, J.B.; McDowell, E.N.; Martin, K.A.; Gaido, K.W.; Johnson, K.; Boekelheide, K. Human Fetal Testis Xenografts Are Resistant to Phthalate-Induced Endocrine Disruption. *Environ. Health Perspect.* **2012**, *120*, 1137–1143. [CrossRef]
8. Jenardhanan, P.; Panneerselvam, M.; Mathur, P.P. Effect of environmental contaminants on spermatogenesis. *Semin. Cell Dev. Biol.* **2016**, *59*, 126–140. [CrossRef]
9. Sidorkiewicz, I.; Zaręba, K.; Wolczynski, S.; Czerniecki, J. Endocrine-disrupting chemicals—Mechanisms of action on male reproductive system. *Toxicol. Ind. Health* **2017**, *33*, 601–609. [CrossRef]
10. Zhang, S.; Mo, J.; Wang, Y.; Ni, C.; Li, X.; Zhu, Q.; Ge, R.S. Endocrine disruptors of inhibiting testicular 3beta-hydroxysteroid dehydrogenase. *Chem. Biol. Interact.* **2019**, *303*, 90–97. [CrossRef]

11. Gore, A.C.; Chappell, V.A.; Fenton, S.E.; Flaws, J.A.; Nadal, A.; Prins, G.S.; Toppari, J.; Zoeller, R.T. EDC-2: The Endocrine Society's Second Scientific Statement on Endocrine-Disrupting Chemicals. *Endocr. Rev.* **2015**, *36*, E1–E150. [PubMed]

12. Walker, C.; Garza, S.; Papadopoulos, V.; Culty, M. Impact of endocrine-disrupting chemicals on steroidogenesis and consequences on testicular function. *Mol. Cell. Endocrinol.* **2021**, *527*, 111215. [CrossRef] [PubMed]

13. Thuillier, R.; Manku, G.; Wang, Y.; Culty, M. Changes in MAPK pathway in neonatal and adult testis following fetal estrogen exposure and effects on rat testicular cells. *Microsc. Res. Tech.* **2009**, *72*, 773–786. [CrossRef] [PubMed]

14. Thuillier, R.; Wang, Y.; Culty, M. Prenatal exposure to estrogenic compounds alters the expression pattern of platelet-derived growth factor receptors alpha and beta in neonatal rat testis: Identification of gonocytes as targets of estrogen exposure. *Biol. Reprod.* **2003**, *68*, 867–880. [CrossRef] [PubMed]

15. Wang, Y.; Thuillier, R.; Culty, M. Prenatal Estrogen Exposure Differentially Affects Estrogen Receptor-Associated Proteins in Rat Testis Gonocytes1. *Biol. Reprod.* **2004**, *71*, 1652–1664. [CrossRef] [PubMed]

16. Rozman, K.K.; Bhatia, J.; Calafat, A.M.; Chambers, C.; Culty, M.; Etzel, R.; Flaws, J.; Hansen, D.K.; Hoyer, P.B.; Jeffery, E.; et al. NTP-CERHR Expert Panel Report on the reproductive and developmental toxicity of soy formula. *Birth Defects Res. Part B Dev. Reprod. Toxicol.* **2006**, *77*, 280–397. [CrossRef] [PubMed]

17. Martinez-Arguelles, D.; Campioli, E.; Culty, M.; Zirkin, B.; Papadopoulos, V. Fetal origin of endocrine dysfunction in the adult: The phthalate model. *J. Steroid Biochem. Mol. Biol.* **2013**, *137*, 5–17. [CrossRef]

18. Buñay, J.; Larriba, E.; Moreno, R.D.; del Mazo, J. Chronic low-dose exposure to a mixture of environmental endocrine disruptors induces microRNAs/isomiRs deregulation in mouse concomitant with intratesticular estradiol reduction. *Sci. Rep.* **2017**, *7*, 3373. [CrossRef]

19. Zarean, M.; Keikha, M.; Poursafa, P.; Khalighinejad, P.; Amin, M.; Kelishadi, R. A systematic review on the adverse health effects of di-2-ethylhexyl phthalate. *Environ. Sci. Pollut. Res.* **2016**, *23*, 24642–24693. [CrossRef]

20. Edwards, L.; McCray, N.L.; VanNoy, B.N.; Yau, A.; Geller, R.J.; Adamkiewicz, G.; Zota, A.R. Phthalate and novel plasticizer concentrations in food items from U.S. fast food chains: A preliminary analysis. *J. Expo. Sci. Environ. Epidemiol.* **2021**, *32*, 366–373. [CrossRef]

21. Culty, M.; Thuillier, R.; Li, W.; Wang, Y.; Martinez-Arguelles, D.B.; Benjamin, C.G.; Triantafilou, K.M.; Zirkin, B.R.; Papadopoulos, V. In Utero Exposure to Di-(2-ethylhexyl) Phthalate Exerts Both Short-Term and Long-Lasting Suppressive Effects on Testosterone Production in the Rat1. *Biol. Reprod.* **2008**, *78*, 1018–1028. [CrossRef] [PubMed]

22. Li, H.; Spade, D.J. REPRODUCTIVE TOXICOLOGY: Environmental exposures, fetal testis development and function: Phthalates and beyond. *Reproduction* **2021**, *162*, F147–F167. [CrossRef] [PubMed]

23. Jones, S.; Boisvert, A.; Francois, S.; Zhang, L.; Culty, M. In Utero Exposure to Di-(2-Ethylhexyl) Phthalate Induces Testicular Effects in Neonatal Rats That Are Antagonized by Genistein Cotreatment1. *Biol. Reprod.* **2015**, *93*, 92. [CrossRef] [PubMed]

24. Mitchell, R.T.; Childs, A.J.; Anderson, R.A.; Driesche, S.V.D.; Saunders, P.T.K.; McKinnell, C.; Wallace, W.H.B.; Kelnar, C.J.H.; Sharpe, R.M. Do Phthalates Affect Steroidogenesis by the Human Fetal Testis? Exposure of Human Fetal Testis Xenografts to Di-n-Butyl Phthalate. *J. Clin. Endocrinol. Metab.* **2012**, *97*, E341–E348. [CrossRef]

25. Muczynski, V.; Cravedi, J.; Lehraiki, A.; Levacher, C.; Moison, D.; Lécureuil, C.; Messiaen, S.; Perdu, E.; Frydman, R.; Habert, R.; et al. Effect of mono-(2-ethylhexyl) phthalate on human and mouse fetal testis: In vitro and in vivo approaches. *Toxicol. Appl. Pharmacol.* **2012**, *261*, 97–104. [CrossRef]

26. Walker, C.; Ghazisaeidi, S.; Collet, B.; Boisvert, A.; Culty, M. In utero exposure to low doses of genistein and di-(2-ethylhexyl) phthalate (DEHP) alters innate immune cells in neonatal and adult rat testes. *Andrology* **2020**, *8*, 943–964. [CrossRef]

27. Sen Sharma, S.; Vats, A.; Majumdar, S. Regulation of Hippo pathway components by FSH in testis. *Reprod. Biol.* **2019**, *19*, 61–66. [CrossRef]

28. Takase, H.M.; Nusse, R. Paracrine Wnt/beta-catenin signaling mediates proliferation of undifferentiated spermatogonia in the adult mouse testis. *Proc. Natl. Acad. Sci. USA* **2016**, *113*, E1489–E1497. [CrossRef]

29. Griswold, M.D. Spermatogenesis: The Commitment to Meiosis. *Physiol. Rev.* **2016**, *96*, 1–17. [CrossRef]

30. Roumaud, P.; Martin, L.J. Transcriptomic analysis of overexpressed SOX4 and SOX8 in TM4 Sertoli cells with emphasis on cell-to-cell interactions. *Biochem. Biophys. Res. Commun.* **2019**, *512*, 678–683. [CrossRef]

31. Behr, R.; Sackett, S.D.; Bochkis, I.M.; Le, P.P.; Kaestner, K.H. Impaired male fertility and atrophy of seminiferous tubules caused by haploinsufficiency for Foxa3. *Dev. Biol.* **2007**, *306*, 636–645. [CrossRef] [PubMed]

32. Garon, G.; Bergeron, F.; Brousseau, C.; Robert, N.M.; Tremblay, J.J. FOXA3 Is Expressed in Multiple Cell Lineages in the Mouse Testis and Regulates Pdgfra Expression in Leydig Cells. *Endocrinology* **2017**, *158*, 1886–1897. [CrossRef] [PubMed]

33. Anjum, S.; Krishna, A.; Sridaran, R.; Tsutsui, K. Localization of Gonadotropin-Releasing Hormone (GnRH), Gonadotropin-Inhibitory Hormone (GnIH), Kisspeptin and GnRH Receptor and Their Possible Roles in Testicular Activities From Birth to Senescence in Mice. *J. Exp. Zool. Part A Ecol. Genet. Physiol.* **2012**, *317*, 630–644. [CrossRef] [PubMed]

34. Zhu, Q.; Dong, Y.; Li, X.; Ni, C.; Huang, T.; Sun, J.; Ge, R.S. Dehydroepiandrosterone and Its CYP7B1 Metabolite 7alpha-Hydroxydehydroepiandrosterone Regulates 11beta-Hydroxysteroid Dehydrogenase 1 Directions in Rat Leydig Cells. *Front. Endocrinol.* **2019**, *10*, 886. [CrossRef] [PubMed]

35. Papadopoulos, V.; Baraldi, M.; Guilarte, T.R.; Knudsen, T.B.; Lacapère, J.-J.; Lindemann, P.; Norenberg, M.D.; Nutt, D.; Weizman, A.; Zhang, M.-R.; et al. Translocator protein (18kDa): New nomenclature for the peripheral-type benzodiazepine receptor based on its structure and molecular function. *Trends Pharmacol. Sci.* **2006**, *27*, 402–409. [CrossRef] [PubMed]

36. Yang, L.; Ma, T.; Zhao, L.; Jiang, H.; Zhang, J.; Liu, D.; Zhang, L.; Wang, X.; Pan, T.; Zhang, H.; et al. Circadian regulation of apolipoprotein gene expression affects testosterone production in mouse testis. *Theriogenology* **2021**, *174*, 9–19. [CrossRef] [PubMed]

37. Teng, M.; Zhou, S.; Cai, C.; Lupien, M.; He, H.H. Pioneer of prostate cancer: Past, present and the future of FOXA1. *Protein Cell* **2020**, *12*, 29–38. [CrossRef]

38. Benayoun, B.A.; Caburet, S.; Veitia, R.A. Forkhead transcription factors: Key players in health and disease. *Trends Genet.* **2011**, *27*, 224–232. [CrossRef]

39. Griswold, S.; Behringer, R. Fetal Leydig Cell Origin and Development. *Sex. Dev.* **2009**, *3*, 1–15. [CrossRef]

40. Shima, Y. Development of fetal and adult Leydig cells. *Reprod. Med. Biol.* **2019**, *18*, 323–330. [CrossRef]

41. Lacagnina, S. The Developmental Origins of Health and Disease (DOHaD). *Am. J. Lifestyle Med.* **2019**, *14*, 47–50. [CrossRef]

42. Beck, D.; Nilsson, E.E.; Ben Maamar, M.; Skinner, M.K. Environmental induced transgenerational inheritance impacts systems epigenetics in disease etiology. *Sci. Rep.* **2022**, *12*, 5452. [CrossRef] [PubMed]

43. Fatima, S.; Wagstaff, K.M.; Loveland, K.L.; Jans, D.A. Interactome of the negative regulator of nuclear import BRCA1-binding protein 2. *Sci. Rep.* **2015**, *5*, 9459. [CrossRef] [PubMed]

44. Mohler, J.L.; Morris, T.L.; Ford, O.H., 3rd; Alvey, R.F.; Sakamoto, C.; Gregory, C.W. Identification of differentially expressed genes associated with androgen-independent growth of prostate cancer. *Prostate* **2002**, *51*, 247–255. [CrossRef] [PubMed]

45. Masood, M.; Grimm, S.; El-Bahrawy, M.; Yagüe, E. TMEFF2: A Transmembrane Proteoglycan with Multifaceted Actions in Cancer and Disease. *Cancers* **2020**, *12*, 3862. [CrossRef] [PubMed]

46. Fan, L.; Zhu, Q.; Liu, L.; Zhu, C.; Huang, H.; Lu, S.; Liu, P. CXCL13 is androgen-responsive and involved in androgen induced prostate cancer cell migration and invasion. *Oncotarget* **2017**, *8*, 53244–53261. [CrossRef] [PubMed]

47. Cargnelutti, F.; Di Nisio, A.; Pallotti, F.; Sabovic, I.; Spaziani, M.; Tarsitano, M.G.; Paoli, D.; Foresta, C. Effects of endocrine disruptors on fetal testis development, male puberty, and transition age. *Endocrine* **2021**, *72*, 358–374. [CrossRef]

48. Lymperi, S.; Giwercman, A. Endocrine disruptors and testicular function. *Metabolism* **2018**, *86*, 79–90. [CrossRef]

49. Manku, G.; Culty, M. Regulation of Translocator Protein 18 kDa (TSPO) Expression in Rat and Human Male Germ Cells. *Int. J. Mol. Sci.* **2016**, *17*, 1486. [CrossRef]

50. Midzak, A.; Rone, M.; Aghazadeh, Y.; Culty, M.; Papadopoulos, V. Mitochondrial protein import and the genesis of steroidogenic mitochondria. *Mol. Cell. Endocrinol.* **2011**, *336*, 70–79. [CrossRef]

51. Rozman, K.K.; Bhatia, J.; Calafat, A.M.; Chambers, C.D.; Culty, M.; Etzel, R.; Flaws, J.; Hansen, D.K.; Hoyer, P.B.; Jeffery, E.; et al. NTP-CERHR expert panel report on the reproductive and developmental toxicity of genistein. *Birth Defects Res. Part B Dev. Reprod. Toxicol.* **2006**, *77*, 485–638. [CrossRef] [PubMed]

52. Kavlock, R.; Barr, D.; Boekelheide, K.; Breslin, W.; Breysse, P.; Chapin, R.; Marcus, M. NTP-CERHR Expert Panel Update on the Reproductive and Developmental Toxicity of di(2-ethylhexyl) phthalate. *Reprod. Toxicol.* **2006**, *22*, 291–399. [PubMed]

53. Jarrell, J.; Foster, W.G.; Kinniburgh, D.W. Phytoestrogens in Human Pregnancy. *Obstet. Gynecol. Int.* **2012**, *2012*, 850313. [CrossRef] [PubMed]

54. Foster, W.G.; Chan, S.; Platt, L.; Hughes, C.L. Detection of phytoestrogens in samples of second trimester human amniotic fluid. *Toxicol. Lett.* **2002**, *129*, 199–205. [CrossRef]

55. Latini, G.; De Felice, C.; Presta, G.; Del Vecchio, A.; Paris, I.; Ruggieri, F.; Mazzeo, P. In utero exposure to di-(2-ethylhexyl)phthalate and duration of human pregnancy. *Environ. Health Perspect.* **2003**, *111*, 1783–1785. [CrossRef]

56. Manku, G.; Hueso, A.; Brimo, F.; Chan, P.; Gonzalez-Peramato, P.; Jabado, N.; Gayden, T.; Bourgey, M.; Riazalhosseini, Y.; Culty, M. Changes in the expression profiles of claudins during gonocyte differentiation and in seminomas. *Andrology* **2015**, *4*, 95–110. [CrossRef]

57. Li, Y.; Chen, L.; Li, L.; Sottas, C.; Petrillo, S.K.; Lazaris, A.; Metrakos, P.; Wu, H.; Ishida, Y.; Saito, T.; et al. Cholesterol-binding translocator protein TSPO regulates steatosis and bile acid synthesis in nonalcoholic fatty liver disease. *Iscience* **2021**, *24*, 102457. [CrossRef]

58. Manku, G.; Wing, S.S.; Culty, M. Expression of the Ubiquitin Proteasome System in Neonatal Rat Gonocytes and Spermatogonia: Role in Gonocyte Differentiation1. *Biol. Reprod.* **2012**, *87*, 44. [CrossRef]

International Journal of
Molecular Sciences

Review

Germline Mutations in Steroid Metabolizing Enzymes: A Focus on Steroid Transforming Aldo-Keto Reductases

Andrea J. Detlefsen [1,†] , Ryan D. Paulukinas [2,3,†] and Trevor M. Penning [2,3,*]

1   Department of Biochemistry & Biophysics, Perelman School of Medicine, University of Pennsylvania, Philadelphia, PA 19104, USA
2   Department of Systems Pharmacology & Translational Therapeutics, Perelman School of Medicine, University of Pennsylvania, Philadelphia, PA 19104, USA
3   Center of Excellence in Environmental Toxicology, Perelman School of Medicine, University of Pennsylvania, Philadelphia, PA 19104, USA
*   Correspondence: penning@upenn.edu; Tel.: +1-215-898-9445; Fax: +1-215-573-0200
†   These authors contributed equally to this work.

Abstract: Steroid hormones synchronize a variety of functions throughout all stages of life. Importantly, steroid hormone-transforming enzymes are ultimately responsible for the regulation of these potent signaling molecules. Germline mutations that cause dysfunction in these enzymes cause a variety of endocrine disorders. Mutations in *SRD5A2*, *HSD17B3*, and *HSD3B2* genes that lead to disordered sexual development, salt wasting, and other severe disorders provide a glimpse of the impacts of mutations in steroid hormone transforming enzymes. In a departure from these established examples, this review examines disease-associated germline coding mutations in steroid-transforming members of the human aldo-keto reductase (AKR) superfamily. We consider two main categories of missense mutations: those resulting from nonsynonymous single nucleotide polymorphisms (nsSNPs) and cases resulting from familial inherited base pair substitutions. We found mutations in human AKR1C genes that disrupt androgen metabolism, which can affect male sexual development and exacerbate prostate cancer and polycystic ovary syndrome (PCOS). Others may be disease causal in the AKR1D1 gene that is responsible for bile acid deficiency. However, given the extensive roles of AKRs in steroid metabolism, we predict that with expanding publicly available data and analysis tools, there is still much to be uncovered regarding germline AKR mutations in disease.

Keywords: aldo-keto reductase; bile acid; deficiency; congenital adrenal hyperplasia; hydroxysteroid dehydrogenase; prostate cancer; single nucleotide polymorphism; structure-function; steroid reductase; pseudohermaphroditism

Citation: Detlefsen, A.J.; Paulukinas, R.D.; Penning, T.M. Germline Mutations in Steroid Metabolizing Enzymes: A Focus on Steroid Transforming Aldo-Keto Reductases. *Int. J. Mol. Sci.* **2023**, *24*, 1873. https://doi.org/10.3390/ijms24031873

Academic Editor: Jacques J. Tremblay

Received: 9 November 2022
Revised: 15 January 2023
Accepted: 16 January 2023
Published: 18 January 2023

1. Introduction

Steroid hormones drive essential processes throughout all stages of life, including fetal gonadal differentiation, childhood and pre-adolescent development, adulthood, menopause, andropause, and cognitive decline. They regulate a diverse set of functions, including sexual differentiation, reproduction, water retention, electrolyte balance, and stress responses [1]. Underpinning this action are steroid hormone biosynthetic and metabolizing enzymes, which act as puppet masters and are responsible for the production and metabolism of these potent signaling molecules. However, when germline mutations cause dysfunction in these enzymes, the strings are cut, and a variety of endocrine-dependent disorders can ensue.

Over the years, there have been many well-documented cases of mutations in steroid hormone-transforming enzymes leading to a variety of disorders. Mutations in *SRD5A2* that result in the rare autosomal recessive 5α-reductase type 2 deficiency is a well-studied example. Normally, 5α-reductase type 2 converts Testosterone (T) to 5α-dihydrotestosterone (DHT), which is critical for male external genitalia differentiation and prostate development [2]. First documented in the 1970s in isolated clusters of individuals with 46,XY disordered sexual development (DSD), a biochemical catalog of mutations in *SRD5A2* and their impact on protein function and clinical presentation has expanded over the following decades [2]. Batista et al. provide a comprehensive evaluation of 451 identified cases to date of 5α-reductase type 2 deficiency in 48 countries, stemming from 151 allelic variants.

Relatedly, *HSD17B3* encodes for 17β-hydroxysteroid dehydrogenase type 3, which is highly expressed in the testes where it converts Δ^4-androstene-3,17-dione (Δ^4-AD) to T [3]. Mutations in this enzyme that lead to decreased T production underlie 17β-hydroxysteroid dehydrogenase type 3 deficiency, another rare autosomal recessive cause of 46,XY DSD. Due to a lack of T production in fetal testes, patients typically present at birth with a spectrum of phenotypes from completely female to ambiguous genitalia. However, individuals later experience amenorrhea and severe virilization during puberty [3]. A summary of clinical presentations, epidemiology, and demographic history of mutations causing 17β-hydroxysteroid dehydrogenase type 3 deficiency is detailed in a review from George et al. [3].

Mutations in the *HSD3B2* gene that result in 3β-hydroxysteroid dehydrogenase type 2 deficiency are one rare underlying cause of congenital adrenal hyperplasia (CAH) [4]. 3β-Hydroxysteroid dehydrogenase type 2 is responsible for the conversion of pregnenolone to progesterone, 17α-hydroxypregnenolone to 17α-hydroxyprogesterone, and dehydroepiandrosterone (DHEA) to Δ^4-AD [4]. Consequently, its dysfunction in adrenal glands and gonads leads to DHEA accumulation, lowered cortisol and aldosterone levels, and dysregulation of other downstream and intermediate metabolites. Due to its extensive metabolic influence, the clinical presentation, diagnostic window, and phenotypic severity of 3β-hydroxysteroid dehydrogenase type 2 deficiency are heavily dependent on the location of the mutation within the gene and its outcome. Mutations that result in frameshift or protein truncation that completely eradicate enzyme activity cause severe salt-wasting that is diagnosed in the first weeks of life and is fatal if left untreated [5]. Missense mutations that leave some enzyme activity intact present as over-virilization in females and under-masculinization of males at birth. Even mutations that have a mild effect on enzyme activity can contribute to premature puberty, acne, and menstrual disorders in females, including polycystic ovary syndrome (PCOS) [5]. A detailed analysis of CAH caused by 3β-hydroxysteroid dehydrogenase type 2 deficiency can be found in reviews from Al Alawi et al. [4] and Sahakitrungruang [5].

Recently, Chang et al. detailed a unique scenario where a nonsynonymous single nucleotide polymorphism (nsSNP) in *HSD3B1* results in the gain of function N367T substitution that stabilizes its encoded protein, 3β-hydroxysteroid dehydrogenase type 1, leading to increased DHT production from adrenal-derived DHEA that could ultimately drive castration-resistant prostate cancer (CRPC) [6]. The authors describe a mode of selective adaptation in CRPC tumors that operates through loss-of-heterozygosity of the wild-type copy of *HSD3B1* in individuals who are germline heterozygous with wild-type/N367T alleles. This phenomenon leads to overexpression of the N367T allele due to increased resistance to ubiquitination and degradation, which allows for elevated intratumoral DHT production that can drive androgen receptor signaling and, ultimately, CRPC tumor growth [6]. In contrast to the previous examples where mutations were the source of severe disorders, the N367T nsSNP is an illustration of a mutant whose impact only becomes apparent when it exacerbates a disease.

The stories of *SRD5A2*, *HSD17B3*, and *HSD3B2* are flagship examples that establish the relevance of germline mutations in steroid metabolizing enzymes in disease. Each has been thoroughly documented, from mutation identification in patients to functional genomic studies that provide evidence of the protein dysfunction that is either disease causal or exacerbates disease. Comprehensive reviews that catalog known cases and consider what can be learned from the amassed data of each deficiency already exist and are referenced above. In a departure from these well-studied examples, this review will focus on the association of disease-associated germline coding mutations in members of the human aldo-keto reductase (AKR) superfamily that transform steroids. While there are many types of mutations that can lead to disease, our analysis considers two main categories: nsSNPs that results in a missense mutation and familial inherited single base pair transitions/transversions that result in rarer missense mutations. The main differentiation between these groups is that single nucleotide polymorphisms (SNPs) are classically defined as occurring in a significant portion of the population (minor allelic frequency >1%), while other damaging mutations are rarer. The reader is referred to Figure 1, which shows a generalized scheme for sex steroid hormone biosynthesis and metabolism and the genes that have germline mutations.

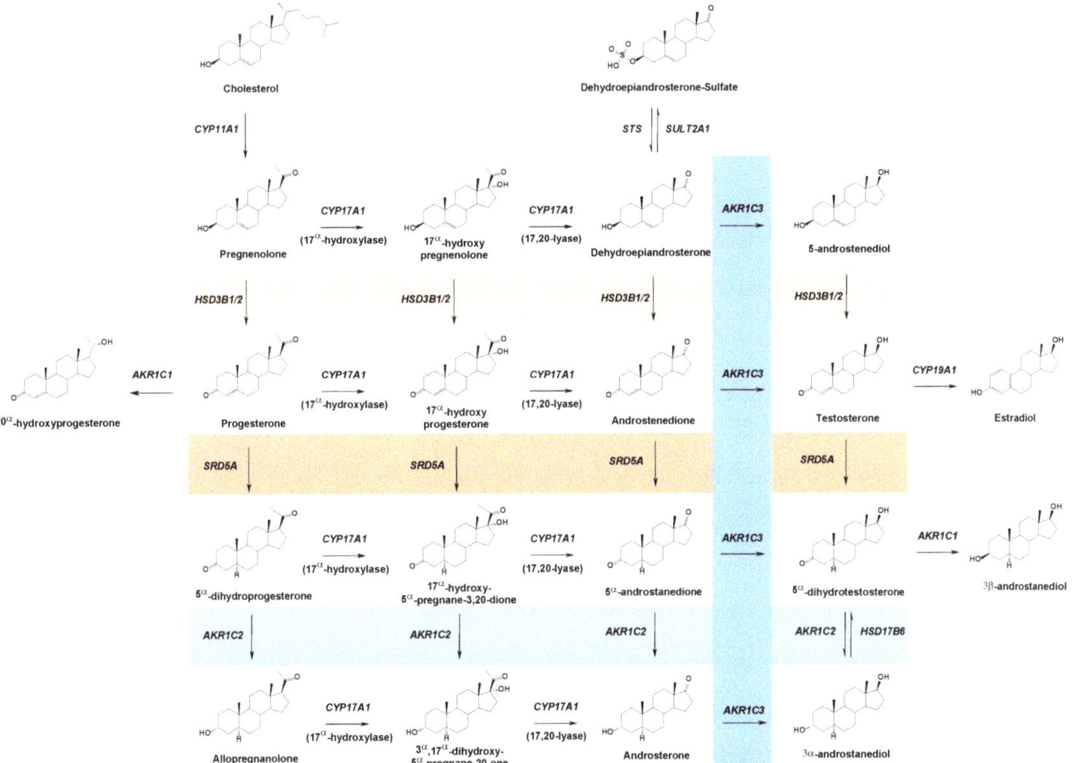

Figure 1. Human sex steroid hormone metabolism with short-chain dehydrogenases/reductase enzymatic conversions influenced by germ-line mutations. AKR family members reviewed in the present work are shaded in blue, while non-AKR family members previously reviewed are shaded in orange.

2. Aldo-Keto Reductases

AKRs are a superfamily of enzymes that reduce carbonyl substrates in an NAD(P)H-dependent reaction by a sequential ordered bi-bi kinetic mechanism [7]. Members share an $(\alpha/\beta)_8$-barrel motif with a highly conserved catalytic tetrad and cofactor binding site and possess three large loops (A, B, and C) at the back of the barrel, which defines substrate specificity [7]. Often these enzymes are pluripotent and can reduce 3-, 17-, and 20-ketosteroids as well as prostaglandins, retinol, or quinones. Several can also oxidize polycyclic aromatic hydrocarbon (PAH) *trans*-dihydrodiols [7]. Rather than discuss all 15 human AKRs, we focus on steroid-metabolizing enzymes, which account for much of the literature on disease-associated AKR mutations. Unlike the preceding examples, where a germline mutation results in a loss or gain of enzyme function that can be related to phenotypic change and disease outcome, the results are less complete for the AKRs. The one notable exception is the link between *AKR1D1* mutations and bile acid deficiency, which will be discussed later.

In our review of germline AKR mutations, we report three major approaches to identify possible links to disease. The first approach involves a survey of the epidemiological literature that identifies a collection of nsSNPs in a panel of AKR genes that may be associated with the disease. In the second approach, nsSNPs, without any formal link to disease, are identified, and biochemical characterization of the mutated protein is performed. The third approach identifies rare inherited mutations with an almost certain link to disease identified in the clinic and biochemically characterized. We found numerous reports claiming that nsSNPs in human *AKR* genes are associated with various aspects of disease without any functional genomic follow-up studies. In another case, the reverse was true, where analysis of several nsSNPs in *AKR1C2* uncovered kinetic differences in the enzyme that could theoretically impact prostate cancer (PCa). Still, no clinical study testing for this association was pursued. Even when a follow-up functional genomic study is conducted, it is possible that the effects of a deleterious mutation could be overlooked depending on whether the assay used is able to identify the impact of the mutation. For example, a study measuring steady-state kinetic parameters would not be suitable for detecting a change in protein stability. Furthermore, many diseases involving AKRs are polygenic, making it difficult to conduct studies that isolate the individual contributions of a single mutation. These efforts are further complicated due to AKRs' involvement in multiple pathways of steroid metabolism where mutations in one step may be compensated for or compounded by nsSNPs in genes responsible for other steps in the same pathway. These obstacles in unraveling the roles of AKR mutations in polygenic diseases would ideally be overcome by studying them in vivo, where multiple mutations could be introduced by CRISPR/cas9. Unfortunately, the use of murine knockouts or transgenic mice to study AKR mutants is not feasible. Velica et al. [8] investigated eight of the nine existing murine AKR isoforms and found that they had largely different substrate and tissue distribution compared to human AKRs and, most importantly, that they were absent from many steroid hormone target tissues. Alternatively, CRISPR/cas9 could be used to create variants in cell models. The challenge of this approach lies in the design of specific guide RNAs, as the AKR1C genes share greater than 86% sequence identity.

We also consider that certain methodological assumptions may be made when searching for nsSNPs with a connection to the disease. Biases built into current systems, such as the origin of reference genomes, could, for example, tip the balance in determining which allele should be considered wild-type. As is evident from studies such as the 1000 genomes project, "wild-type" can shift between two alleles depending on which continental population is under consideration [9,10]. However, with the continually expanding publicly available data such as the recent additions to the UK biobank, the cancer genome atlas (TCGA), and more specific resources like the stand up to cancer (SU2C) prostate cancer foundation, these biases will likely dissipate [11–13]. This expansion also creates the potential for the categorization of mutations to be upgraded to an SNP status if the frequency in a newly evaluated population shifts the global minor allelic frequency (MAF). Additionally, programs that predict mutational impact in silico, such as PolyPhen and Sift, are useful in a first-pass evaluation of new mutations [14,15]. These programs provide a prediction of whether mutations are deleterious vs. benign, which often corresponds to amino acids that are evolutionarily conserved or non-conserved throughout the AKR superfamily respectively. Furthermore, the addition of new developments in computational analysis methods and tools like AlphaFold makes it increasingly more accessible to make an initial evaluation of the impact of mutations on structure in silico [16]. However, while these programs should be able to accurately predict the presence of the $(\alpha/\beta)_8$ barrel motif in mutant proteins, they may be challenged to predict the effect of mutations on the conformation of the disordered loops A, B, and C, which change upon ligand binding [17].

With help from advancements in available data and analysis tools, we predict that there is still much to be uncovered with respect to germline AKR mutations in disease. In addition to the examples detailed below, there is a web of connections linking AKRs to disease through their activity in key metabolic pathways, altered expression in female reproductive diseases such as endometriosis and PCOS, and general participation in a variety of other disorders and processes [18,19]. Given their wide reach, it is likely that there are other layers of genetic involvement for these enzymes that have yet to be discovered. Our goal is to provide a synopsis of germline coding mutations in steroid-transforming AKRs to reinvigorate the exploration of their connection to the disease.

3. Steroid Metabolizing AKR Enzymes

Members of the AKR1C subfamily display varying levels of 3-, 17-, and 20-ketosteroid reductase activities and metabolize several major classes of steroids. These reductive enzymes work in the opposite direction to oxidative hydroxysteroid dehydrogenases, where the complementary oxidoreductase activity of these enzyme pairs ultimately controls ligand availability and occupancy for steroid hormone receptors, including the androgen, estrogen, and progesterone receptors [18]. AKR1Cs also play a role in the metabolism of exogenous environmental toxicants that drive cancer initiation and progression. AKR1D1 is an exception and is the only steroid 5β-reductase in humans that catalyzes the 5β-reduction of the double-bond in Δ^4-3-ketosteorids to produce 5β-dihydrosteroids, which are essential intermediates in bile acid biosynthesis. Importantly, several of these enzymes work in concert one with one another. For example, AKR1C3 is a major peripheral enzyme that converts Δ^4-AD to T, which is a direct precursor to DHT. In contrast, AKR1C1 and AKR1C2 can metabolize DHT into the inactive 3β- and 3α-androstanediols, respectively. In another example, AKR1D1 precedes AKR1C4 in hepatic bile acid biosynthesis. Moreover, AKR1D1 works sequentially with AKR1C1 and AKR1C2 to provide a source of 3β,5β- and 3α,5β-tetrahydrosteroids, respectively. Genetic mutation of these critical enzymes may lead to disease at various stages of life. Understanding how these variants alter protein structure and function will improve the identification of at-risk populations and precision therapies. It is with this in mind that we chose to review current research and the remaining knowledge gaps in the study of germline mutations in human steroid metabolizing AKRs: AKR1C1-4 and AKR1D1.

3.1. AKR1C1

AKR1C1 (20α(3α)-hydroxysteroid dehydrogenase) is known to play an essential role in progesterone metabolism via its 20-ketosteroid reduction activity, where it converts progesterone to the inactive 20α-hydroxyprogesterone. Knockout of the murine homolog AKR1C18 leads to a delay in parturition due to the reduced ability to metabolize progesterone to 20α-hydroxyprogesterone [20]. The NCBI SNP database reveals no nsSNPs with missense outcomes that occur with an MAF greater than 0.01. A sequence alignment of the mutations detailed below and their locations with respect to conserved catalytic, cofactor, and substrate binding sites can be found in Table 1 and Figure 2. Recently a mutation in AKR1C1 was found to be associated with lipedema, a disease of subcutaneous fat accumulation [21]. In the context of inappropriate lipid accumulation, AKR1C1 is highly expressed in fat and liver tissues. Michelini et al. [21] hypothesized that an inherited AKR1C1 loss-of-function mutation could underpin nonsyndromic primary lipedema in one family, based on a similar accumulation of progesterone that promotes lipogenesis and lipid accumulation.

Table 1. Familial inherited missense mutations in steroid transforming human AKRs.

AKR Enzyme	Missense Mutation	References	Disorder
AKR1C1	L213Q	Michelini et al., 2020 [21]	Nonsyndromic Primary Lipedema
AKR1C2	I79V	Flück et al., 2011 [22]	46,XY DSD
	H90Q		
	N300T		
	H222Q		
AKR1D1	P198L	Lemonde et al., 2003, Drury et al., 2010 [23,24]	Bile Acid Deficiency
	L106F		
	P133R	Gonzales et al., 2004, Drury et al. 2010, Chen et al., 2020 [24–26]	
	R261C	Gonzales et al., 2004, Seki et al., 2013, Drury et al., 2010 [24,25,27]	
	G223E	Ueki et al., 2009, Seki et al., 2013, Drury et al., 2010 [24,27,28]	
	R266Q	Seki et al., 2013, Chen et al., 2020 [26,27]	
	T25I	Chen et al., 2020 [26]	

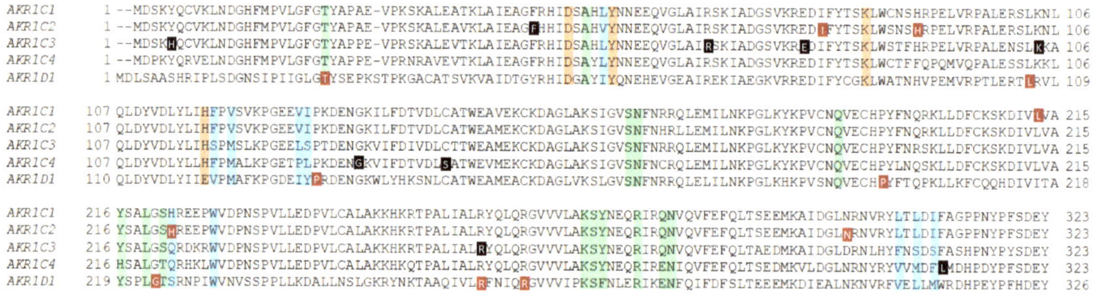

```
AKR1C1    1 --MDSKYQCVKLNDGHFMPVLGFGTYAPAE-VPKSKALEATKLAIEAGFRHIDSAHLYNNEEQVGLAIRSKIADGSVKREDIFYTSKLWCNSHRPELVRPALERSLKNL 106
AKR1C2    1 --MDSKYQCVKLNDGHFMPVLGFGTYAPAE-VPKSKALEAVKLAIEAGFHHIDSAHVYNNEEQVGLAIRSKIADGSVKREDIFYTSKLWSNSHRPELVRPALERSLKNL 106
AKR1C3    1 --MDSKHQCVKLNDGHFMPVLGFGTYAPPE-VPRSKALEVTKLAIEAGFRHIDSAHLYNNEEQVGLAIRSKIADGSVKREDIFYTSKLWSTFHRPELVRPALENSLKKA 106
AKR1C4    1 --MDPKYQRVELNDGHFMPVLGFGTYAPPE-VPRNRAVEVTKLAIEAGFRHIDSAYLYNNEEQVGLAIRSKIADGSVKREDIFYTSKLWCTFFQPQMVQPALESSLKKL 106
AKR1D1    1 MDLSAASHRIPLSDGNSIPIIGLGTYSEPKSTPKGACATSVKVAIDTGYRHIDGAYIYQNEHEVGEAIREKIAEGKVRREDIFYCGKLWATNHVPEMVRPTLERTLRVL 109

AKR1C1  107 QLDYVDLYLIHFPVSVKPGEEVIPKDENGKILFDTVDLCATWEAVEKCKDAGLAKSIGVSNFNRRQLEMILNKPGLKYKPVCNQVECHPYFNQRKLLDFCKSKDIVLVA 215
AKR1C2  107 QLDYVDLYLIHFPVSVKPGEEVIPKDENGKILFDTVDLCATWEAMEKCKDAGLAKSIGVSNFNHRLLEMILNKPGLKYKPVCNQVECHPYFNQRKLLDFCKSKDIVLVA 215
AKR1C3  107 QLDYVDLYLIHSPMSLKPGEELSPTDENGKVIFDIVDLCTTWEAMEKCKDAGLAKSIGVSNFNRRQLEMILNKPGLKYKPVCNQVECHPYFNRSKLLDFCKSKDIVLVA 215
AKR1C4  107 QLDYVDLYLLHFPMALKPGETPLPKDENGKVIFDTVDLSATWEVMEKCKDAGLAKSIGVSNFNCRQLEMILNKPGLKYKPVCNQVECHPYLNQSKLLDFCKSKDIVLVA 215
AKR1D1  110 QLDYVDLYIIEVPMAFKPGDEIYPRDENGKWLYHKSNLCATWEAMEACKDAGLVKSLGVSNFNRRQLELILNKPGLKHKPVSNQVECHPYFTQPKLLKFCQQHDIVITA 218

AKR1C1  216 YSALGSHREEPWVDPNSPVLLEDPVLCALAKKHHKRTPALIALRYQLQRGVVVLAKSYNEQRIRQNVQVFEFQLTSEEMKAIDGLNRNVRYLTLDIFAGPPNYPFSDEY 323
AKR1C2  216 YSALGSHREEPWVDPNSPVLLEDPVLCALAKKHHKRTPALIALRYQLQRGVVVLAKSYNEQRIRQNVQVFEFQLTSEEMKAIDGLNRNVRYLTLDIFAGPPNYPFSDEY 323
AKR1C3  216 YSALGSQRDKRWVDPNSPVLLEDPVLCALAKKHHKRTPALIALRYQLQRGVVVLAKSYNEQRIRQNVQVFEFQLTAEDMKAIDGLDRNLHYFNSDSFASHPNYPYSDEY 323
AKR1C4  216 HSALGTQRHKLWVDPNSPVLLEDPVLCALAKKHKQTPALIALRYQLQRGVVVLAKSYNEQRIRENIQVFEFQLTSEDMKVLDGLNRNYRYVVMDFLMDHPDYPFSDEY 323
AKR1D1  219 YSPLGTSRNPIWVNVSSPPLLKDALLNSLGKRYNKTAAQIVLRFNIQRGVVVIPKSFNLERIKENFQIFDFSLTEEEMKDIEALNKNVRFVELLMWRDHPEYPFHDEY 326
```

Figure 2. AKR sequence alignment shows the position of major nsSNPs and inherited mutations. Catalytic tetrad is shown in orange, steroid binding residues in blue, cofactor binding residues in green, nsSNPs with an MAF greater than 0.01 in black boxes, and familial mutations in red boxes.

A 638 T > A transversion that results in a one amino acid substitution of L213 to Q in AKR1C1 was identified in a family with three members afflicted by sex-limited autosomal dominant nonsyndromic lipedema. Disease occurred in heterozygous carriers of the mutation. While the mutation did not affect protein expression in carriers, the L213Q variant was determined to be a loss-of-function mutation. L213 is not located on the active site. However, it is within the core of the protein, and its mutation appears to influence structure and function. Using molecular dynamic simulations, residue L213 was shown to participate in hydrophobic interactions. Mutation to glutamine disrupts these interactions due to the new polar side chain forming new hydrogen bonds. This mutation was found to affect the solvent accessibility of substrates with the steroid and cofactor binding pockets of AKR1C1. As a result of the structural changes, both steroid substrate and cofactor are more solvent-exposed, leading to a decrease in interaction energy between steroid substrate and cofactor that likely stems from a loss of non-covalent interactions. Quantitative structure-activity relationship (QSAR) modeling was performed to predict enzymatic parameters, where the mutant reduced both turnover number and catalytic efficiency by 50%.

This is a good example of the identification of a potentially clinically relevant mutation followed by the pursuit of functional genomic analysis to determine how it affects the protein of interest. However, the structure–function relationships were only performed with computational modeling and predictive relationships to determine reaction kinetics. These predictions were only performed with one steroid (20α-hydroxyprogesterone) and one cofactor (NADP⁺). The mutation may affect only one type of reaction, whereas conversion of other steroids may not be as drastically affected. Furthermore, the studies were not performed with the preferred substrates for AKR1C1, namely NADPH and progesterone. The QSAR modeling should have been ideally performed with additional steroid substrates to increase the robustness of the model's predictions. Beyond progesterone conversion, the authors raise the idea that prostaglandin F 2alpha (PGF2α) is known to inhibit adipogenesis, whereas AKR1C1 catalyzes its synthesis [29,30]. Decreased AKR1C1 activity could release this brake on adipogenesis that is normally carried out by prostaglandins. However, we are unconvinced that the catalytic efficiency of this reaction would support this functional relationship since AKR1C3 appears to be the major PGF2a synthase [31]. The authors suggest that mutation of AKR1C1 could result in less PGF2α and stimulation of adipogenesis, which was described in their discussion. However, the group was unable to perform predictive kinetics on this reaction. These studies all point to the need to conduct kinetic analysis on each of the reactions of interest using recombinant enzymes bearing this mutation.

Lipid production is also known to be androgen-dependent in prostate cancer cells and PCOS adipocytes [32,33]. AKR1C1 inactivates the potent androgen receptor ligand DHT to 3β-androstanediol. Androgens have been shown to increase de novo lipogenesis and decrease lipid breakdown in adipocytes. Therefore, mutation of AKR1C1 would reduce DHT inactivation, allowing it to promote lipedema through androgen receptor signaling. Collectively, mutations in AKR1C1 may be able to promote lipedema through progesterone, androgen, or prostaglandin pathways. Therefore, it is important to perform additional functional genomic studies to determine how the mutation affects the reaction kinetics of progesterone conversion to 20α-hydroxyprogesterone, DHT to 3β-androstanediol, or prostaglandin E2 to PGF2α. These cases illustrate how mutations in a protein could have diverse effects that are ultimately mediated through the same pathway, from a delay in parturition in mice to dysregulated lipid accumulation in adults, both the result of decreased progesterone metabolism.

3.2. AKR1C2

AKR1C2 (type 3 3α-hydroxysteroid dehydrogenase) demonstrates preferred 3-ketosteroid reductase activity. Importantly, it inactivates the potent androgen DHT to 3α-androstanediol and converts dihydroprogesterone (DHP) to allopregnanolone, an important neurosteroid that modulates the GABA receptor. There are no k/o mice available since a murine equivalent of this enzyme does not exist [8]. The NCBI SNP database reveals one nsSNP with a missense outcome that occurs with an MAF greater than 0.01. It should be noted that several of the nsSNPs investigated by Takahashi et al. [34], detailed below, fall below the MAF threshold we used in this search. A sequence alignment of the mutations and nsSNPs is detailed below, and also their locations with respect to conserved catalytic, cofactor, and substrate binding sites can be found in Tables 1 and 2, Figure 2. The impact of mutations in AKR1C2 on hormone-dependent diseases and disorders stems primarily from its involvement in DHT metabolism. As a potent androgen receptor ligand, DHT regulates sexual differentiation in embryonic males, promotes secondary sexual characteristics in adult males, and in some cases, causes androgen-dependent disorders. Mutations in AKR1C2 that result in an under or oversupply of DHT have been associated with several disorders.

Table 2. Steroid-transforming human AKR nsSNPs with an MAF greater than 0.01, * indicates that nsSNPs with a frequency of <0.01 exist but are not included in the table.

AKR Enzyme	Missense nsSNP	NCBI Identifier	MAF
AKR1C1 *	-	-	-
AKR1C2	F46Y	rs2854482	0.0649
	H5Q	rs12529	0.4203
	K104D	rs12387	0.1518
AKR1C3	E77G	rs11551177	0.0367
	R258C	rs62621365	0.0325
	R66Q	rs35961894	0.0230
	C145S	rs3829125	0.1028
AKR1C4	L311V	rs17134592	0.1024
	G135E	rs11253043	0.0270
AKR1D1 *	-	-	-

Flück et al. described how several germline mutations in AKR1C2 underlie the dysregulation of male sexual differentiation observed in a Swiss family in 1972, beginning with the development of the fetal gonad driven by DHT [22]. Several members of a Swiss family exhibited varying degrees of under-virilization at birth, resulting in female sex assignment and 46,XY DSD [35]. Years later, Flück et al. sequenced DNA from the original family as well as one other family with similarly presenting individuals and identified a total of four germline inherited mutations in AKR1C2 that they further investigated for functional differences [22]. The authors proposed that fetal DHT synthesis proceeds through a backdoor pathway ending with the oxidation of 3α-androstanediol to DHT and that the identified mutations in AKR1C2 are the main cause of this pathway deficiency in affected individuals. They also identified a mutation in AKR1C4 that resulted in aberrant splicing, consequently eliminating its supplementary role to AKR1C2 metabolism and exacerbating the deficiency in the backdoor pathway to DHT. A kinetic analysis of wild-type and the three AKR1C2 variants identified in family one (I79V, H90Q, and N300T) showed reduced catalytic activity for two key reactions in the alternative pathway to DHT: the reduction of 5α-dihydroprogesterone (5α-DHP) to allopregnanolone, and the oxidation of 3α-androstanediol to DHT. The authors note that while AKR1C2 wild-type was able to oxidize 3α-androstanediol to DHT in vitro, it acts primarily as a 3-ketosteroid reductase in vivo based on its high affinity for NADPH. Therefore, the impact of these mutations most likely manifests in AKR1C2's decreased ability to catalyze the reduction of 5α-DHP to allopregnanolone and 17α-hydroxyprogesterone to 17α-hydroxyallopregnanolone, both of which feed into the backdoor pathway [22], see Figure 3. Another AKR1C2 mutation, H222Q, was identified in the second family. In an assay performed in COS1 cells, the H222Q variant resulted in significantly reduced DHT production compared to wild-type [22]. The authors noted that while all four mutations identified in AKR1C2 resulted in reduced catalytic activity, it was not to the degree typically associated with recessive disorders of steroidogenesis. Ultimately, a combination of mutations in AKR1C2 and AKR1C4 is necessary to stunt fetal DHT synthesis and cause disordered sexual development in these two families.

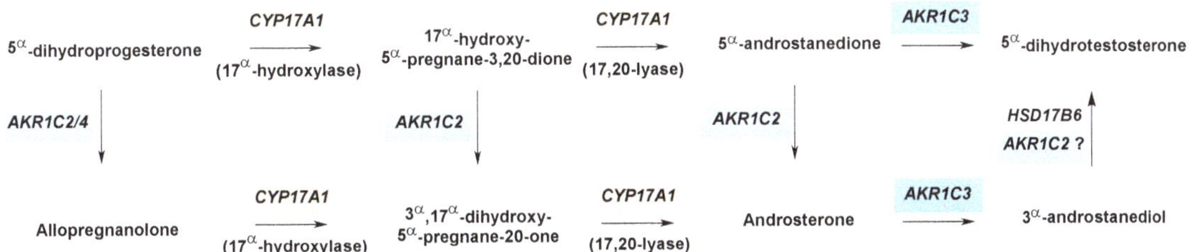

Figure 3. Role of AKR1C2 and AKR1C4 in the fetal backdoor pathway for DHT. Arrows depict reactions carried out by the indicated enzymes. The involvement of AKR1C2 is unconfirmed (indicated by a question mark) in the reaction converting 3α-androstanediol to 5α-dihydrotestosterone.

In contrast to the previous study, where AKR1C2 mutations caused an undersupply of DHT that resulted in a sex development disorder, Takahashi et al. highlight how mutations in AKR1C2 that lead to an accumulation of DHT could exacerbate PCa. DHT is a potent AR ligand and a driver of PCa. As a mechanism to regulate its occupancy on the AR, DHT is reduced to 3α-androstanediol by AKR1C2, where disruption of this shunt could lead to an accumulation of DHT that exacerbates PCa [34]. Takahashi et al. examined how five AKR1C2 nsSNPs might affect protein function and disrupt DHT synthesis in PCa using transient transfection followed by enzyme assays in Sf9 cell lysates. They found that two variants, F46Y and L172Q, had decreased V_{max} compared to wild-type in the reduction of DHT to 3α-androstanediol [34]. They also reported that three variants, L172Q, K185E, and R258C, had significantly lower apparent K_m values compared to wild-type. None of the nsSNPs resulted in protein stability differences. The authors note that the frequency of the F46Y variant in several continental-based populations corresponds with the occurrence and severity of PCa in these groups. The F46Y allele is found in 15% of African and 5.9% of European populations and is undetected in Asian populations [34]. Interestingly, this mirrors the occurrence and severity of PCa in decreasing order for these ethnic groups. The authors highlight the parallel between F46Y occurrence and PCa and suggest that this or other AKR1C2 mutations could be a contributing factor to genetic PCa risk. They proposed that an association between the occurrence of the F46Y allele and elevated DHT serum levels in PCa patients may predict the severity of the disease; however, to our knowledge, this study has not been conducted. A weakness of the Takahasi study is that, as a control, they mutated the catalytic tyrosine and found that the Y55F mutant still had one-tenth the catalytic efficiency of wild-type AKR1C2. This calls into question the reliability of kinetic data obtained using Sf9 cell lysates as opposed to those obtained using purified recombinant mutant proteins.

In a follow-up study, Arthur et al. [36] used homology modeling to predict and analyze the structural impact of the mutations investigated by the Takahashi group. With respect to the F46Y variant, they report that the substitution of tyrosine introduces a polar residue into a previously hydrophobic environment. This tyrosine is predicted to form a hydrogen bond with a water molecule that could change the local environment and destabilize the hydrogen bond that normally forms nearby between N280 and the cofactor [36]. The authors suggest that this destabilization in cofactor binding could account for a reduced maximum velocity rather than F46Y directly disrupting the interaction with substrate DHT. They also report that the L172Q substitution might result in a similar disruption of hydrogen bonding between N167 and cofactor that leads to a decreased V_{max} reported by Takahashi [35]. Together, these studies suggest that the F46Y nsSNP decreases AKR1C2's catalytic ability due to disrupted cofactor binding, which could result in an accumulation of DHT. However, in the absence of a more rigorous kinetic analysis of the mutant proteins and a clinical study to confirm the impact of this finding in PCa patients, this example serves to highlight how a more comprehensive investigation spanning several disciplines is necessary to provide a complete story of the role of these nsSNPs in disease.

3.3. AKR1C3

AKR1C3 (type 2 3α(17β)-hydroxysteroid dehydrogenase) is the only steroid metabolizing AKR that preferentially displays 17-ketosteroid reductase activity that can convert Δ^4-AD to T and 11-oxo-Δ^4-AD to 11-keto-T [19]. It is also known as prostaglandin F2α synthase and oxidizes PAH-*trans*-dihydrodiols [31]. There are no k/o mice available since a murine equivalent of this enzyme does not exist [8]. The NCBI SNP database reveals five nsSNPs with missense outcomes that occur with an MAF greater than 0.01. A sequence alignment of the nsSNPs detailed below and their locations with respect to conserved catalytic, cofactor, and substrate binding sites can be found in Table 2, Figure 2.

AKR1C3 drives prostate cancer and other hormone-dependent disorders, e.g., PCOS, where coding nsSNPs are thought to further modify their role in various aspects of the disease [37]. Until recently, most studies that implicated AKR1C3 nsSNPs were epidemiological and lacked functional genomic experiments to identify any change in protein function that could explain the connection to the disease. The AKR1C3 nsSNP rs12529, which corresponds to the H5Q mutation, has been flagged by many groups without further investigation. However, recent work from our group showed no major differences between AKR1C3 wild-type and the top four most frequently occurring variants, calling into question the weight of these associations.

Because of the role of AKR1C3 in the peripheral synthesis of T, there are numerous epidemiological studies that associate AKR1C3 nsSNPs, mainly H5Q, with PCa detection, prognosis, and treatment effectiveness. One study conducted in a New Zealand PCa cohort by Karunasinghe et al. suggested that the presence of the Q5 mutation in combination with smoking is associated with unusually low serum PSA levels that belie the severity of disease and lead to late detection by current prostate-specific antigen (PSA) diagnostic benchmarks [38]. The same group suggested that the Q5 mutation and smoking are associated with an increased age at which PCa is diagnosed by PSA serum levels [39]. Karunasinghe et al. also report that in patients receiving androgen deprivation therapy (ADT), the Q5 variant is associated with increased hormone treatment-related symptoms [40]. They proposed that these adverse drug effects could be avoided if individuals were genotyped for the Q5 variant to achieve precise treatment monitoring.

Three groups have associated AKR1C3 nsSNPs with deviations in serum T levels. Shiota et al. associated the presence of Q at position 5 with higher serum T levels in patients receiving ADT [41]. Inversely, the presence of H at position 5 is associated with a better prognosis. Relatedly, Jakobsson et al. described that in healthy Swedish subjects, the E77G variant is associated with lower serum T levels. However, they found no significant difference in enzyme activity of this variant compared to wild-type [42]. In the third study, Ju et al. report that Q5 is associated with increased serum T levels in a cohort of Chinese women with PCOS, which may indicate that the variant contributes to hyperandrogenism [43]. Notably, these three studies predict that mutations in AKR1C3 may affect T biosynthesis without performing any supporting biochemical characterization of the mutant proteins. Relatedly, several groups highlight AKR1C3 nsSNPs that might play a role in non-lethal disorders in men. Roberts et al. found that the Q5 mutation is associated with decreased risk of prostate enlargement in benign prostate hyperplasia [44]. Additionally, Soderhall et al. report the identification of a novel AKR1C3 nsSNP resulting in A215T substitution that is unique to a boy with penile hypospadias compared to a control group [45].

Studies implicating AKR1C3 nsSNPs extend beyond T production and PCa. The effects of AKR1C3 nsSNPs in the metabolic activation of PAHs and other carcinogens have also been considered. A preliminary study from one group suggests that the Q5 variant could be involved in the molecular pathogenesis of urinary bladder cancer [46]. In contrast, Figueroa et al. report an inverse association of Q5 with bladder cancer risk with no connection to smoking in a Caucasian population [47]. Interestingly, Lan et al. report the Q5 variant to be associated with a significant risk of developing lung cancer in a Chinese population exposed to high levels of PAH-rich coal combustions from cooking and heating [48]. Together, these studies highlight how AKR1C3 nsSNPs might modify their contribution to the production of ultimate carcinogens that can lead to cancer.

Given the large number of reports relating AKR1C3 variants with various positive and negative aspects of the disease, there are disproportionately fewer functional genomic studies examining how Q5 and other nsSNPs affect AKR1C3 function. To our knowledge, there are only two studies that biochemically characterize AKR1C3 nsSNPs. The first analysis, conducted by Platt et al., evaluated wild-type and five AKR1C3 variants in their ability to metabolize exemestane, an aromatase inhibitor used to treat breast cancer, to 17β-dihydroexemestane [49]. Their findings indicate that H5Q, E77G, K104D, P180S, and R258C are 17-250-fold less catalytically active compared to wild-type in this reaction, which could significantly affect exemestane metabolism in breast cancer patients with different variants and warrant the consideration of AKR1C3 genotype in treatment protocols [49]. However, in our own work, where we evaluated wild-type and the top four most frequently occurring variants (H5Q, K104D, E77G, and R258C), we found no significant kinetic differences in the ability of the variants to metabolize Δ^4-AD to T, progesterone to 20α-hydroxyprogesterone, or exemestane to 17β-dihydroexemestane compared to wild-type [10]. Additionally, while the K104D variant was less stable than WT, the presence of cofactors NAD(P)$^+$ diminished this effect. In contrast to associative and biochemical studies that implicate H5/Q5 and other AKR1C3 nsSNPs in disease, our findings indicate that none of the variants we examined have significant differences that would likely manifest in patient prognosis.

While several of the associative studies discussed above convey contradictory impacts of the H5Q variant on disease prognosis, the sizable number of reports that flag this variant makes it difficult to excuse these associations completely. However, our own work is in opposition to the idea that H5Q is an influential variant, as we found no significant impact of this nsSNP or any other on AKR1C3 protein function or stability. H5Q is the most commonly occurring AKR1C3 variant with a global MAF of 0.42 [9,10]. However, in certain continental populations, the distribution is reversed, and Q5 is the major allele. This raises the concept that depending on the population observed in a given study, Q5 could be the major allele in contrast to the global MAF. This could lead to a distorted enrichment of the Q5 variant in certain populations. Furthermore, other factors, such as the upregulation of AKR1C3 via ADT or other modifications of its expression that occur in PCa and other disease states, could obscure the influence of the H5Q variant and account for the widespread identification of this variant in associative studies.

3.4. AKR1C4

AKR1C4 (type 1 3α-hydroxysteroid dehydrogenase) is predominantly expressed in the liver, where it displays 3-ketosteroid reductase activity and is responsible for making 3α-hydroxysteroids. However, alteration of AKR1C4 has been associated with mood disorders. Progesterone is converted into an intermediate (5α-dihydroprogesterone) by SRD5A1, which is then available for conversion by AKR1C4 into allopregnanolone, which is implicated in negative mood changes due to dysregulation of GABAergic signaling of glutamatergic neurons [50]. However, it is thought that the main AKR involved in central nervous system (CNS) regulation via allopregnanolone metabolism is AKR1C2 [51]. There are no k/o mice available since a murine equivalent of this enzyme does not exist [8]. The NCBI SNP database reveals six nsSNPs with missense outcomes that occur with an MAF greater than 0.01. A sequence alignment of the mutations and nsSNPs detailed below and their locations with respect to conserved catalytic, cofactor, and substrate binding sites can be found in Table 2, Figure 2.

One study identified the C145S AKR1C4 nsSNP due to a C to G transversion in a purely associative study [52]. The majority of patients in this study had type 1 bipolar disorder and exhibited an irritable mood during mania/hypomania based on affective disorders evaluation (ADE). The SNP was associated with increased manic or hypomanic irritability, where men with the SNP were 5.44 times more likely to experience manic or hypomanic irritability compared to those without the SNP. This effect was not seen in women. Paradoxically, as increased irritability correlated with the C145S variant in men, there was a corresponding decrease in serum progesterone levels. The authors speculated why this could occur but did not follow up with any mechanistic studies. The same group published follow-up studies in another population of men and women with bipolar disorder [53]. There was a correlation between men with paranoid ideation and both DHEA-S and progesterone levels, where the mean levels of both steroids were lower in men with paranoid ideation compared to those without. In terms of the C145S variant in AKR1C4, women had a reduced likelihood of exhibiting paranoid ideation, indicating that the mutation may have the reverse outcome and yield a protective effect in women. However, in order to determine whether the C145S mutation causes a change in AKR1C4 catalytic activity that reflects decreases in DHEA-S and progesterone levels, it would be necessary to conduct a functional genomic analysis of the protein. It is also uncertain whether these changes would be related to changes in systemic steroid metabolism in the liver or CNS. For example, the NCBI database only shows transcript expression in the liver and gall bladder.

AKR1C4 was also linked to breast cancer in a population of postmenopausal women who were receiving estrogen or combined estrogen and progesterone therapy [54]. The group observed that carriers who were heterozygous or homozygous for L311V SNP correlated with a 16.7 and 29.3% increase, respectively, in mammographic percentage density (MPD), a risk factor for breast cancer in women who were receiving combined estrogen and progesterone as hormone replacement therapy. However, the sample size was small for these groups, with an N of seven and one, respectively, so these findings should be considered with caution. It was previously reported that L311V does have a 66–80% decrease in AKR1C4 catalytic activity, so this residue may be important for substrate binding since it resides in the C-terminal loop [55]. Since functional data about the mutant was already known, the group was able to relate this SNP to the MPD risk factor in postmenopausal women who were receiving combined estrogen and progesterone replacement therapy.

3.5. AKR1D1

AKR1D1 (Δ^4-3-oxosteroid-5β-reductase) is a key enzyme for bile acid synthesis, and disruption of this critical pathway leads to bile acid deficiency. AKR1D1 reduces Δ^4-cholesten-7-ol-3-one and Δ^4-cholesten-7,12-diol-3-one to their respective 5β-dihydrosteroid forms. AKR1D1 works sequentially with AKR1C4 to produce the 3α,5β-configuration in the A-ring of the steroid, which is essential for the proper emulsification of fats, Figure 4. Interestingly, steroid 5b-reductase k/o mice retained some ability to synthesize bile acids. Still, the bile acid levels were reduced, and composition differed in males and females, where the former had significantly reduced 12a-hydroxylated bile acids [56]. If undetected in humans, bile acid deficiency can be a fatal disorder in the neonate due to the resulting inability to emulsify fat and absorb fat-soluble vitamins. In addition, these mutations lead to diversion in the metabolism of 7α-hydroxy-4-cholesten-3-one and 7α,12α-dihydroxy-4-cholesten-3-one to the 5α-reduced (allo)-bile acids which are hepatotoxic. Historically the treatment of this deficiency focused on relieving symptoms rather than identifying the underlying genetic causes of the disorder, leading to the possibility that the frequency of this genetic disorder may be underestimated. A sequence alignment of the mutations detailed below and their locations with respect to conserved catalytic, cofactor, and substrate binding sites can be found in Table 1 and Figure 2.

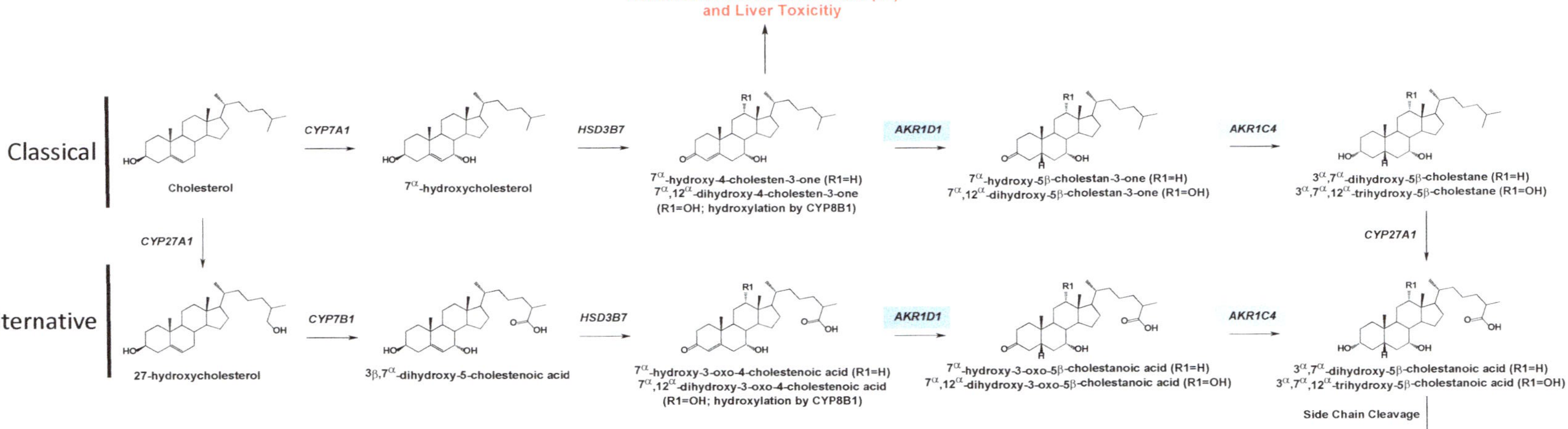

Figure 4. Role of AKR1C4 and AKR1D1 in Classical (neutral) and Alternative (acidic) Bile Acid Metabolism to regulate farnesoid X receptor (FXR). The FXR is required to regulate bile acid homeostasis. 5β-reductase deficiency, such as mutation of AKR1D1, shunts the pathway away from proper bile acid metabolism and leads to the accumulation of allo-bile acids resulting in liver toxicity.

There are several mutations in AKR1D1 associated with bile acid deficiency. Lemonde identified two transitions in infants with cholestatic liver disease [23]. One patient was homozygous for a 662 C > T transition that resulted in a P198L substitution, while their parents were both heterozygous for the mutation. This mutation was not found in a control population of 100 chromosomes. The group also identified a second mutation (385 C > T), resulting in an L106F substitution where the patient was similarly homozygous for the mutation and had parents that were both heterozygous, Table 1.

Gonzales et al. described a case report where a pair of 8-month-old sisters with progressive cholestasis and liver failure were compound heterozygous for two missense mutations in AKR1D1: P133R (467 C > G) in exon 4 and R261C (850 C > T) in exon 7 [25]. Each parent was heterozygous for one of the mutations, and it was confirmed that both girls had two mutated alleles. Two patients with chronic cholestasis had a heterozygous mutation resulting in a G223E (737 G > A) amino acid substitution in exon 6 [28]. Finally, another case report identified two infants with neonatal cholestasis [27]. One was heterozygous for a novel R266Q mutation, which was detected in the heterozygous mother but not in the father. The second patient was compound heterozygous for the G223E and R261C mutations. G223E was detected as heterozygous in the patient's father and absent in the mother, while R261C was heterozygous in the patient's mother and absent in the father. Recently, another group identified more inborn errors of bile acid metabolism in three infants [26]. One patient had a compound mutation consisting of R50X, where X indicates an early stop codon and the R266Q mutation. The second had a chromosomal mutation (74 C > T), resulting in a T25I mutation. These studies present the connection between AKR1D1 deficiency and its effect on bile acid metabolism and the development of cholestasis. However, it is necessary to evaluate relevant biochemical studies on the structure–function relationship to determine why these mutations result in bile acid deficiency.

Our group conducted an extensive functional genomic analysis of how these mutations affect AKR1D1 structure and function. The point mutations L106F, P133R, P198L, G223E, and R261C were tested to determine how they affect AKR1D1 [24]. All mutations are highly conserved across AKR1D1 homologs in other mammalian species except for P133R. None of the mutations were in direct contact with the catalytic tetrad, cofactor, or substrate binding sites. Interestingly, only P133R could be successfully purified while the other four accumulated in inclusion bodies indicating protein aggregation, misfolding, or instability. G223E degraded within 24 h in cycloheximide pulse-chase experiments conducted in transfected HEK293 cells, while L106F and R261C were poorly expressed and degraded within 6 h. 5β-Reduction of T was assessed for each mutant, and very low activity was seen with L106F and R261C within 24 h, which is consistent with their poor expression. Additionally, no conversion was observed with the G223E or P198L mutations over 60 h. However, as reaction times increased, background conversion of T to other androgens occurred, making it difficult to quantify residual 5β-reduction of T. Protein expression was measured for each of the five mutations, revealing that L106F, R261, and G223E all had reduced stability. P198L remained in cells for up to 24 h; however, the mutation may still affect its enzymatic activity or expression since it could not be successfully purified.

The P133R mutation was the only recombinant enzyme to be successfully purified, and reactions could be conducted to determine steady-state kinetic parameters. When using T as a substrate, the K_m increased from 2.7 to 12.7 μM, and the k_{cat} decreased from 7.1 to 2.7, resulting in an over 10-fold decrease in the catalytic efficiency. When using cortisone as a substrate, the K_m decreased from 15.1 to 1.3 μM. However, the k_{cat} also decreased from 9.9 to 0.6, ultimately resulting in only a small reduction in the catalytic efficiency. This illustrates how kinetic parameters for two different substrates could be drastically different for the same mutation in an enzyme. The substrate of interest Δ^4-cholesten-7α-ol-3-one has a longer C17 side chain, similar to that of cortisone. However, the K_m for this substrate could not be accurately measured due to saturation at the lowest substrate concentration. At saturation, K_m was significantly lower than 0.8 μM (the value for wild-type), while the k_{cat} was reduced 7-fold. These differences were more like those observed with cortisone and reframe AKR1D1 from a low affinity, high-capacity enzyme to a high affinity, low-capacity enzyme in this scenario. This could indicate that AKR1D1 binds tightly to bile acid substrates resulting in insufficient turnover to the 5β-reduced bile acid precursors for proper emulsification, which could account for bile acid deficiency from this mutant.

Thermal stability studies showed that 50% of mutant enzymatic activity was lost at 42 °C and less than 5–10% remained at 46.5 °C, while >60% of the enzymatic activity of wild-type remained. This indicated that the instability of the mutants is exacerbated at temperatures above 37 °C. Thus, while the mutants are less stable than the wild-type enzyme, this observation is not relevant at physiological temperatures, and it is thought that differences in kinetic parameters may be more important.

Transient kinetics were performed with the P133R mutant, where it was shown to result in a 40-fold increase in the K_d value for NADPH and an increased rate of release of NADP$^+$ by two orders of magnitude compared to wild-type. The reduced affinity for the cofactor suggests that the enzyme exists in a cofactor-free form. Impaired NADPH binding and hydride transfer were found to be the molecular basis for bile acid deficiency in patients with the P133R mutation [57].

Here we see that biochemical studies become critical to understand how a point mutation may affect structure and function. If only evaluated by computational methods, these mutations may have been overlooked as they are largely not near the catalytic tetrad, substrate, or cofactor binding sites. However, the biochemical studies brought to light why these mutations may lead to bile acid deficiency due to disruption of AKR1D1 expression/stability or catalytic activity. However, each point mutation can have a different effect on enzyme function, and the same mutation, P133R, can have different effects on different steroid substrates that can only be revealed by transient kinetic studies. Molecular dynamic studies could have been useful in understanding how these nsSNPs may directly affect protein stability, as we saw with the AKR1C1 analysis. The biochemical studies support the association of the AKR1D1 mutants with cholestasis/liver failure and bile acid metabolism deficiency. It was suspected that AKR1D1 might be the offender; however, the biochemical studies aid in definitively explaining why these nsSNPs alter the structure and function of AKR1D1.

4. Conclusions

AKR steroid metabolizing enzymes are critical players in proper physiological growth and development. Germline mutations in these enzymes can disrupt androgen, progesterone, and bile acid metabolism and lead to debilitating pathological disease states that greatly diminish the quality of life and could ultimately lead to death, especially in young individuals born with these variants. These coding-region point mutations ultimately alter the proper expression levels and function of these key metabolizing enzymes. In many cases, epidemiological studies link a variant to a specific disease state; however, the functional genomic studies to support the association and understand the structure and functional impact of the mutation are often incomplete.

Most often, a patient is treated for a phenotype. However, there may be different underlying dysfunctions that lead to the same disorder. This is evident in the examples highlighted in the introduction, where mutations in *SRD5A2*, *HSD17B3*, and *HSD3B2* all lead to some form of disordered sexual development. Only with a close examination of which key steroid metabolites were in excess in patients in combination with sequencing can these disorders be teased apart. In rare cases, when the diagnosis is based on a phenotype without investigation of the underlying cause of the disease, a patient could even be misdiagnosed. This is where precision medicine is helpful in recognizing that patients with the same categorical disease may need different types of treatment. Additionally, mutations in a protein should be considered in the greater context of the entire metabolic pathway of which the enzyme is a part. This is especially true in the case of human AKRs, which typically catalyze several reactions in alternative routes to the same end steroid. If one portion of the pathway is shunted due to a mutation that hinders the biological activity of the enzyme, this may drive the pathway through another arm. Disruption of this metabolic pathway could even lead to the formation of novel steroids, making it important to study the steroid metabolome in the context of a mutation to understand how the disease state arises. The accumulation of SNPs in the same pathway should also be considered, as the combination of mutations may have debilitating cumulative effects.

When examining the impact of a SNP, it could also be important to consider how a mutation may precipitate disease in men and women differently. As can be seen in the proteins highlighted in the introduction, certain deficiencies cause severe phenotypes in one gender over the other due to the nature of steroid signaling. This is also evident in the case of AKR1C3, where T production can affect men and women differently through PCa and PCOS, respectively. Therefore, mutations may have different outcomes depending on the role of the enzyme's target product in different sex-dependent developmental and reproductive processes.

With the recent additions to large genome databases such as the UK biobank and increasing accessibility to analytical tools like AlphaFold, there are new opportunities to unearth connections between AKR SNPs and disease. It will be essential to draw from both associative and biochemical studies in order to uncover the full story of a mutation and evaluate how this knowledge may be incorporated into medical intervention. As is evident from the diversity of research groups and the expanse of time in some of the more developed stories, this work often requires perseverance by many different groups with complementary expertise and resources. With this perspective, we might consider some of these narratives pending rather than permanently incomplete.

Author Contributions: A.J.D. and R.D.P. reviewed the literature and wrote drafts of the manuscript, and approved the submitted version. T.M.P. edited all drafts of the manuscript, provided intellectual guidance, obtained funding for the work, and approved the submitted version. All authors have read and agreed to the published version of the manuscript.

Funding: This work was supported in part by the following grants from the National Institutes of Health P30 ES013508 (T.M.P.), T32-ES0199851 (R.D.P.), and T32-GM133398 (A.J.D.).

Institutional Review Board Statement: Not applicable.

Informed Consent Statement: Not applicable.

Data Availability Statement: No new data were created or analyzed in this study. Data sharing is not applicable to this article.

Conflicts of Interest: T.M.P. is the Founder of Pennzymes, is a consultant for Propella Therapeutics and Sage Pharmaceuticals, and is a member of the Expert Panel for Research on Fragrance Materials. A.J.D. and R.D.P. declare no conflict of interest.

References

1. Hu, J.; Zhang, Z.; Shen, W.J.; Azhar, S. Cellular cholesterol delivery, intracellular processing and utilization for biosynthesis of steroid hormones. *Nutr. Metab.* **2010**, *7*, 47. [CrossRef] [PubMed]
2. Batista, R.L.; Mendonca, B.B. The Molecular Basis of 5alpha-Reductase Type 2 Deficiency. *Sex. Dev.* **2022**, *16*, 171–183. [CrossRef] [PubMed]
3. George, M.M.; New, M.I.; Ten, S.; Sultan, C.; Bhangoo, A. The clinical and molecular heterogeneity of 17betaHSD-3 enzyme deficiency. *Horm. Res. Paediatr.* **2010**, *74*, 229–240. [CrossRef] [PubMed]
4. Al Alawi, A.M.; Nordenstrom, A.; Falhammar, H. Clinical perspectives in congenital adrenal hyperplasia due to 3beta-hydroxysteroid dehydrogenase type 2 deficiency. *Endocrine* **2019**, *63*, 407–421. [CrossRef]
5. Sahakitrungruang, T. Clinical and molecular review of atypical congenital adrenal hyperplasia. *Ann. Pediatr. Endocrinol. Metab.* **2015**, *20*, 1–7. [CrossRef]
6. Chang, K.H.; Li, R.; Kuri, B.; Lotan, Y.; Roehrborn, C.G.; Liu, J.; Vessella, R.; Nelson, P.S.; Kapur, P.; Guo, X.; et al. A gain-of-function mutation in DHT synthesis in castration-resistant prostate cancer. *Cell* **2013**, *154*, 1074–1084. [CrossRef]
7. Penning, T.M. The Aldo-Keto Reductases (AKRs): Overview. *Chem. Biol. Interact.* **2015**, *176*, 139–148. [CrossRef]
8. Velica, P.; Davies, N.J.; Rocha, P.P.; Schrewe, H.; Ride, J.P.; Bunce, C.M. Lack of functional and expression homology between human and mouse aldo-keto reductase 1C enzymes: Implications for modelling human cancers. *Mol. Cancer* **2009**, *8*, 121. [CrossRef]

9. Clarke, L.; Zheng-Bradley, X.; Smith, R.; Kulesha, E.; Xiao, C.; Toneva, I.; Vaughan, B.; Preuss, D.; Leinonen, R.; Shumway, M.; et al. The 1000 Genomes Project: Data management and community access. *Nat. Methods* **2012**, *9*, 459–462. [CrossRef]

10. Detlefsen, A.J.; Wangtrakuldee, P.; Penning, T.M. Characterization of the major single nucleotide polymorphic variants of aldo-keto reductase 1C3 (type 5 17beta-hydroxysteroid dehydrogenase). *J. Steroid Biochem. Mol. Biol.* **2022**, *221*, 106121. [CrossRef]

11. Halldorsson, B.V.; Eggertsson, H.P.; Moore, K.H.S.; Hauswedell, H.; Eiriksson, O.; Ulfarsson, M.O.; Palsson, G.; Hardarson, M.T.; Oddsson, A.; Jensson, B.O.; et al. The sequences of 150,119 genomes in the UK Biobank. *Nature* **2022**, *607*, 732–740. [CrossRef] [PubMed]

12. Rosenthal, E.T. Second Stand Up To Cancer Prostate Cancer Dream Team. *Oncology Times* **2012**, *34*, 38. [CrossRef]

13. Cancer Genome Atlas Research, N.; Weinstein, J.N.; Collisson, E.A.; Mills, G.B.; Shaw, K.R.; Ozenberger, B.A.; Ellrott, K.; Shmulevich, I.; Sander, C.; Stuart, J.M. The Cancer Genome Atlas Pan-Cancer analysis project. *Nat. Genet.* **2013**, *45*, 1113–1120.

14. Adzhubei, I.; Jordan, D.M.; Sunyaev, S.R. Predicting functional effect of human missense mutations using PolyPhen-2. *Curr. Protoc. Hum. Genet.* **2013**, *76*, 7–20. [CrossRef]

15. Sim, N.L.; Kumar, P.; Hu, J.; Henikoff, S.; Schneider, G.; Ng, P.C. SIFT web server: Predicting effects of amino acid substitutions on proteins. *Nucleic Acids Res.* **2012**, *40*, W452–W457. [CrossRef]

16. Jumper, J.; Evans, R.; Pritzel, A.; Green, T.; Figurnov, M.; Ronneberger, O.; Tunyasuvunakool, K.; Bates, R.; Zidek, A.; Potapenko, A.; et al. Highly accurate protein structure prediction with AlphaFold. *Nature* **2021**, *596*, 583–589. [CrossRef]

17. Cooper, W.C.; Jin, Y.; Penning, T.M. Elucidation of a complete kinetic mechanism for a mammalian hydroxysteroid dehydrogenase (HSD) and identification of all enzyme forms on the reaction coordinate: The example of rat liver 3alpha-HSD (AKR1C9). *J. Biol. Chem.* **2007**, *282*, 33484–33493. [CrossRef]

18. Penning, T.M.; Wangtrakuldee, P.; Auchus, R.J. Structural and Functional Biology of Aldo-Keto Reductase Steroid-Transforming Enzymes. *Endocr. Rev.* **2019**, *40*, 447–475. [CrossRef]

19. Paulukinas, R.D.; Mesaros, C.A.; Penning, T.M. Conversion of Classical and 11-Oxygenated Androgens by Insulin-Induced AKR1C3 in a Model of Human PCOS Adipocytes. *Endocrinology* **2022**, *163*, bqac068. [CrossRef]

20. Piekorz, R.P.; Gingras, S.; Hoffmeyer, A.; Ihle, J.N.; Weinstein, Y. Regulation of progesterone levels during pregnancy and parturition by signal transducer and activator of transcription 5 and 20alpha-hydroxysteroid dehydrogenase. *Mol. Endocrinol.* **2005**, *19*, 431–440. [CrossRef]

21. Michelini, S.; Chiurazzi, P.; Marino, V.; Dell'Orco, D.; Manara, E.; Baglivo, M.; Fiorentino, A.; Maltese, P.E.; Pinelli, M.; Herbst, K.L.; et al. Aldo-Keto Reductase 1C1 (AKR1C1) as the First Mutated Gene in a Family with Nonsyndromic Primary Lipedema. *Int. J. Mol. Sci.* **2020**, *21*, 6264. [CrossRef] [PubMed]

22. Flück, C.E.; Meyer-Boni, M.; Pandey, A.V.; Kempna, P.; Miller, W.L.; Schoenle, E.J.; Biason-Lauber, A. Why boys will be boys: Two pathways of fetal testicular androgen biosynthesis are needed for male sexual differentiation. *Am. J. Hum. Genet.* **2011**, *89*, 201–218. [CrossRef]

23. Lemonde, H.A.; Custard, E.J.; Bouquet, J.; Duran, M.; Overmars, H.; Scambler, P.J.; Clayton, P.T. Mutations in SRD5B1 (AKR1D1), the gene encoding delta(4)-3-oxosteroid 5beta-reductase, in hepatitis and liver failure in infancy. *Gut* **2003**, *52*, 1494–1499. [CrossRef]

24. Drury, J.E.; Mindnich, R.; Penning, T.M. Characterization of disease-related 5beta-reductase (AKR1D1) mutations reveals their potential to cause bile acid deficiency. *J. Biol. Chem.* **2010**, *285*, 24529–24537. [CrossRef] [PubMed]

25. Gonzales, E.; Cresteil, D.; Baussan, C.; Dabadie, A.; Gerhardt, M.F.; Jacquemin, E. SRD5B1 (AKR1D1) gene analysis in delta(4)-3-oxosteroid 5beta-reductase deficiency: Evidence for primary genetic defect. *J. Hepatol.* **2004**, *40*, 716–718. [CrossRef] [PubMed]

26. Chen, J.Y.; Wu, J.F.; Kimura, A.; Nittono, H.; Liou, B.Y.; Lee, C.S.; Chen, H.S.; Chiu, Y.C.; Ni, Y.H.; Peng, S.S.; et al. AKR1D1 and CYP7B1 mutations in patients with inborn errors of bile acid metabolism: Possibly underdiagnosed diseases. *Pediatr. Neonatol.* **2020**, *61*, 75–83. [CrossRef]

27. Seki, Y.; Mizuochi, T.; Kimura, A.; Takahashi, T.; Ohtake, A.; Hayashi, S.; Morimura, T.; Ohno, Y.; Hoshina, T.; Ihara, K.; et al. Two neonatal cholestasis patients with mutations in the SRD5B1 (AKR1D1) gene: Diagnosis and bile acid profiles during chenodeoxycholic acid treatment. *J. Inherit. Metab Dis.* **2013**, *36*, 565–573. [CrossRef]

28. Ueki, I.; Kimura, A.; Chen, H.L.; Yorifuji, T.; Mori, J.; Itoh, S.; Maruyama, K.; Ishige, T.; Takei, H.; Nittono, H.; et al. SRD5B1 gene analysis needed for the accurate diagnosis of primary 3-oxo-Delta4-steroid 5beta-reductase deficiency. *J. Gastroenterol. Hepatol.* **2009**, *24*, 776–785. [CrossRef]

29. Taketani, Y.; Yamagishi, R.; Fujishiro, T.; Igarashi, M.; Sakata, R.; Aihara, M. Activation of the prostanoid FP receptor inhibits adipogenesis leading to deepening of the upper eyelid sulcus in prostaglandin-associated periorbitopathy. *Invest. Ophthalmol. Vis. Sci.* **2014**, *55*, 1269–1276. [CrossRef]

30. Lepak, N.M.; Serrero, G. Prostaglandin F2 alpha stimulates transforming growth factor-alpha expression in adipocyte precursors. *Endocrinology* **1995**, *136*, 3222–3229. [CrossRef]

31. Komoto, J.; Yamada, T.; Watanabe, K.; Takusagawa, F. Crystal structure of human prostaglandin F synthase (AKR1C3). *Biochemistry* **2004**, *43*, 2188–2198. [CrossRef] [PubMed]

32. O'Reilly, M.W.; Kempegowda, P.; Walsh, M.; Taylor, A.E.; Manolopoulos, K.N.; Allwood, J.W.; Semple, R.K.; Hebenstreit, D.; Dunn, W.B.; Tomlinson, J.W.; et al. AKR1C3-Mediated Adipose Androgen Generation Drives Lipotoxicity in Women With Polycystic Ovary Syndrome. *J. Clin. Endocrinol. Metab.* **2017**, *102*, 3327–3339. [CrossRef] [PubMed]

33. Swinnen, J.V.; Esquenet, M.; Goossens, K.; Heyns, W.; Verhoeven, G. Androgens stimulate fatty acid synthase in the human prostate cancer cell line LNCaP. *Cancer Res.* **1997**, *57*, 1086–1090.
34. Takahashi, R.H.; Grigliatti, T.A.; Reid, R.E.; Riggs, K.W. The effect of allelic variation in aldo-keto reductase 1C2 on the in vitro metabolism of dihydrotestosterone. *J. Pharmacol. Exp. Ther.* **2009**, *329*, 1032–1039. [CrossRef]
35. Zachmann, M.; Vollmin, J.A.; Hamilton, W.; Prader, A. Steroid 17,20-desmolase deficiency: A new cause of male pseudo-hermaphroditism. *Clin. Endocrinol.* **1972**, *1*, 369–385. [CrossRef] [PubMed]
36. Arthur, J.W.; Reichardt, J.K. Modeling single nucleotide polymorphisms in the human AKR1C1 and AKR1C2 genes: Implications for functional and genotyping analyses. *PLoS ONE* **2010**, *5*, e15604. [CrossRef]
37. Penning, T.M.; Detlefsen, A.J. Intracrinology-revisited and prostate cancer. *J. Steroid Biochem. Mol. Biol.* **2020**, *196*, 105499. [CrossRef]
38. Karunasinghe, N.; Ambs, S.; Wang, A.; Tang, W.; Zhu, S.; Dorsey, T.H.; Goudie, M.; Masters, J.G.; Ferguson, L.R. Influence of lifestyle and genetic variants in the aldo-keto reductase 1C3 rs12529 polymorphism in high-risk prostate cancer detection variability assessed between US and New Zealand cohorts. *PLoS ONE* **2018**, *13*, e0199122. [CrossRef]
39. Karunasinghe, N.; Symes, E.; Gamage, A.; Wang, A.; Murray, P.; Zhu, S.; Goudie, M.; Masters, J.; Ferguson, L.R. Interaction between leukocyte aldo-keto reductase 1C3 activity, genotypes, biological, lifestyle and clinical features in a prostate cancer cohort from New Zealand. *PLoS ONE* **2019**, *14*, e0217373. [CrossRef]
40. Karunasinghe, N.; Zhu, Y.; Han, D.Y.; Lange, K.; Zhu, S.; Wang, A.; Ellett, S.; Masters, J.; Goudie, M.; Keogh, J.; et al. Quality of life effects of androgen deprivation therapy in a prostate cancer cohort in New Zealand: Can we minimize effects using a stratification based on the aldo-keto reductase family 1, member C3 rs12529 gene polymorphism? *BMC Urol.* **2016**, *16*, 48. [CrossRef]
41. Shiota, M.; Endo, S.; Fujimoto, N.; Tsukahara, S.; Ushijima, M.; Kashiwagi, E.; Takeuchi, A.; Inokuchi, J.; Uchiumi, T.; Eto, M. Polymorphisms in androgen metabolism genes with serum testosterone levels and prognosis in androgen-deprivation therapy. *Urol. Oncol.* **2020**, *38*, 849.e11–849.e18. [CrossRef] [PubMed]
42. Jakobsson, J.; Palonek, E.; Lorentzon, M.; Ohlsson, C.; Rane, A.; Ekstrom, L. A novel polymorphism in the 17beta-hydroxysteroid dehydrogenase type 5 (aldo-keto reductase 1C3) gene is associated with lower serum testosterone levels in caucasian men. *Pharmacogenomics J.* **2007**, *7*, 282–289. [CrossRef] [PubMed]
43. Ju, R.; Wu, W.; Fei, J.; Qin, Y.; Tang, Q.; Wu, D.; Xia, Y.; Wu, J.; Wang, X. Association analysis between the polymorphisms of HSD17B5 and HSD17B6 and risk of polycystic ovary syndrome in Chinese population. *Eur. J. Endocrinol.* **2015**, *172*, 227–233. [CrossRef] [PubMed]
44. Roberts, R.O.; Bergstralh, E.J.; Farmer, S.A.; Jacobson, D.J.; Hebbring, S.J.; Cunningham, J.M.; Thibodeau, S.N.; Lieber, M.M.; Jacobsen, S.J. Polymorphisms in genes involved in sex hormone metabolism may increase risk of benign prostatic hyperplasia. *Prostate* **2006**, *66*, 392–404. [CrossRef] [PubMed]
45. Soderhall, C.; Korberg, I.B.; Thai, H.T.; Cao, J.; Chen, Y.; Zhang, X.; Shulu, Z.; van der Zanden, L.F.; van Rooij, I.A.; Frisen, L.; et al. Fine mapping analysis confirms and strengthens linkage of four chromosomal regions in familial hypospadias. *Eur. J. Hum. Genet.* **2015**, *23*, 516–522. [CrossRef] [PubMed]
46. Tiryakioglu, N.O.; Tunali, N.E. Association of AKR1C3 Polymorphisms with Bladder Cancer. *Urol J* **2016**, *13*, 2615–2621.
47. Figueroa, J.D.; Malats, N.; Garcia-Closas, M.; Real, F.X.; Silverman, D.; Kogevinas, M.; Chanock, S.; Welch, R.; Dosemeci, M.; Lan, Q.; et al. Bladder cancer risk and genetic variation in AKR1C3 and other metabolizing genes. *Carcinogenesis* **2008**, *29*, 1955–1962. [CrossRef]
48. Lan, Q.; Mumford, J.L.; Shen, M.; Demarini, D.M.; Bonner, M.R.; He, X.; Yeager, M.; Welch, R.; Chanock, S.; Tian, L.; et al. Oxidative damage-related genes AKR1C3 and OGG1 modulate risks for lung cancer due to exposure to PAH-rich coal combustion emissions. *Carcinogenesis* **2004**, *25*, 2177–2181. [CrossRef]
49. Platt, A.; Xia, Z.; Liu, Y.; Chen, G.; Lazarus, P. Impact of nonsynonymous single nucleotide polymorphisms on in-vitro metabolism of exemestane by hepatic cytosolic reductases. *Pharmacogenet Genomics* **2016**, *26*, 370–380. [CrossRef]
50. Andreen, L.; Nyberg, S.; Turkmen, S.; van Wingen, G.; Fernandez, G.; Backstrom, T. Sex steroid induced negative mood may be explained by the paradoxical effect mediated by GABAA modulators. *Psychoneuroendocrinology* **2009**, *34*, 1121–1132. [CrossRef]
51. Liang, J.J.; Rasmusson, A.M. Overview of the Molecular Steps in Steroidogenesis of the GABAergic Neurosteroids Allopreg-nanolone and Pregnanolone. *Chronic Stress* **2018**, *2*, 2470547018818555. [CrossRef] [PubMed]
52. Johansson, A.G.; Nikamo, P.; Schalling, M.; Landen, M. AKR1C4 gene variant associated with low euthymic serum progesterone and a history of mood irritability in males with bipolar disorder. *J. Affect. Disord.* **2011**, *133*, 346–351. [CrossRef] [PubMed]
53. Johansson, A.G.; Nikamo, P.; Schalling, M.; Landen, M. Polymorphisms in AKR1C4 and HSD3B2 and differences in serum DHEAS and progesterone are associated with paranoid ideation during mania or hypomania in bipolar disorder. *Eur. Neuropsy-chopharmacol.* **2012**, *22*, 632–640. [CrossRef] [PubMed]
54. Lord, S.J.; Mack, W.J.; Van Den Berg, D.; Pike, M.C.; Ingles, S.A.; Haiman, C.A.; Wang, W.; Parisky, Y.R.; Hodis, H.N.; Ursin, G. Polymorphisms in genes involved in estrogen and progesterone metabolism and mammographic density changes in women randomized to postmenopausal hormone therapy: Results from a pilot study. *Breast Cancer Res.* **2005**, *7*, R336–R344. [CrossRef] [PubMed]

55. Kume, T.; Iwasa, H.; Shiraishi, H.; Yokoi, T.; Nagashima, K.; Otsuka, M.; Terada, T.; Takagi, T.; Hara, A.; Kamataki, T. Characterization of a novel variant (S145C/L311V) of 3alpha-hydroxysteroid/dihydrodiol dehydrogenase in human liver. *Pharmacogenetics* **1999**, *9*, 763–771. [CrossRef]
56. Gathercole, L.L.; Nikolaou, N.; Harris, S.E.; Arvaniti, A.; Poolman, T.M.; Hazlehurst, J.M.; Kratschmar, D.V.; Todorcevic, M.; Moolla, A.; Dempster, N.; et al. AKR1D1 knockout mice develop a sex-dependent metabolic phenotype. *J. Endocrinol.* **2022**, *253*, 97–113. [CrossRef]
57. Chen, M.; Jin, Y.; Penning, T.M. In-Depth Dissection of the P133R Mutation in Steroid 5beta-Reductase (AKR1D1): A Molecular Basis of Bile Acid Deficiency. *Biochemistry* **2015**, *54*, 6343–6351. [CrossRef]

International Journal of
Molecular Sciences

Article

DHCR24, a Key Enzyme of Cholesterol Synthesis, Serves as a Marker Gene of the Mouse Adrenal Gland Inner Cortex

Huifei Sophia Zheng, Yuan Kang, Qiongxia Lyu, Kristina Junghans, Courtney Cleary, Olivia Reid, Greer Cauthen, Karly Laprocina and Chen-Che Jeff Huang *

Department of Anatomy, Physiology & Pharmacology, College of Veterinary Medicine, Auburn University, Auburn, AL 36849, USA
* Correspondence: jeff.huang@auburn.edu

Abstract: Steroid hormones are synthesized through enzymatic reactions using cholesterol as the substrate. In steroidogenic cells, the required cholesterol for steroidogenesis can be obtained from blood circulation or synthesized de novo from acetate. One of the key enzymes that control cholesterol synthesis is 24-dehydrocholesterol reductase (encoded by *DHCR24*). In humans and rats, DHCR24 is highly expressed in the adrenal gland, especially in the zona fasciculata. We recently reported that DHCR24 was expressed in the mouse adrenal gland's inner cortex and also found that thyroid hormone treatment significantly upregulated the expression of *Dhcr24* in the mouse adrenal gland. In the present study, we showed the cellular expression of DHCR24 in mouse adrenal glands in early postnatal stages. We found that the expression pattern of DHCR24 was similar to the X-zone marker gene 20αHSD in most developmental stages. This finding indicates that most steroidogenic adrenocortical cells in the mouse adrenal gland do not synthesize cholesterol locally. Unlike the 20αHSD-positive X-zone regresses during pregnancy, some DHCR24-positive cells remain present in parous females. Conditional knockout mice showed that the removal of *Dhcr24* in steroidogenic cells did not affect the overall development of the adrenal gland or the secretion of corticosterone under acute stress. Whether DHCR24 plays a role in conditions where a continuous high amount of corticosterone production is needed requires further investigation.

Keywords: adrenal gland inner cortex; X-zone; DHCR24; seladin-1

Citation: Zheng, H.S.; Kang, Y.; Lyu, Q.; Junghans, K.; Cleary, C.; Reid, O.; Cauthen, G.; Laprocina, K.; Huang, C.-C.J. DHCR24, a Key Enzyme of Cholesterol Synthesis, Serves as a Marker Gene of the Mouse Adrenal Gland Inner Cortex. *Int. J. Mol. Sci.* **2023**, 24, 933. https://doi.org/10.3390/ijms24020933

Academic Editor: Jacques J. Tremblay

Received: 19 September 2022
Revised: 2 December 2022
Accepted: 13 December 2022
Published: 4 January 2023

1. Introduction

The mitochondrial cytochrome P450 cholesterol side-chain cleavage enzyme (P450scc, encoded by *CYP11A1*) controls the first step of the steroidogenesis pathway, which converts cholesterol into pregnanolone (Figure 1) [1]. Steroidogenic cells can obtain the required cholesterol for steroidogenesis by taking up circulating cholesterol via receptors on the cell membrane, using the cholesterol stored in lipid droplets in the cytoplasm, or synthesizing cholesterol de novo from acetate (Figure 1) [2]. Similar to many other steroid-producing cells, adrenocortical cells can synthesize cholesterol locally [3]. In human adrenal glands, de novo synthesized cholesterol contributes to 20% of cortisol production [4]. One of the key enzymes that control cholesterol synthesis is DHCR24, which is highly expressed in the adrenal gland in both humans [5] and rats [6], especially in the zona fasciculata. *DHCR24* is also named Selective Alzheimer disease indicator 1 (seladin-1) because it was first identified using neuronal cells from Alzheimer's disease (AD) patients [7]. In humans, the adrenal gland is the tissue with the highest expression of DHCR24 [7]. The expression of DHCR24 has been reported to be altered in human adrenal adenomas and carcinomas [8,9]. Patients carrying mutations in DHCR24 show the accumulation of cholesterol precursor, desmosterol, and cause desmosterolosis, which is a disorder characterized by multiple congenital anomalies, neurological complications, and developmental delays [10]. We recently reported that DHCR24 was expressed in the mouse adrenal gland inner cortex partially colocalized with the inner cortex marker gene 20αHSD [11]. In mouse adrenal glands, the 20αHSD-positive zone is also the X-zone [12], a structure that undergoes regression during postnatal development [13]. A recent study used single-cell approaches and chronic stress challenges to demonstrate the existence and the recruitment of a subpopulation localized in the adrenal gland inner cortex [14]. This finding suggests that the inner part of the adrenal cortex might be critical for optimal stress response. To characterize the significance of the zonal-restricted marker gene, DHCR24, in the mouse adrenal gland inner cortex, we used X-gal staining, immunostaining, and quantitative polymerase chain reaction (qPCR) for the time-course expression of DHCR24. A conditional knockout (cKO) mouse model was used to define the possible role of DHCR24 in adrenal gland development and function at the baseline level (without treatments/challenges other than CO_2 euthanization).

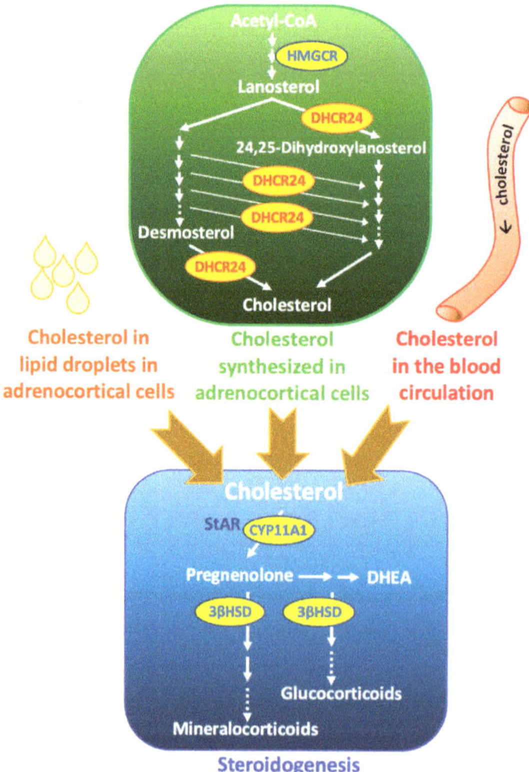

Figure 1. Cholesterol synthesis and steroidogenesis in the adrenal gland.

2. Results

2.1. Cellular Expression of DHCR24 in Mouse Adrenal Glands

A previous finding of the expression of DHCR24 in the mouse adrenal gland inner cortex [11] was confirmed by X-gal staining (Figure 2A). Three commercially available antibodies against DHCR24 were able to detect DHCR24 on formalin-fixed, paraffin-embedded adrenal gland sections (Figure 2B) from mice that were treated with T3, which is known to significantly induce DHCR24 expression [11,15,16] through thyroid hormone receptor (TR) binding sites on the promotor region [17]. Under the euthyroid condition, the DHCR24-positive cells were mainly in the 3βHSD low-expressing inner cortex (Figure 2C), which is also the X-zone in mouse adrenal glands [12]. To obtain the spatial and temporal expression profile of DHCR24 in mouse adrenal glands, both male and female mice were analyzed by immunostaining (Figure 2D) and qPCR (Figure 2E). Both results showed that DHCR24 was expressed in postnatal adrenal glands with a sexually dimorphic pattern. The sexually dimorphic cellular expression and the expression timeline were similar to those of the X-zone marker gene 20αHSD [11,18]. At P14, immunostaining results showed that a few DHCR24-positive cells were found at the cortical–medullary boundary in both sexes. A clear DHCR24-positive zone existed in the inner-cortical region at P21. At P28 and P35, the DHCR24-positive zone remained in adrenal glands in female mice. However, in male mice, only a few DHCR24-positive cells were found at P28; immunostaining did not detect any DHCR24-positive cells in males at P35 (Figure 2D). Despite the strong increase seen by immunostaining from P14 to P21 in both sexes, it is interesting that the qPCR result only showed an increase of the *Dhcr24* expression in female mice when the expression is normalized with *Actb*. It is possible that a housekeeping gene is differentially expressed across different postnatal stages [19], thus masking the differential expression of *Dhcr24*. Unlike the 20αHSD-positive zone that regresses soon during the first pregnancy [13], immunostaining showed that the DHCR24-positive zone remained present in parous females (Figure 3), suggesting that pregnancy does not lead to loss of the DHCR24-positive zone in the adrenal glands of female mice.

2.2. Deletion of Dhcr24 in Steroidogenic Cells Did Not Affect the Zonal Structure of the Adrenal Gland

To understand the role of DHCR24 in the adrenal gland, we used *Sf1-Cre* to remove *Dhcr24* in steroidogenic cells. The expression of *Dhcr24* in the whole adrenal gland shown by qPCR was dramatically reduced in cKO mice (*Dhcr24$^{flox/flox}$; Sf1Cre*) in both sexes (Figure 4A). The expression level of *Dhcr24* in P35 cKO males was 24.6% of that in wild-type males, whereas in female cKO mice at the same age, the expression was only 10.2% of that in wild-type females. An RNA-seq result in a previous study showed that the expression levels of *Dhcr24* in the whole adrenal gland in P35 mice were 33.62 FPKM (fragments per kilobase of exon per million mapped fragments) in males and 111.96 FPKM in females [11]. The reduction rates of the *Dhcr24* expression in cKO mice shown by the qPCR suggested that the expression levels of *Dhcr24* in the adrenal glands in cKO mice could be around 10 FPKM in both sexes. This finding indicates that *Dhcr24* is mainly expressed in SF1-positive cortical cells and that *Sf1-Cre* could remove most *Dhcr24* in the adrenal gland. Indeed, immunostaining showed that no DHCR24-positive cells were detected in adrenal glands in cKO mice of both sexes in the euthyroid condition. Even after thyroid hormone treatment, the adrenal glands in cKO mice remained DHCR24-negative for immunostaining signals (Figure 4B), confirming the deletion of DHCR24 in cKO mice adrenal glands. To determine if the structure of the concentric cortical layers was affected in cKO adrenal glands, we performed immunostaining to label different cortical layers, including the X-zone and the CYP2F2-positive zone, which is a newly identified inner subzone in mouse adrenal zona fasciculata [11]. The lack of DHCR24 did not lead to any significant change in adrenal cortex zonation at P35 (Figure 4C,D). The qPCR results showed that expressions of three major marker genes of the inner cortex, *Akr1c18* (encodes 20αHSD), *Pik3c2g*, and *Thrb1* were not significantly altered in cKO adrenal glands (Figure 5). Interestingly, the *Cyp2f2* expression was slightly reduced in female cKO adrenal glands (WT vs. cKO: 1 ± 0.30 vs. 0.53 ± 0.20, *p* = 0.020). Expressions of *Cyp11a1* and *Star*, the two key factors that control the rate-limiting step of steroidogenesis, were not affected by the deletion of *Dhcr24*. The mRNA level of the steroidogenic enzyme 11β-hydroxylase (encoded by *Cyp11b1*) was also unchanged. The enzyme that controls the rate-limiting step of cholesterol synthesis (HMG-CoA reductase, encoded by *Hmgcr*), was slightly reduced in cKO females (WT vs. cKO: 1 ± 0.26 vs. 0.45 ± 0.43, *p* = 0.038).

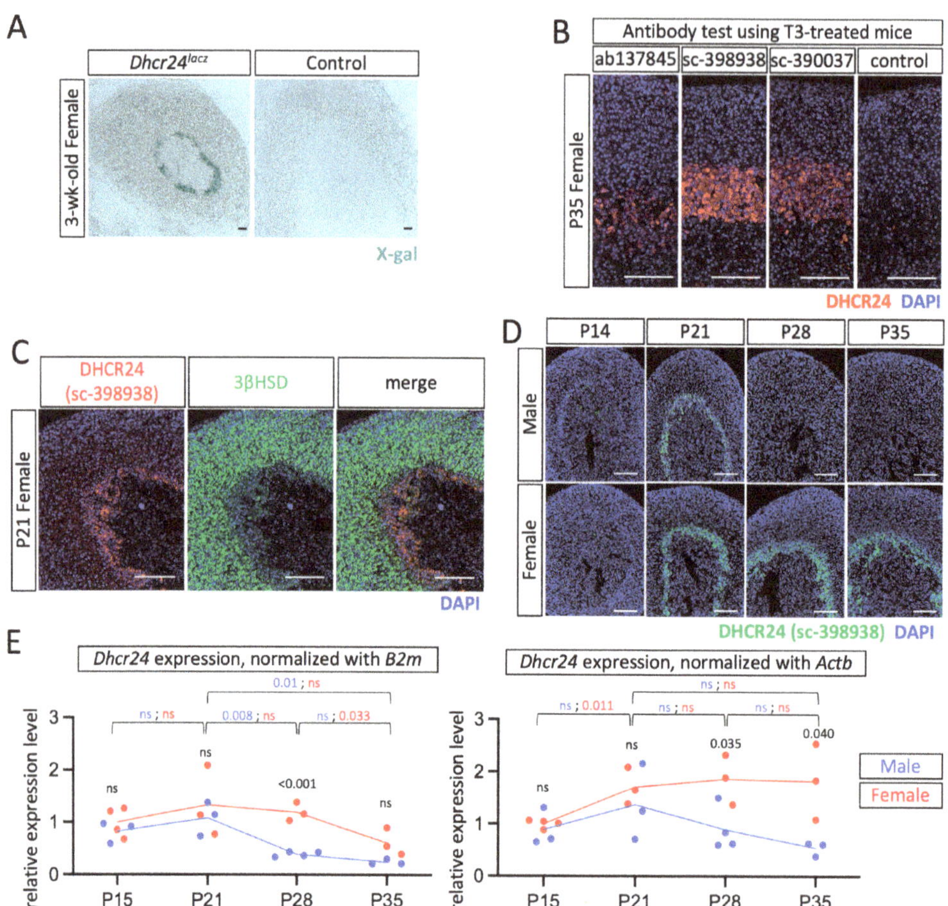

Figure 2. DHCR24 expression. (**A**) X-gal staining of tissues from 3-week-old female mice that were either heterozygous for *Dhcr24^lacz^* or the wild-type from the *Dhcr24^lacz^* strain. (**B**) Immunostaining of DHCR24 using three different commercially available antibodies. Mice were treated with T3 water for 10 days to increase the expression level of DHCR24. The negative control sample was stained using the same method except for incubating with the primary antibody. (**C**) Double immunostaining using DHCR24 antibody #sc-398938 and 3βHSD on euthyroid wild-type B6 mice. (**D**) Immunostaining using DHCR24 antibody #sc-398938 on euthyroid wild-type B6 mice. (**E**) Quantitative polymerase chain reaction was performed to detect the relative expression levels of *Dhcr24*, normalized with either *B2m* or *Actb* to P15 males (set as 1), in whole adrenal glands. *p*-values are shown for the comparison of the same sex between adjacent time points (font color red: female, font color blue: male), and the comparison of male vs. female within each time point (font color black). The trends are shown with the mean. Each data point contains pooled samples from at least three mice. The cell nuclei were counterstained with DAPI (blue). Scale bars, 130 μm. ns, no significant difference.

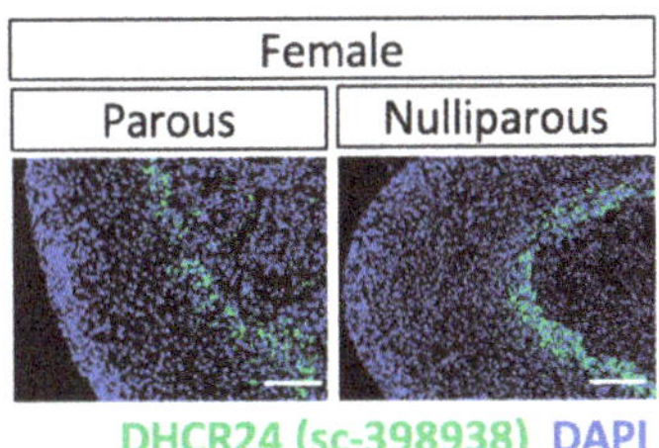

Figure 3. Immunostaining using DHCR24 antibody #sc-398938 of tissue from parous and nulliparous euthyroid wild-type B6 mice. The cell nuclei were counterstained with DAPI (blue). Scale bars, 130 μm.

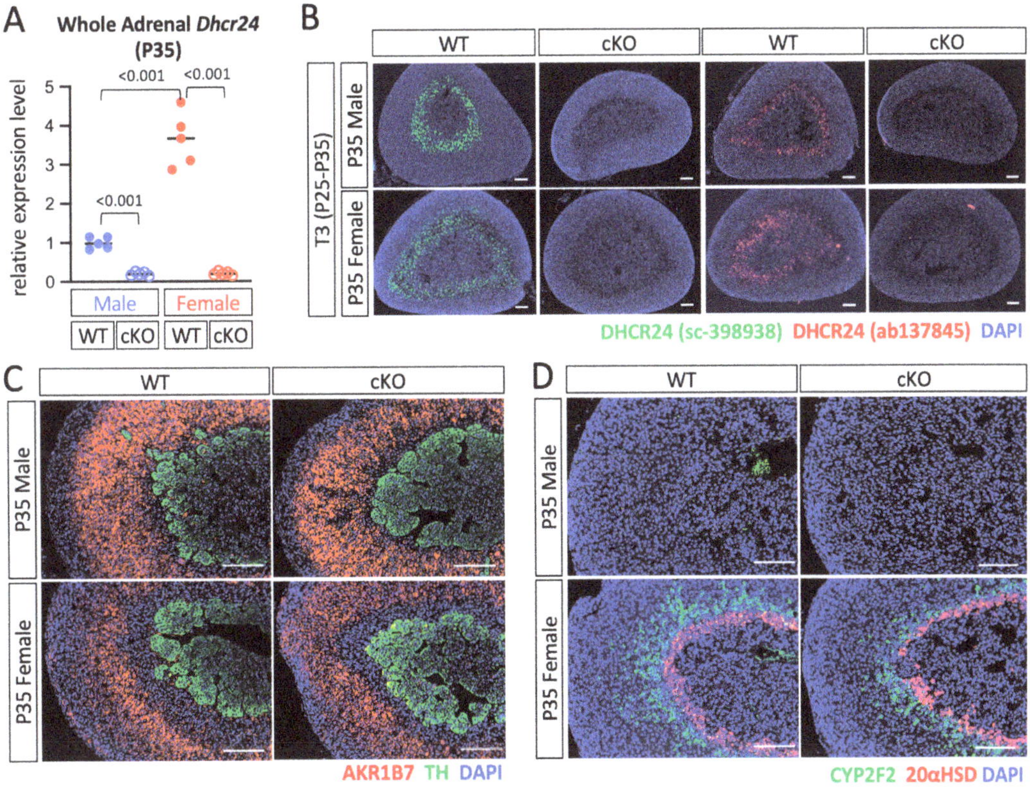

Figure 4. Phenotypic analyses of *Dhcr24* cKO mice. (**A**) Quantitative polymerase chain reaction was performed to detect the relative expression levels of *Dhcr24* in whole adrenal glands from cKO and WT littermates at P35. The relative expression levels of *Dhcr24* in cKO mice were normalized to those of the male WT mice. Each dot (WT) and circle (cKO) represents data from one mouse. Data are shown with the mean. (**B**) Immunostaining using DHCR24 antibodies #sc-398938 and #ab137845. The mice were treated with T3 drinking water for 10 days to increase the expression level of DHCR24. The cell nuclei were counterstained with DAPI (blue). (**C,D**) Immunostaining showed areas of zona fasciculata (AKR1B7-positive), medulla (tyrosine hydroxylase (TH)-positive), and inner cortex (CYP2F2- or 20αHSD-positive) in euthyroid P35 mice. The cell nuclei were counterstained with DAPI (blue). WT: wild-type mice, cKO: *Dhcr24* conditional knockout mice. Scale bars, 130 μm.

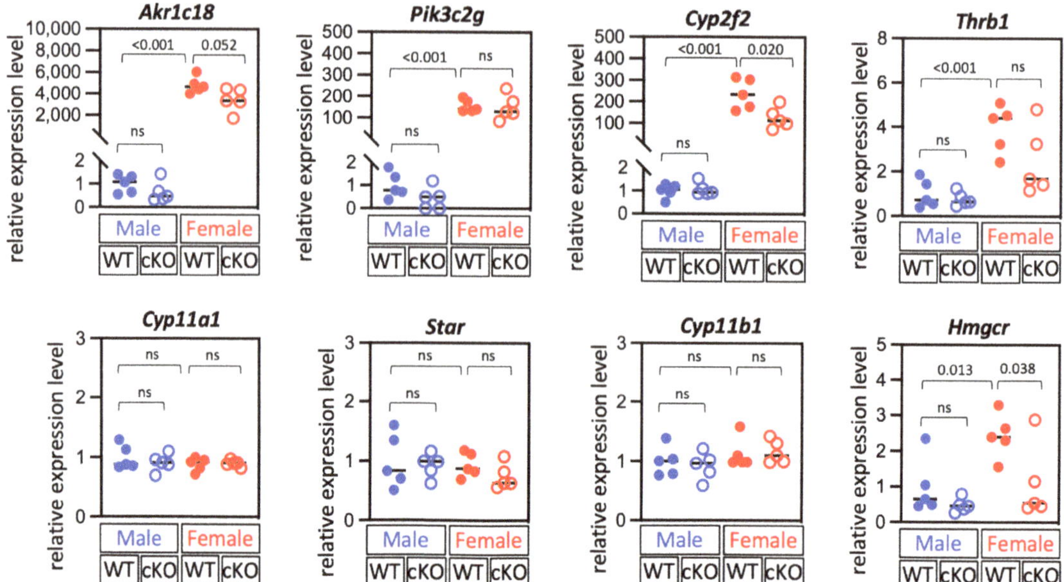

Figure 5. Quantitative polymerase chain reaction showed the relative expressions of key marker genes in the adrenal cortex. The relative expression levels of each gene in cKO mice were normalized to those of the male WT mice. Each dot (WT) and circle (cKO) represents data from one mouse. Data are shown with the mean. WT: wild-type mice, cKO: *Dhcr24* conditional knockout mice. ns, no significant difference.

2.3. Deletion of Dhcr24 in Steroidogenic cells Did Not Affect the Lipid-Droplet Accumulation in Adrenocortical Cells or the Blood Corticosterone and ACTH Levels

Because DHCR24 is the key enzyme that controls cholesterol synthesis, it is possible that the de novo synthesis of cholesterol is affected in adrenocortical cells in *Dhcr24* cKO mice and then further alters the subsequent steroidogenesis. Phenotypic analyses using Oil Red O staining and a hormone assay showed that the oil accumulation was not affected in adrenal glands at the histological level in cKO mice (Figure 6A); the corticosterone and ACTH levels in blood circulation and the ACTH/corticosterone ratio were also unchanged in both sexes of the cKO mice (Figure 6B). It is important to note that these mice were euthanized by carbon dioxide inhalation, which is considered to be a stressor that leads to elevated corticosterone secretion [20]. The comparable stimulated corticosterone and ACTH levels in cKO mice suggest that the lack of de novo synthesized cholesterol does not significantly affect steroidogenesis under certain stress levels, such as the stress caused by CO_2 euthanasia. This result is consistent with the expression characteristic of DHCR24 in which it is specifically expressed in the inner cortex or X-zone, the area that is not considered a primary contributor to the steroidogenic function of the adrenal gland cortex. However, the de novo synthesized cholesterol might still be detrimental to chronic stress response, especially some genes involved in cholesterol synthesis (e.g., *Sqle* and *Hsd17b7*) has been reported to be upregulated in the mouse adrenal gland after long-term ACTH administration [21], with the recent identification of an *Abcb1b*-positive subpopulation in the adrenal gland inner cortex [14].

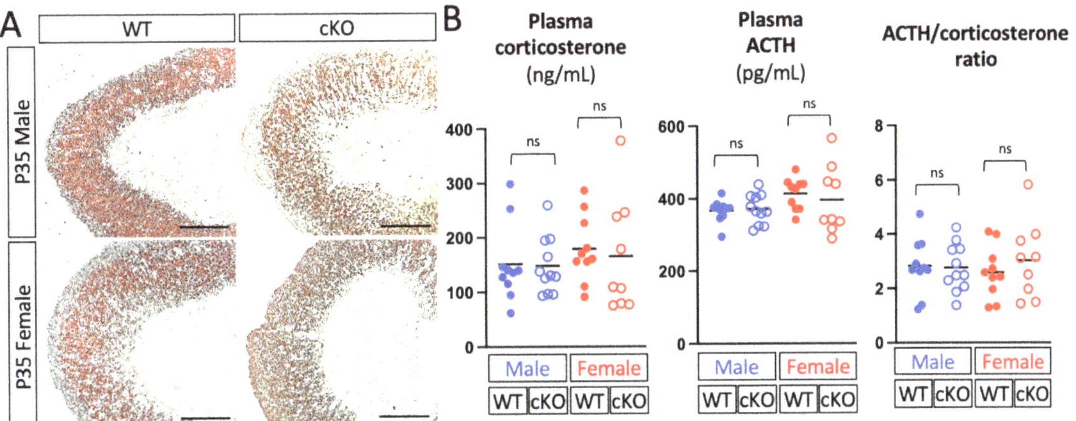

Figure 6. Phenotypic analyses of Dhcr24 cKO mice. (**A**) Oil Red O staining showing lipid droplets in the adrenal glands of euthyroid P35 mice. (**B**) Plasma levels of corticosterone, ACTH, and the ACTH/corticosterone ratio in euthyroid P35 mice. Each dot (WT) and circle (cKO) represents data from one mouse. Data are shown with the mean. Scale bars, 210 μm. WT: wild-type mice, cKO: Dhcr24 conditional knockout mice. ns, no significant difference.

3. Discussion

Although the *Dhcr24* cKO mice did not show any strong phenotypes that were examined in this study, our data demonstrated that in the mouse adrenal gland, DHCR24 was specifically expressed in the inner cortex, with temporal and spatial expression patterns that were very similar to those of the X-zone marker gene 20αHSD, which has a sexually dimorphic expression pattern. The major difference between DHCR24 and 20αHSD is that there are DHCR24-positive cells in the adrenal glands of parous female mice, but there is no X-zone in these mice because it regresses during pregnancy [13]. This zonal-restricted expression of DHCR24 indicates that most adrenocortical cells in the mouse adrenal gland do not synthesize cholesterol locally because of the lack of DHCR24. We also found that the deletion of DHCR24 in steroidogenic cells did not lead to major histological and functional changes in the adrenal gland, at least under euthyroid conditions. Although a few subpopulations in the adrenal gland inner cortex have been reported [11,14] along with evidence showing the origin [22–24] and the developmental timeline [18] of the X-zone, more studies are needed to discover the physiological function of cells in the mouse adrenal gland inner cortex.

Different sources or supply routes of cholesterol ensure the high demand for cholesterol for steroidogenesis in steroid hormone-producing cells. Other than (1) taking up high- and low-density lipoproteins from the blood circulation and (2) deriving cholesterol from cholesterol esters stored as lipid droplets, synthesizing cholesterol de novo in the endoplasmic reticulum is another way steroidogenic cells obtain cholesterol [2,25]. The restricted spatial expression of DHCR24 in the mouse adrenal inner cortex with the finding of normal corticosterone levels in the *Dhcr24* cKO mice suggest that the de novo synthesis of cholesterol is not the major source of cholesterol used for steroid hormone production in the mouse adrenal gland at least for the stress response caused by CO_2 euthanization. Because the lack of DHCR24 could lead to the accumulation of its substrate desmosterol [26] and the P450scc enzyme is predicted also to use desmosterol to initiate steroid hormone synthesis [27], the lack of DHCR24 may not affect steroidogenesis if desmosterol is used for steroidogenesis as cholesterol. Whether the de novo cholesterol synthesis is important under severe and chronic stress where continuous high amounts of corticosterone production are needed requires further study. It is possible that the *Dhcr24* cKO will start showing phenotypes under certain conditions (e.g., dexamethasone-induced cortex regression, fasting, long-term ACTH administration, etc.).

In addition to its involvement in the mevalonate pathway that controls cholesterol synthesis, DHCR24 has also been linked to many cellular functions and diseases. DHCR24 is also named 'selective Alzheimer's disease indicator-1' or Seladin-1 because of its connection with AD [7]. Patients with AD suffer from massive neuronal death due to apoptosis in both neurons and glial cells [28]. In the brain, DHCR24 is less abundant in the areas affected by AD [7,29]. Over-expression of *DHCR24* in neurons prevents β-amyloid accumulation and oxidative stress. This neuroprotective function that prevents neuronal loss has been seen both in vitro and in vivo [30–34]. Additionally, it has been demonstrated that DHCR24 has a protective effect against apoptosis by inhibiting caspase-3 activity [35,36]. However, the molecular mechanism for the DHCR24-mediated cell protective effect is not fully understood. Our previous study showed that TRβ1 is specifically expressed in the adrenal gland inner cortex, especially in the X-zone [18]. Because T3 treatment has been shown to increase the size of inner-cortical cells and delay their regression [18,37,38], the T3-mediated delayed regression of the X-zone is possibly a direct effect of the T3-induced high expression of DHCR24 in the inner cortex. We are currently using *Dhcr24* cKO mice to determine if DHCR24 contributes to T3-mediated delayed regression.

Because *Dhcr24* is highly expressed in the inner cortex even under euthyroid conditions, it is also possible that DHCR24 confers cell protective effects under euthyroid conditions [27]. Although no significant difference was noticed in *Dhcr24* cKO mice at P35 in both sexes (Figure 6), we are currently examining some earlier developmental stages to determine if deletion of DHCR24 leads to early regression of the X-zone in euthyroid mice. For example, compared with wild-type mice at the same age who usually retain a thin X-zone, male *Dhcr24* cKO mice may lose all 20αHSD-positive cells at P28 [18]. A time-course analysis will show whether or not *Dhcr24* has a role in controlling the normal regression process of the X-zone under euthyroid conditions.

4. Materials and Methods

4.1. Animals

C57BL/6J mice were purchased from the Jackson Lab. The $Dhcr24^{tm1a(EUCOMM)Wtsi}$ (C57BL/6 background) mutant mice ($Dhcr24^{lacz}$) were obtained from The European Conditional Mouse Mutagenesis Program (EUCOMM). To generate the conditional knockout mice, the $Dhcr24^{lacz}$ mice were first crossed with the Flp deleter strain (C57BL/6 background, #7089 from Taconic Biosciences, Germantown, NY, USA) to generate the 'conditional ready' strain ($Dhcr24^{flox}$). Then, the mice were crossed with C57BL/6J mice to remove the Flp. The $Dhcr24^{flox/flox}$ mice were then crossed with the SF1-Cre mice (gift from Dr. Keith Parker, has been back-crossed to C57BL/6 for more than 10 generations) to obtain the cKO mice ($Dhcr24^{flox/flox}$; *Sf1-Cre*). All mice were housed in a 12:12 h light-dark cycle (lights on at 6 am) with free access to regular rodent chow and water until sample collection. Mice were euthanized between 2 pm and 4 pm using carbon dioxide, followed by decapitation. Tissues were collected immediately and fixed in ice-cold 4% (*v/v*%) paraformaldehyde (PFA) in 1X phosphate-buffered saline (PBS) or frozen by liquid nitrogen. All procedures followed the protocols approved by the Institutional Animal Care and Use Committees at Auburn University.

4.2. X-Gal Staining

Tissues were fixed in 2% (*v/v*) paraformaldehyde in phosphate-buffered saline (PBS) on ice for 20 min and then rinsed with PBS three times. Samples were immersed in 30% (m/v) sucrose at 4 °C on a shaker until tissues were sunk to the vial's bottom. Samples were embedded into Tissue-Tek O.C.T. compound (Sakura Finetek, Torrance, CA, USA), and 8 μm sections were collected using positive-charged slides. Cryosections were stained using X-gal staining solution [2 mM $MgCl_2$, 5 mM potassium ferricyanide (Sigma-Aldrich,

St. Louis, MO, USA), 5 mM potassium ferrocyanide (Sigma-Aldrich), 1 mg/mL X-gal (Teknova, Hollister, CA, USA) in PBS] with incubation at 37 °C overnight.

4.3. Immunohistochemistry

Tissues were fixed at 4 °C overnight and processed according to standard immunostaining procedures [11]. In short, paraffin-embedded sections were incubated with primary antibodies (DHCR24, #sc-398938, RRID: AB_2832944, 1:100; DHCR24, #sc-390037, RRID: AB_2923495, 1:100; DHCR24, #ab137845, RRID: AB_2923496, 1:100; Tyrosine Hydroxylase (TH), RRID: AB_628422, 1:500; 20αHSD, RRID: AB_2832956, 1:500; CYP2F2, #sc-374540, RRID: AB_10987684, 1:250) followed by appropriate fluorescein-conjugated secondary antibodies. DHCR24 was detected by a biotinylated secondary antibody followed by a fluorescence tyramide [39]. Fluorescent images were obtained using a Revolve 4 microscope (ECHO). ImageJ (https://imagej.nih.gov/ij/, accessed on 14 December 2022) was used for adjusting the brightness and contrast.

4.4. Oil Red O Staining

After overnight fixation, tissues were rinsed by ice-cold PBS and then immersed in 30% ($w/v\%$) sucrose in PBS at 4 °C on a shaker until tissues were sunk to the vial's bottom. Samples were embedded into Tissue-Tek O.C.T. compound (Sakura Finetek, Torrance, CA, USA), and stored at −80 °C until cryosectioning. Sections (8 μm) were collected using positive-charged slides, rinsed in PBS, and then incubated in Oil Red O solution [0.18% ($w/v\%$) Oil Red O powder in 60% ($v/v\%$) isopropanol/ddH$_2$O] for 5 min. After washing with ddH$_2$O twice for 5 min each, the slides were covered with an aqueous mounting medium. For each group, at least 3 adrenal glands from 3 mice were analyzed.

4.5. Quantitative Real-Time RT-PCR (qPCR)

Total RNA from snape-freezing adrenal glands was isolated using the TRIzol reagent (Thermo Fisher, Waltham, MA, USA) according to the manufacturer's instructions. The reverse transcription was performed with total RNA and SuperScript IV reverse transcriptase (Thermo Fisher, Waltham, MA, USA) with oligo dT primers. The qPCR analysis was performed as described in a previously published study using PowerUP SYBR Green Master Mix (Thermo Fisher, Waltham, MA, USA) [11]. The relative gene expressions were calculated using relative standard curves with *B2m* or *Actb* (Figure 2E, right panel) as the internal control. Primers used for qPCR are listed below: GCCCTTGGTGTC-TATGGGTC (forward) and AGCTCGTAGGCAGTGCAAAT (reverse) for *Dhcr24*; GATAG-GCCAGGCCATTCTAAGC (forward) and CATTCCCTGGCTTCAGAGACAC (reverse) for *Akr1c18*; CCATTTGTGGACCCAGGTGA (forward) and GGGTCAGTGCATTTTG-GAACA (reverse) for *Pik3c2g*; AAGTGCAACGCTTTGCTGAC (forward) and TGAACTC-CTGAGGCGTCTTG for *Cyp2f2*; CCTGGATCCTGACGATGTGAA (forward) and ACAGGT-GATGCAGCGATAGT (reverse) for *Thrb1*; CTGCCTCCAGACTTCTTTCG (forward) and TTCTTGAAGGGCAGCTTGTT (reverse) for *Cyp11a1*; TATTGACCTGAAGGGGTGGC (forward) and CAGGTGGTTGGCGAACTCTAT (reverse) for *Star*; CAGTGTTCCCAAGGCCT-GAACG (forward) and GGCCATCCGCACATCCTCTTTC (reverse) for *Cyp11b1*; GGAGGC-CTTTGATAGCACCA (forward) and TTCAGCAGTGCTTTCTCCGT (reverse) for *Hmgcr*; TGCTACGTAACACAGTTCCACCC (forward) and CATGATGCTTGATCACATGTCTCG (reverse) for *B2m*; ATGGAGGGGAATACAGCCC (forward) and TTCTTTGCAGCTC-CTTCGTT (reverse) for *Actb*.

4.6. Hormone Assays

Mice were exposed to carbon dioxide (~2 min) until they stopped breathing. The blood was then collected at the decapitation site using EDTA-coated tubes. Plasma was stored at −80 °C until use. Hormones were measured using the corticosterone EIA kit (K014-H1, Arbor Assays, Ann Arbor, MI, USA) and the mouse ACTH ELISA kit (NBP3-14759, Novus Biologicals, Centennial, CO, USA) according to manufacturers' instructions. Five (corticosterone) and 25 (ACTH) microliters of plasma from each mouse were measured for each data point.

4.7. Statistical Analysis

The two-tailed unpaired Student's *t*-test function in Microsoft Excel was used to calculate p values. p values less than 0.05 were considered statistically significant.

Author Contributions: Conceptualization, C.-C.J.H. and H.S.Z.; formal analysis, H.S.Z., Y.K., Q.L., K.J., C.C., O.R., G.C. and K.L.; writing—original draft preparation, C.-C.J.H. and H.S.Z.; writing—review and editing, H.S.Z., Y.K., K.J. and C.-C.J.H.; funding acquisition, C.-C.J.H. All authors have read and agreed to the published version of the manuscript.

Funding: This work was supported by NIH R00HD082686, R03AG065518, and Auburn University Animal Health and Disease Research Program.

Institutional Review Board Statement: The animal study protocol [#2020-3790 (1 October 2020), #2020-3828 (7 December 2020)] was approved by the Institutional Animal Care and Use Committees at Auburn University.

Informed Consent Statement: Not applicable.

Data Availability Statement: Not applicable.

Acknowledgments: Q.L. was supported by the Chinese government scholarship (from China Scholarship Council) as a visiting scholar in the Lab. Current address: College of Animal Science & Technology, Henan University of Science and Technology, LuoYang, Henan 471000, China

Conflicts of Interest: The authors declare no conflict of interest.

References

1. Miller, W.L. Steroidogenic enzymes. *Endocr. Dev.* **2008**, *13*, 1–18. [CrossRef] [PubMed]
2. Miller, W.L.; Bose, H.S. Early steps in steroidogenesis: Intracellular cholesterol trafficking. *J. Lipid. Res.* **2011**, *52*, 2111–2135. [CrossRef] [PubMed]
3. Mason, J.I.; Rainey, W.E. Steroidogenesis in the human fetal adrenal: A role for cholesterol synthesized de novo. *J. Clin. Endocrinol. Metab.* **1987**, *64*, 140–147. [CrossRef] [PubMed]
4. Borkowski, A.J.; Levin, S.; Delcroix, C.; Mahler, A.; Verhas, V. Blood cholesterol and hydrocortisone production in man: Quantitative aspects of the utilization of circulating cholesterol by the adrenals at rest and under adrenocorticotropin stimulation. *J. Clin. Investig.* **1967**, *46*, 797–811. [CrossRef]
5. Battista, M.C.; Roberge, C.; Martinez, A.; Gallo-Payet, N. 24-dehydrocholesterol reductase/seladin-1: A key protein differentially involved in adrenocorticotropin effects observed in human and rat adrenal cortex. *Endocrinology* **2009**, *150*, 4180–4190. [CrossRef]
6. Battista, M.C.; Roberge, C.; Otis, M.; Gallo-Payet, N. Seladin-1 expression in rat adrenal gland: Effect of adrenocorticotropic hormone treatment. *J. Endocrinol.* **2007**, *192*, 53–66. [CrossRef]
7. Greeve, I.; Hermans-Borgmeyer, I.; Brellinger, C.; Kasper, D.; Gomez-Isla, T.; Behl, C.; Levkau, B.; Nitsch, R.M. The human DIMINUTO/DWARF1 homolog seladin-1 confers resistance to Alzheimer's disease-associated neurodegeneration and oxidative stress. *J. Neurosci.* **2000**, *20*, 7345–7352. [CrossRef]
8. Luciani, P.; Ferruzzi, P.; Arnaldi, G.; Crescioli, C.; Benvenuti, S.; Nesi, G.; Valeri, A.; Greeve, I.; Serio, M.; Mannelli, M.; et al. Expression of the novel adrenocorticotropin-responsive gene selective Alzheimer's disease indicator-1 in the normal adrenal cortex and in adrenocortical adenomas and carcinomas. *J. Clin. Endocrinol. Metab.* **2004**, *89*, 1332–1339. [CrossRef]
9. Sarkar, D.; Imai, T.; Kambe, F.; Shibata, A.; Ohmori, S.; Siddiq, A.; Hayasaka, S.; Funahashi, H.; Seo, H. The human homolog of Diminuto/Dwarf1 gene (hDiminuto): A novel ACTH-responsive gene overexpressed in benign cortisol-producing adrenocortical adenomas. *J. Clin. Endocrinol. Metab.* **2001**, *86*, 5130–5137. [CrossRef]
10. Rohanizadegan, M.; Sacharow, S. Desmosterolosis presenting with multiple congenital anomalies. *Eur. J. Med. Genet.* **2018**, *61*, 152–156. [CrossRef]
11. Lyu, Q.; Wang, H.; Kang, Y.; Wu, X.; Zheng, H.S.; Laprocina, K.; Junghans, K.; Ding, X.; Huang, C.J. RNA-Seq Reveals Sub-Zones in Mouse Adrenal Zona Fasciculata and the Sexually Dimorphic Responses to Thyroid Hormone. *Endocrinology* **2020**, *161*, bqaa126. [CrossRef] [PubMed]
12. Hershkovitz, L.; Beuschlein, F.; Klammer, S.; Krup, M.; Weinstein, Y. Adrenal 20 alpha-hydroxysteroid dehydrogenase in the mouse catabolizes progesterone and 11-deoxycorticosterone and is restricted to the X-zone. *Endocrinology* **2007**, *148*, 976–988. [CrossRef]

13. Huang, C.C.; Kang, Y. The transient cortical zone in the adrenal gland: The mystery of the adrenal X-zone. *J. Endocrinol.* **2019**, *241*, R51–R63. [CrossRef] [PubMed]

14. Lopez, J.P.; Brivio, E.; Santambrogio, A.; De Donno, C.; Kos, A.; Peters, M.; Rost, N.; Czamara, D.; Brückl, T.M.; Roeh, S.; et al. Single-cell molecular profiling of all three components of the HPA axis reveals adrenal ABCB1 as a regulator of stress adaptation. *Sci. Adv.* **2021**, *7*, eabe4497. [CrossRef] [PubMed]

15. Benvenuti, S.; Luciani, P.; Cellai, I.; Deledda, C.; Baglioni, S.; Saccardi, R.; Urbani, S.; Francini, F.; Squecco, R.; Giuliani, C.; et al. Thyroid hormones promote cell differentiation and up-regulate the expression of the seladin-1 gene in in vitro models of human neuronal precursors. *J. Endocrinol.* **2008**, *197*, 437–446. [CrossRef] [PubMed]

16. Patyra, K.; Löf, C.; Jaeschke, H.; Undeutsch, H.; Zheng, H.; Tyystjärvi, S.; Pulawska, K.; Doroszko, M.; Chrusciel, M.; Loo, B.M.; et al. Congenital hypothyroidism and hyperthyroidism alters adrenal gene-expression, development, and function. *Thyroid* **2022**, *32*, 459–471. [CrossRef]

17. Ishida, E.; Hashimoto, K.; Okada, S.; Satoh, T.; Yamada, M.; Mori, M. Crosstalk between thyroid hormone receptor and liver X receptor in the regulation of selective Alzheimer's disease indicator-1 gene expression. *PLoS ONE* **2013**, *8*, e54901. [CrossRef]

18. Huang, C.-C.J.; Kraft, C.; Moy, N.; Ng, L.; Forrest, D. A Novel Population of Inner Cortical Cells in the Adrenal Gland That Displays Sexually Dimorphic Expression of Thyroid Hormone Receptor-beta 1. *Endocrinology* **2015**, *156*, 2338–2348. [CrossRef]

19. Ho, K.H.; Patrizi, A. Assessment of common housekeeping genes as reference for gene expression studies using RT-qPCR in mouse choroid plexus. *Sci. Rep.* **2021**, *11*, 3278. [CrossRef]

20. Boivin, G.P.; Hickman, D.L.; Creamer-Hente, M.A.; Pritchett-Corning, K.R.; Bratcher, N.A. Review of CO_2 as a Euthanasia Agent for Laboratory Rats and Mice. *J. Am. Assoc. Lab. Anim. Sci.* **2017**, *56*, 491–499.

21. Menzies, R.I.; Zhao, X.; Mullins, L.J.; Mullins, J.J.; Cairns, C.; Wrobel, N.; Dunbar, D.R.; Bailey, M.A.; Kenyon, C.J. Transcription controls growth, cell kinetics and cholesterol supply to sustain ACTH responses. *Endocr. Connect.* **2017**, *6*, 446–457. [CrossRef] [PubMed]

22. Zubair, M.; Ishihara, S.; Oka, S.; Okumura, K.; Morohashi, K. Two-step regulation of Ad4BP/SF-1 gene transcription during fetal adrenal development: Initiation by a Hox-Pbx1-Prep1 complex and maintenance via autoregulation by Ad4BP/SF-1. *Mol. Cell Biol.* **2006**, *26*, 4111–4121. [CrossRef] [PubMed]

23. Zubair, M.; Parker, K.L.; Morohashi, K.-i. Developmental Links between the Fetal and Adult Zones of the Adrenal Cortex Revealed by Lineage Tracing. *Mol. Cell Biol.* **2008**, *28*, 7030–7040. [CrossRef] [PubMed]

24. Freedman, B.D.; Kempna, P.B.; Carlone, D.L.; Shah, M.S.; Guagliardo, N.A.; Barrett, P.Q.; Gomez-Sanchez, C.E.; Majzoub, J.A.; Breault, D.T. Adrenocortical zonation results from lineage conversion of differentiated zona glomerulosa cells. *Dev. Cell* **2013**, *26*, 666–673. [CrossRef] [PubMed]

25. Gallo-Payet, N.; Battista, M.C. Steroidogenesis-adrenal cell signal transduction. *Compr. Physiol.* **2014**, *4*, 889–964. [CrossRef]

26. Kanuri, B.; Fong, V.; Ponny, S.R.; Weerasekera, R.; Pulakanti, K.; Patel, K.S.; Tyshynsky, R.; Patel, S.B. Generation and validation of a conditional knockout mouse model for desmosterolosis. *J. Lipid. Res.* **2021**, *62*, 100028. [CrossRef]

27. Zerenturk, E.J.; Sharpe, L.J.; Ikonen, E.; Brown, A.J. Desmosterol and DHCR24: Unexpected new directions for a terminal step in cholesterol synthesis. *Prog. Lipid. Res.* **2013**, *52*, 666–680. [CrossRef]

28. Martiskainen, H.; Paldanius, K.M.A.; Natunen, T.; Takalo, M.; Marttinen, M.; Leskelä, S.; Huber, N.; Mäkinen, P.; Bertling, E.; Dhungana, H.; et al. DHCR24 exerts neuroprotection upon inflammation-induced neuronal death. *J. Neuroinflammation* **2017**, *14*, 215. [CrossRef]

29. Iivonen, S.; Hiltunen, M.; Alafuzoff, I.; Mannermaa, A.; Kerokoski, P.; Puoliväli, J.; Salminen, A.; Helisalmi, S.; Soininen, H. Seladin-1 transcription is linked to neuronal degeneration in Alzheimer's disease. *Neuroscience* **2002**, *113*, 301–310. [CrossRef]

30. Cecchi, C.; Rosati, F.; Pensalfini, A.; Formigli, L.; Nosi, D.; Liguri, G.; Dichiara, F.; Morello, M.; Danza, G.; Pieraccini, G.; et al. Seladin-1/DHCR24 protects neuroblastoma cells against Abeta toxicity by increasing membrane cholesterol content. *J. Cell Mol. Med.* **2008**, *12*, 1990–2002. [CrossRef]

31. Crameri, A.; Biondi, E.; Kuehnle, K.; Lütjohann, D.; Thelen, K.M.; Perga, S.; Dotti, C.G.; Nitsch, R.M.; Ledesma, M.D.; Mohajeri, M.H. The role of seladin-1/DHCR24 in cholesterol biosynthesis, APP processing and Abeta generation in vivo. *EMBO J.* **2006**, mboxemph25, 432–443. [CrossRef] [PubMed]

32. Kuehnle, K.; Crameri, A.; Kälin, R.E.; Luciani, P.; Benvenuti, S.; Peri, A.; Ratti, F.; Rodolfo, M.; Kulic, L.; Heppner, F.L.; et al. Prosurvival effect of DHCR24/Seladin-1 in acute and chronic responses to oxidative stress. *Mol. Cell Biol.* **2008**, *28*, 539–550. [CrossRef] [PubMed]

33. Lu, X.; Li, Y.; Wang, W.; Chen, S.; Liu, T.; Jia, D.; Quan, X.; Sun, D.; Chang, A.K.; Gao, B. 3 β-hydroxysteroid-Δ 24 reductase (DHCR24) protects neuronal cells from apoptotic cell death induced by endoplasmic reticulum (ER) stress. *PLoS ONE* **2014**, *9*, e86753. [CrossRef]

34. Wu, C.; Miloslavskaya, I.; Demontis, S.; Maestro, R.; Galaktionov, K. Regulation of cellular response to oncogenic and oxidative stress by Scladin-1. *Nature* **2004**, *432*, 640–645. [CrossRef] [PubMed]

35. Lu, X.; Kambe, F.; Cao, X.; Kozaki, Y.; Kaji, T.; Ishii, T.; Seo, H. 3beta-Hydroxysteroid-delta24 reductase is a hydrogen peroxide scavenger, protecting cells from oxidative stress-induced apoptosis. *Endocrinology* **2008**, *149*, 3267–3273. [CrossRef] [PubMed]

36. Sarajärvi, T.; Haapasalo, A.; Viswanathan, J.; Mäkinen, P.; Laitinen, M.; Soininen, H.; Hiltunen, M. Down-regulation of seladin-1 increases BACE1 levels and activity through enhanced GGA3 depletion during apoptosis. *J. Biol. Chem.* **2009**, *284*, 34433–34443. [CrossRef]

37. Preston, M.I. Effects of thyroxin injections on the suprarenal glands of the mouse. *Endocrinology* **1928**, *12*, 323–334. [CrossRef]
38. Gersh, I.; Grollman, A. The nature of the X-zone of the adrenal gland of the mouse. *Anat. Rec.* **1939**, *75*, 131–153. [CrossRef]
39. Lyu, Q.; Zheng, H.S.; Laprocina, K.; Huang, C.C. Microwaving and Fluorophore-Tyramide for Multiplex Immunostaining on Mouse Adrenals—Using Unconjugated Primary Antibodies from the Same Host Species. *J. Vis. Exp.* **2020**, *156*, e60868. [CrossRef]

International Journal of
Molecular Sciences

Article

Development of Human Adrenocortical Adenoma (HAA1) Cell Line from Zona Reticularis

Hans K. Ghayee [1,2,*], Yiling Xu [1], Heather Hatch [3], Richard Brockway [3], Asha S. Multani [4], Tongjun Gu [5], Wendy B. Bollag [6,7], Adina Turcu [8], William E. Rainey [8], Juilee Rege [8], Kazutaka Nanba [8], Vikash J. Bhagwandin [9], Fiemu Nwariaku [10], Victor Stastny [11], Adi F. Gazdar [11], Jerry W. Shay [12], Richard J. Auchus [8] and Sergei G. Tevosian [3,*]

1   Division of Endocrinology & Metabolism, University of Florida, Gainesville, FL 32610, USA
2   Malcom Randall VAMC, Gainesville, FL 32608, USA
3   Department of Physiological Sciences, University of Florida, Gainesville, FL 32603, USA
4   Department of Genetics, The University of Texas MD Anderson Cancer Center, Houston, TX 77030, USA
5   Department of Bioinformatics, University of Florida, Gainesville, FL 32610, USA
6   Charlie Norwood VA Medical Center, Augusta, GA 30904, USA
7   Department of Physiology, Augusta University, Augusta, GA 30912, USA
8   Division of Endocrinology & Metabolism, University of Michigan, Ann Arbor, MI 48109-5624, USA
9   OncoStemyx, Inc., Palo Alto, CA 94401, USA
10 Department of Surgery, University of Texas Southwestern Medical Center, Dallas, TX 75390, USA
11 Hamon Center for Therapeutic Oncology, University of Texas Southwestern Medical Center, Dallas, TX 75390-8593, USA
12 Department of Cell Biology, University of Texas Southwestern Medical Center, Dallas, TX 75390-9039, USA
*   Correspondence: hans.ghayee@medicine.ufl.edu (H.K.G.); stevosian@ufl.edu (S.G.T.)

Citation: Ghayee, H.K.; Xu, Y.; Hatch, H.; Brockway, R.; Multani, A.S.; Gu, T.; Bollag, W.B.; Turcu, A.; Rainey, W.E.; Rege, J.; et al. Development of Human Adrenocortical Adenoma (HAA1) Cell Line from Zona Reticularis. *Int. J. Mol. Sci.* **2023**, *24*, 584. https://doi.org/10.3390/ijms24010584

Academic Editor: Jacques J. Tremblay

Received: 28 October 2022
Revised: 8 December 2022
Accepted: 20 December 2022
Published: 29 December 2022

Abstract: The human adrenal cortex is composed of distinct zones that are the main source of steroid hormone production. The mechanism of adrenocortical cell differentiation into several functionally organized populations with distinctive identities remains poorly understood. Human adrenal disease has been difficult to study, in part due to the absence of cultured cell lines that faithfully represent adrenal cell precursors in the early stages of transformation. Here, Human Adrenocortical Adenoma (HAA1) cell line derived from a patient's macronodular adrenocortical hyperplasia and was treated with histone deacetylase inhibitors (HDACis) and gene expression was examined. We describe a patient-derived HAA1 cell line derived from the zona reticularis, the innermost zone of the adrenal cortex. The HAA1 cell line is unique in its ability to exit a latent state and respond with steroidogenic gene expression upon treatment with histone deacetylase inhibitors. The gene expression pattern of differentiated HAA1 cells partially recreates the roster of genes in the adrenal layer that they have been derived from. Gene ontology analysis of whole genome RNA-seq corroborated increased expression of steroidogenic genes upon HDAC inhibition. Surprisingly, HDACi treatment induced broad activation of the Tumor Necrosis Factor (TNF) alpha pathway. This novel cell line we developed will hopefully be instrumental in understanding the molecular and biochemical mechanisms controlling adrenocortical differentiation and steroidogenesis.

Keywords: adrenocortical carcinoma; histone deacetylase inhibitor; cell differentiation; gene expression

1. Introduction

In mammals, the adrenal cortex is composed of concentric cellular zones that surround an inner medulla (M) and are anatomically and functionally distinct. Three major zones are distinguished in the human adrenal cortex: (a) zona glomerulosa (ZG), (b) zona fasciculata (ZF), and (c) zona reticularis (ZR). Steroidogenic cells present in these three zones synthesize steroid hormones: mineralocorticoids, glucocorticoids, and androgens, respectively. In the murine adrenal cortex, ZG and ZF can be distinguished, but in contrast to humans, the presence of ZR in mice is controversial (e.g., [1–4]) and no adrenal androgens are produced. Furthermore, rodent adrenals do not express *Cyp17a1* (a gene encoding 17α-hydroxylase/17,20-lyase), and their ZF cells secrete corticosterone, while the main glucocorticoid produced by human adrenals is cortisol [5]. As a result of these notable dissimilarities, mouse models of several human adrenal diseases have been difficult to establish. A thin capsule (C) that surrounds the gland provides the structural support and serves as a source of cortical stem cells [6,7].

Successful production of essential steroidogenic hormones in the adrenal gland relies on the combination of the universal and unique steroidogenic regulatory proteins. Accordingly, master regulators of steroidogenesis confer the common hormone-producing characteristics, while zone-specific enzymes act as refining molecular coordinators for adrenal cell specificity. Steroidogenic factor 1 (SF1/NR5A1) is the transcriptional master regulator of steroidogenic cell identity in several endocrine organs, including the adrenal gland. *Sf1* expression serves as a molecular marker of steroidogenic cell identity [8]. Mice lacking *Sf1* do not develop steroidogenic cells and fail to form gonads or adrenals [9]. In humans, *SF1* mutations cause adrenal failure and a 46,XY-sex reversal [10,11]. As we described previously, GATA4 and GATA6 transcription factors are necessary for the expression of steroidogenic enzymes (e.g., cytochrome P450 family 11 subfamily A (CYP11A1, cholesterol side-chain cleavage enzyme), steroid 3β-hydroxysteroid dehydrogenase/$\Delta^{5/4}$-isomerase type 2 (HSD3B2), cytochrome P450 family 11 subfamily B member 1 (CYP11B1, 11β-hydroxylase), cytochrome P450 family 11 subfamily B member 2 (CYP11B2, aldosterone synthase), and cell surface receptors (e.g., melanocortin receptor type 2 (MC2R, the adrenocorticotropin receptor)) [12,13]. Tissue- and zone-specific gene expression patterns serve as unique molecular signatures for human adrenal cell populations. For example, CYP11B2 required for aldosterone synthesis is expressed by ZG cells, while cells residing in the ZF and ZR express CYP11B1 required for the synthesis of glucocorticoids [5].

Immortalized cell lines are critical tools for understanding disease mechanisms. Unlike many other tissues (e.g., muscle, adipose, breast or colon), few or no cultured cell models that faithfully recapitulate adrenal differentiation states currently exist. The armamentarium currently available for understanding human adrenal differentiation in vitro is patently scarce [14,15]. NCI-H295R, the sole human adrenocortical cell line currently in wide use, was derived from a late-stage aggressive carcinoma [16]. This lineage provides a valid and clinically relevant target for drug therapy. However, these cells are not representative of any specific adrenal lineage and, being late-stage cancer, have limitations for studies of adrenal differentiation and neoplasia.

Epigenetic regulation modulates gene expression through modification of nucleosomes (DNA and histones), without altering the DNA nucleotide sequence. Histone acetylation (hyperacetylation) by histone acetyltransferases results in a relatively open chromatin arrangement that is favorable for DNA transcription. In contrast, histone deacetylases (HDACs) catalyze the removal of the acetyl group from the lysine on the target proteins. Their main function is to balance the acetylation level of histones (and other proteins, most notably transcription factors) by opposing the action of histone acetyltransferases. A total of 18 HDAC enzymes that employ zinc- or NAD+-dependent mechanisms to deacetylate acetyl lysine substrates are known in humans. Small molecules that specifically target these epigenetic regulators have been identified. HDACis (<u>H</u>istone <u>d</u>eacetylase <u>i</u>nhibitor<u>s</u>) are natural and synthetically produced compounds that interfere with HDAC function [17]. Here, we describe a non-secretory SF1-positive human adrenocortical adenoma (HAA) derived cell line HAA-1 from a patient with a ZR tumor, which produced dehydroepiandrosterone sulfate (DHEAS). When these cells were placed in culture, they dedifferentiated and no longer produced hormones. We demonstrate here that, upon treatment with HDACis, HAA1 cells undergo steroidogenic differentiation and highly up-regulate the expression of steroidogenic genes and enzymes. We propose that HAA1 represents an early stage in the differentiation of adrenocortical cells and provide a valuable tissue culture model of adrenal differentiation and disease.

2. Results

2.1. Developing the Novel HAA1 Cell Line from a Benign Neoplasm

The HAA1 cell line has been in culture for over eight years and propagated for over 1000 passage doublings. The cells have an epithelial and cuboidal appearance with variably granular cytoplasm (Figure 1A). Upon addition of 10 µM forskolin, the cell line showed evidence of morphologic changes (rounding up, not shown). This cAMP-dependent protein kinase-mediated characteristic has historically been seen with adrenocortical NCI-H295R cells [18].

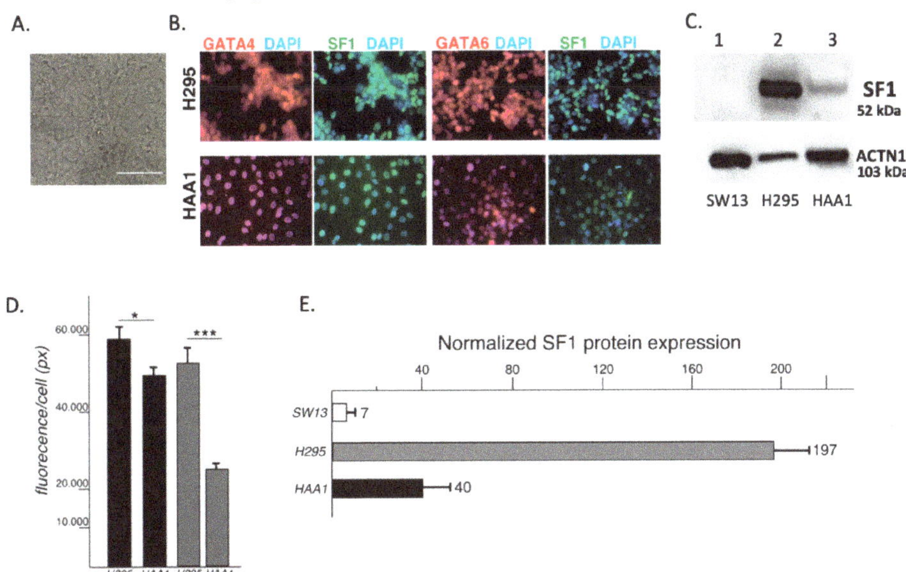

Figure 1. (**A**) HAA1 cells in culture. (**B**) Both master-regulator NR5A1/SF1 (green) and the key adrenal transcription GATA4 and GATA6 proteins (red) are expressed at a lower level in the HAA1 cell line (bottom panels) than in the control NCI-H295R (top panels) human adrenal cortical carcinoma cell line. Scale bar, 50 µ. (**C**) A representative Western blot analysis of protein expression; NR5A1/SF1 (top panel) and ACTIN1 (bottom panel); L1. SW13 adrenal carcinoma cells (these cells produce no steroids and serve as a negative control) L2. NCI-H295.L3. HAA1 (**D**) Quantitative analysis of GATA4 and GATA6 protein expression for the experiment shown in (**B**); *, $p < 05$; ***, $p < 0.001$. (**E**) Quantitative analysis of NR5A1/SF1 expression for the experiment shown in (**C**). All differences are significant, $p < 0.05$.

2.2. HAA1 Cells Exhibit Some Characteristics of the Progenitor Cells of the Adrenal

To characterize the phenotype of the HAA1 cells, we examined the expression of several genes and proteins normally present in adrenocortical cells. Expression of master-regulator SF1 and the key adrenal transcription factors GATA4 and GATA6 in the HAA1 line was present, but lower than that in the control NCI-H295R human adrenal cortical carcinoma cell line (Figure 1B,C). In contrast to the steroidogenic NCI-H295R cells, the HAA1 cell line showed low RNA levels of most steroidogenic enzymes, which were comparable to the non-steroidogenic SW13 cells. However, steroidogenic factor-1 (*SF1*, *NR5A1*) and *STAR* RNA expression levels in HAA1 cells were intermediate between SW13 and NCI-H295R (Supplementary Figure S1). Since the HAA1 line did not exhibit overt signs of lineage-specific steroidogenic differentiation, we examined the cells for the expression of genes specific to progenitor cells present in the adrenal gland. Sonic Hedgehog (SHH) and GLI1 proteins have been described as markers of the major progenitor cell populations present in the adrenal cortex [7,19,20]. HAA1 cells express nuclear-localized GLI1 protein that is also notably present in the sub-population of the adrenal stem cells [7,20–22], but not SHH (Figure 2).

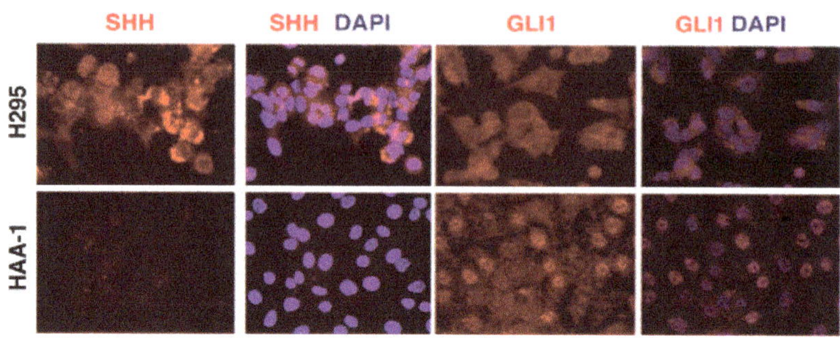

Figure 2. HAA1 cells exhibit some characteristics of the progenitor cells of the adrenal. HAA1 cells (**bottom panels**), but not NCI-H295 cells (**top panels**), express GLI1 protein that is also expressed in the sub-population of the adrenal stem cells, but not SHH. H295 cells express SHH protein. Scale bar, 50 μ.

2.3. Increased Expression of Master Regulators alone Does Not Induce Re-Differentiation in HAA1 Cells

HAA1 cells retain the expression of master steroidogenic regulators *SF1*, *GATA4* and *GATA6*, but at a lower level than steroidogenically active NCI-H295R cells (Figure 1B,C) and, similar to the progenitor adrenal cells, HAA1 cells express key steroidogenic pathway genes at a low level [23]. One explanation for the absence of steroidogenesis would be de-differentiation and loss of adrenocortical gene expression upon 2D culture conditions over an extended time. Thus, we sought to reestablish steroidogenesis in HAA1 cells. Over-expression of SF1 during embryogenesis is sufficient to cause robust ectopic steroidogenesis in fetal mice [24]. Hence, we tested whether simply boosting the expression of master regulators SF1, GATA4, and GATA6, to the levels of NCI-H295R cells, would be sufficient for re-differentiation. Transient transfection (either separately or combined) of plasmid DNA in which *SF1*, *GATA4* or *GATA6* cDNAs were driven by a strong CMV promoter did not lead to detectable increases in steroidogenic enzyme expression in HAA1 cells (Supplementary Figure S2). Other established approaches for inducing differentiation (e.g., treating cells with the adenylate cyclase activator forskolin or 5-Azacytidine, a DNA methyltransferase inhibitor) were equally unsuccessful.

2.4. HAA1 Cells Respond to HDACis by Reprogramming Their Gene Expression

It has been known for a long time that the butyrate ion is a potent inducer of terminal erythroid differentiation in cultured erythroleukemic cells [25]. The mechanistic basis for this phenomenon, based on the ability of butyrate to inhibit histone deacetylation, has been proposed [26]. However, unlike undifferentiated hematopoietic cells that undergo differentiation, cells derived from solid tumors normally respond to HDACis by apoptosis (e.g., [27] and references therein). Nonetheless, we sought to attempt this differentiation-inducing protocol for HAA1 cells. HAA1 cells were treated with various concentrations of sodium butyrate (SoBu), and cells were harvested after 2, 4 or 6 days. RNA was isolated, converted to cDNA, and subjected to qRT-PCR analysis.

We determined that SoBu treatment led to a prominent adrenocortical differentiation in HAA1 cells (Figure 3). Gene expression for several key enzymes involved in steroidogenic hormone synthesis was highly upregulated (Figure 3a); longer treatment times resulted in higher gene expression (Figure 3b). Since another HDACi, Trichostatin A, was equally effective in inducing steroidogenic gene expression in the HAA1 cells (Figure 3c), we concluded that it is HDAC inhibition that is capable of promoting steroidogenic re-differentiation in these cells. Interestingly, derived from zona reticularis (ZR), HAA1 cells respond to SoBu by robustly inducing *CYP11B1* mRNA, which encodes the 11β-hydroxylase that catalyzes the final step of cortisol (corticosterone) biosynthesis (Figure 3c) This gene is normally expressed in the ZR and ZF but not the ZG [28]. We also noted that not all zonal-specific steroidogenic gene expression is completely restored in HAA1 cells upon SoBu treatment. For example, CYP17A1 required for the synthesis of adrenal androgens is normally present in human ZR cells. qRT-PCR analysis did not detect a notable increase in *CYP17A1* RNA expression in the cell line upon sodium butyrate or Trichostatin A treatment (Figure 3). In contrast, HDACi treatment of NCI-H295R cells did not result in a robust induction of steroidogenic gene expression (Supplementary Figure S3A).

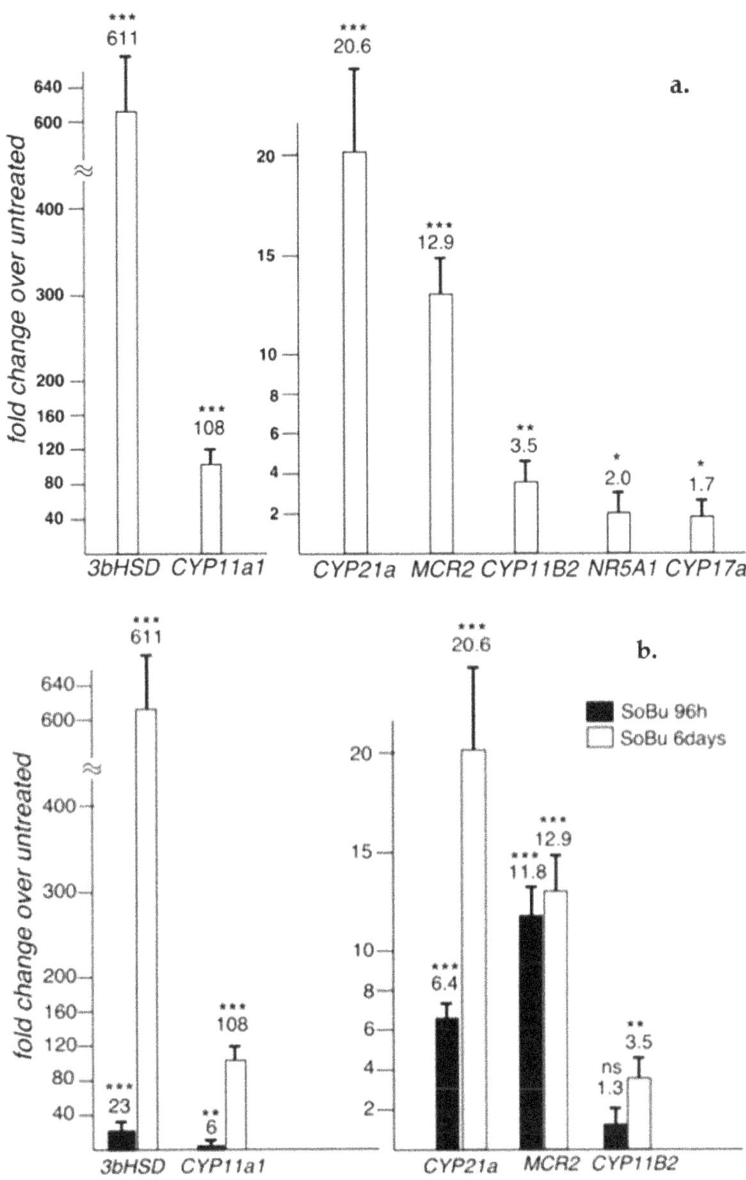

Figure 3. *Cont.*

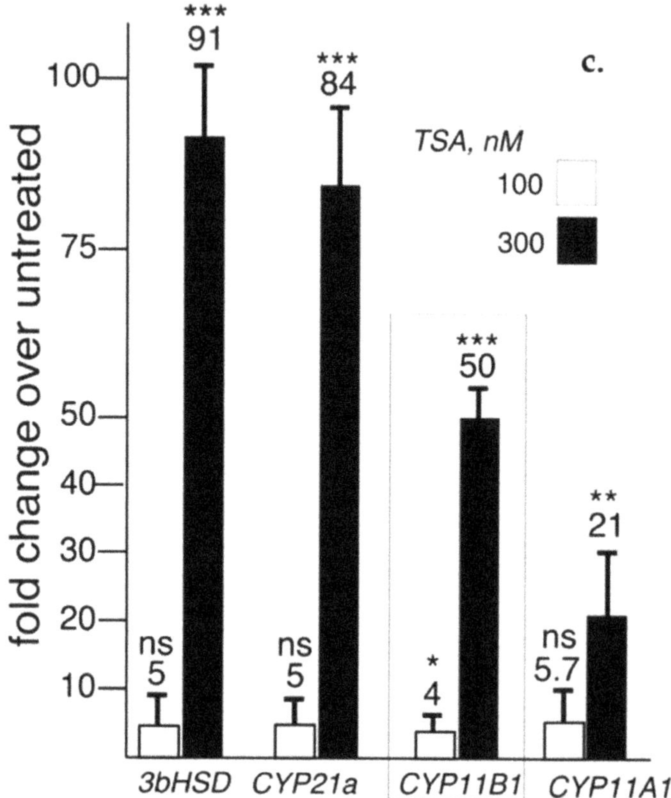

Figure 3. (**a**) qRT-PCR analysis of steroidogenic gene expression in the HAA1 cells. Gene expression in treated cells compared to the untreated is shown. Sodium Butyrate treatment leads to a prominent adrenocortical differentiation in HAA1 cell lines and gene expression for several key enzymes involved in steroidogenic hormone synthesis is highly up-regulated. ***, $p < 0.001$; **, $p < 0.01$; *, $p < 0.05$. (**b**) A comparison of gene expression in HAA1 cells upon 4 (black) or 6 (white) day sodium butyrate treatment. Longer treatment times increase steroidogenic gene expression. ***, $p < 0.001$; **, $p < 0.01$; ns, no significant. The data for 6-day treatment are also shown in 3A. (**c**) qRT-PCR analysis of gene expression in HAA1 cells upon treatment with another HDAC inhibitor, Trichostatin A (TSA). TSA, similarly, is effective in inducing steroidogenic gene expression in the HAA1 cells as sodium butyrate. Notice that CYP11B1 gene expression is efficiently induced in HAA1 cells derived from ZR cells. ***, $p < 0.001$; **, $p < 0.01$; *, $p < 0.05$; ns, no significant.

To confirm the RNA expression data at the protein level, we performed immunofluorescence staining analysis using antibodies on CYP11A1 and HSD3B2. CYP11A1 and HSD3B2 immunostaining is prominent only in the HDACi-treated cells (Figure 4). To examine protein expression in HDACi-treated cells in more detail, we performed a Western blot for StAR. The StAR gene encodes for the steroidogenic acute regulatory protein that regulates cholesterol transport within the mitochondria, a rate-limiting step in the production of all steroid hormones. The expression of StAR was elevated upon HDACi treatment in HAA1 cells (Supplementary Figure S4).

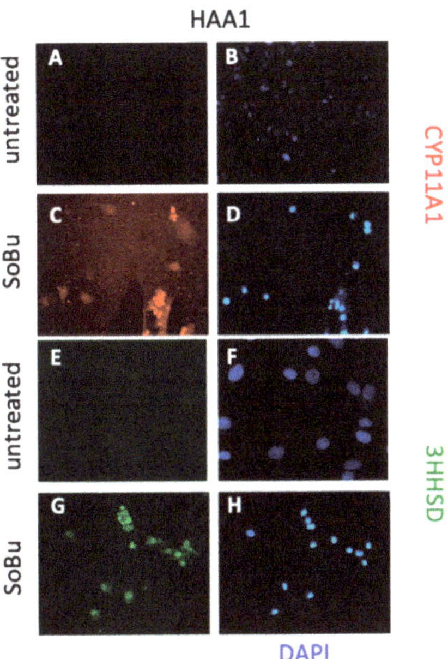

Figure 4. SoBu treatment induces expression of steroidogenic enzymes in HAA1 cells. (**A–H**) Untreated (**A,B,E,F**) and SoBu-treated (**C,D,G,H**) HAA1 cells were stained either for CYP11A1 (**A–D**) or 3BHSD (**E–H**) protein. DAPI staining shows that cells are present in all fields. Scale bars, 100 μ (**A–D,G,H**) and 50 μ (**E,F**).

2.5. Multiple HDACs Contribute to Repression of Differentiation in the HAA1 Cells

Both SoBu and trichostatin A are pan-HDAC inhibitors; therefore, it is not clear which HDAC(s) is/are repressing the differentiation of HAA1 cells. To identify the HDACs involved in this process, we treated HAA1 cells with additional HDAC inhibitors with both broad and selective activity (Figure 5). We noted that, in HAA1 cells, CI994 inhibitor (Selleck Chemicals, Houston, TX, USA), which is reported to inhibit HDACs 1, 3, 6, and 8, is the most effective in inducing steroidogenic differentiation, followed by pan-inhibitors SAHA and SoBu, whereas specific inhibitors PCI-34051 (Selleck Chemicals, Houston, TX, HDAC8-specific) and RGFP966 (Selleck Chemicals, HDAC3-specific) were less potent. Overall, these results suggest that multiple HDACs may contribute to repressing differentiation in the HAA1 cells.

2.6. RNA-Seq Analysis of Gene Expression in HAA1 Cells

To gain a better understanding of the differentiation process in HAA1 cells upon HDACi treatment, we performed an RNA-seq analysis of untreated and SoBu-treated cells (Figure 6A). Ingenuity pathway analysis revealed that the steroidogenic (cholesterol biosynthetic) pathway is the top canonical pathway activated in SoBu-treated HAA1 cells (Table 1).

Additionally, TNF-alpha appears as a top activated regulator (Table 1), with increased expression of numerous inflammation-related genes upon HDACi treatment (Figure 6B). To confirm activation of the TNF-alpha pathway by a different assay, we performed qRT-PCR analysis of several genes associated with this pathway. We demonstrated a profound activation of TNF-alpha pathway genes, confirming the RNA-seq results (Figure 6C). Similarly, the TNF-alpha pathway genes were not induced by the HDACi treatment of NCI-H295 cells (Supplementary Figure S3B).

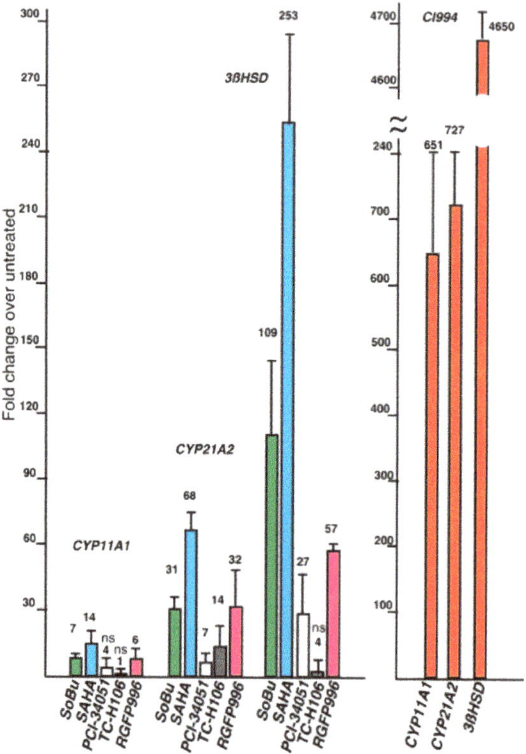

Figure 5. Several HDAC inhibitors were compared for their ability to induce differentiation in the HAA1 cells. CI-994 is the most potent inducer of steroidogenic differentiation in HAA1 cells. All values are significant ($p < 0.05$ or less) except those indicated with "ns".

Table 1. Ingenuity Pathway Analysis of gene expression in HAA1 cells. Summary of Analysis-HAA1 for Ingenuity.

Top Canonical Pathways		
Name	**p-Value**	**Overlap**
Superpathway of Cholesterol Biosynthesis	6.72×10^{-15}	53.6% 15/28
LXR/RXR Activation	1.89×10^{-12}	20.7% 25/121
Hepatic Fibrosis/Hepatic Stellate Cell Activation	2.10×10^{-12}	16.6% 31/187
Cholesterol Biosynthesis I	7.59×10^{-11}	69.2% 9/13
Cholesterol Biosynthesis II (via 24,25, dihydrolanosterol)	7.59×10^{-11}	69.2% 9/13
Top Upstream Regulators	p value of overlap	Predicted activation
TNF	1.47×10^{-55}	Activated
IL1B	9.44×10^{-35}	Activated
Cg	6.48×10^{-34}	Activated
TGFB1	1.32×10^{-32}	Activated
Beta-estradiol	1.73×10^{-30}	

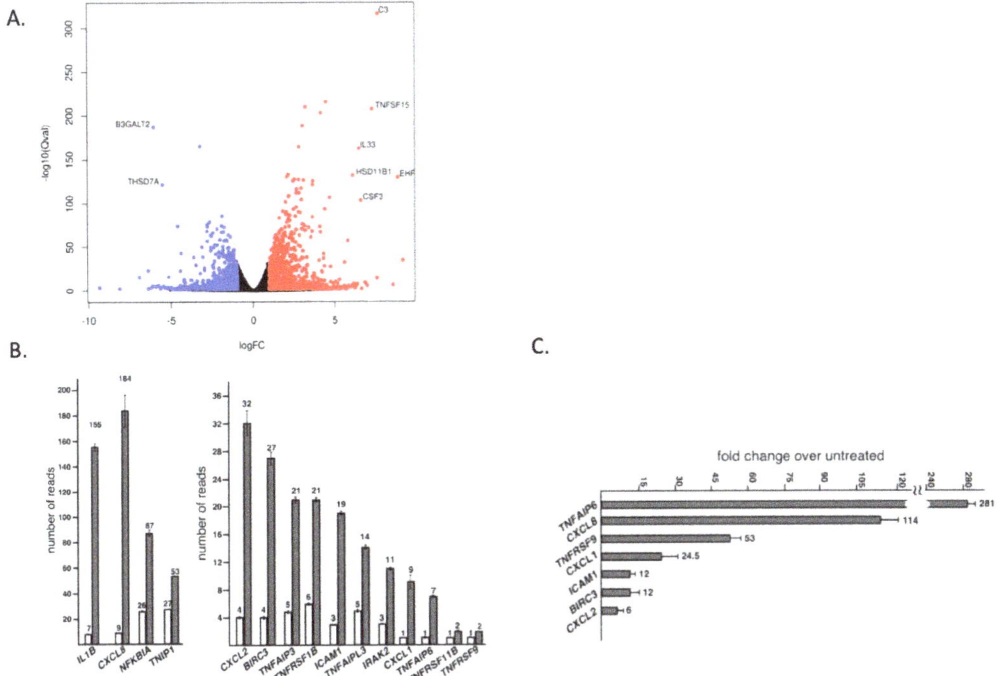

Figure 6. (**A**) A volcano plot of RNAseq analysis of HAA1 cells upon HDACi treatment. The genes with Qval < 4.78 × 10^{-95} and log2 (fold change) > 5 are labeled. (**B**) RNA-seq analysis of gene expression in HAA1 cells after HDACi treatment showing increased expression of a number of TNF-alpha pathway related genes. All values are highly significant, $p < 0.001$. (**C**) qRT-PCR analysis of gene expression in HAA1 cells after HDAC-inhibitor treatment, confirming increased expression of TNF-alpha pathway genes. All values are significant, $p < 0.05$.

2.7. Chromosomal Characteristics

This is a male-derived cell line with a chromosome number ranging from 45 to 92 and a modal chromosome number of 46. Representative karyotypes of this cell line at passages 25 and 155 are shown in Supplementary Figure S5A,B, respectively. Five consistent clonal markers were present in this cell line. Tentative identification of these markers are M1 = t (8q; ?), M2 = dup (11q), M3 = t (14p; ?), M4 = t (15q; 10q), M5 = t (16q; ?) and M6 = t (17p; 1q). Marker M4 was not present in every metaphase, as shown in Supplementary Figure S5B.

2.8. Quantitative Characterization of the Steroidogenic Synthesis in HAA1 Cells upon HDACi Treatment

To examine the steroidogenic potential of the cell line, we measured a set of 23 steroids by LC-MS/MS. We compared the steroid output in media from HAA1 cells under basal conditions and after HDACi treatment using the pan-HDACi SoBu, as well as the optimal HDACi for the cells, CI994. No notable hormone production was observed in untreated or treated cells, despite a dramatic elevation of RNA messages for several key steroidogenic enzymes. To examine the limiting step in steroid hormone biosynthesis in HAA1 cells, we attempted to bypass the first two biosynthetic steps mediated by StAR and CYP11A1. To this end, we supplemented the media with 22(*R*)-hydroxy-cholesterol (22ROH) and pregnenolone. HDACi treatment resulted in the appearance of modest amounts of progesterone in the treated, but not untreated, cells (not shown). We concluded that HDACi-induced differentiation does not induce a complete steroidogenic capacity to a sufficient extent for HAA1 cells to synthesize steroid hormones.

2.9. Short Tandem Repeat (STR) Analysis

The HAA1 cells were compared by STR analysis with the H295R (adrenal cortical carcinoma) cell line-ATCC website (https://www.atcc.org/products/all/CRL-2128.aspx# specifications (accessed on 17 October 2019). The HAA1 cells are a unique cell line (Table 2).

Table 2. Short tandem repeat profile of HAA1.

	Amelogenin	CSF1PO	D13S317	D16S539	D18S51	D19S433	D21S11	D2S1338
HAA1	X, Y	11	12, 13	9, 11	14, 18	14, 15	27, 32.2	24, 25
	D3S1358	**D5S818**	**D7S820**	**D8S1179**	**FGA ***	**TH01**	**TPOX**	**vWA**
HAA1	14, 18	12	8	10, 14	21, 22, 24	7. 9.3	8, 10	16, 18

*, multiple low-level peaks were also observed.

3. Discussion

In this study, we describe a non-secretory SF1-positive cell line, HAA1, derived from a human DHEAS-producing tumor. We demonstrate that, upon treatment with HDACis, HAA1 cells undergo further steroidogenic differentiation and up-regulate the expression of steroidogenic genes and enzymes. Epigenetic regulation modulates gene expression through the alteration of nucleosomes (by modifying DNA and histones), without changing DNA nucleotide sequence. One of the established regulatory mechanisms is through the control of histone acetylation. The degree of histone acetylation in a cell is mostly determined by opposing activities of two types of enzymes: histone acetyltransferases and HDACs. HDACs' main function is to balance the acetylation level of histones by opposing the action of histone acetyltransferases. HDACs catalyze the removal of the acetyl group from lysines on target proteins. Intuitively, one may think that HDACs, by promoting chromatin condensation, should be dedicated to gene repression; however, recent evidence points to HDAC function in highly transcribed genes, where they regulate the turnover of acetylated histones and reset chromatin after transcription [29]. On the other hand, histones are not the sole target of HDAC action and hypoacetylation can certainly result in down-regulation of gene expression, and HDAC-dependent down-regulation of key tumor suppressor genes, such as gatekeepers *TP53* and *RB1*, has been reported [30,31]. Information available with respect to HDAC expression and function in adrenocortical cells is very limited. It has been previously reported that HDACis inhibit steroidogenesis through ubiquitination and degradation of steroidogenic factor 1 (SF1, NR5A1) in Y-1 murine cultured cells [32]; however, as the results reported here demonstrate, this observation does not appear to hold true in the human HAA1 cell line. HDAC function in adrenocortical cells in humans remains to be understood.

HDACis include both natural and synthetically produced compounds that interfere with the function of HDACs [17]. While they have important additional targets, the key substrates for these enzymes are the core DNA histones H2A, H2B, H3 and H4 [33]. Histone acetylation (hyperacetylation) by histone acetyltransferases neutralizes the positive charge of the histone tail and destabilizes binding to the negatively charged DNA. This weakened affinity results in a relatively open chromatin arrangement that is favorable for DNA transcription. Acetylation of nucleosomes residing in the vicinity of transcription start sites (TSSs) is thought to promote the binding of chromatin remodeling factors at promoter regions and/or destabilize chromatin structure [34,35], which may lead to decreased nucleosome occupancy immediately upstream of TSSs and facilitate RNA Pol II binding and transcription [36].

Our data show that HDACis induce expression of steroidogenic genes in HAA1 cells. The results also suggest that multiple HDACs may be involved in the suppression of HAA1 steroidogenesis (Figure 5). Furthermore, HDACi treatment of HAA1 cells demonstrates that they preserve some memory of their cell of origin. We found that *CYP11B1* and *MCR2 MC2R* expression is induced in HAA1 cells (Figure 3B,C). We have also determined that untreated HAA1 cells express GLI1 protein that is normally present in the stem cell population in the adrenal cortex (Figure 2). However, a complete pattern of gene expression corresponding to the original adrenal layer was not restored. For example, CYP17A1 was not expressed in HAA1 cells upon HDACi treatment (Figure 3A).

It is not entirely surprising that a complete roster of lineage-specific expression in HAA1 cells is not restored upon HDACi treatment. For example, some of the genes may require a qualitative or quantitative blend of transcription factors that was not recreated as a result of the treatment, or these genes could be controlled through other pathways. It is also possible that these genes were deregulated in the primary tumors that the HAA1 cells were derived from, as adrenal cancer cells are known to exhibit disorganized steroidogenic gene expression [37]. It has been also postulated that early stage, immature steroidogenesis is a characteristic of adrenal tumors. Exploring methylation patterns and chromatin configuration for genes that failed to re-activate in the HAA1 cells upon treatment should be informative and will be the subject of further studies.

Our data further suggest that one of the functions for HDACs could be regulation and suppression of the TNF-alpha and inflammation pathway in adrenocortical cells. Previous research convincingly demonstrated that TNF-alpha is a potent indirect activator of steroid secretion through its ability to stimulate ACTH production [38]. A dedicated role for the intra-adrenal TNF-alpha pathway has also been proposed based on its presence in adrenal cell lines [39,40] and adrenal tumors [39,41]. The addition of TNF-alpha to HAA1 cells did not result in activation of steroidogenic gene expression, so at this point we can only speculate on the role of this pathway in adrenocortical differentiation of HAA1 cells and whether it is a contributing factor or a bystander in their steroidogenic differentiation. In agreement with previous studies, we favor the hypothesis that a chronically active TNF-alpha pathway could be important for activating steroidogenic gene expression in HAA1 cells. In addition, we observed cell death in HDACi-treated cultures of HAA1 cells, consistent with TNF-alpha's ability to induce apoptosis [42,43]. TNF-alpha can also be a contributing factor to adrenal cancer development and its drug resistance. In conclusion, we have observed that HDAC inhibition partially restores steroidogenic genes in a human cell line derived from an adrenal adenoma.

4. Materials and Methods

4.1. The HAA1 Cell Line Derivation

Human Adrenal Adenoma Line 1 (HAA1) cell line was derived from a 29-year-old man incidentally found to have bilateral macronodular adrenal hyperplasia. Hormonal work-up revealed elevated DHEA and DHEAS levels, which normalized following bilateral adrenalectomy [23]. Tissue samples were procured under an IRB-approved protocol after obtaining written informed consent. To derive cell line HAA1, tissue from DHEAS-producing macronodular hyperplasia was surgically excised, minced, dispersed with collagenase and DNAse I, and passaged in ACL-4 medium. Primary HAA1 cells were infected with lentiviruses containing human telomerase reverse transcriptase (hTERT) and the human papilloma virus early genes E6/E7, and cells were selected with G418. The resultant line was characterized with phase-contrast microscopy. Later, passage cells were adapted to grow in RPMI 1640 medium supplemented with 10% fetal bovine serum (FBS).

4.2. Ectopic Gene Expression in HAA1 Cells

The pCS2_SF1_IRES_EGFP, pCS2_GATA4 and pCS2_GATA6 plasmids were introduced into HAA1 cells by lipofection using Lipofectamine 3000 (Invitrogen, Waltham, MA, USA). For lipofection, 2 µg of plasmid DNA mix was combined with Lipofectamine 3000 reagent as recommended by the manufacturer and added to the mini-chamber slide (Lab-Tek, Grand Rapids, MI, USA). The cells were incubated with the plasmids for 48 h, washed with phosphate-buffered saline (PBS), and fixed with 4% paraformaldehyde.

4.3. Immunocytochemistry

Immunofluorescence analysis of HAA1 cells was performed, essentially as previously described [12]. Briefly, cells were grown on chamber slides (Lab-Tek) and used for plasmid DNA transfection, HDACi treatment or as controls. Cells were gently washed with PBS and fixed with 4% paraformaldehyde or cold methanol for 7 min on ice. Fixed cells were washed twice with PBS, and blocked for 30 min in PBS, 5% bovine serum albumin (BSA), and 0.1% Triton X-100. Cells were incubated with primary antibodies diluted in PBS, 1% BSA, and 0.1% Triton X-100. After 1 h of incubation, cells were washed twice with PBS and incubated with Alexa Fluor-conjugated secondary antibodies diluted in PBS/1% BSA/0.1% Triton X-100 for 1 h. Cells were washed and mounted in medium containing 4,6-diamidine-2-phenylidole-dihydrochloride (DAPI, Vector Laboratories, Newark, CA, USA). The following antibody combinations were used: goat anti-GATA4 antibody (R&D Systems, Minneapolis, MN, USA), followed by donkey anti-goat Alexa Fluor 555-conjugated antibodies (Alexa); rabbit anti-GATA6 antibody (Cell Signaling Technology, Danvers, MA, USA) and rabbit anti-GLI1 antibody (Santa Cruz Biotechnology, Dallas, TX, USA) followed by goat anti-rabbit Alexa Fluor 488-conjugated antibodies (Invitrogen); and goat anti-SHH antibody, goat anti-HSD3B2, and goat anti-CYP11A1 (all Santa Cruz Biotechnology) antibodies followed by donkey anti-goat Alexa Fluor 555-conjugated antibodies (Invitrogen). All primary antibodies were diluted 1:300, and all secondary (conjugated) antibodies were diluted 1:500. Images of cells were obtained and photographed using an Olympus BX-51 microscope and an Olympus DP72 digital camera. Images were overlaid in Photoshop and assembled and labeled in Canvas, CorelDraw or Powerpoint. To quantify the GATA4; GATA6 levels, immunofluorescence staining for GATA4 and GATA6 was converted to grayscale and analyzed in a minimum of 20 cells in four sections. The lasso tool was used for nucleus contouring, and the integrated density immunofluorescence for each nucleus was calculated; the background was subtracted from each image. The Mann–Whitney test was performed, and the data were plotted in Excel and presented as corrected total cell fluorescence (CTCF) for nucleus $\pm$ SEM.

4.4. Western Blot Analysis

Whole cell lysates were prepared using a sodium deoxycholate lysis buffer. Nuclear protein extracts were prepared using a dual buffer method, with the first buffer containing a detergent and the second containing glycerol. Samples were isolated from well washed cell pellets from control or sodium butyrate-treated HAA1 cells and BJ fibroblasts (negative control), which were flash-frozen and kept at $-80\ ^{\circ}\mathrm{C}$. The protein in each sample was measured using a NanoDrop Lite spectrophotometer (ThermoFisher Scientific Inc., Waltham, MA, USA). A 100 µL aliquot was separated from the original sample and boiled with 4X LDS sample buffer (Invitrogen) for 5 min. A total of 30 µg of protein for each sample was loaded and resolved on a 12% SDS-PAGE gel along with a BenchMark Protein Ladder (Invitrogen), followed by electroblotting onto PVDF (BioRad, Hercules, CA, USA) membrane. The membranes were incubated with anti-SF1 antibody (Perseus Proteomics, Tokyo, Japan) followed by horseradish peroxidase (HRP)–conjugated anti-mouse secondary (BioRad), and anti-StAR antibody (Santa Cruz Biotechnology) followed by anti-rabbit HRP secondary antibody (BioRad). The HRP signal was developed using Clarify Western ECL substrate (BioRad) and detected using a Li-Cor scanner and Image Studio Digits version 3.1. Sample loading was confirmed through incubation with anti-beta actin antibody (Novus, Centennial, CO, USA), followed by anti-mouse HRP antibody and ECL development and detection. To quantify the protein levels, the staining was converted to grayscale and the images were inverted. The marquee tool was used for band contouring, and the integrated density for each band was calculated; the background was subtracted from each image. The Mann–Whitney test was performed, and the data were plotted in Excel and presented as mean $\pm$ SEM.

4.5. Chromosome Analysis

HAA-1 cells were cultured in RPMI 1640 medium supplemented with 10% FBS, and chromosome preparations were made at passages 25 and 155 following the standard air-drying technique. Aged slides were G-banded by Trypsin Giemsa technique. G-banded metaphase spreads were photographed using 80i Nikon Microscope and Applied Spectral Imaging (ASI) Karyotyping system. A minimum of ten metaphases were karyotyped.

4.6. HDACi Treatment

HAA1 cells were plated onto 60 mm plates at a density of 8.0×10^5 in RPMI 1640 medium supplemented with 10% FBS. Then, 24 h after plating, cells were either left untreated or treated with sodium butyrate (3 mM, Thermo) or trichostatin A (100–300 nM, Sigma-Aldrich, Saint Louis, MO, USA). Cells were maintained in the presence of the HDACi for a defined number of days, at which time the media was removed and cells harvested directly on the plate with TRI® reagent (Sigma-Aldrich, St. Louis, MO, USA). Experiments with HDAC isoform-specific inhibitors and suberoylanilide hydroxamic acid (SAHA) were performed in a similar fashion, except that HAA1 cells were grown in triplicate in 6-well plates (Corning) and either left untreated or treated for 4–6 days with sodium butyrate, Trichostatin A, SAHA (5 µM, #SML0061, Sigma-Aldrich, St. Louis, MO), CI994 (50 µM, #S2818, SelleckChem, Houston, TX, USA), PC-34051 (50 µM, #S2012, SelleckChem) or RGFP996 (50 µM, #S7229, SelleckChem, Houston, TX, USA). RNA from untreated and treated cells was prepared and analyzed as described below.

4.7. Total RNA Extraction, First cDNA Synthesis and Quantitative RT-PCR (qPCR)

Total RNA was isolated with the TRI® reagent (Sigma-Aldrich), following the manufacturer's recommendations, and treated with DNase I (Roche Diagnostics Corporation, Indianapolis, IN, USA), according to the vendor's instructions. DNase I-treated RNA was purified with Qiagen Mini columns (Qiagen, Germantown, MD, USA), and the quantity and quality of RNA were determined spectrophotometrically with a NanoDrop Lite spectrophotometer. Equal concentrations of total RNA were reverse transcribed using an M-MLV (Moloney Murine Leukemia Virus) Reverse Transcriptase kit (Invitrogen, Thermo), following the manufacturer's specifications. Quantitative RT-PCR experiments were performed in an ABI 7500 instrument (Applied Biosystems, Foster City, CA, USA) using SYBR Green PCR master mix (Applied Biosystems) under the following conditions: 40 cycles of 95 °C for 15 s and 60 °C for 1 min in a 2-step thermal cycle, preceded by two initial steps: 2 min at 50 °C and 10 min at 95 °C. The primer sequences are shown in Table 3.

Table 3. qRT-PCR primers.

Gene Name	Primers	Reference
1. *CYP11B1*	CYP11B1_For: GGCAGAGGCAGAGATGCTG CYP11B1_REV: TCTTGGGTTAGTGTCTCCACCTG	[44]
2. *CYP11B2*	CYP11B2_FOR: GGCAGAGGCAGAGATGCTG CYP11B2_REV: CTTGAGTTAGTGTCTCCACCAGGA	[44]
3. *CYP17A1*	CYP17A1_For: TGTGGACAAGGGCACAGAAG CYP17A1_Rev: GGATTCAAGAAACGCTCAGGC	[45]
4. *CYP11A1*	CYP11A1_FOR: AGCTAGAGATGACCATCTTCC CYP11A1_REV: GGCATCAGAATGAGGTTGAATG	[45]
5. CYP21A2	Cyp21a2_FOR: ACCTGTCCTTGGGAGACTAC Cyp21a2_REV: TGCGCTCACAGAACTCCTGGGT	[46]
6. HSD3B2	HSD3B2_FOR: AGAAGAGCCTCTGGAAAACACATG HSD3B2_REV: CGCACAAGTGTACAAGGTATCACCA	[47]
7. NR5A1	NR5A1_For: TGGCTACCTCTACCCTGCCTTTCC NR5A1_Rev: GCCTTCTCCTGAGCGTCTTTCACC	[48]
8. StAR	hStAR For: AAGACCAAACTTACGTGGC hStAR Rev: GTGGTTGGCAAAATCCACC	[45]

Table 3. *Cont.*

Gene Name	Primers	Reference
9. MC2R	MC2R_For: AGCCTGTCTGTGATTGCTG MC2R_Rev: AGATGACCGTAAGCACCACC	[45]
10. SULT2A1	SULT2A1_For: TGATGTCAGACTATAATTGGTTTGAAGGC SULT2A1_Rev: GGTTATGAGTCGTGGTCCTTCCTTATTG	[49]
11. AKR1C3	AKR1C3_For: GAGAAGTAAAGCTTTGGAGGTCACA AKR1C3_Rev: CAACCTGCTCCTCATTATTGTATAAATGA	[50]
12. CYB5A	CYB5A_For: CCAAAGTTAAACAAGCCTCCG CYB5A_Rev: TGTTCAGTCCTCTGCCATG	[51]
13. CYPA	CYPA_For: TATCTGCACTGCCAAGACTGAGTG CYPA_Rev: CTTCTTGCTGGTCTTGCCATTCC	[52]
14. ACTN	hbAct_For: TCACCATTGGCAATGAGCG hbAct_Rev: TGGAGTTGAAGGTAGTTTCGTG	[45]

For the initial analysis of the HAA1 cells, standardization was performed relative to cyclophilin A (*PPIA*) RNA, and the expression was compared to that in the NCI-H295R and the non-steroidogenic SW13 cell lines. For the analysis of the HDACi-treated cells, standardization of the qPCR data was performed with the endogenous reference *ACTB* (human beta actin) gene RNA. The samples were analyzed in triplicate from at least 3 biological replicates (independent experiments), and the fold change was calculated using the $\Delta\Delta Ct$ method. Statistical analysis (Student's *t*-test; two-tailed) was performed on the $\Delta\Delta Ct$ values, and the results were considered significant at $p < 0.05$. The results were graphed as fold-change differences relative to wild-type controls using GraphPad Prism®, San Diego, CA (6.02 version) software. Fold change equal to 1 represents no change in gene expression.

4.8. RNAseq Analysis

HAA1 cells were grown in triplicate and either left untreated or treated with sodium butyrate (3 mM). After six days, cells were harvested and total RNA isolated as described above. For quality control, RNA concentration was determined on a Qubit® 2.0 Fluorometer (ThermoFisher/Invitrogen, Grand Island, NY, USA). RNA quality was assessed using the Agilent 2100 Bioanalyzer (Agilent Technologies, Inc., Santa Clara, CA, USA). Only total RNA with 28S/18S > 1 and RNA integrity number (RIN) $\geq$ 7 were used for RNA-seq library construction. The RINs of RNA ranged between 7.5 and 9.4.

RNA library construction was performed at the Interdisciplinary Center for Biotechnology Research (ICBR) Gene Expression Core, University of Florida, and sequencing runs were performed in the NextGen core. RNA-seq library preparation was performed with 2 µL of 1:200 diluted RNA spike-in External RNA Controls Consortium (ERCC; $0.5\times$ of the amount suggested in the ERCC user guide: Cat# 4456740) and 1000 ng of total RNA, followed by mRNA isolation using NEBNext Poly(A) mRNA Magnetic Isolation module (New England Biolabs, catalog # E7490) and RNA library construction with NEBNext Ultra RNA Library Prep Kit for Illumina (New England Biolabs, catalog # E7530) according to the manufacturer's instructions. RNA fragmenting time was adjusted according to the RIN of total RNA. Briefly, 1000 ng of total RNA together with 2 µL of 1/200 diluted ERCC were incubated with 15 µL of NEBNext Magnetic Oligo d(T)25 and fragmented in an NEBNext First Strand Synthesis Buffer by heating at 94 °C for the desired time. First strand cDNA synthesis was performed using reverse transcriptase and random primers, and the synthesis of double-stranded DNA was completed using the second-strand master mix provided in the kit. The resulting double-stranded DNA was end-repaired, dA-tailed and ligated with NEBNext adaptors. Finally, the synthesized libraries were enriched by 13 cycles of amplification and purified by Meg-Bind RxnPure Plus beads (Omega Biotek, Norcross, GA, catalog # M1386). For library quality control and pooling, barcoded libraries were sized on the bioanalyzer, quantitated by QUBIT and qPCR (Kapa Biosystems, Wilmington, MA, catalog number: KK4824). A total of 12 individual libraries were pooled at equal molar value of 20 nM, and a total of 2 lanes of HiSeq 000 were run. Differentially expressed genes were plotted as a volcano plot and the genes with Qval < 4.78×10^{-95} and log2 (fold change) > 5 were assigned and labeled. Differentially expressed genes were further analyzed using Illumina Pathways analysis and PANTHER. Differential expression of genes belonging to the TNF alpha pathway was analyzed by qRT-PCR as described above.

4.9. Liquid Chromatography—Tandem Mass-Spectrometry (LC-MS/MS) Analysis

For the LC-MS/MS experiment, HAA1 cells were grown in 6-well plates in the RPMI media with 10% FBS. Treatments included 15 µM 22*R*-hydroxycholesterol, 15 µM pregnenolone, 10 µM forskolin, without (control) or with HDAC inhibitor, 50 µM of CI-994, to induce steroidogenic differentiation. On day six of the experiment, the media was replaced with the same media as above, except that FBS was omitted. Then, 1 mL aliquots of the media were collected at 0, 4, 8 and 24 h and frozen at −20 °C. Steroid quantitation of 20 3-keto-Δ^4 (Δ4) and three 3β-hydroxy-Δ^5 (Δ5) steroids was performed by LC-MS/MS as described previously [53,54].

4.10. Short Tandem Repeat (STR) Analysis

Genomic DNA from HAA1 cells was extracted and DNA profiling was performed using the AmpFLSTR Identifier PCR Amplification kit (Thermo Fisher Scientific) and subsequently analyzed on a 3730XL DNA analyzer (Thermo Fisher Scientific). The kit amplified 15 tetranucleotide repeat loci and Amelogenin gender-determining marker. The results were analyzed using GeneMapper v3.7 (Applied Biosystems, Waltham, MA, USA).

Supplementary Materials: The following supporting information can be downloaded at: https://www.mdpi.com/article/10.3390/ijms24010584/s1.

Author Contributions: For H.K.G.: designed and performed experiments, manuscript development, and analysis; Y.X.: performed key experiments; H.H.: performed key experiments; R.B.: performed key experiments; A.S.M.: performed key experiments; T.G.: analysis; W.B.B.: designed experiments; A.T.: performed key experiments; W.E.R.: designed experiments; J.R.: performed experiments; K.N.: performed key experiments; V.J.B.: experiment design; F.N.: experiment design; V.S.: performed key experiments; A.F.G.: experimental design; J.W.S.: experimental design; R.J.A.: experimental design; S.G.T.: designed and performed experiments, manuscript development, and analysis. All authors have read and agreed to the published version of the manuscript.

Funding: This research was supported by the Gatorade Trust Fund from the Department of Medicine, University of Florida Gatorade (Project# 00122235), Southeast Center for Integrated Metabolomics, University of Florida, and Malcom Randall VA Pilot Award Program (H.K.G).

Institutional Review Board Statement: The study was conducted in accordance with the Declaration of Helsinki, and approved by the Institutional Review Board at the University of Texas Southwestern Medical Center (IRB File 052004-044).

Informed Consent Statement: Informed consent was obtained from the subject involved in the study.

Data Availability Statement: All materials described in this publication are available upon request.

Acknowledgments: The authors are grateful to Sen Pathak, from the University of Texas MD Anderson Cancer Center for helpful discussions and guidance in the development of this manuscript. In addition, services from the University of Michigan Advanced Genomics Core for the performance of short tandem repeat analysis are appreciated.

Conflicts of Interest: The authors declare no competing interest.

References

1. Tanaka, S.; Matsuzawa, A. Comparison of adrenocortical zonation in C57BL/6J and DDD mice. *Exp. Anim.* **1995**, *44*, 285–291. [CrossRef] [PubMed]
2. Tanaka, S.; Nishimura, M.; Kitoh, J.; Matsuzawa, A. Strain difference of the adrenal cortex between A/J and SM/J mice, progenitors of SMXA recombinant inbred group. *Exp. Anim.* **1995**, *44*, 127–130. [CrossRef] [PubMed]
3. Inomata, A.; Sasano, H. Practical approaches for evaluating adrenal toxicity in nonclinical safety assessment. *J. Toxicol. Pathol.* **2015**, *28*, 125–132. [CrossRef] [PubMed]
4. Gannon, A.L.; O'Hara, L.; Mason, J.I.; Jorgensen, A.; Frederiksen, H.; Milne, L.; Smith, S.; Mitchell, R.T.; Smith, L.B. Androgen receptor signalling in the male adrenal facilitates X-zone regression, cell turnover and protects against adrenal degeneration during ageing. *Sci. Rep.* **2019**, *9*, 10457. [CrossRef]
5. Ghayee, H.K.; Auchus, R.J. Basic concepts and recent developments in human steroid hormone biosynthesis. *Rev. Endocr. Metab. Disord.* **2007**, *8*, 289–300. [CrossRef]
6. Simon, D.P.; Hammer, G.D. Adrenocortical stem and progenitor cells: Implications for adrenocortical carcinoma. *Mol. Cell. Endocrinol.* **2012**, *351*, 2–11. [CrossRef]
7. Wood, M.A.; Acharya, A.; Finco, I.; Swonger, J.M.; Elston, M.J.; Tallquist, M.D.; Hammer, G.D. Fetal adrenal capsular cells serve as progenitor cells for steroidogenic and stromal adrenocortical cell lineages in M. musculus. *Development* **2013**, *140*, 4522–4532. [CrossRef]
8. Morohashi, K.; Honda, S.; Inomata, Y.; Handa, H.; Omura, T. A common trans-acting factor, Ad4-binding protein, to the promoters of steroidogenic P-450s. *J. Biol. Chem.* **1992**, *267*, 17913–17919. [CrossRef]
9. Luo, X.; Ikeda, Y.; Parker, K.L. A cell-specific nuclear receptor is essential for adrenal and gonadal development and sexual differentiation. *Cell* **1994**, *77*, 481–490. [CrossRef]
10. Achermann, J.C.; Ozisik, G.; Ito, M.; Orun, U.A.; Harmanci, K.; Gurakan, B.; Jameson, J.L. Gonadal determination and adrenal development are regulated by the orphan nuclear receptor steroidogenic factor-1, in a dose-dependent manner. *J. Clin. Endocrinol. Metab.* **2002**, *87*, 1829–1833. [CrossRef]
11. Achermann, J.C.; Ito, M.; Hindmarsh, P.C.; Jameson, J.L. A mutation in the gene encoding steroidogenic factor-1 causes XY sex reversal and adrenal failure in humans. *Nat. Genet.* **1999**, *22*, 125–126. [CrossRef] [PubMed]
12. Tevosian, S.; Jimenez, E.; Hatch, H.M.; Jiang, T.; Morse, D.A.; Fox, S.; Padua, M.B. Adrenal Development in Mice Requires GATA4 and GATA6 Transcription Factors. *Endocrinology* **2015**, *156*, 2503–2517. [CrossRef] [PubMed]
13. Padua, M.B.; Jiang, T.; Morse, D.A.; Fox, S.C.; Hatch, H.M.; Tevosian, S.G. Combined Loss of the GATA4 and GATA6 Transcription Factors in Male Mice Disrupts Testicular Development and Confers Adrenal-Like Function in the Testes. *Endocrinology* **2015**, *156*, 1873–1886. [CrossRef] [PubMed]
14. Rainey, W.E.; Bird, I.M.; Mason, J.I. The NCI-H295 cell line: A pluripotent model for human adrenocortical studies. *Mol. Cell. Endocrinol.* **1994**, *100*, 45–50. [CrossRef]
15. Wang, T.; Rainey, W.E. Human adrenocortical carcinoma cell lines. *Mol. Cell. Endocrinol.* **2012**, *351*, 58–65. [CrossRef]
16. Gazdar, A.F.; Oie, H.K.; Shackleton, C.H.; Chen, T.R.; Triche, T.J.; Myers, C.E.; Chrousos, G.P.; Brennan, M.F.; Stein, C.A.; La Rocca, R.V. Establishment and characterization of a human adrenocortical carcinoma cell line that expresses multiple pathways of steroid biosynthesis. *Cancer Res.* **1990**, *50*, 5488–5496.
17. Newbold, A.; Falkenberg, K.J.; Prince, M.H.; Johnstone, R.W. How do tumor cells respond to HDAC inhibition? *FEBS J.* **2016**, *283*, 4032–4046. [CrossRef]
18. Gallo-Payet, N. 60 YEARS OF POMC: Adrenal and extra-adrenal functions of ACTH. *J. Mol. Endocrinol.* **2016**, *56*, T135–T156. [CrossRef]
19. King, P.; Paul, A.; Laufer, E. SHH signaling regulates adrenocortical development and identifies progenitors of steroidogenic lineages. *Proc. Natl. Acad. Sci. USA* **2009**, *106*, 21185–21190. [CrossRef]
20. Bandiera, R.; Vidal, V.P.; Motamedi, F.J.; Clarkson, M.; Sahut-Barnola, I.; von Gise, A.; Pu, W.T.; Hohenstein, P.; Martinez, A.; Schedl, A. WT1 maintains adrenal-gonadal primordium identity and marks a population of AGP-like progenitors within the adrenal gland. *Dev. Cell* **2013**, *27*, 5–18. [CrossRef]
21. Huang, C.C.; Miyagawa, S.; Matsumaru, D.; Parker, K.L.; Yao, H.H. Progenitor cell expansion and organ size of mouse adrenal is regulated by sonic hedgehog. *Endocrinology* **2010**, *151*, 1119–1128. [CrossRef] [PubMed]
22. Ching, S.; Vilain, E. Targeted disruption of Sonic Hedgehog in the mouse adrenal leads to adrenocortical hypoplasia. *Genesis* **2009**, *47*, 628–637. [CrossRef] [PubMed]
23. Ghayee, H.K.; Rege, J.; Watumull, L.M.; Nwariaku, F.E.; Carrick, K.S.; Rainey, W.E.; Miller, W.L.; Auchus, R.J. Clinical, biochemical, and molecular characterization of macronodular adrenocortical hyperplasia of the zona reticularis: A new syndrome. *J. Clin. Endocrinol. Metab.* **2011**, *96*, E243–E250. [CrossRef] [PubMed]
24. Zubair, M.; Oka, S.; Parker, K.L.; Morohashi, K. Transgenic expression of Ad4BP/SF-1 in fetal adrenal progenitor cells leads to ectopic adrenal formation. *Mol. Endocrinol.* **2009**, *23*, 1657–1667. [CrossRef] [PubMed]
25. Riggs, M.G.; Whittaker, R.G.; Neumann, J.R.; Ingram, V.M. n-Butyrate causes histone modification in HeLa and Friend erythroleukaemia cells. *Nature* **1977**, *268*, 462–464. [CrossRef] [PubMed]
26. Simpson, R.T. Modification of chromatin with acetic anhydride. *Biochemistry* **1971**, *10*, 4466–4470. [CrossRef]

27. Easwaran, H.; Tsai, H.C.; Baylin, S.B. Cancer epigenetics: Tumor heterogeneity, plasticity of stem-like states, and drug resistance. *Mol. Cell* **2014**, *54*, 716–727. [CrossRef]

28. Rainey, W.E. Adrenal zonation: Clues from 11beta-hydroxylase and aldosterone synthase. *Mol. Cell. Endocrinol.* **1999**, *151*, 151–160. [CrossRef]

29. Shahbazian, M.D.; Grunstein, M. Functions of site-specific histone acetylation and deacetylation. *Annu. Rev. Biochem.* **2007**, *76*, 75–100. [CrossRef]

30. Zhao, Y.; Tan, J.; Zhuang, L.; Jiang, X.; Liu, E.T.; Yu, Q. Inhibitors of histone deacetylases target the Rb-E2F1 pathway for apoptosis induction through activation of proapoptotic protein Bim. *Proc. Natl. Acad. Sci. USA* **2005**, *102*, 16090–16095. [CrossRef]

31. Juan, L.J.; Shia, W.J.; Chen, M.H.; Yang, W.M.; Seto, E.; Lin, Y.S.; Wu, C.W. Histone deacetylases specifically down-regulate p53-dependent gene activation. *J. Biol. Chem.* **2000**, *275*, 20436–20443. [CrossRef] [PubMed]

32. Chen, W.Y.; Weng, J.H.; Huang, C.C.; Chung, B.C. Histone deacetylase inhibitors reduce steroidogenesis through SCF-mediated ubiquitination and degradation of steroidogenic factor 1 (NR5A1). *Mol. Cell. Biol.* **2007**, *27*, 7284–7290. [CrossRef] [PubMed]

33. Falkenberg, K.J.; Johnstone, R.W. Histone deacetylases and their inhibitors in cancer, neurological diseases and immune disorders. *Nat. Rev. Drug Discov.* **2014**, *13*, 673–691. [CrossRef] [PubMed]

34. Boeger, H.; Griesenbeck, J.; Strattan, J.S.; Kornberg, R.D. Nucleosomes unfold completely at a transcriptionally active promoter. *Mol. Cell* **2003**, *11*, 1587–1598. [CrossRef]

35. Reinke, H.; Horz, W. Histones are first hyperacetylated and then lose contact with the activated PHO5 promoter. *Mol. Cell* **2003**, *11*, 1599–1607. [CrossRef]

36. Schones, D.E.; Cui, K.; Cuddapah, S.; Roh, T.Y.; Barski, A.; Wang, Z.; Wei, G.; Zhao, K. Dynamic regulation of nucleosome positioning in the human genome. *Cell* **2008**, *132*, 887–898. [CrossRef]

37. Uchida, T.; Nishimoto, K.; Fukumura, Y.; Asahina, M.; Goto, H.; Kawano, Y.; Shimizu, F.; Tsujimura, A.; Seki, T.; Mukai, K.; et al. Disorganized Steroidogenesis in Adrenocortical Carcinoma, a Case Study. *Endocr. Pathol.* **2017**, *28*, 27–35. [CrossRef]

38. Bernardini, R.; Kamilaris, T.C.; Calogero, A.E.; Johnson, E.O.; Gomez, M.T.; Gold, P.W.; Chrousos, G.P. Interactions between tumor necrosis factor-alpha, hypothalamic corticotropin-releasing hormone, and adrenocorticotropin secretion in the rat. *Endocrinology* **1990**, *126*, 2876–2881. [CrossRef]

39. Hantel, C.; Ozimek, A.; Lira, R.; Ragazzon, B.; Jackel, C.; Frantsev, R.; Reincke, M.; Bertherat, J.; Mussack, T.; Beuschlein, F. TNF alpha signaling is associated with therapeutic responsiveness to vascular disrupting agents in endocrine tumors. *Mol. Cell. Endocrinol.* **2016**, *423*, 87–95. [CrossRef]

40. Mikhaylova, I.V.; Kuulasmaa, T.; Jaaskelainen, J.; Voutilainen, R. Tumor necrosis factor-alpha regulates steroidogenesis, apoptosis, and cell viability in the human adrenocortical cell line NCI-H295R. *Endocrinology* **2007**, *148*, 386–392. [CrossRef]

41. Murakami, M.; Yoshimoto, T.; Nakano, Y.; Tsuchiya, K.; Minami, I.; Bouchi, R.; Fujii, Y.; Nakabayashi, K.; Hashimoto, K.; Hata, K.I.; et al. Expression of inflammation-related genes in aldosterone-producing adenomas with KCNJ5 mutation. *Biochem. Biophys. Res. Commun.* **2016**, *476*, 614–619. [CrossRef] [PubMed]

42. Heyninck, K.; Beyaert, R. Crosstalk between NF-kappaB-activating and apoptosis-inducing proteins of the TNF-receptor complex. *Mol. Cell Biol. Res. Commun.* **2001**, *4*, 259–265. [CrossRef] [PubMed]

43. Fotin-Mleczek, M.; Henkler, F.; Samel, D.; Reichwein, M.; Hausser, A.; Parmryd, I.; Scheurich, P.; Schmid, J.A.; Wajant, H. Apoptotic crosstalk of TNF receptors: TNF-R2-induces depletion of TRAF2 and IAP proteins and accelerates TNF-R1-dependent activation of caspase-8. *J. Cell Sci.* **2002**, *115 Pt 13*, 2757–2770. [CrossRef]

44. Ye, P.; Nakamura, Y.; Lalli, E.; Rainey, W.E. Differential effects of high and low steroidogenic factor-1 expression on CYP11B2 expression and aldosterone production in adrenocortical cells. *Endocrinology* **2009**, *150*, 1303–1309. [CrossRef] [PubMed]

45. Xu, B.; Yang, W.H.; Gerin, I.; Hu, C.D.; Hammer, G.D.; Koenig, R.J. Dax-1 and steroid receptor RNA activator (SRA) function as transcriptional coactivators for steroidogenic factor 1 in steroidogenesis. *Mol. Cell. Biol.* **2009**, *29*, 1719–1734. [CrossRef]

46. Witchel, S.F.; Lee, P.A.; Suda-Hartman, M.; Trucco, M.; Hoffman, E.P. Evidence for a heterozygote advantage in congenital adrenal hyperplasia due to 21-hydroxylase deficiency. *J. Clin. Endocrinol. Metab.* **1997**, *82*, 2097–2101. [CrossRef]

47. Doi, M.; Takahashi, Y.; Komatsu, R.; Yamazaki, F.; Yamada, H.; Haraguchi, S.; Emoto, N.; Okuno, Y.; Tsujimoto, G.; Kanematsu, A.; et al. Salt-sensitive hypertension in circadian clock-deficient Cry-null mice involves dysregulated adrenal Hsd3b6. *Nat. Med.* **2010**, *16*, 67–74. [CrossRef]

48. Morales, A.; Vilchis, F.; Chavez, B.; Morimoto, S.; Chan, C.; Robles-Diaz, G.; Diaz-Sanchez, V. Differential expression of steroidogenic factors 1 and 2, cytochrome p450scc, and steroidogenic acute regulatory protein in human pancreas. *Pancreas* **2008**, *37*, 165–169. [CrossRef]

49. Shimizu, C.; Fuda, H.; Yanai, H.; Strott, C.A. Conservation of the hydroxysteroid sulfotransferase SULT2B1 gene structure in the mouse: Pre- and postnatal expression, kinetic analysis of isoforms, and comparison with prototypical SULT2A1. *Endocrinology* **2003**, *144*, 1186–1193. [CrossRef]

50. Mantel, A.; Carpenter-Mendini, A.B.; Vanbuskirk, J.B.; De Benedetto, A.; Beck, L.A.; Pentland, A.P. Aldo-keto reductase 1C3 is expressed in differentiated human epidermis, affects keratinocyte differentiation, and is upregulated in atopic dermatitis. *J. Investig. Derm.* **2012**, *132*, 1103–1110. [CrossRef]

51. Rhoads, K.; Sacco, J.C.; Drescher, N.; Wong, A.; Trepanier, L.A. Individual variability in the detoxification of carcinogenic arylhydroxylamines in human breast. *Toxicol Sci* **2011**, *121*, 245–256. [CrossRef] [PubMed]

52. Batalha, V.L.; Ferreira, D.G.; Coelho, J.E.; Valadas, J.S.; Gomes, R.; Temido-Ferreira, M.; Shmidt, T.; Baqi, Y.; Buee, L.; Muller, C.E.; et al. The caffeine-binding adenosine A2A receptor induces age-like HPA-axis dysfunction by targeting glucocorticoid receptor function. *Sci. Rep.* **2016**, *6*, 31493. [CrossRef] [PubMed]
53. Turcu, A.F.; Rege, J.; Chomic, R.; Liu, J.; Nishimoto, H.K.; Else, T.; Moraitis, A.G.; Palapattu, G.S.; Rainey, W.E.; Auchus, R.J. Profiles of 21-Carbon Steroids in 21-hydroxylase Deficiency. *J. Clin. Endocrinol. Metab.* **2015**, *100*, 2283–2290. [CrossRef] [PubMed]
54. Turcu, A.F.; Nanba, A.T.; Chomic, R.; Upadhyay, S.K.; Giordano, T.J.; Shields, J.J.; Merke, D.P.; Rainey, W.E.; Auchus, R.J. Adrenal-derived 11-oxygenated 19-carbon steroids are the dominant androgens in classic 21-hydroxylase deficiency. *Eur. J. Endocrinol.* **2016**, *174*, 601–609. [CrossRef]

International Journal of
Molecular Sciences

Article

Inclusion of 11-Oxygenated Androgens in a Clinical Routine LC-MS/MS Setup for Steroid Hormone Profiling

Robert Zeidler, Ronald Biemann, Uta Ceglarek, Jürgen Kratzsch, Berend Isermann and Alexander Gaudl *

Institute of Laboratory Medicine, Clinical Chemistry and Molecular Diagnostics, Leipzig University, Liebigstraße 27, 04103 Leipzig, Germany
* Correspondence: alexander.gaudl@medizin.uni-leipzig.de

Abstract: 11-Oxygenated androgens (11-OAs) are being discussed as potential biomarkers in diagnosis and therapy control of disorders with androgen excess such as congenital adrenal hyperplasia and polycystic ovary syndrome. However, quantification of 11-OAs by liquid chromatography-tandem mass spectrometry (LC-MS/MS) still relies on extensive sample preparation including liquid–liquid extraction, derivatization and partial long runtimes, which is unsuitable for high-throughput analysis under routine laboratory settings. For the first time, an established online-solid-phase extraction-LC-MS/MS (online-SPE-LC-MS/MS) method for the quantitation of seven serum steroids in daily routine use was extended and validated to include 11-ketoandrostenedione, 11-ketotestosterone, 11β-hydroxyandrostenedione and 11β-hydroxytestosterone. Combining a simple protein precipitation step with fast chromatographic separation and ammonium fluoride-modified ionization resulted in a high-throughput method (6.6 min run time) featuring lower limits of quantification well below endogenous ranges (63–320 pmol/L) with recoveries between 85% and 117% (CVs ≤ 15%). Furthermore, the ability of this method to distinguish between adrenal and gonadal androgens was shown by comparing 11-OAs in patients with hyperandrogenemia to healthy controls. Due to the single shot multiplex design of the method, potential clinically relevant ratios of 11-OAs and corresponding androgens were readily available. The fully validated method covering endogenous concentration levels is ready to investigate the diagnostic values of 11-OAs in prospective studies and clinical applications.

Keywords: 11-oxygenated androgens; androgens; steroid hormones; LC-MS/MS; method validation

Citation: Zeidler, R.; Biemann, R.; Ceglarek, U.; Kratzsch, J.; Isermann, B.; Gaudl, A. Inclusion of 11-Oxygenated Androgens in a Clinical Routine LC-MS/MS Setup for Steroid Hormone Profiling. *Int. J. Mol. Sci.* **2023**, *24*, 539. https://doi.org/10.3390/ijms24010539

Academic Editor: Jacques J. Tremblay

Received: 1 November 2022
Revised: 21 December 2022
Accepted: 23 December 2022
Published: 29 December 2022

1. Introduction

The adrenal gland is the source of 11-hydroxylated androgens, which are primary synthesized in the *zona reticularis* by 11β-hydroxylase (CYP11B1) from androstenedione and testosterone under regulation of adrenocorticotropic hormone [1]. The low potent androgens 11β-hydroxyandrostenedione (11-OHA4) and 11β-hydroxytestosterone (11-OHT) are precursors for the higher potent 11-ketoandrostenedione (11-KA4) and 11-ketotestosterone (11-KT) which are formed by 11β-HSDB2 (11β-hydroxysteroid dehydrogenase type 2) in adrenal glands and kidneys (Figure 1) [2–5]. In adipose tissues, the synthesis of 11-KT from 11-KA, as well as the metabolization of the higher potent 11-ketoandrogens to 11-hydroxylated androgens by 11β-HSDB1 (11β-hydroxysteroid dehydrogenase type 1), is suggested [3,6,7]. Due to their origin, 11-oxygenated androgens (11-OAs) allow the differentiation of adrenal- and gonadal-produced androgens and provide a potential diagnostic tool to reliably assign contributions of these organs to disorders of androgen synthesis, such as polycystic ovarian syndrome (PCOS), androgen-producing or -dependent tumors, and for therapy control in congenital adrenal hyperplasia (CAH) [3,8–14]. Furthermore, the ratios of 11-OAs to testosterone or androstenedione are discussed as potential biomarkers regarding disorders of androgen synthesis [12,13,15].

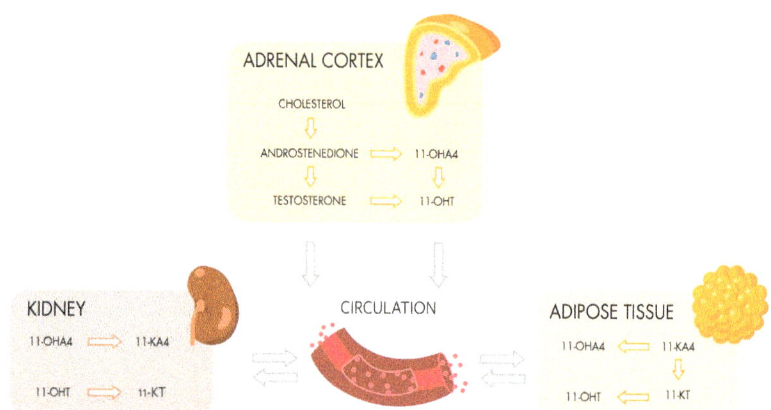

Figure 1. Metabolism of 11-OAs and their circulation in peripheral tissues. Circulating 11-OHA4 and 11-OHT is de novo synthesized from cholesterol via androstenedione and testosterone in adrenal glands. 11-OHA4 and 11-OHT are metabolized to 11-KA4 and 11-KT in kidneys and vice versa converted as well as metabolized to 11-KA4 from 11-KT in adipose tissues. 11-KA4, 11-ketoandrostenedione; 11-KT, 11-ketotestosterone; 11-OHA4, 11β-hydroxyandrostenedione; 11-OHT, 11β-hydroxytestosterone.

Considering their diagnostic potential, 11-OAs are not yet established as parameters in clinical routine analysis, due to low serum concentration levels, high laborious effort and interferences in routine immunoassays. First mass spectrometry-based methods to reliably quantify 11-OAs were published between 2008 and 2011 and were limited by, e.g., lack of corresponding internal standards, long runtimes, low sensitivity or missing validation for human serum samples as well as a lack of certified calibrators and quality controls [16–19]. In recent years, several newly developed LC-MS/MS methods have been published partially overcoming those limitations [10,13,15,20–32]. These analytical methods, including validation data as well as advantages and disadvantages, are described in detail in Caron et al. (2021) [21]. Overall, published methods rely on extensive sample preparation including liquid–liquid extraction and derivatization, which is unsuitable for high-throughput analysis in clinical routine diagnostics. Therefore, the major objective of this work was the inclusion of 11-oxygenated androgens in an established routine online-SPE-LC-MS/MS setup for profiling of seven clinically relevant steroid hormones including 17α-hydroxyprogesterone (17-OHP), aldosterone (A), androstenedione (A4), cortisol (F), cortisone (E), dehydroepiandrosterone sulfate (DHEAS), estradiol (E2), progesterone (P) and testosterone (T). Details of the method are described in Gaudl et al., (2016) [33].

2. Results

Multiple reaction monitoring of the four 11-OAs was integrated into the established LC-MS/MS setup as shown in Figure 2. Limits of detection (LODs) were calculated at 15 pmol/L for 11-KA4, 18 pmol/L for 11-KT, 32 pmol/L for 11-OHA4 and 19 pmol/L for 11-OHT (see Table S2). Linearity was proven between 0.08 and 3.3 nmol/l for 11-KA4, 11-KT, 11-OHT, and between 0.8 and 33 nmol/l for 11-OHA4. Relative standard deviations of the slopes of the calibration curves were below 4% with $R^2 > 0.999$. Among the 11-OAs, multiple interferences were observed (Table S1). Interferences above 1% of the original signal intensity were chromatographically separated (R > 1.5) except for 11-OHA4-d7 interfering with 11-OHT, adding 1.3% of its original signal intensity to the analyte. In serum samples, post column infusion showed an intensity loss by the first fraction of matrix constituents reaching the mass spectrometer and affected all analytes starting with the red dashed line in example chromatograms for 11-OAs in Figure S1. The ion suppression by the matrix of serum samples was observed across the detection window affecting all mass transitions of all analytes and internal standards (intensity loss in serum: 40% for 11-KA4, 50% for 11-KT and 11-OHA4 and 30% for 11-OHT). Specific ion-suppressing effects at the retention times of the analytes were not detected. Example chromatograms are shown in Figure S1.

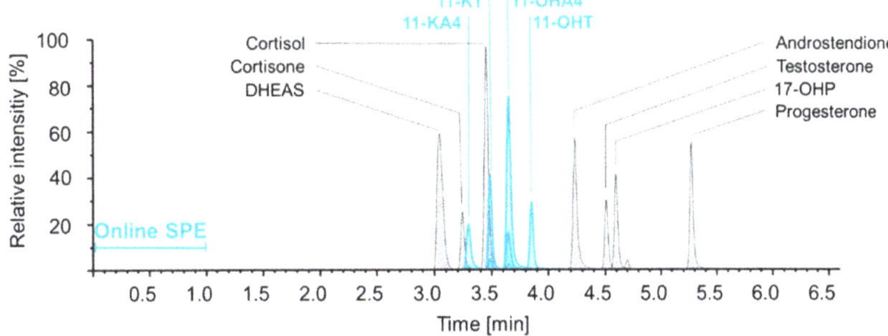

Figure 2. Exemplary chromatogram for single shot analysis of 11 steroid hormones. Chromatographic separation with retention times and relative intensities of 11-OAs (blue; 11-KA4, 11-KT, 11-OHA4 and 11-OHT) and established steroid panel (grey; 17-OHP, androstenedione, cortisol, cortisone, DHEAS, testosterone and progesterone) [34]. 11-KA4, 11-ketoandrostenedione; 11-KT, 11-ketotestosterone; 11-OHA4, 11β-hydroxyandrostenedione; 11-OHT, 11β-hydroxytestosterone; 17-OHP, 17α-hydroxyprogesterone; DHEAS, dehydroepiandrosterone sulfate; SPE, solid phase extraction.

Inter- and intra-assay coefficients of variation (CVs) in spiked serum and spiked controls were between 2% and 13% for 11-KA4, 2% and 15% for 11-KT, 2% and 7% for 11-OHA4, 2% and 10% for 11-OHT. Mean recovery ranges in spiked serum and spiked controls were between 102% and 115% for 11-KA4, 85% and 105% for 11-KT, 100% and 114% for 11-OHA4 and 99% and 117% for 11-OHT (Table 1). Lower limits of quantification (LLOQs) were determined at 63 pmol/L for 11-KA4 (CV 20%, s/n = 13), 100 pmol/L for 11-KT (CV 9%, s/n = 17), 320 pmol/L for 11-OHA4 (CV 3.9%, s/n = 30) and 83 pmol/L for 11-OHT (CV 5.7%, s/n = 13). In freeze/thaw stability experiments, the concentrations of 11-OAs were stable across five cycles with a reproducibility within the acceptable limit of 20% and without increasing or decreasing trend (Figure 3).

Table 1. Inter-assay imprecision and recovery of 11-oxygenated androgens. Means and coefficients of variation for spiked serum and quality controls at low, moderate and high concentration levels.

	11-KA4			11-KT			11-OHA4			11-OHT		
	Mean [nmol/L]	CV	Recovery	Mean [nmol/L]	CV	Recovery	Mean [nmol/L]	CV	Recovery	Mean [nmol/L]	CV	Recovery
Serum Level 1	0.34	13%	102%	1.5	7%	91%	6.0	5%	113%	0.44	10%	108%
Serum Level 2	0.9	10%	107%	1.9	4%	86%	12	6%	114%	0.62	9%	113%
Serum Level 3	19	7%	115%	16	3%	88%	188	4%	111%	1.2	6%	116%
QK Level 1	0.18	7%	109%	0.14	10%	85%	1.8	5%	107%	0.18	13%	109%
QK Level 2	0.7	10%	108%	0.6	4%	85%	7.1	3%	107%	0.69	5%	105%
QK Level 3	19	8%	113%	15	5%	90%	165	6%	100%	16	3%	99%

QK, quality control; 11-KA4, 11-ketoandrostenedione; 11-KT, 11-ketotestosterone; 11-OHA4, 11β-hydroxyandrostenedione; 11-OHT, 11β-hydroxytestosterone; CV, coefficient of variation.

For clinical verification, comparison of CAH patients and healthy controls revealed elevated levels of 11-KA4 (p = 0.016), 11-OHA4 (p = 0.006), 11-OHT (p = 0.08), 11-KT (p = 0.001), 17 OHP (p < 0.001), T (p = 0.015) and DHEAS (p = 0.012) as expected (Figure 4 and Table S6).

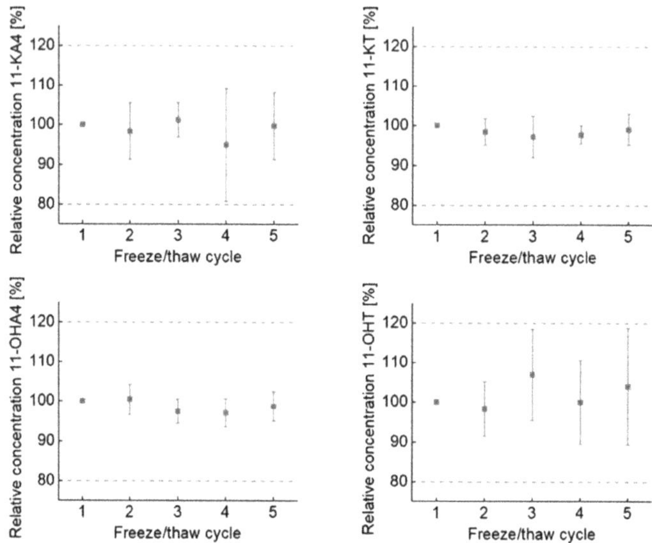

Figure 3. 11-OAs are stable across repeated freeze/thaw cycles. Dot plots show mean and coefficient of variation of 5 individually sample preparations. Data are presented relative to the concentration of first freeze/thaw cycle. Analyte ranges were from 3.1 nmol/L to 8.1 nmol/L for 11-OHA4, 0.3 nmol/L to 0.8 nmol/L for 11-KA4, 0.3 nmol/L to 1.2 nmol/L for 11-OHT and 0.6 nmol/L to 2.6 nmol/L for 11-KT. Error bars represent standard derivations, dashed lines indicate the acceptable limit of change ($\pm$20%). 11-KA4, 11-ketoandrostenedione; 11-KT, 11-ketotestosterone; 11-OHA4, 11β-hydroxyandrostenedione; 11-OHT, 11β-hydroxytestosterone;.

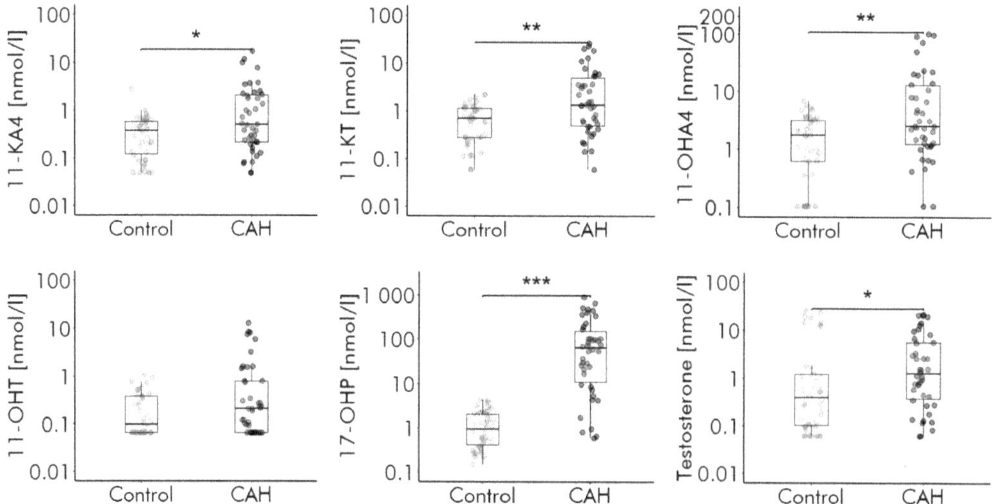

Figure 4. Levels of 17-OHP, testosterone and 11-OAs are elevated in CAH. Boxplots show median and interquartile range, whiskers indicate 95% confidence interval; * $p < 0.01$, ** $p < 0.05$, *** $p < 0.001$. CAH, congential adrenal hyperplasia; 11-KA4, 11-ketoandrostenedione; 11-KT, 11-ketotestosterone; 11-OHA4, 11β-hydroxyandrostenedione; 11-OHT, 11β-hydroxytestosterone; 17-OHP, 17α-hydroxyprogesterone.

Accordingly, ratios of A/11-OHT (p = 0.004), DHEAS/11-OAs (p < 0.001), T/11-OHT (p = 0.042) were decreased in CAH patients, reflecting a higher adrenal versus gonadal androgen synthesis. 11-OA levels in the 42 healthy individuals were comparable to published concentration levels for 11-OAs (Table S3).

3. Discussion

11-OAs are currently being discussed as biomarkers to improve diagnosis and therapy of diseases that are associated with disturbed androgen production, including CAH, PCOS, premature adrenarche, metabolic syndrome or obesity [3,8–14]. However, the quantification of 11-OA is labor intensive and remains a challenge for routine clinical applications. By implementing 11-OAs into an established routine online solid phase extraction online-SPE-LC-MS/MS setup for steroids, these challenges were already resolved and we enabled the simultaneous analysis of 11-OAs and seven relevant steroids within a single run using only 100 µL serum. The presented method covers 11 clinically relevant steroid hormones with fast and simple sample preparation protocol and a short runtime of 6.6 min. Compared to run times of published high-throughput LC-MS/MS methods (4 to 15 min) the presented method can be ranked within the fast ones [14,15,20,23,25,27,28,30]. While these methods utilize liquid–liquid extraction, including evaporation and re-suspension, the single protein precipitation step of the described method generates minimal hands-on time prior to LC-MS/MS measurement, which has been proven to be reliable during five years of routine application [33,34].

Similar to the referenced routine steroid hormones method, imprecision as well as recoveries are within the acceptable limits given by CLSI guideline C62-A, confirming the validity of the method [35]. The obtained sensitivity competes with the most sensitive methods utilizing complex derivatization and multi-stage liquid–liquid extraction [10,24,27,30,32]. In contrast to others, LLOQs and LODs were determined in native sera, reflecting genuine patient samples instead of diluted standards or serum with decreasing matrix influence. Since the experimentally determined LLOQs of the current method are at least three times below the published endogenous ranges of 11-OAs (Table S3), the proposed method is sufficiently sensitive to improve the assessment of clinically relevant hyperandrogenism [7].

Through previous experience in steroid analysis, interferences between the analytes with similar masses ranging between 301.2 g/mol and 311.2 g/mol and similar fragments were expected (Table S5) as well as interfering masses by incomplete deuteration of internal standards and proton substitution within the ion source. Sufficient chromatographic separated interferences were non-relevant and neglected. The interference of 11-OHA-d7 with 11-OHT delivers an 1.3% of 4e5 intense signal which is negligible below the noise of 2e3 for 11-OHT in endogenous samples.

In post-column experiments, the loss of sensitivity in serum compared to methanol and calibrator is most probably caused by the serum matrix affecting standards and internal standards equally. Therefore, a negative effect on the determined concentration of any given analyte can be neglected. The negative effect on LLOQs, however, is considered as a price of compromising between sensitivity, selectivity, and high-throughput capability. As repeated freeze/thaw cycles (n = 5) have no effect on the analytical stability of 11-OAs, the method is also suitable for batchwise analysis of clinical studies.

To improve clinical diagnostics, the method must reliably discriminate between patients with and without disturbed androgen production. The determined levels for 11-OAs in 13 healthy individuals are comparable to reported concentration levels for 11-OAs in healthy individuals, thus indicating a correct determination of 11-OAs by the method [7,32]. Elevated levels of 11-KA4, 11-KT and 11-OHA4 in CAH patients similar to recently reported findings in adults are proving the clinical verification of the method (Figure 4 and Table S3). The non-significant elevation of 11-OHT indicates a lack of power due to small sample size of n = 42. Decreased ratios of DHEAS and T to 11-OAs confirm the expected increased adrenal versus gonadal androgen synthesis in CAH patients [36]. Despite a fast and easy sample preparation protocol combined with short runtimes, the method delivers results for

CAH patients as well as healthy individuals that are similar to recently reported methods utilizing liquid–liquid extraction and derivatization. Since commercial calibrators and traceable controls for 11-OAs are not accessible yet, the fully validated method containing in-house spiked calibrators and controls is ready to investigate the diagnostic values of 11-OAs in prospective studies and clinical applications. A minimal sample volume of 100 µL makes it highly relevant for children as well.

In conclusion, we present a robust high-throughput method with high sensitivity for simultaneous quantification of four 11-OAs and seven relevant steroid hormones using minimal sample preparation. By multiplex-design in single shot analysis, ratios of clinically relevant steroids to 11-OAs are instantly accessible, allowing the application of 11-OAs in future routine diagnostics and therapy control in disorders associated with androgen excess. In future studies, the method will be used for determination of reference intervals (from birth to 80 years) as well as to investigate the influence of 11-OAs on obesity, metabolic syndrome, and puberty.

4. Materials and Methods

4.1. Chemicals and Reagents

Deionized water was produced in-house using a Barnstead Nanopure from Thermo Scientific, Waltham, MA, USA. Zinc sulfate heptahydrate was obtained from Merck, Darmstadt, Germany, ammonium fluoride from Sigma Aldrich, St. Louis, MO, USA, LC-MS grade methanol from Biosolve, Valkenswaard, The Netherlands. 11-KA4, 11-KT, 11-OHA4, 11-OHT were purchased from Steraloids, Inc., Newport, RI, USA, 11-KT-16,16,17-d3 from Eurisotop GmbH, Saarbrücken, Germany, 11-OHT-2,2,4,6,6-d5 from CDN Isotopes, Pointe-Claire, Quebec, Canada, 11-KA4-3,3,6,6,7,7,9,10,10,17-d10 from LGC Group, Luckenwalde, Germany and 11-OHA4-2,2,4,6,6,16,16-d7 from EQ Laboratories GmbH, Augsburg, Germany.

Methanolic working standards were produced for all analytes and internal standards based on 1 mg/mL (3.3 mmol/L) stock solutions. 6PLUS1 Multilevel serum calibrator levels 1–5 as well as MassCheck® steroid serum control levels 1–3 were obtained from Chromsystems Instruments and Chemicals GmbH, Munich, Germany and were used to generate calibrators and quality controls (QC) and serum covering expected endogenous levels by spiking with working standards of 11-KA4, 11-KT, 11-OHA4, 11-OHT (Table S4) [7]. Spiked Calibrators ranged from 0.08 nmol/L to 33.3 nmol/L for 11-KA4, 11-KT and 11-OHT, and from 0.83 nmol/L to 331 nmol/L for 11-OHA4. Spiked QC's and serum controls ranged from 0.16 nmol/L to 16.6 nmol/L for 11-KA4, 11-KT and 11-OHT, and from 1.7 nmol/L to 165 nmol/L for 11-OHA4 (Table S4).

4.2. Human Samples

Residual serum taken from routine diagnostics of patients with treated CAH was used for method verification. The study was approved by the ethics committee of the University Hospital Leipzig (082 10 190-42010) according to the declaration of Helsinki ethical principles. Serum samples of healthy individuals with normal 17-OHP concentrations were obtained from the LIFE Child study (Leipzig Research Centre for Civilization Diseases) approved by the Ethical Committee of the University of Leipzig (reference number: Reg. No. 264-10-19042010) and is registered at ClinicalTrials.gov (NCT02550236).

4.3. Sample Preparation

Aliquots of calibrators, quality controls, blank, and serum (100 µL) were treated with 200 µL precipitating agent (ZnSO4 in water (0.3 mol/L)/methanol 1/4 v/v, including the internal standards (3.3 nmol/L)), thoroughly mixed and centrifuged for 10 min at 14,000× g. The supernatant was transferred to autosampler vials with 250 µL inserts.

4.4. LC-MS/MS

A Prominence UFLC system from Shimadzu (Duisburg, Germany) was coupled to a QTRAP® 6500plus from SCIEX (Framingham, MA, USA). A PAL3 RSI autosampler from CTC Analytics (Zwingen, Switzerland) handled sample injection. Injection volume was 100 µL. Online solid phase extraction was performed on a POROS® column (30 × 2.1 mm) from Applied Biosystems (Foster City, CA, USA) at a flow rate of 3 mL/min. For chromatographic separation, a Chromolith® High Resolution column (RP-18, endcapped, 100 × 4.6 mm) from Merck (Darmstadt, Germany) was used. The mobile phase consisted of 50% eluent A (0.2 mmol/L ammonium fluoride (NH4F) in water/methanol 97/3 *v/v*) and 50% eluent B (0.2 mmol/L NH4F in water/methanol 3/97 *v/v*) and was adjusted as follows: 0–1 min 50% B, 1–5.5 min 50% to 95% B, 5.5–6.5 min 100% B, 6.5–6.6 min 50% B. Flow rate was 1.5 mL/min and the column oven was set to 35 °C. Electrospray ionization (ESI) was applied in positive mode and detection was carried out using multiple reaction monitoring. Mass transitions of the 11-OAs as well as their corresponding internal standards are listed in Table S5. The concentrations were determined using calibration curves which were obtained via ratios of analyte peak area/deuterated standard peak area. Furthermore, estradiol (E2) and aldosterone (A) (ESI negative mode) can additionally be determined without extra sample preparation [33,34].

4.5. Validation

Serum samples with known low concentrations of androgens were used for the determination of LLOQ and LOD. LOD was calculated at *s/n* = 3. LLOQ was defined as the lowest concentration at which a triplicate measurement resulted in CV $\leq$ 20% with signal to noise (*s/n*) $\geq$ 10. Linear range of calibration was determined by regression analysis. Means of slopes and regression coefficients of ten 5-point calibrations were determined for robustness of linearity. Potential interferences between 11-OAs, 9 established steroid hormones (17-OHP, A, A4, F, E, E2, DHEAS, T, P), as well as their corresponding internal standards were investigated by measuring highly concentrated standard solutions (28–37 nmol/L; 27 µmol/L for DHEAS/DHEAS-d6). Matrix effects were investigated by post column infusion of 11-OHA4, 11-KA4, 11-OHT and 11-KT (50 nmol/L, 10 µL/min) during the measurement of methanol, calibrator and serum. Imprecision (intra- and interassay) and recoveries were determined by measurement of 10 replicates of spiked quality controls as well as spiked serum samples at three concentration levels each as described above. To asses potential effects of freeze/thaw cycles, five serum samples were frozen at 80 °C within 2 h after blood sampling, thawed and refrozen on 4 individual days prior to threefold measurement. Changes above 20% indicated that repeated freeze/thaw cycles affect analytical stability of 11-OAs.

4.6. Clinical Verification

Residual serum samples of 42 patients with treated CAH (19 males, 2 to 79 years) and expected elevated concentrations of 11-OAs were compared to a control group of 42 healthy individuals matched for sex and age. In both groups, concentrations of 11-OAs as well as the routine steroid hormone panel (17-OHP, A, A4, F, E, E2, DHEAS, T, P) and ratios of 17-OHP, A, A4, P or T to 11-OAs were compared.

Supplementary Materials: The supporting information can be downloaded at: https://www.mdpi.com/article/10.3390/ijms24010539/s1.

Author Contributions: Conceptualization, R.B., U.C. and A.G.; Methodology, R.Z. and A.G.; Project administration, U.C.; Resources, B.I.; Supervision, R.B. and A.G.; Validation, R.Z., R.B. and A.G.; Visualization, R.Z.; Writing—original draft, R.Z., R.B. and A.G.; Writing—review and editing, U.C., J.K. and B.I. All authors were involved in writing the article and approved the submitted version. All authors have read and agreed to the published version of the manuscript.

Funding: This work was supported by 'Stiftung Pathobiochemie und Molekulare Diagnostik' (SPMD).

Institutional Review Board Statement: The study is approved by the ethics committee of the University Hospital Leipzig (082 10 190-42010). The LIFE Child study (Leipzig Research Centre for Civilization Diseases) is approved by the Ethical Committee of the University of Leipzig (reference number: Reg. No. 264-10-19042010) and is registered at ClinicalTrials.gov (NCT02550236). The study was conducted in accordance with the Declaration of Helsinki.

Informed Consent Statement: Informed consent was obtained from participants of the LIFE Child study.

Data Availability Statement: Not applicable.

Acknowledgments: This publication is supported by LIFE—Leipzig Research Center for Civilization Diseases, University of Leipzig. LIFE is funded by means of the European Union, by means of the European Social Fund (ESF), by the European Regional Development Fund (ERDF), and by means of the Free State of Saxony within the framework of the excellence initiative. The authors thank the participants for volunteering for these studies.

Conflicts of Interest: The authors declare no conflict of interest.

References

1. Mornet, E.; Dupont, J.; Vitek, A.; White, P.C. Characterization of Two Genes Encoding Human Steroid 11β-Hydroxylase (P-450(11β)). *J. Biol. Chem.* **1989**, *264*, 20961–20967. [CrossRef] [PubMed]
2. Schiffer, L.; Arlt, W.; Storbeck, K.H. Intracrine Androgen Biosynthesis, Metabolism and Action Revisited. *Mol. Cell. Endocrinol.* **2018**, *465*, 4–26. [CrossRef] [PubMed]
3. Turcu, A.F.; Rege, J.; Auchus, R.J.; Rainey, W.E. 11-Oxygenated Androgens in Health and Disease. *Nat. Rev. Endocrinol.* **2020**, *16*, 284–296. [CrossRef] [PubMed]
4. Turcu, A.F.; Nanba, A.T.; Auchus, R.J. The Rise, Fall, and Resurrection of 11-Oxygenated Androgens in Human Physiology and Disease. *Horm. Res. Paediatr.* **2018**, *89*, 284–291. [CrossRef]
5. Rege, J.; Nakamura, Y.; Wang, T.; Merchen, T.D.; Sasano, H.; Rainey, W.E. Transcriptome profiling reveals differentially expressed transcripts between the human adrenal zona fasciculata and zona reticularis. *J. Clin. Endocrinol. Metab.* **2014**, *99*, E518–E527. [CrossRef]
6. Kelly, D.M.; Jones, T.H. Testosterone and Obesity. *Obes. Rev.* **2015**, *16*, 581–606. [CrossRef]
7. Davio, A.; Woolcock, H.; Nanba, A.T.; Rege, J.; O'day, P.; Ren, J.; Zhao, L.; Ebina, H.; Auchus, R.; Rainey, W.E.; et al. Sex Differences in 11-Oxygenated Androgen Patterns across Adulthood. *J. Clin. Endocrinol. Metab.* **2020**, *105*, e2921–e2929. [CrossRef]
8. Conway, G.; Dewailly, D.; Diamanti-Kandarakis, E.; Escobar-Morreale, H.F.; Franks, S.; Gambineri, A.; Kelestimur, F.; Macut, D.; Micic, D.; Pasquali, R.; et al. The polycystic ovary syndrome: A position statement from the European society of endocrinology. *Eur. J. Endocrinol.* **2014**, *171*, P1–P29. [CrossRef]
9. Turcu, A.F.; Nanba, A.T.; Chomic, R.; Upadhyay, S.K.; Giordano, T.J.; Shields, J.J.; Merke, D.P.; Rainey, W.E.; Auchus, R.J. Adrenal-derived 11-oxygenated 19-carbon steroids are the dominant androgens in classic 21-hydroxylase deficiency. *Eur. J. Endocrinol.* **2016**, *174*, 601–609. [CrossRef]
10. Turcu, A.F.; Rege, J.; Chomic, R.; Liu, J.; Nishimoto, H.K.; Else, T.; Moraitis, A.G.; Palapattu, G.S.; Rainey, W.E.; Auchus, R.J. Profiles of 21-carbon steroids in 21-hydroxylase deficiency. *J. Clin. Endocrinol. Metab.* **2015**, *100*, 2283–2290. [CrossRef]
11. Rege, J.; Garber, S.; Conley, A.J.; Elsey, R.M.; Turcu, A.F.; Auchus, R.J.; Rainey, W.E. Circulating 11-oxygenated androgens across species. *J. Steroid Biochem. Mol. Biol.* **2019**, *190*, 242–249. [CrossRef] [PubMed]
12. Carmina, E.; Stanczyk, F.Z.; Chang, L.; Miles, R.A.; Lobo, R.A. The ratio of androstenedione: 11β-Hydroxyandrostenedione Is an important marker of adrenal androgen excess in women. *Fertility and Sterility* **1992**, *58*, 148–152. [CrossRef] [PubMed]
13. Nanba, A.T.; Rege, J.; Ren, J.; Auchus, R.J.; Rainey, W.E.; Turcu, A.F. 11-Oxygenated C19 steroids do not decline with age in women. *J. Clin. Endocrinol. Metab.* **2019**, *104*, 2615–2622. [CrossRef] [PubMed]
14. du Toit, T.; Bloem, L.M.; Quanson, J.L.; Ehlers, R.; Serafin, A.M.; Swart, A.C. Profiling adrenal 11β-Hydroxyandrostenedione metabolites in prostate cancer cells, tissue and plasma: UPC2-MS/MS quantification of 11β-Hydroxytestosterone, 11keto-testosterone and 11keto-dihydrotestosterone. *J. Steroid Biochem. Mol. Biol.* **2017**, *166*, 54–67. [CrossRef] [PubMed]
15. Han, B.; Zhu, H.; Yao, H.; Ren, J.; O'Day, P.; Wang, H.; Zhu, W.; Cheng, T.; Auchus, R.J.; Qiao, J. Differences of adrenal-derived androgens in 5α-reductase deficiency versus androgen insensitivity syndrome. *Clin. Transl. Sci.* **2021**, *15*, 658–666. [CrossRef] [PubMed]
16. Blasco, M.; Carriquiriborde, P.; Marino, D.; Ronco, A.E.; Somoza, G.M. A Quantitative HPLC-MS Method for the Simultaneous Determination of Testosterone, 11-Ketotestosterone and 11-β Hydroxyandrostenedione in Fish Serum. *J. Chromatogr. B: Anal. Technol. Biomed. Life Sci.* **2009**, *877*, 1509–1515. [CrossRef] [PubMed]
17. Flores-Valverde, A.M.; Hill, E.M. Methodology for profiling the steroid metabolome in animal tissues using ultraperformance liquid chromatography-electrospray-time-of-flight mass spectrometry. *Anal. Chem.* **2008**, *80*, 8771–8779. [CrossRef]
18. Schloms, L.; Storbeck, K.H.; Swart, P.; Gelderblom, W.C.A.; Swart, A.C. The influence of aspalathus linearis (rooibos) and dihydrochalcones on adrenal steroidogenesis: Quantification of steroid intermediates and end products in H295R Cells. *J. Steroid Biochem. Mol. Biol.* **2012**, *128*, 128–138. [CrossRef]
19. Xing, Y.; Edwards, M.A.; Ahlem, C.; Kennedy, M.; Cohen, A.; Gomez-Sanchez, C.E.; Rainey, W.E. The effects of ACTH on steroid metabolomic profiles in human adrenal cells. *J. Endocrinol.* **2011**, *209*, 327–335. [CrossRef]

20. Zheng, J.; Islam, R.M.; Skiba, M.A.; Zheng, J.; Islam, R.M.; Skiba, M.A.; Bell, R.J.; Davis, S.R. Associations between androgens and sexual function in premenopausal women: A cross-sectional study. *Lancet Diabetes Endocrinol.* **2020**, *8*, 693–702. [CrossRef]
21. Caron, P.; Turcotte, V.; Guillemette, C. A quantitative analysis of total and free 11-oxygenated androgens and its application to human serum and plasma specimens using liquid-chromatography tandem mass spectrometry. *J. Chromatogr. A* **2021**, *1650*, 462228. [CrossRef] [PubMed]
22. Zhang, X.; Zhou, C.; Xu, H.; Feng, Y.; Yang, P.; Zhai, S.; Song, J.; Yang, L. A Sensitive HPLC-DMS/MS/MS method for multiplex analysis of androgens in human serum without derivatization and its application to PCOS patients. *J. Pharm. Biomed. Anal.* **2021**, *192*, 113680. [CrossRef] [PubMed]
23. Hawley, J.M.; Adaway, J.E.; Owen, L.J.; Keevil, B.G. Development of a Total Serum Testosterone, Androstenedione, 17-Hydroxyprogesterone, 11β-Hydroxyandrostenedione and 11-Ketotestosterone LC-MS/MS Assay and Its Application to Evaluate Pre-Analytical Sample Stability. *Clin. Chem. Lab. Med.* **2020**, *58*, 741–752. [CrossRef]
24. Wright, C.; O'Day, P.; Alyamani, M.; Sharifi, N.; Auchus, R.J. Abiraterone acetate treatment lowers 11-Oxygenated androgens. *Eur. J. Endocrinol.* **2020**, *182*, 413–421. [CrossRef] [PubMed]
25. Häkkinen, M.R.; Murtola, T.; Voutilainen, R.; Poutanen, M.; Linnanen, T.; Koskivuori, J.; Lakka, T.; Jääskeläinen, J.; Auriola, S. Simultaneous analysis by LC–MS/MS of 22 ketosteroids with hydroxylamine derivatization and underivatized estradiol from human plasma, serum and prostate tissue. *J. Pharm. Biomed. Anal.* **2019**, *164*, 642–652. [CrossRef]
26. Houghton, L.C.; Howland, R.E.; Wei, Y.; Ma, X.; Kehm, R.D.; Chung, W.K.; Genkinger, J.M.; Santella, R.M.; Hartmann, M.F.; Wudy, S.A.; et al. The steroid metabolome and breast cancer risk in women with a family history of breast cancer: The novel role of adrenal androgens and glucocorticoids. *Cancer Epidemiol. Biomark. Prevention* **2021**, *30*, 89–96. [CrossRef] [PubMed]
27. Quanson, J.L.; Stander, M.A.; Pretorius, E.; Jenkinson, C.; Taylor, A.E.; Storbeck, K.H. High-throughput analysis of 19 endogenous androgenic steroids by ultra-performance convergence chromatography tandem mass spectrometry. *J. Chromatogr. B Anal. Technol. Biomed. Life Sci.* **2016**, *1031*, 131–138. [CrossRef]
28. O'Reilly, M.W.; Kempegowda, P.; Jenkinson, C.; Taylor, A.E.; Quanson, J.L.; Storbeck, K.H.; Arlt, W. 11-oxygenated C19 steroids are the predominant androgens in polycystic ovary syndrome. *J. Clin. Endocrinol. Metab.* **2017**, *102*, 840–848. [CrossRef]
29. du Toit, T.; Stander, M.A.; Swart, A.C. A High-Throughput UPC2-MS/MS method for the separation and quantification of C19 and C21 steroids and Their C11-Oxy steroid metabolites in the classical, alternative, backdoor and 11OHA4 steroid pathways. *J. Chromatogr. B Anal. Technol. Biomed. Life Sci.* **2018**, *1080*, 71–81. [CrossRef]
30. Yoshida, T.; Matsuzaki, T.; Miyado, M.; Saito, K.; Iwasa, T.; Matsubara, Y.; Ogata, T.; Irahara, M.; Fukami, M. 11-Oxygenated C19 steroids as circulating androgens in women with polycystic ovary syndrome. *Endocr. J.* **2018**, *65*, 979–990. [CrossRef]
31. Skiba, M.A.; Bell, R.J.; Islam, R.M.; Handelsman, D.J.; Desai, R.; Davis, S.R. Androgens during the reproductive years: What is normal for women? *J. Clin. Endocrinol. Metab.* **2019**, *104*, 5382–5392. [CrossRef] [PubMed]
32. Rege, J.; Turcu, A.F.; Kasa-Vubu, J.Z.; Lerario, A.M.; Auchus, G.C.; Auchus, R.J.; Smith, J.M.; White, P.C.; Rainey, W.E. 11-ketotestosterone is the dominant circulating bioactive androgen during normal and premature adrenarche. *J. Clin. Endocrinol. Metab.* **2018**, *103*, 4589–4598. [CrossRef] [PubMed]
33. Gaudl, A.; Kratzsch, J.; Bae, Y.J.; Kiess, W.; Thiery, J.; Ceglarek, U. Liquid chromatography quadrupole linear ion trap mass spectrometry for quantitative steroid hormone analysis in Plasma, Urine, Saliva and Hair. *J. Chromatogr. A* **2016**, *1464*, 64–71. [CrossRef] [PubMed]
34. Gaudl, A.; Kratzsch, J.; Ceglarek, U. Advancement in Steroid Hormone Analysis by LC–MS/MS in clinical routine diagnostics—A three year recap from serum cortisol to dried blood 17α-hydroxyprogesterone. *J. Steroid Biochem. Mol. Biol.* **2019**, *192*, 105389. [CrossRef] [PubMed]
35. *CLSI Liquid Chromatography-Mass Spectrometry Methods*; Approved Guideline; CLSI document C62-A 2014; Clinical and Laboratory Standards Institute: Wayne, PA, USA, 2014.
36. Auer, M.K.; Paizoni, L.; Neuner, M.; Lottspeich, C.; Schmidt, H.; Hawley, J.; Keevil, B.; Reisch, N. Elevated 11-oxygenated androgens are not a major contributor to HPG-Axis disturbances in adults with congenital adrenal hyperplasia due to 21-Hydroxylase deficiency. *medRxiv* **2021**. [CrossRef]

International Journal of
Molecular Sciences

Review

Transgenic Mouse Models to Study the Development and Maintenance of the Adrenal Cortex

Nour Abou Nader, Gustavo Zamberlam and Alexandre Boyer *

Centre de Recherche en Reproduction et Fertilité, Faculté de Médecine Vétérinaire, Université de Montréal, Saint-Hyacinthe, QC J2S 7C6, Canada
* Correspondence: alexandre.boyer.1@umontreal.ca; Tel.: +1-450-773-8521 (ext. 8345)

Abstract: The cortex of the adrenal gland is organized into concentric zones that produce distinct steroid hormones essential for body homeostasis in mammals. Mechanisms leading to the development, zonation and maintenance of the adrenal cortex are complex and have been studied since the 1800s. However, the advent of genetic manipulation and transgenic mouse models over the past 30 years has revolutionized our understanding of these mechanisms. This review lists and details the distinct Cre recombinase mouse strains available to study the adrenal cortex, and the remarkable progress total and conditional knockout mouse models have enabled us to make in our understanding of the molecular mechanisms regulating the development and maintenance of the adrenal cortex.

Keywords: transgenic mice; adrenal cortex; development; maintenance

1. Introduction

The adrenal gland is an organ formed of two main regions: the centrally located medulla that produces catecholamines, and the adrenal cortex that produces steroid hormones essential for body homeostasis in mammals. The adrenal cortex is further organized into concentric zones that produce distinct steroid hormones. The outermost zone of the adrenal cortex, the zona glomerulosa (zG), secretes aldosterone. The intermediate zone, the zona fasciculata (zF), secretes corticosterone or cortisol depending on the species. Finally, the inner zone that is absent in rodents, the zona reticularis (zR), secretes dehydroepiandrosterone and dehydroepiandrosterone sulfate (DHEA/DHEAS).

The development and maintenance of the mammalian adrenal cortex is complex and includes five main sequential steps: (1) formation of a common primordium with the gonads known as the adrenogonadal primordium (AGP); (2) separation of the AGP into the gonadal primordium (GP) and the adrenal primordium (AP), with the latter being responsible for the formation of the fetal adrenal cortex; (3) encapsulation of the fetal adrenal cortex, following the invasion of neural crest-derived cells that will form the future chromaffin cells of the medulla; (4) replacement of the fetal cortex by the definitive adrenal cortex; and (5) establishment of zonation (zG, zF and zR) and subsequent maintenance of these zones.

Foundations for the study of the adrenal cortex were laid 139 years ago, when it was first suggested that cells forming all three of the above-mentioned adrenocortical zones originate from the outer capsule and migrate inwards to ultimately die at the boundary between the adrenal cortex and the medulla [1]. Although this theory was reinforced five decades later by the study of cellular renewal in adrenal injury models [2,3], it took transgenic mouse technology to begin to understand the molecular mechanisms regulating the development and maintenance of the adrenal cortex. The objectives of the present review are: (1) to detail the mouse Cre strains available to study adrenal cortex development and maintenance; and (2) to present the main findings acquired from these different models, from AGP formation up to the postnatal maintenance of the adrenal cortex.

Citation: Abou Nader, N.; Zamberlam, G.; Boyer, A. Transgenic Mouse Models to Study the Development and Maintenance of the Adrenal Cortex. *Int. J. Mol. Sci.* **2022**, *23*, 14388. https://doi.org/10.3390/ijms232214388

Academic Editor: David W. Walker

Received: 2 October 2022
Accepted: 15 November 2022
Published: 19 November 2022

Publisher's Note: MDPI stays neutral with regard to jurisdictional claims in published maps and institutional affiliations.

2. Mouse Models Used to Study the Adrenal Cortex Development and Maintenance

In the last three decades, gene targeting approaches have been the most insightful technique to comprehend the mechanisms regulating the development and maintenance of the adrenal cortex. Aside from traditional total knockout (KO) mouse models, several Cre recombinase mouse strains have been generated and used to perform fate mapping studies and to conditionally (cKO) inactivate genes of interest in the adrenal cortex (Table 1, Supplementary Table S1). Before presenting the results obtained with these mouse models, it is important to first present the specificities and the limitations of the Cre recombinase mouse strains models used in these studies, to better understand the relevance of the results acquired from them so far.

Table 1. Cre mouse strains used to study adrenal cortex development and maintenance.

Mouse Strains	Ref (Mouse Dev.)	Targeted Tissues/Cells during Adrenal Development and Maintenance	Targeted Genes (Inactivated)/Tracing Cell Populations	Ref
CAG-CreER	[4]	Global inactivation	*Gata4* *Rspo3*	[5] [6]
Osr1[eGFP-CreERt2]	[7]	Coelomic epithelium	*Gata4* Tracing intermediate mesoderm descendants	[5] [8]
Tbx18[Cre]	[9]	Coelomic epithelium	*Fgfr2*	[10]
Gata4[CreERt2]	[11]	Coelomic epithelium/AGP		
Wt1[CreERt2/+]	[12]	Coelomic epithelium/AGP Subpopulation of capsular cells	*Gata4* Tracing (capsular population) Tracing intermediate mesoderm descendants	[5] [13] [8]
Nr5a1-Cre[high] (High transgene copy numbers)	[14]	AGP/fetal cortex, definitive cortex	*Apc* *Ctnnb1* *Ctnnb1ex3* *Dicer* *Ezh2* *Fgfr2* *Gata4* *Gata6* *Porcn* *Prkar1a* *Rnf43* *Shh* *Smo* *Wnt4* *Wt1 (activation)* *Znrf3*	[15] [16,17] [17] [17,18] [19] [20] [21,22] [21,22] [23] [24–26] [23] [17,27–29] [28] [6,25] [13] [23]
Nr5a1-Cre[low] (low transgene copy numbers)	[14]	AGP/fetal cortex, definitive cortex, few cells affected	*Apc* *Ctnnb1* *H19*	[15] [15,16] [15]
Nr5a1-Cre	[30]	AGP/fetal cortex, definitive cortex	*Gata6* *Yap/Taz* *Lats1/Lats2* *Mst1/Mst2*	[31] [32] [33] [34]

Table 1. *Cont.*

Mouse Strains	Ref (Mouse Dev.)	Targeted Tissues/Cells during Adrenal Development and Maintenance	Targeted Genes (Inactivated)/Tracing Cell Populations	Ref
FAdE/Nr5a1-Cre	[35]	Fetal cortex	Tracing fetal adrenocortical cells descendants	[35,36]
FAdE/Nr5a1-CreERT2	[35]	Fetal cortex	*Prkar1a* Tracing fetal adrenocortical cells descendants	[37] [35,37]
Nr5a1 ^{eGFP-CreERt2}	-	AGP/fetal cortex, definitive cortex		
hCyp11a1-iCre	[38]	Fetal cortex/definitive cortex	*Insr/Igf1r*	[39]
mCyp11a1-iCre	[40]	Fetal cortex/definitive cortex	*Ctnnb1*^{ex3} *Nr5a1*	[41] [40]
Cyp11a1^{Gfp,Cre/+}	[42]	Fetal cortex/definitive cortex	*AR*	[43,44]
Akr1b7-Cre	[45]	Fetal cortex/definitive cortex	*Ctnnb1*^{ex3} *Prkar1a* *Prkaca*	[24,25,46,47] [25,26] [25]
Cyp11b2^{Cre}/AS^{Cre}	[48]	Aldosterone producing zG cells and their zF descendants	*Ctnnb1* *Ctnnb1*^{ex3} *Ffg2r* *Prkar1a* *Nr5a1* *Nr0b1* *Znrf3* Tracing zG cell descendants	[49] [49,50] [49] [37] [48] [48] [23] [23,37,48,50]
Cyp11b1^{eGFP-Cre}	[51]	zF cells	*Cth*	[51]
Gli1^{CreERt2}	[52]	Capsular stem cells	*Rspo3* *Smo (activation)* *Tracing capsular stem cell descendants*	[6] [53] [27,28,36,53–55]
Shh^{Cre}	[56]	Subcapsular progenitor zG cells	Tracing subcapsular progenitor cell descendants	[28]
Shh^{CreERt2}	[56]	Subcapsular progenitor zG cells	Tracing subcapsular progenitor cell descendants	[28,53]
Axin2^{CreERt2}	[57]	WNT signaling activated zG cells	*Ctnnb1* Tracing zG cells descendants (including subcapsular progenitor cells)	[53] [53,54]
Wnt4 ^{CreERt2}	[58]	WNT signaling activated zG cells	Tracing zG cells descendants (including subcapsular progenitor cells)	[54]
Nes-CreERt2	[59]	Stress induced adrenocortical progenitor cells	Tracing stress induced progenitor cell descendants	[60]

2.1. Mouse Strains to Study AGP Formation

Adrenal development is first initiated with the thickening of the coelomic epithelium, followed by subsequent delamination of a group of cells that migrate inwards to form the AGP beginning at around e9.5. Most information obtained for this stage of development comes from studies employing KO animals. Identifying the best promoter to drive Cre expression and generate cKO to specifically study this step of development have proven difficult because genes expressed in the coelomic epithelium are also expressed in several other tissues. For example, *Gata4* (gene names/abbreviations are listed at the end of this review) and *Wt1* are two key transcription factors expressed in the coelomic epithelium and involved in the early steps of AGP formation [5,61]. However, both genes present broad expression, including in the developing heart, and using them to drive a Cre recombinase and generate cKO could lead to the death of the embryo at a time point that precedes the thickening of the coelomic epithelium. *Gata4*-Cre [62,63] and *Wt1-Cre* [12,64,65] strains have therefore not been employed to generate cKO models and evaluate early stages of adrenal development. Tamoxifen inducible models have provided a more promising solution to understand this stage of development, as demonstrated by the study of the role of GATA4 in the coelomic epithelium [5]. In this study, three mouse strains were used to inactivate *Gata4* at e8.75: the *CAG-CreER* [4]; the *Osr1$^{eGFP\text{-}CreERt2}$* [7]; and the *Wt1^{CreERt2}* [12] (mouse strains expressing, respectively, Cre following tamoxifen injections ubiquitously [4] or in the intermediate mesoderm/early coelomic epithelium [7,8,12,66–68] among other tissues). It was further suggested that creating mutants with both *Osr1-* and *Wt1-*driven Cre alleles in the same animal might be the best solution to study the earliest steps of the AGP formation [5], since recombination with both *Osr1$^{eGFP\text{-}CreERt2/+}$* and *Wt1$^{CreERt2/+}$* strains has shown variant efficiency while the *CAG-CreER* strain increases the risk of an indirect effect [5]. Two other mouse strains have been generated that could be useful to evaluate the AGP. First, the *Tbx18^{Cre}* mouse strain targets, among other tissues, the adrenal precursors in the anterior coelomic epithelium [9,10]. This mouse strain has only been used once to evaluate the AGP [10]. Finally, the generation of a *Gata4^{CreERt2}* strain was recently reported [11]. However, this strain has not yet been used to study the development of the AGP.

2.2. Mouse Strains Using Nr5a1 Regulatory Sequences to Drive Cre Expression

Concomitantly to coelomic epithelium thickening and delamination, *Nr5a1* expression rapidly increases in the forming AGP. Contrary to *Gata4* and *Wt1*, *Nr5a1* expression is maintained in the AP, and the fact that its expression is mainly restricted to endocrine/steroidogenic tissues, makes its regulatory region an interesting driver of Cre expression. Indeed, two *Nr5a1-Cre* mouse strains have been generated [14,30] and, to date, they remain the most common strains used to study the development and maintenance of the adrenal glands. A less efficient version of one of these *Nr5a1-Cre* strains called *Nr5a1-Cre^{low}* (in opposition to *Nr5a1-Cre^{high}*) [14] has also been reported, but has only been used seldomly since its driven recombination occurs in fewer cells. Several aspects must be considered when analyzing cKO models generated using the *Nr5a1-Cre* strains. First, *Nr5a1* is expressed in the AGP, the fetal cortex and the adult cortex, making it difficult to determine if a phenotype observed at a certain time point indicates a role for the deleted floxed target genes at this particular time point, or if the observed phenotype is actually associated with an alteration that has started in a previous step of development. For example, the fetal cortex initially contributes to adult cortex formation [35,36] and inactivating a gene important for the development of the fetal cortex could also indirectly affect the formation of the adult cortex. Secondly, conditional deletion of the gene of interest will also be performed in other steroidogenic and endocrine cells including the Leydig and Sertoli cells in the testis, the granulosa and theca cells in the ovary, the gonadotropes in the pituitary and the neurons of the ventromedial hypothalamus [14,30]. Considering the main hormones produced by each of these cells, it is likely that the loss of the target gene expression in any of these cells might affect the maintenance of the adrenal cortex indirectly. Androgens [37,54] and, to a lesser degree, estrogen [69] have been shown to affect homeostasis of the adrenal cortex, while luteinizing hormone (LH) has been shown to induce the transdifferentiation of adrenal hyperplastic spindle-shaped cells into sex-steroid producing cells in gonadectomized mice [70,71]. Furthermore, it was demonstrated that lesions of the ventromedial hypothalamus in rats increased the adrenal weight and inhibited corticosterone/basal adrenocorticotropic hormone (ACTH) diurnal rhythm feedback [72]. Gonadal and gonadotropic hormones (and potentially corticosterone circadian rhythm) should therefore be evaluated, to determine if an observed abnormal phenotype following gene inactivation is exclusively due to alterations happening in the adrenal cortex and/or depends on hormones secreted from other tissues. Finally, it was recently demonstrated that conditional deletion of the genes of interest could also occur in a subset of dermal fibroblast progenitors when using the *Nr5a1-Cre*^{high} model [73]. Although inactivation of a gene of interest in these cells is unlikely to indirectly affect the adrenal cortex, this expression must be considered when characterizing the phenotype of mouse models. This is particularly true for mouse models that attempt to simulate a complex syndrome affecting multiple organs like Carney complex [73].

Three other mouse strains using *Nr5a1* regulatory sequences to drive Cre expression have been reported. First, a tamoxifen inducible *Nr5a1*$^{eGFP-CreERt2}$ mouse strain was created by the Welcome Trust Sanger Institute. That strain, which has not yet been used, could technically target *Nr5a1+* cells at specific time points to only study postnatal steroidogenic cells, for example. Secondly, two strains, the *FAdE/Nr5a1-Cre* and the *FAdE/Nr5a1-CreERT2*, have been reported [35]. These strains use the enhancer that selectively drives *Nr5a1* expression in the fetal adrenal cortex to specifically express Cre in the fetal adrenal cortex [35]. These two models have been used mainly to perform lineage experiments and determine the fate of the fetal adrenal cortex [35,36]. However, the *FAdE/Nr5a1-CreERT2* was used once for gene inactivation [37]. Interestingly, the *FAdE/Nr5a1-CreERT2* model can specifically target the fetal adrenal cortex without subsequently affecting the definitive cortex if the recombinase Cre is activated after e14.5 [35].

2.3. Mouse Strains Using Regulatory Sequence of Genes Coding for Steroidogenic Enzymes to Drive Cre Expression

Three strains using the promoter of the steroidogenic enzyme *Cyp11a1* to drive Cre expression have also been employed to inactivate genes in the adrenal cortex. The first two of these models used either 4.4 Kb of the human *CYP11A1* promoter [38] or 2.8 Kb of the mouse *Cyp11a1* promoter [40]. In both strains, Cre expression is detected in the fetal and adult adrenal cortex and in Leydig cells. However, Cre expression is also detected in the theca cells and corpus luteum of the postnatal ovaries [38,40] and, for the human *Cyp11a1-iCre* strain, at lower levels in the female gonads and in the diencephalon and midbrain [38]. Expression in the brain was not evaluated in the mouse *Cyp11a1-iCre* model [40]. More recently, a third model (called *Cyp11a1*^{Gfp,Cre}) was generated to drive Cre expression under the endogenous *Cyp11a1* promoter [42]. In this model, the integration of a GFP/Cre cassette was used to disrupt the *Cyp11a1* exon containing the *ATG* site [42]. Again, Cre expression was detected in the fetal and adult testis, adrenal cortex and adult ovary (as well as in the cerebellum) [42]. Although these models are a little bit more specific than the *Nr5a1-Cre* strains, similar problems will arise since sexual hormones will/could also be affected in them.

Other strains have been generated/used to target recombination in the adrenal cortex in a more specific matter. First, a *Cyp11b2*^{Cre} knocking allele (better known as AS^{Cre}) has been created to target zG cells. Using tracing experiments, it was demonstrated that a few zG cells were marked between e16.5 and 1dpp with all zG cells marked at 6 weeks after birth [48]. Tracing experiments further demonstrated that all cells of the adrenal cortex were eventually marked due to centripetal migration and lineage conversion of zG cells into zF cells [48]. Although the AS^{Cre} is specific to the adrenal cortex and has the potential to inactivate a gene of interest in all the zones of the adrenal cortex, it was demonstrated that the zF cell population can be maintained independently of the zG cell population when the capacity of zG cells to differentiate into zF cells is affected [48]. This finding suggests that this strain might not be useful to study the function of a gene in both the zF and zG when genes essential for zG cell survival or differentiation are inactivated [48,50]. Furthermore, loss of zG cells or zG functions can potentially lead to an increase in the proliferation of the non-recombined progenitor and stem cell populations, and their subsequent differentiation into zG steroidogenic cells leading to a mosaic of recombined steroidogenic zG cells and consequently to a weaker phenotype. Similarly, lineage conversion of zG cells into zF cells could also be accelerated if the deleted gene is particularly important for zF functions as the tissue tries to replace/replenish the zF with functional cells. Cells might therefore spend insufficient time in their zG state to allow efficient recombination of some floxed alleles before their transdifferentiation into zF cells. This could also be exacerbated in females in which complete adrenal cortex turnover is three-times faster than in males [54]. More recently, a strain in which a P2A-eGFP-Cre cassette was integrated before the 3′ UTR region of the *Cyp11b1* gene, was generated using Crispr technology to specifically target the zF. However, to date, only one scientific paper (written in Chinese) has been published using this model [51].

2.4. Other Mouse Strains Used to Target the Adrenal Cortex

Other strains have been generated/used to target recombination in the adrenal cortex. In the *Akr1b7-Cre* strain, recombination is observed in about 80% of the adult adrenocortical cells [45]. Recombination is also observed in the adrenal cortex starting at e14.5 and persists in the presumptive X-zone cells until 10 dpp [45]. This suggests that Cre recombination occurs in both the fetal and definitive adrenocortical cells. As by e14.5, the fetal adrenocortical cells do not contribute anymore to the formation of the adult cortex [35]; this also suggests that an abnormal phenotype observed postnatally will directly come from the inactivation of the gene of interest in the definitive cortex. Interestingly, Cre activity is not detected in the gonads (though recombination can be observed in some structures of the kidney) facilitating the interpretation of the phenotype [45].

The *Gli1* ^{CreERt2} strain [52] targets, among other tissues, the capsular adrenal stem cells. This strain has been used for both tracing experiments and for gene inactivation. However, it is important to note that while the capsular stem cells contribute to the adrenocortical steroidogenic cell lineage in juvenile males and females, their contribution is limited to the females in adult mice [53,54]. The dimorphic contribution of the stem cell population and the timing of the inactivation should therefore be accounted for when this model is used. The importance of the Hedgehog signaling in numerous tissues limits the potential usage of this strain. Similarly, a few other mouse strains targeting different cell populations in the adrenal cortex have mostly been used for tracing experiment due to concomitant recombination in several tissues. These strains, respectively, target a subpopulation of capsular cells (the previously mentioned *Wt1*^{CreERt2} [12]), the subcapsular progenitor cells (*Shh*^{Cre} and *Shh*^{CreERt2} [56]), all zG cells (*Axin2*^{CreERt2} [57], *Wnt4* ^{CreERt2} [58]) or a stress induced adrenocortical progenitor cell population (*Nes*-CreERt2) [59,60].

3. AGP Development

As previously mentioned, the adrenal cortex and the gonads arise from the thickening of the coelomic epithelium. However, the genesis of these organs is initiated at an earlier time point of embryonic development. Indeed, recent fate mapping studies have demonstrated that the coelomic epithelium could derive from the posterior intermediate mesoderm which emerges from the primitive streak [8,74]. In mice, it was demonstrated that mesenchymal cells originating from the early primitive streak (and subsequent early posterior intermediate mesoderm/coelomic epithelium) contribute to both the adrenal and the anterior gonad formation, while cells emerging from the late primitive streak (and subsequent late posterior intermediate mesoderm/coelomic epithelium) contribute solely to the gonad formation [8] (Figure 1A). In humans and monkeys the adrenal and gonad arise from two distinct regions of the coelomic epithelium (anterior and posterior regions, respectively) [74], suggesting that complete segregation of both tissues arises earlier in these species. Furthermore, it was suggested that a *Hox* gene code is involved in the anterior/posterior regionalization of the coelomic epithelium [8,74]. Finally, in the chicken, the adrenal gland seems to arise from the inner layer of the coelomic epithelium while the gonad arises from the outer layer [75]. This regionalization of the coelomic epithelium could explain the discrepancies observed following the inactivation of genes considered critical for the thickening of the coelomic epithelium and AGP formation. Indeed, while inactivation of *Osr1* [76] or *Wt1* [61,77] leads to adrenal and gonadal agenesis (Table 2, list of mouse models), inactivation of *Emx2* or *Lhx9* leads to agenesis of the gonads without affecting adrenal development [78–80].

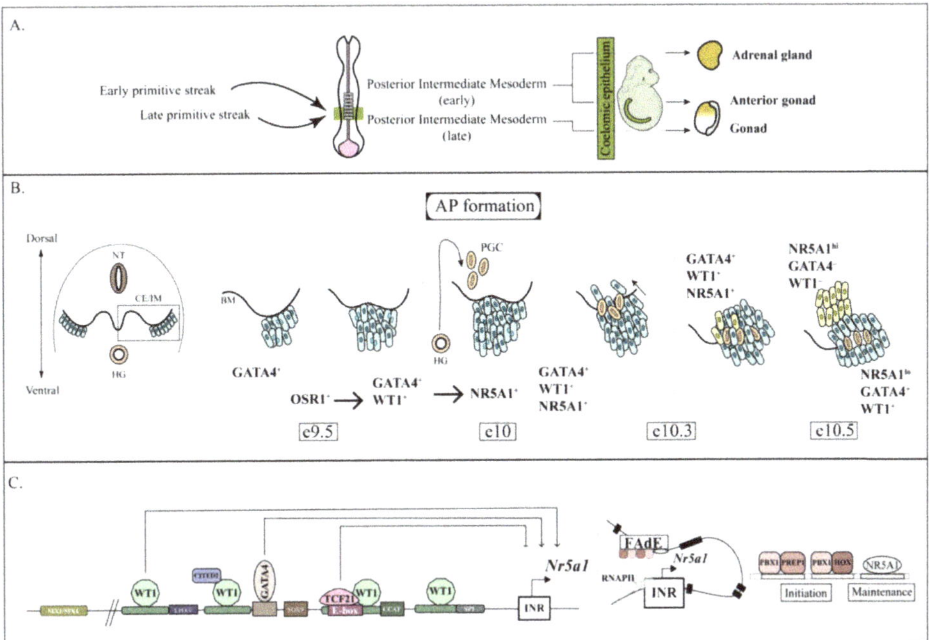

Figure 1. Key events in for the early stages of the development of the fetal adrenal cortex in mice: (**A**) Schematic model for the anterior–posterior regionalization of the posterior intermediate mesoderm, the coelomic epithelium and the AGP in mouse. Fate mapping experiments suggest that cells from the early (anterior) primitive streak migrate to form the early posterior intermediate mesoderm, which is followed by subsequent formation of both the adrenal and the gonad, while cells from the late (posterior) primitive streak migrate to form the late posterior intermediate mesoderm, which is followed by formation of the gonad but not the adrenal gland. (**B**) Schematic model for the development of the AP in mouse. The schematic representation of a transverse view of an embryo shows that the AGP arises from the thickening and delamination of the coelomic epithelium, a process that is initiated around 9.5 and that necessitates the contribution of GATA4 OSR1, WT1 and NR5A1. Once AGP is formed it is invaded by the PGCs (around e10.0), which leads to the separation of the AP from the GP (at e10.5 when a population of cells expressing high levels of NR5A1 begins to migrate dorsomedially). (**C**) Overview of the regulation of the transcription of *Nr5a1* during AGP and AP formation. In the coelomic epithelium/AGP, *Nr5a1* expression is initiated by the binding of several factors to its proximal promoter (only the transcription factors relevant to adrenal development are depicted). Elevated expression of *Nr5a1* in the AP is initiated by the binding of PREP1, PBX1 and HOXs to the FadE enhancer located in the exon 4. Elevated expression of *Nr5a1* in the AP is further maintained by an autoregulatory loop. AGP = adrenogonadal primordium, AP = adrenal primordium, BM = basement membrane. CE/IM = coelomic epithelium/intermediate mesoderm, FadE = fetal adrenal enhancer, GP = gonadal primordium, HG = hindgut, PGCs = primordial germ cells, NT = neural tube.

Table 2. List of mouse models evaluating the development, zonation and maintenance of the adrenal cortex.

Genes	Mouse Models	Phenotype	Ref
Apc	$Apc^{flox/flox}$; $Nr5a1\text{-}Cre^{high}$	Adrenal hypoplasia	[15]
	$Apc^{flox/flox}$; $Nr5a1\text{-}Cre^{low}$	Adrenal Hyperplasia, rare adenoma in older animals	[15]
	$Apc^{flox/flox}$; $Ctnnb1^{flox/flox}$; $Nr5a1\text{-}Cre^{low}$	Rescue of the hyperplasia	[15]
	$Apc^{flox/flox}$; $H19^{floxDMD/floxDMD}$; $Nr5a1\text{-}Cre^{low}$	Adrenal Hyperplasia with higher incidence of adenoma	[15]
AR	$Ar^{flox/Y}$; $Cyp11a1^{Gfp,Cre/+}$	Abnormal retention of the X-zone, Subcapsular spindle-shaped cell hyperplasia	[44]
	$Ar^{flox/flox}$; $Cyp11a1^{Gfp,Cre/+}$	Reduced expression of the zF markers AKR1B7, Subcapsular spindle-shaped cell hyperplasia	[43]
Cbx2	$Cbx2^{-/-}$	Mild hypoplastic adrenal gland at e18.5	[81]
Cited2	$Cited2^{-/-}$	Adrenal agenesis	[82,83]
	$Cited2^{+/-}$; $Wt1^{+/-}$	Adrenal and gonadal hypoplasia	[83]
Ctnnb1	$Ctnnb1^{flox/flox}$; $Nr5a1\text{-}Cre^{high}$	Adrenal aplasia	[16,17]
	$Ctnnb1^{flox/flox}$; $Nr5a1\text{-}Cre^{low}$	Age-dependent adrenal cortex degeneration	[16]
	$Ctnnb1^{flox/flox}$; $Axin2^{CreERt2/+}$	Inefficient regeneration of the adrenal cortex	[53]
	$Ctnnb1^{flox/flox}$; $AS^{Cre/+}$	Impaired rosette formation in the zG	[49]
	$Ctnnb1^{ex3/+}$; $Nr5a1\text{-}Cre^{high}$	Adrenal agenesis (right adrenal), adrenal hypoplasia (left adrenal)	[17,25]
	$Ctnnb1^{ex3/+}$; $Prkar1a^{flox/flox}$; $Nr5a1\text{-}Cre^{high}$	Partial rescue of the adrenal hypoplasia.	[25]
	$Ctnnb1^{ex3/+}$; $Akr1b7\text{-}Cre$	Ectopic expression of zG cells at the expense of zF cells, hyperaldosteronism, Subcapsular spindle-shaped cell hyperplasia, rare adenoma in older animals	[46]
		Increased SUMOylation in the zF	[24]
	$Ctnnb1^{ex3/+}$; $Prkar1a^{flox/flox}$; $Akr1b7\text{-}Cre$	Decreased WNT induced hyperproliferation and ectopic zG differentiation	[25]
	$Ctnnb1^{ex3/+}$; $Prkaca^{+/-}$; $Akr1b7\text{-}Cre$	Accelerated WNT induced tumorigenesis	[25]
	$Ctnnb1^{ex3/+}$; $Akr1b7\text{-}Cre$, $Akr1b7\text{-}Igf2$	Same phenotype as the $Ctnnb1^{ex3/+}$; $Akr1b7\text{-}Cre$ mice	[47]
	$Ctnnb1^{ex3/+}$; $mCyp11a1\text{-}iCre$	Adenoma (Dab2+)	[41]
	$Nr5a1\text{-}Hoxb9$; $Ctnnb1^{ex3/+}$; $mCyp11a1\text{-}iCre$	Adenoma, increase adrenal size in male compared to activation of CTNNB1 alone	[41]
	$Ctnnb1^{ex3/+}$; $AS^{Cre/+}$	Hyperaldosteronism, increased rosette frequency in the zG, block differentiation of zG to zF cells	[49,50]

Table 2. *Cont.*

Genes	Mouse Models	Phenotype	Ref
Dennd1a.V2	*pCMV-BAM hDenndia.V2* (overexpression of the human V2 isoform)	Overexpression of *Cyp17a1*, phenotype not evaluated	[84]
Dicer	*Dicer* $^{flox/flox}$; *Nr5a1-Cre*^{high}	Adrenal hypoplasia at e16.5 and adrenal failure at birth	[17,18]
Ezh2	*Ezh2* $^{flox/flox}$; *Nr5a1-Cre*^{high}	Aberrant zonal differentiation, loss of PKA activity in the zF, expansion of the zG, appearance of subcapsular spindle-shaped cells, phenotype more pronounced in males	[19]
Fgfr2	*Fgfr2* $^{flox/flox}$;*Tbx-Cre*	Adrenal hypoplasia	[10]
	Fgfr2 $^{flox/flox}$;*Nr5a1-Cre*^{high}	Adrenal hypoplasia	[20]
	Fgfr2 IIIb $^{flox/-}$;*K5*$^{Cre/+}$ (global inactivation via recombination in germ cells)	Adrenal hypoplasia at e15.5	[85]
	Fgfr2 $^{flox/flox}$; *AS* $^{Cre/+}$	Impaired rosette formation in the zG	[49]
Gata4 / Gata6	*Gata4* $^{flox/flox}$; *Wt1*$^{CreERt2/+}$	Disruption of coelomic epithelium thickening	[5]
	Gata4 $^{flox/flox}$; *Osr1*$^{eGFP-CreERt2/+}$	Disruption of coelomic epithelium thickening	[5]
	Gata4 $^{flox/flox}$; *Osr1*$^{eGFP-CreERt2/+}$; *Wt1*$^{CreERt2/+}$	Disruption of coelomic epithelium thickening	[5]
	Gata4 $^{flox/flox}$; *CAG-CreER*	Disruption of coelomic epithelium thickening	[5]
	Gata4$^{+/-}$	Reduced subcapsular spindle-shaped cell hyperplasia following gonadectomy	[86]
	Cyp21a1-Gata4	Subcapsular spindle-shaped cell hyperplasia	[71]
	Gata6 $^{flox/flox}$; *Nr5a1-Cre*	Adrenal hypoplasia, absence of an X-zone in postnatal adrenal, Subcapsular spindle-shaped cell hyperplasia	[31]
	Gata4 $^{flox/flox}$; *Gata6* $^{flox/flox}$; *Nr5a1-Cre*^{high}	Adrenocortical like cells in the testes	[22]
		Adrenal agenesis, Adrenocortical like cells in the testes	[21]
Gli3	*Gli3*$^{\Delta699/\Delta699}$	Adrenal aplasia	[87]
		Normal adrenals	[88]
Hoxb9	*Nr5a1-Hoxb9*	Large X-zone	[41]
	Nr5a1-Hoxb9; Ctnnb1$^{ex3/+}$; *mCyp11a1-iCre*	Adrenal tumor formation	[41]
Igf2	*H19* $^{floxDMD/floxDMD}$; *Nr5a1-Cre*^{low}	Normal adrenal	[15]
	Akr1b7-Igf2	Subcapsular spindle-shaped cell hyperplasia	[47]

Table 2. *Cont.*

Genes	Mouse Models	Phenotype	Ref
Insr/ Igf1r	*Insr$^{-/-}$; Igfr1$^{-/-}$* (via recombination of *Insr$^{flox/flox}$; Igfr1$^{flox/flox}$* in germ cells)	Adrenal agenesis and gonadal hypoplasia	[89]
	Insr$^{flox/flox}$; Igfr1$^{flox/flox}$; hCyp11a1-iCre	Abnormal hypoplastic adrenal	[39]
Lats1/ Lats2	*Lats1$^{flox/flox}$; Lats2$^{flox/flox}$; Nr5a1-Cre*	Transdifferentiation of adrenocortical cells into myofibroblast like cells	[33]
Lhcgr	*Lhcgr$^{-/-}$*	Prevention of GATA4 induction and tumor formation in inhα/Tag mice	[70]
Mc2r	*Mc2r$^{-/-}$*	Adrenal hypoplasia limited to the zF, zG still present	[90]
Mst1/ Mst2	*Mst1$^{flox/flox}$; Mst2$^{flox/flox}$; Nr5a1-Cre*	Premature subcapsular spindle-shaped cell hyperplasia	[34]
Mrap	*Mrap$^{-/-}$*	Adrenal hypoplasia limited to the zF (following corticosterone replacement therapy), expansion of WNT/CTNNB1 signaling in the cortex	[91]
Nr0b1	*Nr0b1$^{-/Y}$*	Delayed regression of the X-zone	[92]
		Enhanced subcapsular proliferation in young animals followed by progressive adrenal cortex degeneration in male	[93]
	Nr0b1$^{flox/Y}$; AS$^{Cre/+}$	No effect on the differentiation of zG cells into zF cells	[48]
Nr5a1	*Nr5a1$^{-/-}$*	Gonadal and adrenal agenesis	[94]
	Nr5a1$^{+/-}$	Adrenal hypoplasia	[95,96]
	Nr5a1$^{flox/flox}$; mCyp11a1-iCre	Morphological changes in the shape of steroidogenic cells of the fetal cortex, *Nr5a1-* cells never observed in the definitive cortex	[40]
	FAdE-Nr5a1	Hyperplastic adrenal, ectopic thoracic adrenal tissue, incomplete separation of the AP and GP	[97]
	Nr5a1$^{2KR/2KR}$	Delayed regression of the X-zone	[92]
		Expansion of SHH+ cells in the zF, presence of *Sox9+* (Sertoli-like cells?) in the cortex, delayed regression of the X-zone	[98]
	Nr5a1$^{flox/flox}$; AS$^{Cre/+}$	Loss of zG (and zF maintenance independent of the zG)	[48]
	Nr5a1-TR (overexpression of rat *Nr5a1*)	Subcapsular spindle-shaped cell hyperplasia and nodule formation	[99]
Osr1	*Osr1$^{-/-}$*	Gonadal and adrenal agenesis	[66,76]
Pbx1	*Pbx1$^{-/-}$*	Adrenal agenesis	[100]
	Pbx1$^{+/-}$	Adrenal hypoplasia and smaller X-zone	[101]

Table 2. *Cont.*

Genes	Mouse Models	Phenotype	Ref
Pde8b	*Pde8b$^{-/-}$*	Elevated urinary corticosterone	[102]
		Elevated basal serum corticosterone level in female, Subcapsular spindle-shaped cell hyperplasia	[103]
Pde11a	*Pde11a$^{-/-}$*	Persistence or resurgence of the X-zone, higher cAMP levels, higher incidence of subcapsular spindle-shaped cell hyperplasia, milder phenotype in males	[104]
Porcn	*Porcn$^{flox/flox}$; Nr5a1-Cre*^{high}	Normal adrenal	[23]
Prkar1a	*Prkar1a$^{flox/flox}$; Akr1b7-Cre*	Adrenal hyperplasia, increased PKA signaling, hypercorticosteronemia, appearance of subcapsular spindle-shaped cells, resurgence of an X-zone/presumptive zR (origin not evaluated), milder phenotype in males	[26]
	Prkar1a$^{flox/flox}$; Nr5a1-Cre^{high}	Expansion of the zF at the expense of the zG	[25]
		Repress SUMOylation	[24]
	Prkar1a$^{flox/flox}$; FadE/Nr5a1-CreERT2	Normal adrenal (tamoxifen induction at e14.5)	[37]
	Prkar1a$^{flox/flox}$; AS$^{Cre/+}$	Hypercorticosteronemia, differentiation of lower zF into a presumptive zR, DHEA secretion	[37]
Rnfr3	*Rnfr3$^{flox/flox}$; Nr5a1-Cre*^{high}	Normal adrenal	[105]
Rspo3	*Rspo3$^{flox/flox}$; CAG-CreER*	Progressive adrenal cortex degeneration, loss of zG markers	[6]
	Rspo3$^{flox/flox}$; Gli1$^{CreERt2/+}$	Progressive adrenal cortex degeneration, loss of zG markers	[6]
Siahi1a	*Siahi1a$^{-/-}$*	Smaller X-zone and dysregulation of the zG	[106]
Sfrp2	*Sfrp2$^{-/-}$*	Ectopic expression of CTNNB1+ cells in the zF	[107]
Shh	*Shh$^{flox/flox}$; Nr5a1-Cre*^{high}	Adrenal hypoplasia (more severe on the right side)	[17,27–29]
Six1/ Six4	*Six1$^{-/-}$; Six4$^{-/-}$*	Potential marginal hypoplastic adrenal gland at 1dpp (unconfirmed, suggested in [107])	[108,109]
Smo	*Smo$^{flox/flox}$; Nr5a1-Cre*^{high}	Normal adrenal	[28]
	Rosa^{SmoM2}; Gli1$^{CreERt2/+}$	Enhanced subcapsular WNT/CTNNB1 signaling	[53]
Tcf21	*Tcf21$^{LacZ/LacZ}$*	Improper separation of the AP and GP	[36]
Wnt4	*Wnt4$^{-/-}$*	Reduced aldosterone secretion	[110]
	Wnt4$^{flox/flox}$; Nr5a1-Cre^{high}	Reduction in zG markers	[6]
		Expansion of the zF at the expense of the zG	[25]

Table 2. *Cont.*

Genes	Mouse Models	Phenotype	Ref
Wt1	$Wt1^{-/-}$	Gonadal and adrenal agenesis	[61]
	$Wt1^{-/-}$; *WT280* (WT1 complementation)	Rudimentary hypoplastic adrenal gland at e15.5	[77]
	$Cited2^{+/-}$; $Wt1^{+/-}$	Adrenal hypoplasia	[83]
	$Rosa26^{Wt1+KTS/Wt1+KTS}$; $Nr5a1\text{-}Cre^{high}$	Adrenal hypoplasia, subcapsular spindle-shaped cell hyperplasia	[13]
Yap/Taz	$Yap^{flox/flox}$; $Taz^{flox/flox}$; *Nr5a1-Cre*	Progressive adrenal cortex degeneration in male	[32]
Znrf3	$Znrf3^{flox/flox}$; $Nr5a1\text{-}Cre^{high}$	Adrenal hyperplasia, expansion of the zF, disrupted adrenal organization	[23]
		Development of adrenocortical carcinoma in 78 weeks-old females, activation of androgen-dependent innate antitumor immunity in males	[111]
	$Znrf3^{flox/flox}$; $Rnfr3^{flox/flox}$; $Nr5a1\text{-}Cre^{high}$	Same as the $Znrf3^{flox/flox}$; $Nr5a1\text{-}Cre^{high}$	[23]
	$Znrf3^{flox/flox}$; $Porcn^{flox/flox}$; $Nr5a1\text{-}Cre^{high}$	Rescue the phenotype observed in $Znrf3^{flox/flox}$; $Nr5a1\text{-}Cre^{high}$	[23]
	$Znrf3^{flox/flox}$; $AS^{Cre/+}$	Adrenal hyperplasia, expansion of the zF, disrupted adrenal organization, moderate increased WNT/CTNNB1 signaling in the upper zF	[23]

Three genes, *Gata4* and the aforementioned *Osr1* and *Wt1*, appear to play a central role for AGP thickening (Figure 1B). *Gata4* is detected as early as e8.0 in the coelomic epithelium [112]. Global inactivation of *Gata4* leads to embryonic death before e9.0 [113,114] and cannot be used to study AGP formation. However, its conditional inactivation in the coelomic epithelium at e8.75 completely abolishes its thickening, as well as the subsequent fragmentation of the basement membrane underneath the coelomic epithelium and the proliferation and delamination of the epithelial cells at e10.3 [5]. *Osr1* is also first expressed in the coelomic epithelium/mesenchyme at e8.0-8.5 (66,74,76), and as previously mentioned, its inactivation leads to complete agenesis of the gonads and adrenal glands [76]. Similarly, *Wt1* is also detected around e9.0 in the coelomic epithelium [68,115], and its inactivation also leads to complete agenesis of the gonads and adrenal glands [61,77].

The study evaluating the function of GATA4 in the coelomic epithelium further suggested that GATA4 initiates the thickening process, while OSR1 and WT1 were not essential for the initiation of AGP formation (as a small thickening of the coelomic epithelium was initially observed in $Osr1^{-/-}$ [5] and $Wt1^{-/-}$ animals [61]). However, it is important to note that the functional hierarchy of these genes in the process of AGP formation has not been thoroughly evaluated. *Wt1* appears to be a target of OSR1, at least in some tissues [76], but this was not clearly demonstrated in the AGP. It was also demonstrated that GATA family members can regulate *Wt1* expression in Jurkat and K562 cell lines by binding to a $3'$ enhancer [116]. However, *Wt1* does not appear to be a target of GATA4 in the AGP [5]. GATA4 also does not appear to be a target of WT1 in the AGP [117]. On the other hand, WT1 and GATA4 have been shown to act in synergy to promote the transcription of genes important for sex determination/differentiation [118], suggesting that they could act in synergy in the AGP. More recently, the ontogenic ancestries of the AGP in mouse, human and monkey were evaluated [8,74]. Interestingly, in human and monkey *WT1* expression precedes *GATA4* expression in the AGP [8,74]. Furthermore, it appears that the adrenal primordium specifically originates from a portion of the anterior coelomic epithelium that does not express *GATA4* [74]. Although it was also suggested in these experiments that *Wt1* expression precedes *Gata4* expression in the mouse, the adrenocortical and gonadal lineages (and *Wt1/Gata4* expression) are initially joint in the coelomic epithelium [8,74].

No matter which gene is expressed first, inactivation of *Gata4* and *Wt1* both lead to a decrease in the expression of *Nr5a1* [5,83], and *Nr5a1* inactivation leads to adrenal (and gonadal) agenesis [94] suggesting that *Nr5a1* acts downstream of these factors for AGP formation. It was demonstrated in vitro that both these factors can directly regulate *Nr5a1* transcription by binding to its proximal promoter. WT1 have been shown to bind to four sites located in the first 500 bp of the *Nr5a1* proximal promoter (Figure 1C) [119]. Interestingly, if the mutations of all four WT1 binding sites induce an important decrease in *Nr5a1* promoter activity [83], the introduction of a point mutation in any single WT1 binding site rather increases *Nr5a1* proximal promoter activity. This suggests that WT1 can potentially both activate and repress *Nr5a1* activity depending on the cellular context [119]. GATA4 is also able to bind to the proximal promoter of *Nr5a1* and enhance its activity in vitro [120] (Figure 1C). This activation was achieved in Sertoli and pituitary cell lines, but not in Leydig and adrenal cell lines [120]. Again, these findings reinforce the conclusion that the cellular type/context is important to comprehend the mechanisms regulating *Nr5a1* in the AGP/adrenal cortex.

In addition to *Gata4* and *Wt1*, the inactivation of *Cited2* [82,83], *Tcf21* [36], and *Insr/Igf1r* [89] also leads to early developmental defects of the adrenal gland. Like the inactivation of *Gata4* and *Wt1*, the inactivation of *Cited2* and *Insr/Igf1r* also decreases *Nr5a1* expression [83,89]. CITED2 is first expressed in the coelomic epithelium at e10.0 [83]. Contrary to the inactivation of *Gata4* and *Wt1*, inactivation of *Cited2* has a greater impact on adrenal development than the gonad as the gonads appear to recover from early differentiation defects [83]. CITED2 physically interacts with WT1 and stimulates its transcriptional activity at the *Nr5a1* basal promoter [83]. Interestingly, *Cited2* expression remains high in the AP but decreases in the GP [83], suggesting that *Cited2* might have additional roles in the AP after its separation.

The loss of *Insr/Igf1r*, the receptors for insulin and insulin-like growth factors, leads to a ≈40% reduction of NR5A1+ cells associated with a reduction in the proliferation rate of GATA4+ cells, and an alteration of a quarter of the genes known to be involved in the development of the AGP/bipotential gonad [89]. Despite this more global effect, it is still possible that insulin growth factor (IGF) signaling regulates NR5A1 as INSR/IGF1R have been shown to activate the mitogen-activated protein kinase (MAPK) signaling pathway [121], which promotes phosphorylation-dependent-NR5A1 activation [122].

Finally, contrary to the previously mentioned genes, TCF21 represses *Nr5a1* transcriptional activity by binding to a E-BOX site [123,124] that overlaps with a WT1 binding site [119] (Figure 1C). Interestingly, it was demonstrated that inactivation of *Tcf21* did not affect AGP formation, but rather led to incomplete separation of the AP and the GP [36]. This suggests that *Nr5a1* must be tightly regulated during the development of the AGP/AP.

Other genes expressed in the AGP such as *Six1*, *Six4* and *Cbx2* are also able to regulate the transcriptional activity of *Nr5a1*. SIX1/SIX4 are able to bind and activate the transcriptional activity of *Nr5a1* proximal promoter [108], while CBX2, a component of the mammalian polycomb repressive complex-1 required for chromatin remodeling and histone modification, has been shown to bind several regions of the *Nr5a1* genomic region [81]. However, the inactivation of these genes suggests that they are more important to gonadal development than adrenal development [108,109,125–127] as their inactivation only leads to marginal or mild adrenal hypoplasia [81,108,109,127]. Nonetheless, these factors might fine-tune adrenal development.

4. AP and Fetal Adrenal Development

If GATA4 and WT1 are considered two of the main regulators for the initial formation of the AGP, their expression in mice is switched off in the AP, just after its separation from the GP occurs [13,128]. This suggests that these genes prevent the differentiation of AGP cells into the adrenal steroidogenic cell lineage. Indeed, it was demonstrated that the ectopic expression of high levels of the WT1-KTS isoform (the isoform able to bind DNA and act as a transcription factor) in NR5A1+ cells of the AGP leads to the maintenance of GATA4 expression, reduced NR5A1 expression, and the formation of abnormal small adrenal glands [13]. Furthermore, WT1 has been shown to bind the promoter of *Tcf21* suggesting that WT1 could also inhibit *Nr5a1* expression indirectly [13]. In an initial study using chimeric mice (generated by the injection of $Gata4^{-/-}$ ES cells in blastocysts), it was also suggested that GATA4 was not essential for adrenocortical cell differentiation [129]. However, concomitant inactivation of *Gata4* and *Gata6* (but not *Gata4* alone) in NR5A1+ cells of the AGP leads to adrenal agenesis, suggesting that GATA4/6 have redundant activity in the AGP before AP separation and downregulation of GATA4 expression [21,22]. Interestingly, inactivation of *Gata6* alone in these cells affects adrenal development suggesting that GATA6 is important for later stages of adrenal development [31]. Following *Gata6* inactivation, expression of GATA4 is maintained in the AP. However, residual NR5A1+ cells did not express GATA4, which suggests that GATA4+ cells are unable to commit to the adrenal steroidogenic lineage. The maintenance of GATA4 expression in this model could be due to a compensatory mechanism. It is also possible that GATA6 normally represses *Gata4* expression in later stages of adrenal development, as it was demonstrated in H9 and P19CL6 cell lines and heart development that GATA4 and GATA6 can mutually and directly regulate their transcriptional activity [130–132].

As mentioned previously, *Nr5a1* expression closely follows *Gata4* and *Wt1* expression in the AGP, and global inactivation of *Nr5a1* leads to adrenal and gonadal agenesis [94]. However, contrary to GATA4 and WT1 expression, robust NR5A1 expression is maintained in the AP following its separation from the AGP. Such robust *Nr5a1* transcription is possible due to the activation of the FadE located in the intron 4 of *Nr5a1* [133]. Studies employing both cell line assays and transgenic mouse models have demonstrated that *Pbx/Prep/Hox* binding sites were necessary for the initiation of the transcription by the FadE (highlighted by the fact that deletion of *Pbx1* leads to adrenal agenesis [100]), while *Nr5a1* transcription is further maintained by a NR5A1 positive autoregulatory loop [133] (Figure 1C). It has also been suggested that NR5A1 dosage is critical for AP development, and it is thought that cells expressing higher levels of NR5A1 give rise to cells that will form the AP whereas cells expressing lower levels will form the GP. This was originally proposed because the adrenal glands of $Nr5a1^{+/-}$ heterozygous animals were highly hypoplastic, while the gonads were not [95]. A subsequent study demonstrated that the loss of one *Nr5a1* allele decreased the number of adrenal precursor cells within the AGP but not the gonadal GATA4+ precursor cells [96]. Inactivation of *Nr5a1* using the m*Cyp11a1-Cre* model also demonstrated that Nr5a1- cells adopt a more elongated and flat shape reminiscent of less differentiated cells, but GATA4 and WT1 expression were not evaluated in this model [40]. Furthermore, overexpression of NR5A1 (using a basal *Nr5a1* promoter and the FadE to drive its expression) led to ectopic adrenal tissue formation in the thorax [97], again suggesting that high NR5A1 expression is necessary for the proper differentiation of the AP. Interestingly, separation of the AP and GP was also affected in this model, suggesting that *Nr5a1* dosage is important for this process [97]. As previously mentioned, TCF21 is also important for the separation of the AP and GP and negatively regulates *Nr5a1* expression (36,123,124), suggesting that TCF21 could be essential for *Nr5a1* dosage.

Aside from NR5A1, it was also demonstrated that FGFR2 is important for AP formation. *Fgfr2* inactivation also leads to major adrenal hypoplasia [10,20]. Serial section and 3D reconstruction analyses revealed that the number of NR5A1+ cells was initially normal in AP of e10.5 mutant animals, but that a two-fold reduction was observed at e11.5. It was further demonstrated that cell proliferation of NR5A1+ cells was also reduced by around 50% at e11.5 and e12.5, and that apoptosis increased at e12.5 suggesting that FGFR2 is required for the expansion of the AP by regulating both cell proliferation and apoptosis [10].

As previously mentioned, the late stages of fetal adrenal development are difficult to evaluate as the inactivation of the gene of interest will occur in cells of both the fetal and adult cortex using most Cre strains, and both cortices will be present at these later time points. An indirect manner to confirm the importance of factors involved in later stages of the fetal cortex development would be to evaluate the fetal cortex at birth, which is referred as the X-zone (before its eventual regression at puberty in males or after the first pregnancy in females [134,135]). It could be expected that inactivation of genes that normally regulate the fetal cortex development positively would lead to the absence or a decrease in the size of the X-zone at birth; while inactivation of genes that normally regulate the fetal cortex development negatively would lead to the opposite effect and the presence of a larger X-zone at birth. To illustrate this, a smaller X-zone is observed in animals with *Pbx1* haploinsufficiency [101]. As previously mentioned, *Pbx1* is a gene important for *Nr5a1* transcription from the FadE. On the other hand, *Nr0b1* knockout male mouse present a larger X-zone in young animals and X-zone regression is delayed [92]. A similar phenotype is also observed in animals with a SUMOylation deficient form of NR5A1 [92]. It was further demonstrated that SUMOylation of NR5A1 facilitates the recruitment of NR0B1 to the FadE of *Nr5a1* to inhibit its transcriptional activity [92]. Inactivation of *Gata6*, inactivation of *Siah1a* and overexpression of *Hoxb9*, respectively, lead to animals lacking an X-zone [31], having a smaller X-zone [106] and having a larger X-zone [41]. This suggests that these genes are also involved in the development of the fetal cortex.

Soon after AP separation, a population of peripheral glial stem cells derived from the neural crest will migrate, invade the medulla, and eventually differentiate into chromaffin cells (Figure 2A). Single cell RNAseq experiments suggest that this migrating population of stem cells differentiate into sympathetic neurons (SN) and Schwann cell precursors (SCP), with the SCPs further differentiating into chromaffin cells in the AP [136,137] (although the transition between cell fates in human appears to occur in a different order [138]). Chromaffin cells are not necessary for the development of the adrenal cortex, but their presence is necessary for its proper organization [139].

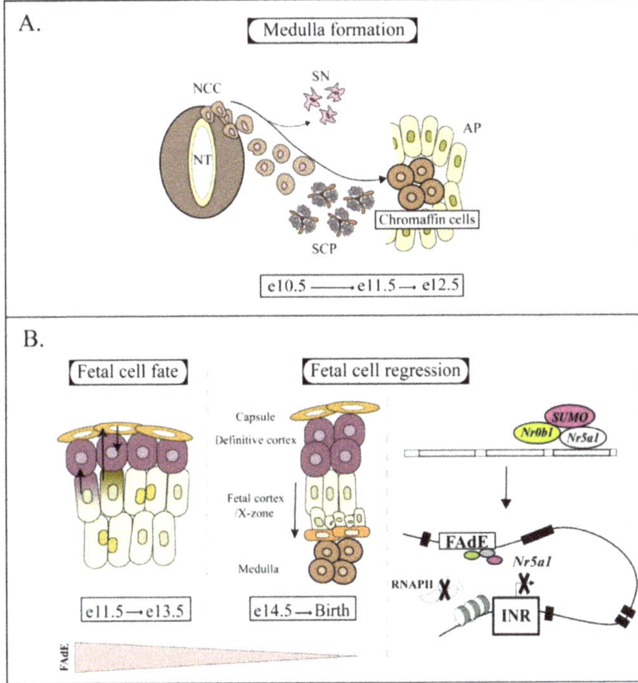

Figure 2. *Cont.*

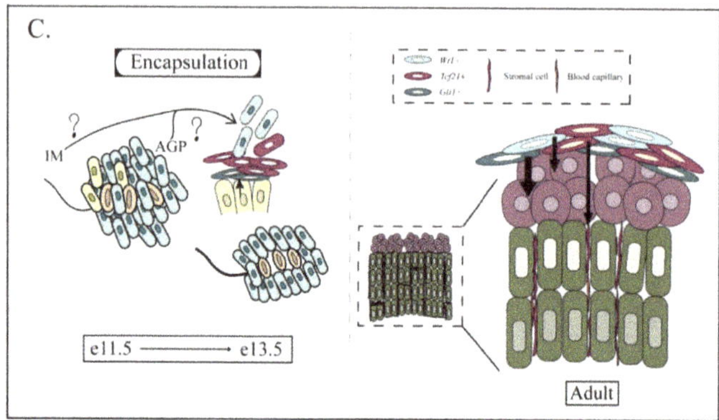

Figure 2. Key events in the late stages of fetal adrenal gland development and fate of the fetal adrenal cortex in mice: (**A**) Schematic model of medulla formation. Chromaffin cell differentiation requires the migration of a subpopulation of neural crest cells and their progressive differentiation into Schwann cell precursors and chromaffin cells. (**B**) Schematic model depicting the fate of the fetal adrenocortical cells. Before e14.5, adrenocortical fetal cells can differentiate into adrenocortical cells of the definitive cortex by first differentiating into capsular stem cells (or potentially by directly differentiating into cells of the definitive cortex), or they can proliferate and maintain the AP. After e14.5, cells from the AP do not contribute anymore to the definitive cortex formation. The progressive inactivation of *Nr5a1* expression via NR5A1 SUMOylation and recruitment of NR0B1 to the FAdE leads to the regression of the fetal cortex/X-zone. (**C**) Schematics model for AP encapsulation. AP encapsulation requires the contribution of cells originating from both the fetal adrenal cortex and cells potentially originating directly from the AGP or the surrounding intermediate mesoderm. Different cell subpopulations have been identified including: *Gli+* cells that are located the nearest to the adrenal cortex and that will eventually differentiate into steroidogenic cells of the definitive adrenal cortex; *Wt1+* cells that have a more limited potential to differentiate into steroidogenic cells; and *Tcf21+* cells that will differentiate into stromal cells of the definitive adrenal cortex. NCC = neural crest cells, NT = neural tube, SN = sympathetic neurons, SCP = Schwann cell precursors.

5. Fate of the Fetal Cortex

As previously mentioned, SUMOylation of NR5A1 followed by NR0B1 binding on the FAdE will eventually inactivate *Nr5a1* transcription in the fetal cortical cells [92] (Figure 2B). This process will eventually lead to the complete regression of the fetal zone that will arise at puberty in male because of androgen action [44,140], or following the first pregnancy in female (or in old virgin females) [141,142] by an unknown mechanism (though it was suggested that progesterone might be involved in this process [143] and that androgen is probably not [43]). However, although some adrenocortical fetal cells follow this above-mentioned path, it does not appear to be the case for all of them. Indeed, tracing experiments using the FAdE driving LacZ demonstrated that if LacZ expression clearly diminished in the outside cortex (where the definitive cortex appears as discussed below) and persisted in the inner fetal adrenal cortex, some cells between the two zones appear to have transient LacZ expression suggesting that cells of the fetal cortex could differentiate into cells of the definitive cortex [35,133]. A subsequent experiment (in which the FAdE was used to drive Cre recombinase in ROSA26-LacZ reporter mice to permanently mark the adrenocortical cells) confirmed that this was indeed the case [35]. Furthermore, a study using a tamoxifen inducible *FAdE-CreERt2*, demonstrated that fetal adrenal cortical cells from e11.5 animals can differentiate into adrenocortical cells of the definitive cortex, but that their potentiality is lost at e14.5 [35]. An additional study further suggested that some adrenocortical fetal cells migrate into the developing capsule and form supporting mesenchymal cells and potential stem cells for the definitive cortex [36] (Figure 2B).

If tracing experiments suggest that the fetal adrenocortical cells can differentiate into cells of the definitive cortex, a recent single cell RNAseq analysis rather suggests that three different cell clusters (adrenal primordium, fetal zone and definitive zone) can be observed in the developing adrenal cortex [137]. Briefly, this study suggests that a subset of cells forming the adrenal primordium cluster first differentiate into cells of the fetal cortex cluster while another subset of cells differentiate into cells of the definitive cortex, suggesting a mutual exclusion of the differential potential of the adrenal primordium cluster across the other two clusters [137]. It is however important to point out that these single cell analyses were performed at time points ranging from e13.5 to P5, a time window that could be considered too late to properly observe the differentiation of the fetal adrenocortical cells into cells of the definitive cortex, as it was previously demonstrated that fetal adrenocortical cells progressively lose their capacity to differentiate into definitive adrenocortical cells between e11.5 and e13.5 before losing this potentiality at e14.5 [35].

6. Encapsulation of the Adrenal Cortex

Following the invasion of the fetal cortex by the chromaffin cells and concomitantly with the beginning of the transition of the fetal cortex/definitive cortex, mesenchymal-like cells will encase the forming adrenal gland and form the capsule; a process fully completed at around e13.5–e14.5. The lineage of these mesenchymal cells is not completely understood, but partially overlapping cell populations have been identified in the capsule (Figure 2C). First, the majority of the capsular cells express NR2F2 and tracing experiments demonstrate that some of these cells originate from the fetal cortex [36]. Three other cell populations have been identified as a *Gli1+* cell population, a *Tcf21+* cell population and a *Wt1+* cell population. The NR2F2+ and *Gli1+* cells partially overlapped, and tracing experiments also suggest that some of the *Gli1+* cells originate from the fetal cortex [36]. *Tcf21+* cells do not arise from the fetal cortex but might arise from the AGP [36] or from other regions of the intermediate mesoderm (Figure 2B). Finally, a *Wt1+* cell population potentially overlaps with the *Tcf21+* cell and with the *Gli1+* cells [13]. It was further demonstrated that WT1 could regulate the transcription of both *Tcf21* and *Gli1* [13]. Single cell RNAseq performed at e13.5 also identified *Gli1*, *Tcf21* and *Wt1* expression in the same cell clusters, while two clusters of capsular cells, *Tcf21^{high}* and *Wt1^{high}*, were identified in late fetal/perinatal adrenal gland suggesting that most *Tcf21+* and *Wt1+* cells belong to distinct capsular cell populations at these time points [137]. However, it is important to note that a limited number of cells (a little over 2000 cells from whole adrenal over six different time points) were used for this latter experiment. The number of capsular cells sequenced was therefore insufficient to truly determine how many cell populations were present in the adrenal capsule [137]. Capsular cells do not express NR5A1, including the cells that originate from the fetal cortex [36] and, interestingly, NR2F2 [144], TCF21 [123,124] and WT1 [13] are all able to negatively regulate *Nr5a1* expression. This could suggest that these genes ensure that capsular cells do not express *Nr5a1*.

7. Development and Maintenance of the Definitive Cortex: The Key Role of Hedgehog and Canonical WNT Signaling Pathways

Fetal adrenal cortical cells can initially differentiate into adrenal cortical cells of the definitive cortex but lose this capacity after e14.5 [35]. This suggests that if the initial cells of the definitive cortex originate from the fetal cortex, the cells necessary for the late stages of the definitive cortex development (and its subsequent maintenance) have a different origin [35]. Again, tracing experiments were essential to better understand the origin of these cells. Using fate mapping it was demonstrated that capsular cells positive for *Gli1+* (the main effector of the Hedgehog signaling pathway) were able to differentiate into steroidogenic cells of the adrenal cortex (Figure 2C), both in the embryo and postnatally, and that marked cells migrate inward while centripetally displacing older cells (27,28,36). It was further demonstrated that some capsular *Wt1+* cells were also able to differentiate into adrenocortical steroidogenic cells (Figure 2C) although the authors suggested that the

Wt1-, Gli1+ cell located in the interior side of the capsule is probably the main capsular stem cell population [13]. Contrary to the *Gli+* cells, the *Tcf21+* cells are only able to form steroidogenic cells before the formation of the capsule and only differentiate into non-steroidogenic stromal adrenocortical cells once the capsule is formed [36] (Figure 2C).

Following *Gli1+* tracing experiments, the importance of the Hedgehog signaling was further confirmed when *Shh*, which is expressed in the subcapsular cell of the adrenal cortex, was inactivated. Conditional deletion of *Shh* in the adrenal cortex using a *Nr5a1-Cre* strain were generated by three different groups that demonstrated that its inactivation led to adrenocortical hypoplasia [27–29]. This phenotype was associated with a thinning of the adrenal capsule [27,28] and a reduction in capsular cell proliferation [27], indicating that Hedgehog signaling affects capsular cells and not steroidogenic cells. This was further confirmed by the fact that the inactivation of *Smo*, a transmembrane protein essential to transduce Hedgehog signaling in the adrenal cortex, did not lead to an apparent phenotype [28]. Furthermore, the number of *Gli1+* cells was dramatically reduced in the capsule following *Shh* inactivation [28], confirming that Hedgehog signaling acts on the capsular cells. *Gli2* and *Gli3* are also important for Hedgehog signaling and are expressed in the capsule [29]. However, the exact role of these molecules for the maintenance of the adrenal cortex is currently unknown. It was originally suggested that GLI3 might have a role in adrenal development, as the expression of a truncated GLI3 with constitutive transcriptional repressor activity leads to the development of Pallister–Hall syndrome in human, which also included adrenal hypoplasia or aplasia in some cases [145]. Mouse bearing a similar *Gli3* allele was also first reported to have adrenal aplasia [87] but a subsequent study using the same model did not observe this phenotype [88].

Although SHH acts on the *Gli+* capsular cells, it was also shown that the *Shh*$^{flox/flox}$; *Nr5a1-Cre* animals had fewer proliferating cells in the outer layer of the adrenal cortex [29]. Moreover, tracing experiments demonstrated that the *Shh+* cells were also able to proliferate and move inward centripetally, indicating that they also are progenitor cells [28]. These results led to the establishment of the two progenitor lineages model in which Gli+ cells (later coined the adrenal stem cells) give rise to both steroidogenic cells and subcapsular progenitor cells, which secrete SHH to allow the proliferation/maintenance of the capsular stem cells. Shh+ cells also proliferate and further differentiate into steroidogenic cells upon their centripetal migration (Figure 3A) [28]. Subsequent experiments demonstrated that both the capsular stem cell and the subcapsular progenitor cell populations were also important to maintain homeostasis of the adrenal cortex in prepubertal animals (males and females) and mature females, while the subcapsular progenitor cells were the main contributors to adrenal homeostasis in mature males (Figure 3A) (though capsular stem cells maintain a role when important regeneration is needed) [53,54]. The importance of sexual dimorphism for the maintenance of the adrenal cortex will be discussed in a later section of this review.

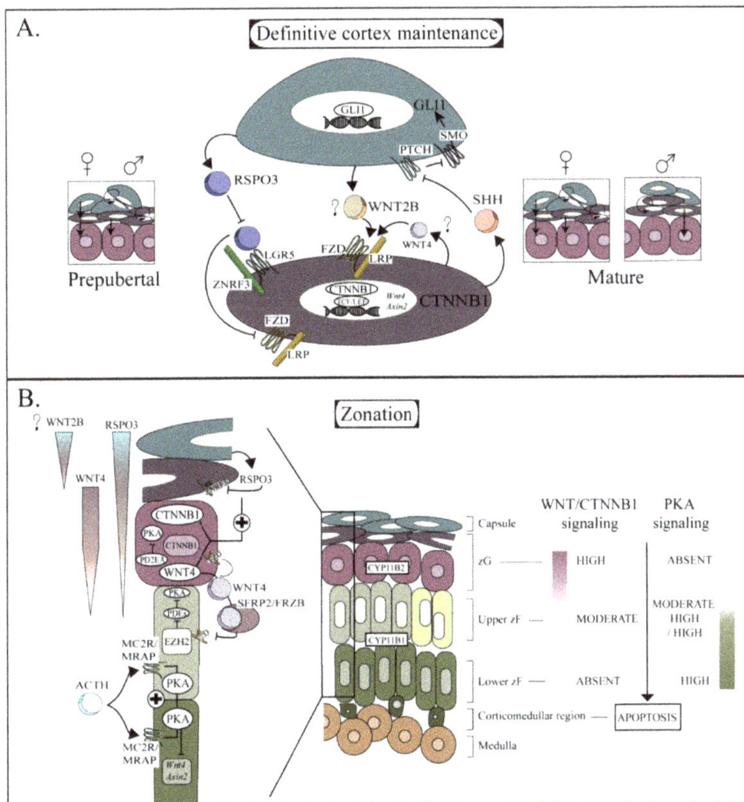

Figure 3. Development, maintenance, and establishment of zonation in the mice definitive adrenal cortex: (**A**) Schematics model for the definitive cortex maintenance. A WNT/SHH double-paracrine mechanism is needed to ensure proper development and maintenance of the adrenal cortex. Capsular stem cells regulate WNT pathway activation in the progenitor zG cells by secreting RSPO3, which in turn induces the clearance from the cell membrane of the WNT signaling inhibitor ZNRF3 and its subsequent degradation. This leads to an increase in canonical WNT signaling in the subcapsular progenitor and it promotes their proliferation (and subsequent differentiation into steroidogenic zG cells). Capsular WNT2B could be the main ligand responsible for the maintenance of the adrenal cortex (as WNT4 expressed in the zG appears to be more important for the differentiation of the steroidogenic zG cells). SHH secreted by the zG progenitor cells promotes the activation of GLI1 in the capsular stem cells and, consequently, their proliferation. Before puberty, both the capsular stem cells and subcapsular progenitor cells contribute to the maintenance of the adrenal cortex. After puberty, both the capsular stem cells and subcapsular progenitor cells contribute to the maintenance of the adrenal cortex in female. In male, only the subcapsular progenitor cells contribute to the maintenance of the adrenal cortex. (**B**) Schematics model for adrenal cortex zonation. Capsular and subcapsular WNTs and RSPO3 secretion create a gradient of WNT/CTNNB1 activity throughout the upper cortex. Elevated WNT/CTNNB1 activity drives steroidogenic zG cells differentiation and inhibits differentiation of zG cells into zF. PDE2A (whose transcription is positively regulated by CTNNB1) could be involved in the inhibition of PKA signaling in the zG by degrading cAMP Moderate/low WNT/CTNNB1 favors the differentiation of zG cells into zF cells and the proliferation of zF cells. Binding of ACTH to its receptor promotes the activation of PKA signaling and antagonizes WNT signaling. Potential expression of FRZB or SFRP2 in the zF could also contribute to the inhibition of WNT signaling in the zF. PKA signaling is also optimized by adrenocortical cell programming by EZH2, which inhibits the transcription of *Pde7b*, *Pde3a* and *Pde1b*.

Aside from the Hedgehog signaling, the canonical WNT signaling also has an important role for the development and the maintenance of the definitive cortex. This role was first suggested when a transgenic mouse model inactivating *Ctnnb1* (the main effector of canonical WNT signaling) in the adrenal cortex was generated. In this model, the adrenal gland initially formed. However, the adrenal cortex was unable to grow properly, leading to the complete atrophy of the adrenal gland by e18.5 [16]. Using a similar model in which *Ctnnb1* recombination was not as efficient, it was demonstrated that CTNNB1 was also essential for the maintenance of the adult adrenal cortex [16]. It was suggested at that time by the authors that the changes observed in the latter model could be associated with the depletion of a population of progenitor cells [16]. The role of CTNNB1 signaling in the progenitor cell population was later confirmed using lineage tracing experiments [53]. Indeed, it was demonstrated that subcapsular cells expressing *Axin2* (a transcriptional target of CTNNB1) contribute to adrenal cortex regeneration following dexamethasone-induced adrenocortical atrophy [53]. Furthermore, inactivation of Ctnnb1 in Axin2+ cells using the *Axin2*^{CreERt2} mouse strain led to inefficient regeneration of the adrenal cortex, reduced *Shh* expression and reduced expression of *Ctnnb1* target genes, demonstrating that at least some progenitor cells were *Shh$^+$*, *Ctnnb1$^+$* [53].

If it was demonstrated that canonical WNT signaling induced SHH signaling in the subcapsular adrenal cortex (which led to the proliferation of the capsular *Gli1*+ cells), it was also demonstrated that the effect was reciprocal and that *Gli1*+ cells induced canonical WNT signaling in the *Shh*+ cells (Figure 3A). Indeed, it was first shown that *Gli1*+ cells express and secrete *Rspo3*, an enhancer of WNT signaling [6]. It was further demonstrated that the inactivation of *Rspo3* at different time points between e11.5 and 12-week-old animals (using either a ubiquitously expressed tamoxifen Cre inducible mouse strain or a *Gli1* regulated tamoxifen Cre inducible mouse strain) results in developmental and maintenance defects of the definitive cortex. These defects were associated with the loss of CTNNB1 target genes and *Shh* expression in the subcapsular cells as well as the loss of capsular *Gli1* expression [6].

It was also demonstrated in other systems that RSPOs promote the clearance of the inhibitor of WNT signaling ZNRF3 from the cell membrane [105,146]. Interestingly, inactivation of *Znrf3* in the developing adrenal cortex (using the *Nr5a1*-cre strain) or in the zG of adult animal (using the AS^{Cre} strain) increases WNT/CTNNB1 signaling and promotes adrenal hyperplasia (Figure 3A) [23]. Together, these findings indicate that inactivation of ZNRF3 by RSPO3 is essential for the development and maintenance of the adrenal cortex. Regulation of the canonical WNT signaling by *Gli1*+ cells was further confirmed as it was demonstrated that CTNNB1 transcriptional activity and adrenocortical regeneration were improved in a mouse model in which a constitutively activated form of SMO is expressed in capsular stem cells following dexamethasone-induced adrenocortical atrophy [53]. Previously mentioned experiments suggested that canonical WNT signaling play a key role in the maintenance of the adrenal cortex. However, 19 WNTs have been identified in the mice suggesting that numerous WNTs could be involved in this process. WNT4 is strongly expressed in the subcapsular region of the adrenal cortex as early as e14.5 [110]. Interestingly, inactivation of *Wnt4* either globally [110] or specifically in the adrenocortical cells [6,25] affects steroidogenic zG cells, CYP11B2 expression and aldosterone secretion [25,110], but only has a marginal effect on adrenal size and does not appear to have an important effect on the proliferation or maintenance of the adrenal cortex. This suggests that WNT4 is more important for the zonation of the adrenal cortex than for the proliferation of the progenitor cell population, and that other WNTs are more important for the proliferation of the progenitor cells (Figure 3A,B). This could also suggest that other WNTs compensate for the loss of WNT4 or act in synergy with WNT4 to regulate the proliferation of the progenitor cells. *Wnt2b* expression has been detected in the adrenal capsule as early as e13.5 [147] and could therefore be the main WNT ligand involved in the crosstalk between the capsular stem cells and the subcapsular progenitor cells. Taken together, all these studies demonstrate that a WNT/SHH double-paracrine mechanism is needed to ensure proper development and maintenance of the adrenal cortex via regulating the capsular stem cell population and the subcapsular progenitor cell population (Figure 3A).

Whether the Hedgehog and WNT signaling pathways appear to be the most important pathways involved in the development and maintenance of the definitive cortex, other biological processes and signaling pathways are also involved. FGFR2 and INSR/IGF1R, which have been shown to respectively have a role during development of the fetal cortex [10] and AGP formation [89], could also participate to the development of the definitive cortex. Inactivation of the *Fgfr2* isoform IIIb (which is expressed in the subcapsular region of the developing adrenal cortex) impairs the development of the adrenal cortex by potentially affecting cell proliferation [85]. On the other hand, inactivation of *Insr1* and *Igf1r* using the h*Cyp11a1-iCre* gives rise to small abnormal adrenal glands in adult animals that produce less corticosterone and that require exogenous sodium supplementation for their survival [39]. MicroRNAs also play an important role in the development of the definitive cortex, as the inactivation of Dicer using a *Nr5a1-Cre* strain leads to adrenal failure at birth associated with the progressive atrophy of the adrenal cortex starting at e16.5 caused by an increase in cellular apoptosis [17,18]. One common denominator with these transgenic mouse models is that development and maintenance defects are associated with a decrease in the expression of NR5A1, or in a decrease in the number of NR5A1+ cells. Although these latter observations could simply indicate that fewer steroidogenic cells are present in these animals, it is likely that the loss of NR5A1 in these models contributes to the observed phenotype. Inactivation of *Nr5a1* could not be achieved in the definitive cortex (contrary to what was observed in the fetal cortex) using a h*Cyp11a1-iCre* strain, suggesting that the lack of *Nr5a1* is incompatible with the definitive cortex development [40]. Furthermore, inactivation of *Nr5a1* using the *AS^{cre}* strain also leads to the loss of the zG suggesting again that *Nr5a1* expression is essential for adrenocortical steroidogenic cell maintenance [48]. Finally, overexpression of rat *Nr5a1* in transgenic mouse models leads to nodule formation [99] while overexpression of *NR5A1* in the human H295R adrenal cell lines increases cellular proliferation [99]. Together, these studies suggest that NR5A1 maintains the definitive adrenal cortex by regulating the proliferation of the steroidogenic cells. Interestingly, loss of NR5A1 SUMOylation also leads to ectopic expression of *Shh* in the developing testes, increased *Shh* expression in the adrenal cortex and the presence of *Shh*+ cells deep into the adrenal cortex [98]. Experiments performed in the embryonic cell line mHypoE-40 further demonstrated that SUMOylation modulated the DNA binding of NR5A1 to the promoter of *Shh*. This suggests that NR5A1 activity must be tightly regulated by SUMOylation to properly regulate the expression of *Shh* in adrenocortical cell populations.

Finally, postnatal impairment of adrenocortical maintenance was also observed following the inactivation of the effectors of the Hippo signaling, *Yap* and *Taz*, in steroidogenic cells [32], or global inactivation of *Nr0b1* [93]. In these two models, degeneration of the adrenal gland was also associated with the appearance of large multinucleated lipoid structure in the adrenal cortex potentially caused by a decrease in the progenitor reserve or progenitor cell reserve [32,93], as suggested by a decrease in the expression of *Shh* following the inactivation of *Yap/Taz* [32]. A potential link between these two models is also suggested by the fact that inactivation of *Yap/Taz* also leads to a decrease in the expression of *Nr0b1* [32]. Interestingly, the degeneration of the adrenal cortex was only observed in male following the inactivation of *Yap/Taz* [32] suggesting that maintenance of the adrenal cortex is sexually dimorphic. Surprisingly, inactivation of the main kinases of Hippo signaling, *Lats1* and *Lats2* (which lead to an increase in YAP and TAZ activity), does not lead to hyperplastic adrenal gland or increased proliferation of the progenitor cells, but rather leads to the transdifferentiation of adrenocortical cells into myofibroblast-like cells [33].

8. Establishment of Zonation: The Opposing Roles of WNT and PKA Signaling Cascades

One key feature of the adrenal cortex is the appearance of concentric zones in which aldosterone (zG) and corticosterone (zF) are synthetized in mice, and aldosterone (zG), cortisol (zF) and DHEA/DHEAS (zR) are synthetized in humans. In mice, functional zonation (as shown by the activity of the *Cyp11b2* promoter) is first observed at e16.5 in rare scattered subcapsular cells (with a similar pattern being also observed at 1dpp) [48]. However, this limited number of cells might reflect a delay in the appearance of the fluorescent reporter marker following recombination. Even though the presence of the enzyme necessary for mineralocorticoid production only appears in late stages of adrenal cortex development, zonation of other genes such as *Wnt4* precedes this time point and can be observed as early as e14.5 in the outer cortex [110]. This suggests that WNT signaling is not only important for adrenal cortex development/maintenance, but also for proper zonation. The importance of WNT signaling for zG development was first suggested when it was demonstrated that *Wnt4* global knockout animals secreted less aldosterone [110]. This result was later confirmed by the inactivation of *Wnt4* using a *Nr5a1-cre* strain in which the expression of zG markers was reduced [6,25], and an expansion of the zF was observed [25]. The loss of *Wnt4* also led to a reduction in the expression of CTNNB1 [25] and its downstream target *Axin2* [6] suggesting that canonical WNT/CTNNB1 signaling is involved in the differentiation of the zG (Figure 3B).

The importance of CTNNB1 in this process was confirmed by the study of transgenic mouse models in which expression of CTNNB1 was stabilized. If the stabilization of CTNNB1 at an early time point of adrenal development (either by inactivating *Apc* or by expressing a constitutively active form of CTNNB1 that lack the phosphorylation sites on exon 3 necessary for its degradation) led to significant adrenal hypoplasia during development [15,17], its stabilization at later time points (either by inactivating *Apc* with a less efficient *Nr5a1-cre* [15] or by expressing the *Ctnnb1^{ex3}* allele in *Akr1b7+* cells [46]) rather led to an increase in cellular proliferation and adrenal dysplasia. Even more important for the zonation process, CTNNB1 stabilization in adrenocortical cells led to the downregulation of the zF marker AKR1B7, expression of ectopic CYP11B2+ cells in the zF and hyperaldosteronism [46]. The importance of CTNNB1 was further confirmed in the H295R cell line as its inactivation also decreased aldosterone production in these cells [107]. Recent studies also demonstrate that ectopic CTNNB1 activation in zG cells blocks their differentiation into zF cells, increases aldosterone production and the number of rosettes (structures adopted by glomerular cells) [49,50]. Furthermore, *Ctnnb1* inactivation in zG cells reduced rosette frequency, though this might be associated with the role of CTNNB1 at cellular junctions rather than in WNT signaling [49] as membranous CTNNB1 is also lost following inactivation of *Fgfr2*, which also leads to the impairment of rosette formation [49]. Finally, global inactivation of the WNT signaling inhibitor *Sfrp2* leads to the appearance of CTNNB1/CYP11B2 positive cells in the zF [107], while the inactivation of the previously mentioned capsular activator of WNT signaling *Rspo3* leads to the loss of all zG markers [6]. Together, these findings clearly indicate that WNT signaling is the key pathway regulating zG cell differentiation (Figure 3B).

If WNT signaling plays a key role in the zG formation and aldosterone production, ACTH/protein kinase A (PKA) signaling is the most important pathway for corticosterone synthesis by the zF. The role for this pathway for the differentiation of the zG cells into zF cells was also recently demonstrated. First, it was demonstrated that zF expansion and expression of zF markers are induced by ACTH treatment in mice, while *Cyp11b2* expression and the activity of a WNT signaling reporter transgene is extinguished [25]. In addition, forskolin (a known pharmacological activator of adenylate cyclase leading to PKA signaling activation) decreases *Wnt4* and *Axin2* expression in H295R cells [25]. This role was further confirmed in a transgenic mouse model inactivating *Prkar1a*, the gene encoding the regulatory subunit type 1a of PKA, which leads to the constitutive activation of PKA in the adrenal cortex. In this model, PKA activation leads to the expansion of zF and an important repression of CTNNB1 activity and *Cyp11b2* expression [25]. Consistent with the role of ACTH for zF maintenance, inactivation of *Mrap* (an accessory protein essential for MC2R, the ACTH receptor) activity leads to neonatal lethality due to the absence of corticosterone secretion [91]. However, surviving animals (following corticosterone replacement treatments) presented hypoplastic adrenals that did not express CYP11B1 [91]. Furthermore, WNT4 and CTNNB1 expression was detected in most cells of the remaining cortex though CYP11B2 was only expressed in a portion of these cells [91]. Similarly, to the inactivation of *Mrap*, the inactivation of *Mc2r* also leads to neonatal lethality in most animals, with rare animals surviving through adulthood [90]. The few surviving animals had hypoplastic adrenal with CYP11B2 expressing zG cells still present (despite lower circulating aldosterone levels) (Figure 3B) [90]. WNT signaling was however not evaluated in this model. Finally, it was demonstrated that global SUMOylation was negatively regulated by PKA signaling and positively by CTNNB1 in the adrenal cortex. However, a possible role for SUMOylation in the zonation process has not yet been identified [24].

Epigenetic factors can also contribute to zF differentiation, as it was demonstrated that the inactivation of the histone methyltransferase *Ezh2* affects zF differentiation leading to an expansion of the zG [19]. The effect of the loss of *Ezh2* on zF differentiation was further associated with a loss of PKA activity that was associated with an increase in the expression of negative regulators of PKA including the *Pde7b*, *Pde3a* and *Pde1b* [19] (Figure 3B). Interestingly, CTNNB1 was also shown to positively regulate the expression of the phosphodiesterase *Pde2a* in the zG, suggesting that phosphodiesterases contribute to the inhibition of PKA signaling in the zG [50]. The contribution of other PDEs for cAMP/PKA signaling inhibition in the adrenal cortex was also demonstrated using both mouse models and adrenal cell lines. Indeed, it was demonstrated that *Pde8b*$^{-/-}$ have elevated levels of serum and urinary corticosterone [102,103], and that inactivation of *Pde8b* potentiates steroidogenesis and corticosterone steroidogenesis in Y1 cells and H295R cells by potentially increasing cAMP levels [102,103]. The cAMP levels were also higher in the adrenal gland of hypomorphic *Pde11a*$^{-/-}$ mice. Furthermore, *Pde11a*$^{-/-}$ mice failed to suppress corticosterone in response to low dose dexamethasone [104].

Whether most studies indicate that WNT signaling is essential for zG differentiation and that PKA signaling is essential for zF differentiation, one study demonstrates that WNT signaling also contributes to zF maintenance. Indeed, inactivation of the WNT signaling inhibitor *Znrf3* promotes the hyperplastic growth of the zF rather than the expected zG growth [23]. It was further confirmed that this hyperplasia is caused by an increase in WNT/CTNNB1 signaling as inactivation of *Porcn* (an O-acyltransferase required for post-translational modification of all WNTs necessary for WNT secretion and activity) concomitantly with *Znrf3*, rescuing the phenotype observed following *Znrf3* inactivation. The Concomitant inactivation of one copy of *Ctnnb1* with *Znrf3* also leads to reduced proliferation and adrenal cortex size following *Znrf3* inactivation [23]. Revisiting the expression of *Wnt4* and CTNNB1, it was demonstrated that a gradient of expression is normally observed in the adrenal cortex going from high expression in the zG to moderate expression in the upper (or outer) zF to no expression in the lower (or inner) zF [23]. Following the deletion of *Znrf3*, moderate *Wnt4* and CTNNB1 expression could be seen throughout the cortex suggesting that not only WNT4/CTNNB1 signaling is responsible for the phenotype observed following the inactivation of *Znrf3*, but that it is also normally involved in the proliferation of upper zF cells [23] (Figure 3B).

The gradient of expression of WNT4/CTNNB1 signaling in the zF from mild expression in the upper zF to its extinction in the lower zF also illustrates the fact that the zF is heterogenous. The presence of different zones (based on gene expression) in the zF was also observed in other studies. For example, by using single-cell transcriptomics analysis (and confirmed by RNAscope analyses) it was demonstrated that a specific cell population formed the lower zF [148]. Furthermore, this cell population, named zFasc1 by the authors, significantly expands in response to chronic stress exposure [148]. Interestingly, it was demonstrated that *Abcb1b*, one of the genes overexpressed in the zFasc1, positively regulates cortisol secretion [148] suggesting that zFasc1 cells could be more potent than cells from the upper zF. Another study that demonstrated that the lower zF differs from the upper zF (based on immunofluorescence for CYP2F2 and DHCR24 and RNAseq data) also suggested that the zF could potentially be divided into even more concentric zones, and that the lower zF shares some similarities with the X-zone [149]. Finally, it was also shown in that study that the lower zF expands in response to T3 treatment in females [149]. Interestingly, *Wnt4* expression decreased in the adrenal cortex of T3 treated mice while expression of *Mrap* was induced [149]. This last result again suggests that the WNT signaling gradient (and potentially PKA signaling) could be key for the establishment of these zF zones and their response to different challenges. Whether the upper and lower zF have partially different functions has, however, not yet been determined.

As mentioned previously, murine adrenal cortex only comprised two zones, the zG and the zF, and lacked the zR that synthetizes androgens in humans. For this reason, very few studies have evaluated the differentiation of zR cells. However, one study demonstrated that PKA signaling activation was involved in its formation. Indeed, inactivation of *Prkar1a* not only affects zG/zF differentiation, but also leads to the formation of a third zone next to the medulla [26,37]. While a study using the *Ark1b7-Cre* model to inactivate *Prkar1a* initially suggested that this zone could correspond to the resurgence of the X-zone [26], a subsequent study combining the inactivation of *Prkar1a* using the AS^{cre} strain and tracing experiments demonstrated that this third zone does not correspond to the resurgence of the X-zone, but rather arises from the differentiation of the lower zF [37]. In this model, the authors demonstrated that expression of CYP17 and its regulator CYB5 could be detected in this zone, and that adrenal cortex from mutant animals could synthetize DHEA/DHEAS [37]. These results indicate that this zone resembles human zR [37]. Interestingly, a persistence or resurgence of the X-zone was also observed in $Pde11a^{-/-}$ mice [104]. As the inactivation of both *Pkrkar1a* and *Pde11a* increase cAMP/PKA signaling, there is a possibility that the observed X-zone in the in $Pde11a^{-/-}$ mice could also resemble the human zR. It would therefore be interesting to determine if the adrenal of $Pde11a^{-/-}$ mice can produce DHEA/DHEAS.

Another study demonstrated that overexpression of human DENND1A.V2 (a truncated isoform of a clathrin-binding protein that has not been detected in rodents but is expressed in human H295A cell line [150]) in a transgenic mouse model leads to an important increase in the expression levels of adrenal *Cyp17a1*, despite only low levels of *DENND1A.V2* being detected [84]. However, it was not determined if a zR was formed in these mice. Nonetheless, these findings correlate with the role of DENND1A.V2 in the hyperandrogenemia associated with polycystic ovarian syndrome (PCOS) in women [151–153]. The exact mechanism of action of DENND1A.V2 is still unknown, but two mechanisms were proposed for its role in PCOS theca cells. First, it was suggested that nuclear DENND1A.V2 could activate the transcriptional activity of *Cyp17a1* either by facilitating the transport of ligand/receptor to the nucleus or by acting as a scaffold for transcription factors [151–153]. Another possibility is that DENND1A.V2, which has a clathrin-binding domain, could regulate (either directly or by interfering with the action of DENND1A.V1) the internalization/endocytosis/recycling of GPCR and therefore increase cAMP/PKA signaling [151–153]. This could lead to the formation of a zR in the mice as observed following *Prkar1a* inactivation [37]. However, again, cAMP/PKA activity was not evaluated in mice overexpressing *DENND1A.V2* [84].

Finally, although it is usually thought that murine adrenal cortex does not produce androgen, it has been demonstrated that the spiny mouse expresses *Cyp17a1* and produces DHEA [154]. More recently, it was also shown that the adrenal gland of C.B.-17 SCID mice also produces a low level of DHEA and its downstream metabolite, suggesting that some strains of mice can actually produce androgens [155]. Again, the presence of a zR in C.B.-17 SCID mouse was not evaluated. Nonetheless, this suggests that some mouse strains could be useful to study zR differentiation and functions.

Tracing experiments (13,27,28,36,37,48,53) have demonstrated that maintenance and zonation are usually linked in the adrenal cortex in a process in which capsular stem cells and subcapsular progenitor cells move inward centripetally and differentiate into zG cells. The latter will then differentiate into zF cells before dying by apoptosis at the junction between the adrenal cortex and the medulla. If this is normally the case, tracing experiments have demonstrated that maintenance of the adrenal cortex and zonation can be separated from one another in certain contexts. For example, inactivation of *Nr5a1* in *Cyp11b2+* zG cells leads to the loss of the zG without affecting zF maintenance [48]. Furthermore, zF cells were no longer derived from zG cells in this model [48]. Overexpression of CTNNB1 in the *Cyp11b2+* zG cells also blocks the differentiation of zG cells into zF cells, leading to the maintenance of a zF no longer derived from the zG [50]. Numerous hypotheses could explain the separation of zG and zF maintenance. First, it is possible that the stem/progenitor cells can bypass their need to differentiate into zG cells before differentiating into zF cells, as recombination does not occur in stem/progenitor cells in these models. On the other hand, it is possible that the differentiation of zG to zF cells is accelerated in some models, and that recombination simply does not have time to occur in zG cells before their differentiation into zF cells. Another possibility is that proliferation of zF cells increases in these models. Residual WNT signaling could be involved in the proliferation of zF cells as observed following the inactivation of *Znrf3* [23]. A fourth possibility is that other populations of normally inactive stem/progenitor cells are present in the adrenal cortex and take over in this context. For example, it was shown in rat that cell expressing POU5F1, a marker of stem cells, could be seen throughout the adrenal cortex before puberty with the number of POU5F1+ cells increasing in the zG after puberty and decreasing in the rest of the cortex [156]. The presence of POU5F1+ cells has not been evaluated in the mouse, but the presence/maintenance/replication of these prepubertal POU5F1+ cells could be maintained in the adult zF in abnormal conditions. Finally, it was demonstrated that a population of Nestin+ cells, mainly located in the subcapsular region (with rare cells also observed in the zF) also has characteristics of stem/progenitor cells and can differentiate into steroidogenic cells of the zG and the zF [60]. These cells lacked co-staining with GLI1 and SHH, suggesting that they are a different population of stem/progenitor cells, though they could potentially still be descendants of GLI1 or SHH positive cells. More interestingly, these cells do not seem to play an important role in the normal maintenance of the adrenal cortex, but their differentiation into steroidogenic cells increases following stress [60], again suggesting that "dormant" progenitor cells could be present in the adrenal cortex and ready to respond if necessary.

9. Sexual Hormones Play an Important Role in the Maintenance of the Adrenal Cortex

As previously stated, the adrenal gland of mouse is sexually dimorphic. Indeed, the adrenal glands of female have a higher weight than their male counterparts [134,157], which could be explained, in part, by the regression of the X-zone at puberty in male and its maintenance in female until the first gestation [134,135], and in part by the fact that the volume of the zF (but not the zG) is higher in females [134]. Furthermore, the higher volume of the zF was also associated with higher levels of circulating corticosterone [134]. Interestingly, it was shown that castration of mature male mice leads to the appearance of a secondary X-zone [158], while testosterone treatment caused the rapid disappearance of the X-zone in females [159]. These findings demonstrate the key role played by male hormones in the adrenal cortex (Figure 4A).

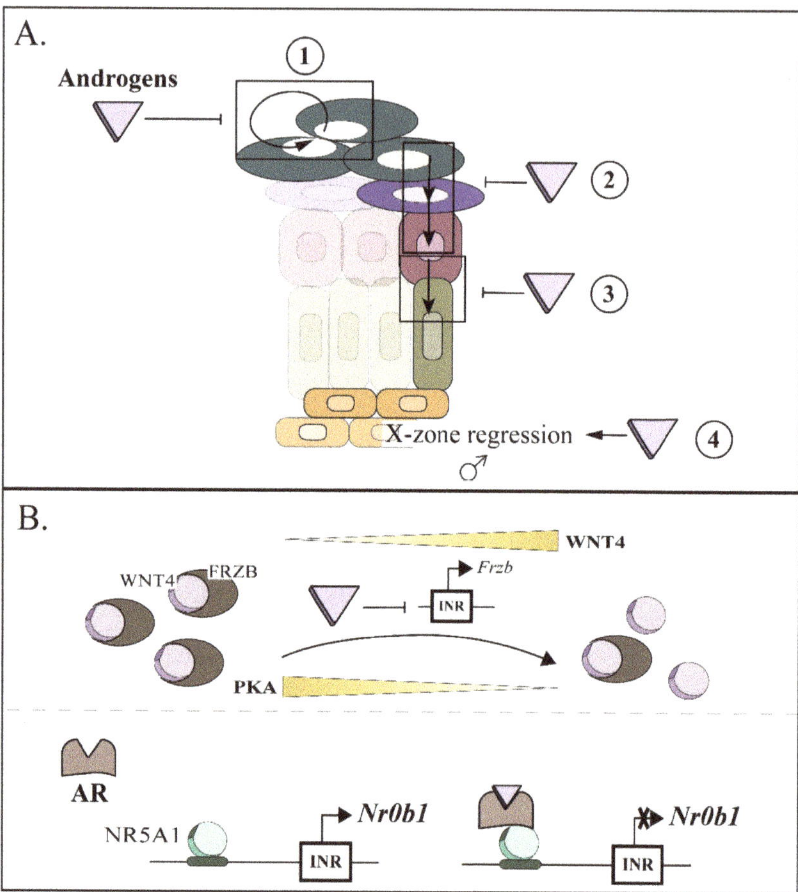

Figure 4. Androgens affect adrenocortical maintenance: (**A**) Androgens act at several levels to regulate the maintenance of the adrenal cortex. ① Androgens can limit cortical cell turnover by reducing the proliferation of the capsular stem cells; ② abolish the contribution of the capsular stem cells to the steroidogenic lineage; ③ limit the differentiation of zG cells into zF cells; and ④ permit the complete regression of the X-zone in male mice. (**B**) Potential molecular action of androgen or its receptor in the adrenal cortex. Global molecular mechanisms regulated by androgens are unclear. It was suggested that androgens could stimulate WNT signaling (and potentially restrain PKA signaling) via the downregulation of the WNT inhibitor *Frzb*. Liganded AR can also negatively regulate the transcriptional activity of *Nr0b1* by binding with NR5A1.

The importance of sexual dimorphism and the roles of androgens for the maintenance of the adrenal cortex were also demonstrated more recently (Figure 4A). By using lineage-tracing experiments of the zG *Axin2*+ or *Wnt4*+ cells, it was first shown that complete renewal of the adrenal cortex was faster in females (approximatively 3 months) than in males (estimated at 9 months) [54]. Furthermore, lineage-tracing experiment of the capsular *Gli1*+ stem cells revealed that these cells contribute to the steroidogenic lineage in both sexes in 3-week-old animals, but that after puberty the contribution of the capsular stem cells to the steroidogenic lineage was almost completely abolished in males but that this contribution remains important in females [54] (Figure 3A, Figure 4A). Again, it was demonstrated that androgens play a key role in this process as recruitment of *Gli1*+ was accelerated in orchiectomized males, while treatment of ovariectomized females with dihydrotestosterone (DHT) had the opposite effect [54]. Adrenal cell renewal was not affected following ovariectomy, suggesting that progesterone and estrogen do not contribute to this process [54]. Interestingly, contribution of *Gli1+* cells to adrenal cell renewal in male could be observed in a model of adrenal regeneration (following dexamethasone treatment), suggesting that these cells can serve as a reservoir in male when important adrenal recovery is needed [53].

The molecular mechanisms of androgen action in adrenal maintenance are not well defined. However, it was demonstrated that gonadectomy decreases the expression of CTNNB1 target genes and dramatically increases the expression of the inhibitor of WNT signaling *Frzb*, while DHT supplementation has the reverse effect [37] suggesting that androgens block the differentiation of zG cells into zF cells or limit the proliferation of the upper zF cells (Figure 4A,B). Another potential target of androgen is *Nr0b1* whose expression is higher in the adrenal cortex of females (compared to males), and whose expression increases in orchiectomized males [160]. Interestingly, it was demonstrated that liganded AR negatively regulates the transcriptional activity of *Nr0b1* by binding with NR5A1 [160] (Figure 4B). YAP have also been shown to bind to AR in prostate cancer [161], suggesting that Hippo signaling could be involved in the regulation of androgen activity in the adrenal cortex. Finally, effects of the loss of *Ar* in the adrenal cortex were evaluated recently. As expected, an abnormal retention of the X-zone could be observed in males [44]. A tendency for higher circulating corticosterone levels in aging animals is also observed in these animals. Interestingly, divergent results could be seen in mutant animals and their orchiectomized counterparts. For example, a reduction in the zF marker ARK1B7 and an increase in apoptosis of zF cells were solely observed in orchiectomized/AR negative animals suggesting that androgens do not exclusively act via AR [44]. Contrary to what is observed in males, ARK1B7 expression was lost in the adrenal cortex of female mutant animals [43] and X-zone regression was independent of AR [43]. Further delineation of the molecular mechanisms of action of AR is still needed to comprehend its function in the adrenal cortex.

Recently, a potential new role for androgen action on adrenal cortex homeostasis has emerged. Indeed, using the *Znrf3* $^{flox/flox}$; *Nr5a1-Cre* ^{high} mouse model, it was first demonstrated that 78-week-old females were developing metastatic adrenocortical carcinoma (ACC) while no such tumors were observed in males [111]. It was further demonstrated that tumor development in males was blunted by the induction of adrenocortical cells senescence, followed by the recruitment and differentiation of phagocytic macrophages [111]. Hyperplasia regression in male was confirmed to be androgen dependent, as the male phenotype (early recruitment of phagocytic macrophages and regression of initial hyperplasia) was recapitulated in females implanted with testosterone pellets [111]. These results might in part explain why ACC incidence is higher in women than in men [162–164]. Interestingly, macrophage number also increases following chronic stress exposure [148]. It was also suggested that adrenal macrophages control lipid metabolism in both sexes, and that macrophage depletion in the whole animals (performed in females) leads to lower local production of aldosterone in stressed animals [165]. Furthermore, sexual dimorphism of the adrenocortical macrophage populations is also observed in wild-type mice as MCH class II^{low} macrophages are solely present in females and are dependent of the X-zone [165]. Together, these studies suggest that macrophages are important to maintain adrenal homeostasis. Further studies are warranted to thoroughly understand their mechanisms of action.

Function of estrogen in adrenal cortex maintenance has not been studied as much as androgens. However, one study using the estrogen-deficient aromatase knockout mouse models suggests that estrogen deficiency leads to the inhibition of telomerase activity, telomere shortening of cortical cells, and a decrease in cell proliferation in the female adrenal cortex [69], while another study also suggests that estrogen might influence *Nr0b1* regulation [160]. Again, further studies will be needed to evaluate the role of female hormones in adrenal cortex maintenance.

Finally, a sexually dimorphic response to the thyroid hormones T3 is observed in the adrenal cortex of prepubertal animals [149]. While differences in this response could be partially explained by the initiation of the X-zone regression in males, the expansion of the inner zF in females suggests that T3 also contributes to the dimorphism of the definitive adrenal cortex. Estrogen and T3 action demonstrate that although androgen action at puberty is probably the most important factor regulating adrenal gland sexual dimorphism, other factors also contribute to this dimorphism.

10. Emergence of Spindle-Shaped Cells in Aging Mice

One last feature commonly observed in older mice is the accumulation of subcapsular non-steroidogenic spindle-shaped cells, named A cells. Further differentiation of a few of these cells in large lipid-laden sex-steroid producing cells, named B cells, is also observed in some mouse strains [166–168]; this is a process that is greatly enhanced following gonadectomy [55,169]. Accumulation of these cell types is often considered a gonad-like tumor as these cells express gonad markers such as *Cyp17a1*, *Gli1*, *Lhcgr* and *Amhr2* [13,55]. Premature appearance of the spindle-shaped cells is also observed following the inactivation or overexpression of numerous genes in the adrenal cortex [13,19,26,31,34,43,44,46,47,71,99,103,104]. Because it was demonstrated that the spindle-shaped cells also express *Gata4* [19,31,46,170] and *Wt1* [13,19], it has been suggested that these cells constitute a population of progenitor cells sharing similarities with AGP cells, which accumulate as an attempt to maintain homeostasis in animals with adrenal insufficiency. Further differentiation of A cells to B cells has also been associated with elevated LH secretion [70,170–172]. Evaluation of LH levels (and other sex hormones) in the different mouse models presenting this phenotype is therefore essential to determine if the appearance of these cells is intrinsic to the adrenal or secondary to the inactivation of these genes in other steroidogenic tissues such as the ovary, testis, or pituitary.

Currently, two theories, both based on tracing experiments, have been suggested for the origin of the A cells. The first theory suggests that both capsular *Wt1*+ [13] and *Gli1*+ [55] cells can form A cells following gonadectomy. Blocking GLI1/2 activity with a pharmacological inhibitor following gonadectomy also decreases the expression of gonad markers, further confirming the results obtained with the tracing experiments [55]. More recently, the origin of these cells was also evaluated following the inactivation of *Ezh2* [19]. In this model, tracing experiments rather suggested that *Nr5a1*+ cells dedifferentiate into A cells [19]. Furthermore, in this model, expression of GATA4 preceded the expression of WT1 in the adrenal cortex, suggesting that GATA4+ cells do not originate from cells expressing WT1 in this model [19]. GATA4 expression was also broader than WT1 expression suggesting that GATA4 induction is independent of WT1 induction [19]. The different origin of the spindle-shaped cells could be associated with differences between the models (gonadectomy vs. aging mice) but capsular *Gli*+ cells have also been shown to contribute to A cell formation in aging animals [55]. It is therefore possible that both the differentiation of capsular stem cells and dedifferentiation of *Nr5a1*+ cells contribute to A cell formation. Which *Nr5a1*+ cells (*Nr5a1*+, *Shh*+ subcapsular progenitor cells or *Nr5a1*+ steroidogenic cells) contribute to A cell formation and what would be the exact contribution of capsular cells and adrenocortical *Nr5a1*+ cells to A cell formation also remain to be determined.

No matter the origin of the spindle-shaped cells, GATA4 appears to play a prominent role in their appearance and their differentiation into B cells. Indeed, it was first demonstrated that the appearance of A cells is delayed and that their differentiation into B cells is blocked in ovariectomized *Gata*$^{+/-}$ mice [86]. Conditional deletion of *Gata4* in the nascent forming A cells of ovariectomized mice also limits their proliferation and blocks their differentiation into B cells [86]. Furthermore, GATA4 appears to be the main factor responsible for the appearance of gonad-like tumor observed in *Inha*$^{-/-}$ animals [172]. While inactivation of *Gata4* has been shown to block the formation of these cells, ectopic expression of GATA4 in the adrenal cortex (under the control of the *Cyp21a1* promoter) has also been shown to lead to the appearance of A cells in females and accelerate the appearance of both A and B cells following gonadectomy in males and females [71].

11. Conclusions

In the last decades, transgenic mouse models have been the driving force behind our understanding of the molecular mechanisms regulating the development, zonation and maintenance of the adrenal cortex. Combination of transgenic mouse models with genome-wide profiling of transcriptomes and epigenomes at the cellular levels further offers the possibility to comprehend the interplay between gene expression, transcription factors and chromatin state to uncover the gene networks regulating adrenocortical cell fate commitment. Disruption of the development, zonation and maintenance of the adrenal cortex have been associated with diseases such as Cushing syndrome, Carney complex and both adrenocortical adenoma and carcinoma. To fully understand these mechanisms could therefore lead to new therapeutic strategies to treat these pathologies.

Supplementary Materials: The following supporting information can be downloaded at: https://www.mdpi.com/article/10.3390/ijms232214388/s1.

Author Contributions: A.B. and N.A.N. wrote this manuscript and G.Z. and A.B. reviewed the final draft. All authors have read and agreed to the published version of the manuscript.

Funding: This work was supported by Discovery Grants from NSERC to AB (RGPIN-2020-05230) and GZ (RGPIN-2018-06470) and by a PhD scholarship from FRQNT to NAN.

Institutional Review Board Statement: Not applicable.

Informed Consent Statement: Not applicable.

Data Availability Statement: Not applicable.

Conflicts of Interest: The authors have no conflicts of interest to declare.

Abbreviations

Abcb1b	ATP-binding cassette, sub-family B (MDR/TAP), member 1B
Akr1b7	aldo-keto reductase family 1, member B7
Amhr2	anti-mullerian hormone type 2 receptor
Apc	adenomatosis polyposis coli
Ar	androgen receptor
Axin2	axin2
Cbx2	chromobox 2
Cited2	Glu/Asp-rich carboxy-terminal domain 2
Ctnnb1	beta-catenin
Cyb5	cytochrome b5
Cyp2f2	cytochrome P450, family 2, subfamily f, polypeptide 2
Cyp11a1	cytochrome P450 side chain cleavage enzyme 11a1
Cyp11b1	cytochrome P450, family 11, subfamily b, polypeptide 1
Cyp11b2/As	cytochrome P450, family 11, subfamily b, polypeptide 2/aldosterone synthase
Cyp17a1	cytochrome P450, family 17, subfamily a, polypeptide 1
Cyp21a1	cytochrome P450, family 21, subfamily a, polypeptide 1
Dennd1a	DENN/MADD domain containing 1A
Dhcr24	24-dehydrocholesterol reductase
Emx2	empty spiracles homeobox 2
Ezh2	zeste homolog 2
FAdE/Nr5a1	Fetal adrenal enhancer of nuclear receptor subfamily 5, Group A, member 1
Foxl2	forkhead box L2
Fgfr2	fibroblast growth factor receptor 2
Frzb	Frizzled-related protein
Gata4	GATA binding protein 4
Gata6	GATA binding protein 6

Gli1	GLI-Kruppel family member 1
Gli2	GLI-Kruppel family member 2
Gli3	GLI-Kruppel family member 3
Hoxb9	homeobox B9
IGF	insulin growth factor
Igf1r	insulin-like growth factor 1 receptor
Inha	inhibin a
Insr	insulin receptor
Lats1	large tumor suppressor 1
Lats2	large tumor suppressor 2
Lhcgr	luteinizing hormone/choriogonadotropin receptor
Lhx9	LIM homeobox protein 9
MAPK	mitogen-activate kinase protein
Mc2r	melanocortin 2 receptor
Mrap	melanocortin 2 receptor accessory protein
Nes	nestin
Nr0b1	nuclear receptor subfamily 0, group B, member 1
Nr2f2	nuclear receptor subfamily 2, group F, member 2
Nr5a1	nuclear receptor subfamily 5, group A, member 1
Osr1	odd-skipped related transcription factor 1
Pbx1	pre B cell leukemia homeobox 1
Pde1b	phosphodiesterase 1B, Ca2+ calmodulin dependent
Pde2a	phosphodiesterase 2A, cGMP-stimulated
Pde3a	phosphodiesterase 3A, cGMP-inhibited
Pde7b	phosphodiesterase 7B
Pde8b	phosphodiesterase 8B
Pde11a	phosphodiesterase 11A
PKA	protein kinase A
Porcn	porcupine homolog
Pou5f1	POU domain, class 5, transcription factor 1
Prkar1a	protein kinase, cAMP dependent regulatory, type 1, alpha
Prep1	Pbx-knotted 1 homeobox
Rspo3	R-spondin 3 homolog
Shh	sonic hedgehog
Siah1a	seven in absentia 1A
Six1	sine oculis-related homeobox-1
Six4	sine oculis-related homeobox-4
Sfrp2	Secreted frizzled-related protein 2
Smo	smoothened homolog
Taz	transcriptional co-activator with PDZ-binding motif
Tcf21	transcription factor 21
Tbx18	T-box18
Wnt4	wingless-type MMTV integration site family, member 4
Wnt2b	wingless-type MMTV integration site family, member 2
Wt1	wilms tumor 1
Yap	yes-associated protein
Znrf3	zinc and ring finger 3

References

1. Gottschau, M. Struktur und Embryonale Entwickelung der Nebennieren bei Saugetieren. Archiv fur Anatomie und Entwickelungsgeschichte. *Anat. Abt.* **1883**, *9*, 412–458.
2. Zwemer, R.L. A study of adrenal cortex morphology. *Am. J. Pathol.* **1936**, *84*, 107–114.1. [CrossRef] [PubMed]
3. Salmon, T.N.; Zwemer, R.L. A study of the life history of cortico-adrenal gland cells of the rat by means of trypan blue injections. *Anat. Rec.* **1941**, *80*, 421–429. [CrossRef]
4. Hayashi, S.; McMahon, A.P. Efficient Recombination in Diverse Tissues by a Tamoxifen-Inducible Form of Cre: A Tool for Temporally Regulated Gene Activation/Inactivation in the Mouse. *Dev. Biol.* **2002**, *244*, 305–318. [CrossRef]
5. Hu, Y.-C.; Okumura, L.M.; Page, D.C. Gata4 Is Required for Formation of the Genital Ridge in Mice. *PLoS Genet.* **2013**, *9*, e1003629. [CrossRef]

6. Vidal, V.; Sacco, S.; Rocha, A.S.; da Silva, F.; Panzolini, C.; Dumontet, T.; Doan, T.M.P.; Shan, J.; Rak-Raszewska, A.; Bird, T.; et al. The adrenal capsule is a signaling center controlling cell renewal and zonation through *Rspo3. Genes Dev.* **2016**, *30*, 1389–1394. [CrossRef]

7. Mugford, J.W.; Sipilä, P.; McMahon, J.A.; McMahon, A.P. Osr1 expression demarcates a multi-potent population of intermediate mesoderm that undergoes progressive restriction to an Osr1-dependent nephron progenitor compartment within the mammalian kidney. *Dev. Biol.* **2008**, *324*, 88–98. [CrossRef]

8. Sasaki, K.; Oguchi, A.; Cheng, K.; Murakawa, Y.; Okamoto, I.; Ohta, H.; Yabuta, Y.; Iwatani, C.; Tsuchiya, H.; Yamamoto, T.; et al. The embryonic ontogeny of the gonadal somatic cells in mice and monkeys. *Cell Rep.* **2021**, *35*, 109075. [CrossRef]

9. Trowe, M.-O.; Shah, S.; Petry, M.; Airik, R.; Schuster-Gossler, K.; Kist, R.; Kispert, A. Loss of Sox9 in the periotic mesenchyme affects mesenchymal expansion and differentiation, and epithelial morphogenesis during cochlea development in the mouse. *Dev. Biol.* **2010**, *342*, 51–62. [CrossRef]

10. Häfner, R.; Bohnenpoll, T.; Rudat, C.; Schultheiss, T.M.; Kispert, A. Fgfr2 is required for the expansion of the early adrenocortical primordium. *Mol. Cell. Endocrinol.* **2015**, *413*, 168–177. [CrossRef]

11. Sankoda, N.; Tanabe, W.; Tanaka, A.; Shibata, H.; Woltjen, K.; Chiba, T.; Haga, H.; Sakai, Y.; Mandai, M.; Yamamoto, T.; et al. Epithelial expression of Gata4 and Sox2 regulates specification of the squamous–columnar junction via MAPK/ERK signaling in mice. *Nat. Commun.* **2021**, *12*, 560. [CrossRef] [PubMed]

12. Zhou, B.; Ma, Q.; Rajagopal, S.; Wu, S.M.; Domian, I.; Rivera-Feliciano, J.; Jiang, D.; Von Gise, A.; Ikeda, S.; Chien, K.R.; et al. Epicardial progenitors contribute to the cardiomyocyte lineage in the developing heart. *Nature* **2008**, *454*, 109–113. [CrossRef] [PubMed]

13. Bandiera, R.; Vidal, V.P.; Motamedi, F.J.; Clarkson, M.; Sahut-Barnola, I.; von Gise, A.; Pu, W.T.; Hohenstein, P.; Martinez, A.; Schedl, A. WT1 Maintains Adrenal-Gonadal Primordium Identity and Marks a Population of AGP-like Progenitors within the Adrenal Gland. *Dev. Cell* **2013**, *27*, 5–18. [CrossRef] [PubMed]

14. Bingham, N.C.; Verma-Kurvari, S.; Parada, L.F.; Parker, K.L. Development of a steroidogenic factor 1/Cre transgenic mouse line. *Genesis* **2006**, *44*, 419–424. [CrossRef] [PubMed]

15. Heaton, J.H.; Wood, M.A.; Kim, A.C.; Lima, L.O.; Barlaskar, F.M.; Almeida, M.Q.; Fragoso, M.C.; Kuick, R.; Lerario, A.M.; Simon, D.P.; et al. Progression to Adrenocortical Tumorigenesis in Mice and Humans through Insulin-Like Growth Factor 2 and β-Catenin. *Am. J. Pathol.* **2012**, *181*, 1017–1033. [CrossRef] [PubMed]

16. Kim, A.C.; Reuter, A.L.; Zubair, M.; Else, T.; Serecky, K.; Bingham, N.C.; Lavery, G.G.; Parker, K.L.; Hammer, G.D. Targeted disruption of beta-catenin in Sf1-expressing cells impairs development and maintenance of the adrenal cortex. *Development* **2008**, *135*, 2593–2602. [CrossRef]

17. Huang, C.-C.J.; Liu, C.; Yao, H.H.-C. Investigating the role of adrenal cortex in organization and differentiation of the adrenal medulla in mice. *Mol. Cell. Endocrinol.* **2012**, *361*, 165–171. [CrossRef]

18. Krill, K.T.; Gurdziel, K.; Heaton, J.H.; Simon, D.P.; Hammer, G.D. Dicer Deficiency Reveals MicroRNAs Predicted to Control Gene Expression in the Developing Adrenal Cortex. *Mol. Endocrinol.* **2013**, *27*, 754–768. [CrossRef]

19. Mathieu, M.; Drelon, C.; Rodriguez, S.; Tabbal, H.; Septier, A.; Damon-Soubeyrand, C.; Dumontet, T.; Berthon, A.; Sahut-Barnola, I.; Djari, C.; et al. Steroidogenic differentiation and PKA signaling are programmed by histone methyltransferase EZH2 in the adrenal cortex. *Proc. Natl. Acad. Sci. USA* **2018**, *115*, E12265–E12274. [CrossRef]

20. Kim, Y.; Bingham, N.; Sekido, R.; Parker, K.L.; Lovell-Badge, R.; Capel, B. Fibroblast growth factor receptor 2 regulates proliferation and Sertoli differentiation during male sex determination. *Proc. Natl. Acad. Sci. USA* **2007**, *104*, 16558–16563. [CrossRef]

21. Tevosian, S.G.; Jimenez, E.; Hatch, H.; Jiang, T.; Morse, D.A.; Fox, S.C.; Padua, M.B. Adrenal Development in Mice Requires GATA4 and GATA6 Transcription Factors. *Endocrinology* **2015**, *156*, 2503–2517. [CrossRef] [PubMed]

22. Padua, M.B.; Jiang, T.; Morse, D.A.; Fox, S.C.; Hatch, H.M.; Tevosian, S.G. Combined Loss of the GATA4 and GATA6 Transcription Factors in Male Mice Disrupts Testicular Development and Confers Adrenal-Like Function in the Testes. *Endocrinology* **2015**, *156*, 1873–1886. [CrossRef] [PubMed]

23. Basham, K.J.; Rodriguez, S.; Turcu, A.F.; Lerario, A.M.; Logan, C.Y.; Rysztak, M.R.; Gomez-Sanchez, C.E.; Breault, D.T.; Koo, B.-K.; Clevers, H.; et al. A ZNRF3-dependent Wnt/β-catenin signaling gradient is required for adrenal homeostasis. *Genes Dev.* **2019**, *33*, 209–220. [CrossRef] [PubMed]

24. Dumontet, T.; Sahut-Barnola, I.; Dufour, D.; Lefrançois-Martinez, A.; Berthon, A.; Montanier, N.; Ragazzon, B.; Djari, C.; Pointud, J.; Roucher-Boulez, F.; et al. Hormonal and spatial control of SUMOylation in the human and mouse adrenal cortex. *FASEB J.* **2019**, *33*, 10218–10230. [CrossRef] [PubMed]

25. Drelon, C.; Berthon, A.; Sahut-Barnola, I.; Mathieu, M.; Dumontet, T.; Rodriguez, S.; Batisse-Lignier, M.; Tabbal, H.; Tauveron, I.; Lefrançois-Martinez, A.-M.; et al. PKA inhibits WNT signalling in adrenal cortex zonation and prevents malignant tumour development. *Nat. Commun.* **2016**, *7*, 12751. [CrossRef]

26. Sahut-Barnola, I.; de Joussineau, C.; Val, P.; Lambert-Langlais, S.; Damon, C.; Lefrançois-Martinez, A.M.; Pointud, J.C.; Marceau, G.; Sapin, V.; Tissier, F.; et al. Cushing's syndrome and fetal features resurgence in adrenal cortex-specific Prkar1a knockout mice. *PLoS Genet.* **2010**, *6*, e1000980. [CrossRef]

27. Huang, C.-C.J.; Miyagawa, S.; Matsumaru, D.; Parker, K.L.; Yao, H.H.-C. Progenitor Cell Expansion and Organ Size of Mouse Adrenal Is Regulated by Sonic Hedgehog. *Endocrinology* **2010**, *151*, 1119–1128. [CrossRef]

28. King, P.; Paul, A.; Laufer, E. Shh signaling regulates adrenocortical development and identifies progenitors of steroidogenic lineages. *Proc. Natl. Acad. Sci. USA* **2009**, *106*, 21185–21190. [CrossRef]

29. Ching, S.; Vilain, E. Targeted disruption of Sonic Hedgehog in the mouse adrenal leads to adrenocortical hypoplasia. *Genesis* **2009**, *47*, 628–637. [CrossRef]

30. Dhillon, H.; Zigman, J.M.; Ye, C.; Lee, C.E.; McGovern, R.A.; Tang, V.; Kenny, C.D.; Christiansen, L.M.; White, R.D.; Edelstein, E.A.; et al. Leptin Directly Activates SF1 Neurons in the VMH, and This Action by Leptin Is Required for Normal Body-Weight Homeostasis. *Neuron* **2006**, *49*, 191–203. [CrossRef]

31. Pihlajoki, M.; Gretzinger, E.; Cochran, R.; Kyrönlahti, A.; Schrade, A.; Hiller, T.; Sullivan, L.; Shoykhet, M.; Schoeller, E.L.; Brooks, M.D.; et al. Conditional Mutagenesis of Gata6 in SF1-Positive Cells Causes Gonadal-Like Differentiation in the Adrenal Cortex of Mice. *Endocrinology* **2013**, *154*, 1754–1767. [CrossRef] [PubMed]

32. Levasseur, A.; St-Jean, G.; Paquet, M.; Boerboom, D.; Boyer, A. Targeted Disruption of YAP and TAZ Impairs the Maintenance of the Adrenal Cortex. *Endocrinology* **2017**, *158*, 3738–3753. [CrossRef] [PubMed]

33. Ménard, A.; Nader, N.A.; Levasseur, A.; St-Jean, G.; Roy, M.L.G.-L.; Boerboom, D.; Benoit-Biancamano, M.-O.; Boyer, A. Targeted Disruption of Lats1 and Lats2 in Mice Impairs Adrenal Cortex Development and Alters Adrenocortical Cell Fate. *Endocrinology* **2020**, *161*, bqaa052. [CrossRef] [PubMed]

34. Nader, N.A.; Blais, É.; St-Jean, G.; Boerboom, D.; Zamberlam, G.; Boyer, A. Effect of inactivation of Mst1 and Mst2 in the mouse adrenal cortex. *J. Endocr. Soc.* **2022**, *7*, 143. [CrossRef]

35. Zubair, M.; Parker, K.L.; Morohashi, K.-I. Developmental Links between the Fetal and Adult Zones of the Adrenal Cortex Revealed by Lineage Tracing. *Mol. Cell. Biol.* **2008**, *28*, 7030–7040. [CrossRef]

36. Wood, M.A.; Acharya, A.; Finco, I.; Swonger, J.M.; Elston, M.J.; Tallquist, M.D.; Hammer, G.D. Fetal adrenal capsular cells serve as progenitor cells for steroidogenic and stromal adrenocortical cell lineages in *M. musculus*. *Development* **2013**, *140*, 4522–4532. [CrossRef]

37. Dumontet, T.; Sahut-Barnola, I.; Septier, A.; Montanier, N.; Plotton, I.; Roucher-Boulez, F.; Ducros, V.; Lefrançois-Martinez, A.-M.; Pointud, J.-C.; Zubair, M.; et al. PKA signaling drives reticularis differentiation and sexually dimorphic adrenal cortex renewal. *JCI Insight* **2018**, *3*, e98394. [CrossRef]

38. Wu, H.-S.; Lin, H.-T.; Wang, C.-K.L.; Chiang, Y.-F.; Chu, H.-P.; Hu, M.-C. HumanCYP11A1 promoter drives Cre recombinase expression in the brain in addition to adrenals and gonads. *Genesis* **2007**, *45*, 59–65. [CrossRef]

39. Neirijnck, Y.; Calvel, P.; Kilcoyne, K.R.; Kühne, F.; Stévant, I.; Griffeth, R.J.; Pitetti, J.-L.; Andric, S.A.; Hu, M.-C.; Pralong, F.; et al. Insulin and IGF1 receptors are essential for the development and steroidogenic function of adult Leydig cells. *FASEB J.* **2018**, *32*, 3321–3335. [CrossRef]

40. Buaas, F.W.; Gardiner, J.R.; Clayton, S.; Val, P.; Swain, A. In vivo evidence for the crucial role of SF1 in steroid-producing cells of the testis, ovary and adrenal gland. *Development* **2012**, *139*, 4561–4570. [CrossRef]

41. Francis, J.C.; Gardiner, J.R.; Renaud, Y.; Chauhan, R.; Weinstein, Y.; Gomez-Sanchez, C.; Lefrançois-Martinez, A.M.; Bertherat, J.; Val, P.; Swain, A. HOX genes promote cell proliferation and are potential therapeutic targets in adrenocortical tumours. *Br. J. Cancer* **2021**, *124*, 805–816. [CrossRef] [PubMed]

42. O'Hara, L.; York, J.P.; Zhang, P.; Smith, L.B. Targeting of GFP-Cre to the Mouse Cyp11a1 Locus Both Drives Cre Recombinase Expression in Steroidogenic Cells and Permits Generation of Cyp11a1 Knock Out Mice. *PLoS ONE* **2014**, *9*, e84541. [CrossRef] [PubMed]

43. Gannon, A.-L.; O'Hara, L.; Mason, I.J.; Jørgensen, A.; Frederiksen, H.; Curley, M.; Milne, L.; Smith, S.; Mitchell, R.T.; Smith, L.B. Androgen Receptor Is Dispensable for X-Zone Regression in the Female Adrenal but Regulates Post-Partum Corticosterone Levels and Protects Cortex Integrity. *Front. Endocrinol.* **2021**, *11*, 599869. [CrossRef] [PubMed]

44. Gannon, A.-L.; O'Hara, L.; Mason, J.I.; Jørgensen, A.; Frederiksen, H.; Milne, L.; Smith, S.; Mitchell, R.T.; Smith, L.B. Androgen receptor signalling in the male adrenal facilitates X-zone regression, cell turnover and protects against adrenal degeneration during ageing. *Sci. Rep.* **2019**, *9*, 10457. [CrossRef]

45. Lambert-Langlais, S.; Val, P.; Guyot, S.; Ragazzon, B.; Sahut-Barnola, I.; De Haze, A.; Lefrançois-Martinez, A.-M.; Martinez, A. A transgenic mouse line with specific Cre recombinase expression in the adrenal cortex. *Mol. Cell. Endocrinol.* **2009**, *300*, 197–204. [CrossRef]

46. Berthon, A.; Sahut-Barnola, I.; Lambert-Langlais, S.; de Joussineau, C.; Damon-Soubeyrand, C.; Louiset, E.; Taketo, M.M.; Tissier, F.; Bertherat, J.; Lefrançois-Martinez, A.M.; et al. Constitutive beta-catenin activation induces adrenal hyperplasia and promotes adrenal cancer development. *Hum. Mol. Genet.* **2010**, *19*, 1561–1576. [CrossRef]

47. Drelon, C.; Berthon, A.; Ragazzon, B.; Tissier, F.; Bandiera, R.; Sahut-Barnola, I.; De Joussineau, C.; Batisse-Lignier, M.; Lefrançois-Martinez, A.-M.; Bertherat, J.; et al. Analysis of the Role of Igf2 in Adrenal Tumour Development in Transgenic Mouse Models. *PLoS ONE* **2012**, *7*, e44171. [CrossRef]

48. Freedman, B.D.; Kempna, P.B.; Carlone, D.L.; Shah, M.S.; Guagliardo, N.A.; Barrett, P.Q.; Gomez-Sanchez, C.E.; Majzoub, J.A.; Breault, D.T. Adrenocortical Zonation Results from Lineage Conversion of Differentiated Zona Glomerulosa Cells. *Dev. Cell* **2013**, *26*, 666–673. [CrossRef]

49. Leng, S.; Pignatti, E.; Khetani, R.S.; Shah, M.S.; Xu, S.; Miao, J.; Taketo, M.M.; Beuschlein, F.; Barrett, P.Q.; Carlone, D.L.; et al. β-Catenin and FGFR2 regulate postnatal rosette-based adrenocortical morphogenesis. *Nat. Commun.* **2020**, *11*, 1680. [CrossRef]

50. Pignatti, E.; Leng, S.; Yuchi, Y.; Borges, K.S.; Guagliardo, N.A.; Shah, M.S.; Ruiz-Babot, G.; Kariyawasam, D.; Taketo, M.M.; Miao, J.; et al. Beta-Catenin Causes Adrenal Hyperplasia by Blocking Zonal Transdifferentiation. *Cell Rep.* **2020**, *31*, 107524. [CrossRef]

51. Zhang, N.-N.; Wang, C.-N.; Ni, X. Construction of transgenic mice with specific Cre recombinase expression in the zona fasciculata in adrenal cortex. *Acta Physiol. Sin.* **2020**, *72*, 148–156.

52. Ahn, S.; Joyner, A.L. Dynamic Changes in the Response of Cells to Positive Hedgehog Signaling during Mouse Limb Patterning. *Cell* **2004**, *118*, 505–516. [CrossRef] [PubMed]

53. Finco, I.; Lerario, A.M.; Hammer, G.D. Sonic Hedgehog and WNT Signaling Promote Adrenal Gland Regeneration in Male Mice. *Endocrinology* **2017**, *159*, 579–596. [CrossRef] [PubMed]

54. Grabek, A.; Dolfi, B.; Klein, B.; Jian-Motamedi, F.; Chaboissier, M.-C.; Schedl, A. The Adult Adrenal Cortex Undergoes Rapid Tissue Renewal in a Sex-Specific Manner. *Cell Stem Cell* **2019**, *25*, 290–296.e2. [CrossRef]

55. Dörner, J.; Rodriguez, V.M.; Ziegler, R.; Röhrig, T.; Cochran, R.S.; Götz, R.M.; Levin, M.D.; Pihlajoki, M.; Heikinheimo, M.; Wilson, D.B. GLI1+ progenitor cells in the adrenal capsule of the adult mouse give rise to heterotopic gonadal-like tissue. *Mol. Cell. Endocrinol.* **2017**, *441*, 164–175. [CrossRef]

56. Harfe, B.D.; Scherz, P.J.; Nissim, S.; Tian, H.; McMahon, A.P.; Tabin, C.J. Evidence for an Expansion-Based Temporal Shh Gradient in Specifying Vertebrate Digit Identities. *Cell* **2004**, *118*, 517–528. [CrossRef]

57. Van Amerongen, R.; Bowman, A.N.; Nusse, R. Developmental stage and time dictate the fate of Wnt/β-catenin-responsive stem cells in the mammary gland. *Cell Stem Cell* **2012**, *11*, 387–400. [CrossRef]

58. Kobayashi, A.; Valerius, M.T.; Mugford, J.W.; Carroll, T.J.; Self, M.; Oliver, G.; McMahon, A.P. Six2 Defines and Regulates a Multipotent Self-Renewing Nephron Progenitor Population throughout Mammalian Kidney Development. *Cell Stem Cell* **2008**, *3*, 169–181. [CrossRef]

59. Burns, K.A.; Ayoub, A.E.; Breunig, J.J.; Adhami, F.; Weng, W.L.; Colbert, M.C.; Rakic, P.; Kuan, C.Y. Nestin-CreER mice reveal DNA synthesis by nonapoptotic neurons following cerebral ischemia hypoxia. *Cereb. Cortex* **2007**, *17*, 2585–2592. [CrossRef]

60. Steenblock, C.; de Celis, M.F.R.; Silva, L.F.D.; Pawolski, V.; Brennand, A.; Werdermann, M.; Berger, I.; Santambrogio, A.; Peitzsch, M.; Andoniadou, C.L.; et al. Isolation and characterization of adrenocortical progenitors involved in the adaptation to stress. *Proc. Natl. Acad. Sci. USA* **2018**, *115*, 12997–13002. [CrossRef]

61. Kreidberg, J.A.; Sariola, H.; Loring, J.M.; Maeda, M.; Pelletier, J.; Housman, D.; Jaenisch, R. WT-1 is required for early kidney development. *Cell* **1993**, *74*, 679–691. [CrossRef]

62. Pilon, N.; Raiwet, D.; Viger, R.S.; Silversides, D.W. Novel pre- and post-gastrulation expression of Gata4 within cells of the inner cell mass and migratory neural crest cells. *Dev. Dyn.* **2008**, *237*, 1133–1143. [CrossRef] [PubMed]

63. Rojas, A.; Schachterle, W.; Xu, S.-M.; Martín, F.; Black, B.L. Direct transcriptional regulation of Gata4 during early endoderm specification is controlled by FoxA2 binding to an intronic enhancer. *Dev. Biol.* **2010**, *346*, 346–355. [CrossRef] [PubMed]

64. Vidal, V.P.; Chaboissier, M.-C.; Lützkendorf, S.; Cotsarelis, G.; Mill, P.; Hui, C.-C.; Ortonne, N.; Ortonne, J.-P.; Schedl, A. Sox9 Is Essential for Outer Root Sheath Differentiation and the Formation of the Hair Stem Cell Compartment. *Curr. Biol.* **2005**, *15*, 1340–1351. [CrossRef]

65. Del Monte, G.; Casanova, J.C.; Guadix, J.A.; MacGrogan, D.; Burch, J.B.; Pérez-Pomares, J.M.; de la Pompa, J.L. Differential Notch signaling in the epicardium is required for cardiac inflow development and coronary vessel morphogenesis. *Circ. Res.* **2011**, *108*, 824–836. [CrossRef]

66. So, P.L.; Danielian, P.S. Cloning and expression analysis of a mouse gene related to Drosophila odd-skipped. *Mech. Dev.* **1999**, *84*, 157–160. [CrossRef]

67. Hata, K.; Kusumi, M.; Yokomine, T.; Li, E.; Sasaki, H. Meiotic and epigenetic aberrations inDnmt3L-deficient male germ cells. *Mol. Reprod. Dev.* **2005**, *73*, 116–122. [CrossRef]

68. Armstrong, J.F.; Pritchard-Jones, K.; Bickmore, W.A.; Hastie, N.D.; Bard, J.B. The expression of the Wilms' tumour gene, WT1, in the developing mammalian embryo. *Mech. Dev.* **1993**, *40*, 85–97. [CrossRef]

69. Bayne, S.; Jones, M.E.; Li, H.; Pinto, A.R.; Simpson, E.R.; Liu, J.-P. Estrogen deficiency leads to telomerase inhibition, telomere shortening and reduced cell proliferation in the adrenal gland of mice. *Cell Res.* **2008**, *18*, 1141–1150. [CrossRef]

70. Doroszko, M.; Chrusciel, M.; Stelmaszewska, J.; Slezak, T.; Rivero-Muller, A.; Padzik, A.; Anisimowicz, S.; Wolczynski, S.; Huhtaniemi, I.; Toppari, J.; et al. Luteinizing Hormone and GATA4 Action in the Adrenocortical Tumorigenesis of Gonadectomized Female Mice. *Cell. Physiol. Biochem.* **2017**, *43*, 1064–1076. [CrossRef]

71. Chrusciel, M.; Vuorenoja, S.; Mohanty, B.; Rivero-Müller, A.; Li, X.; Toppari, J.; Huhtaniemi, I.; Rahman, N.A. Transgenic GATA-4 expression induces adrenocortical tumorigenesis in C57Bl/6 mice. *J. Cell Sci.* **2013**, *126*, 1845–1857. [CrossRef] [PubMed]

72. Suemaru, S.; Darlington, D.N.; Akana, S.F.; Cascio, C.S.; Dallman, M.F. Ventromedial Hypothalamic Lesions Inhibit Corticosteroid Feedback Regulation of Basal ACTH during the Trough of the Circadian Rhythm. *Neuroendocrinology* **1995**, *61*, 453–463. [CrossRef] [PubMed]

73. Salpea-Damola, I.; Lefrançois-Martinez, A.-M.; Dufour, D.; Botto, J.-M.; Kamilaris, C.; Faucz, F.R.; Stratakis, C.A.; Val, P.; Martinez, A. Steroidogenic Factor-1 Lineage Origin of Skin Lesions in Carney Complex Syndrome. *J. Investig. Dermatol.* **2022**, *142*, 2949–2957.e9. [CrossRef] [PubMed]

74. Cheng, K.; Seita, Y.; Moriwaki, T.; Noshiro, K.; Sakata, Y.; Hwang, Y.S.; Torigoe, T.; Saitou, M.; Tsuchiya, H.; Iwatani, C.; et al. The developmental origin and the specification of the adrenal cortex in humans and cynomolgus monkeys. *Sci. Adv.* **2022**, *8*, eabn8485. [CrossRef] [PubMed]

75. Saito, D.; Tamura, K.; Takahashi, Y. Early segregation of the adrenal cortex and gonad in chicken embryos. *Dev. Growth Differ.* **2017**, *59*, 593–602. [CrossRef] [PubMed]
76. Wang, Q.; Lan, Y.; Cho, E.-S.; Maltby, K.M.; Jiang, R. Odd-skipped related 1 (Odd1) is an essential regulator of heart and urogenital development. *Dev. Biol.* **2005**, *288*, 582–594. [CrossRef]
77. Moore, A.; McInnes, L.; Kreidberg, J.; Hastie, N.; Schedl, A. YAC complementation shows a requirement for Wt1 in the development of epicardium, adrenal gland and throughout nephrogenesis. *Development* **1999**, *126*, 1845–1857. [CrossRef]
78. Miyamoto, N.; Yoshida, M.; Kuratani, S.; Matsuo, I.; Aizawa, S. Defects of urogenital development in mice lacking Emx2. *Development* **1997**, *124*, 1653–1664. [CrossRef]
79. Kusaka, M.; Katoh-Fukui, Y.; Ogawa, H.; Miyabayashi, K.; Baba, T.; Shima, Y.; Sugiyama, N.; Sugimoto, Y.; Okuno, Y.; Kodama, R.; et al. Abnormal Epithelial Cell Polarity and Ectopic Epidermal Growth Factor Receptor (EGFR) Expression Induced in Emx2 KO Embryonic Gonads. *Endocrinology* **2010**, *151*, 5893–5904. [CrossRef]
80. Birk, O.S.; Casiano, D.E.; Wassif, C.A.; Cogliati, T.; Zhao, L.; Zhao, Y.; Grinberg, A.; Huang, S.; Kreidberg, J.A.; Parker, K.L.; et al. The LIM homeobox gene Lhx9 is essential for mouse gonad formation. *Nature* **2000**, *403*, 909–913. [CrossRef]
81. Katoh-Fukui, Y.; Owaki, A.; Toyama, Y.; Kusaka, M.; Shinohara, Y.; Maekawa, M.; Toshimori, K.; Morohashi, K.-I. Mouse Polycomb M33 is required for splenic vascular and adrenal gland formation through regulating Ad4BP/SF1 expression. *Blood* **2005**, *106*, 1612–1620. [CrossRef] [PubMed]
82. Bamforth, S.; Bragança, J.; Eloranta, J.J.; Murdoch, J.N.; Marques, F.I.; Kranc, K.R.; Farza, H.; Henderson, D.; Hurst, H.C.; Bhattacharya, S. Cardiac malformations, adrenal agenesis, neural crest defects and exencephaly in mice lacking Cited2, a new Tfap2 co-activator. *Nat. Genet.* **2001**, *29*, 469–474. [CrossRef] [PubMed]
83. Val, P.; Martinez-Barbera, J.P.; Swain, A. Adrenal development is initiated by Cited2 and Wt1 through modulation of Sf-1 dosage. *Development* **2007**, *134*, 2349–2358. [CrossRef] [PubMed]
84. Teves, M.E.; Modi, B.P.; Kulkarni, R.; Han, A.X.; Marks, J.S.; Subler, M.A.; Windle, J.; Newall, J.M.; McAllister, J.M.; Strauss, J.F., 3rd. Human DENND1A.V2 Drives Cyp17a1 Expression and Androgen Production in Mouse Ovaries and Adrenals. *Int. J. Mol. Sci.* **2020**, *21*, 2545. [CrossRef]
85. Guasti, L.; Sze, W.C.; McKay, T.; Grose, R.; King, P.J. FGF signalling through Fgfr2 isoform IIIb regulates adrenal cortex development. *Mol. Cell. Endocrinol.* **2013**, *371*, 182–188. [CrossRef]
86. Krachulec, J.; Vetter, M.; Schrade, A.; Löbs, A.-K.; Bielinska, M.; Cochran, R.; Kyrönlahti, A.; Pihlajoki, M.; Parviainen, H.; Jay, P.Y.; et al. GATA4 Is a Critical Regulator of Gonadectomy-Induced Adrenocortical Tumorigenesis in Mice. *Endocrinology* **2012**, *153*, 2599–2611. [CrossRef]
87. Böse, J.; Grotewold, L.; Rüther, U. Pallister-Hall syndrome phenotype in mice mutant for Gli3. *Hum. Mol. Genet.* **2002**, *11*, 1129–1135. [CrossRef]
88. Laufer, E.; Kesper, D.; Vortkamp, A.; King, P. Sonic hedgehog signaling during adrenal development. *Mol. Cell. Endocrinol.* **2012**, *351*, 19–27. [CrossRef]
89. Pitetti, J.-L.; Calvel, P.; Romero, Y.; Conne, B.; Truong, V.; Papaioannou, M.D.; Schaad, O.; Docquier, M.; Herrera, P.L.; Wilhelm, D.; et al. Insulin and IGF1 Receptors Are Essential for XX and XY Gonadal Differentiation and Adrenal Development in Mice. *PLoS Genet.* **2013**, *9*, e1003160. [CrossRef]
90. Chida, D.; Nakagawa, S.; Nagai, S.; Sagara, H.; Katsumata, H.; Imaki, T.; Suzuki, H.; Mitani, F.; Ogishima, T.; Shimizu, C.; et al. Melanocortin 2 receptor is required for adrenal gland development, steroidogenesis, and neonatal gluconeogenesis. *Proc. Natl. Acad. Sci. USA* **2007**, *104*, 18205–18210. [CrossRef]
91. Novoselova, T.V.; Hussain, M.; King, P.J.; Guasti, L.; Metherell, L.A.; Charalambous, M.; Clark, A.J.L.; Chan, L.F. MRAP deficiency impairs adrenal progenitor cell differentiation and gland zonation. *FASEB J.* **2018**, *32*, 6186–6196. [CrossRef] [PubMed]
92. Xing, Y.; Morohashi, K.-I.; Ingraham, H.A.; Hammer, G.D. Timing of adrenal regression controlled by synergistic interaction between Sf1 SUMOylation and Dax1. *Development* **2017**, *144*, 3798–3807. [CrossRef] [PubMed]
93. Scheys, J.O.; Heaton, J.H.; Hammer, G.D. Evidence of Adrenal Failure in Aging Dax1-Deficient Mice. *Endocrinology* **2011**, *152*, 3430–3439. [CrossRef] [PubMed]
94. Luo, X.; Ikeda, Y.; Parker, K.L. A cell-specific nuclear receptor is essential for adrenal and gonadal development and sexual differentiation. *Cell* **1994**, *77*, 481–490. [CrossRef]
95. Bland, M.L.; Jamieson, C.A.; Akana, S.F.; Bornstein, S.R.; Eisenhofer, G.; Dallman, M.F.; Ingraham, H.A. Haploinsufficiency of steroidogenic factor-1 in mice disrupts adrenal development leading to an impaired stress response. *Proc. Natl. Acad. Sci. USA* **2000**, *97*, 14488–14493. [CrossRef]
96. Bland, M.L.; Fowkes, R.C.; Ingraham, H.A. Differential Requirement for Steroidogenic Factor-1 Gene Dosage in Adrenal Development Versus Endocrine Function. *Mol. Endocrinol.* **2004**, *18*, 941–952. [CrossRef]
97. Zubair, M.; Oka, S.; Parker, K.L.; Morohashi, K.-I. Transgenic Expression of *Ad4BP/SF-1* in Fetal Adrenal Progenitor Cells Leads to Ectopic Adrenal Formation. *Mol. Endocrinol.* **2009**, *23*, 1657–1667. [CrossRef]
98. Lee, F.Y.; Faivre, E.J.; Suzawa, M.; Lontok, E.; Ebert, D.; Cai, F.; Belsham, D.D.; Ingraham, H.A. Eliminating SF-1 (NR5A1) Sumoylation In Vivo Results in Ectopic Hedgehog Signaling and Disruption of Endocrine Development. *Dev. Cell* **2011**, *21*, 315–327. [CrossRef]

99. Doghman, M.; Karpova, T.; Rodrigues, G.A.; Arhatte, M.; de Moura, J.; Cavalli, L.R.; Virolle, V.; Barbry, P.; Zambetti, G.P.; Figueiredo, B.C.; et al. Increased Steroidogenic Factor-1 Dosage Triggers Adrenocortical Cell Proliferation and Cancer. *Mol. Endocrinol.* **2007**, *21*, 2968–2987. [CrossRef]

100. Schnabel, C.A.; Selleri, L.; Cleary, M.L. Pbx1 is essential for adrenal development and urogenital differentiation. *Genesis* **2003**, *37*, 123–130. [CrossRef]

101. Lichtenauer, U.D.; Duchniewicz, M.; Kolanczyk, M.; Hoeflich, A.; Hahner, S.; Else, T.; Bicknell, A.B.; Zemojtel, T.; Stallings, N.R.; Schulte, M.M.; et al. Pre-B-Cell Transcription Factor 1 and Steroidogenic Factor 1 Synergistically Regulate Adrenocortical Growth and Steroidogenesis. *Endocrinology* **2007**, *148*, 693–704. [CrossRef] [PubMed]

102. Tsai, L.C.; Shimizu-Albergine, M.; Beavo, J.A. The high-affinity cAMP-specific phosphodiesterase 8B controls steroidogenesis in the mouse adrenal gland. *Mol. Pharmacol.* **2011**, *79*, 639–648. [CrossRef] [PubMed]

103. Leal, L.F.; Szarek, E.; Berthon, A.; Nesterova, M.; Faucz, F.R.; London, E.; Mercier, C.; Abu-Asab, M.; Starost, M.F.; Dye, L.; et al. Pde8b haploinsufficiency in mice is associated with modest adrenal defects, impaired steroidogenesis, and male infertility, unaltered by concurrent PKA or Wnt activation. *Mol. Cell. Endocrinol.* **2020**, *522*, 111117. [CrossRef] [PubMed]

104. Levy, I.; Szarek, E.; Maria, A.G.; Starrost, M.; Sierra, M.D.L.L.; Faucz, F.R.; Stratakis, C.A. A phosphodiesterase 11 (Pde11a) knockout mouse expressed functional but reduced Pde11a: Phenotype and impact on adrenocortical function. *Mol. Cell. Endocrinol.* **2020**, *520*, 111071. [CrossRef] [PubMed]

105. Hao, H.X.; Jiang, X.; Cong, F. Control of Wnt Receptor Turnover by R-spondin-ZNRF3/RNF43 Signaling Module and Its Dysregulation in Cancer. *Cancers* **2016**, *8*, 54. [CrossRef] [PubMed]

106. Scortegagna, M.; Berthon, A.; Settas, N.; Giannakou, A.; Garcia, G.; Li, J.-L.; James, B.; Liddington, R.C.; Vilches-Moure, J.G.; Stratakis, C.A.; et al. The E3 ubiquitin ligase Siah1 regulates adrenal gland organization and aldosterone secretion. *JCI Insight* **2017**, *2*, e97128. [CrossRef] [PubMed]

107. Berthon, A.; Drelon, C.; Ragazzon, B.; Boulkroun, S.; Tissier, F.; Amar, L.; Samson-Couterie, B.; Zennaro, M.-C.; Plouin, P.-F.; Skah, S.; et al. WNT/β-catenin signalling is activated in aldosterone-producing adenomas and controls aldosterone production. *Hum. Mol. Genet.* **2013**, *23*, 889–905. [CrossRef]

108. Fujimoto, Y.; Tanaka, S.S.; Yamaguchi, Y.L.; Kobayashi, H.; Kuroki, S.; Tachibana, M.; Shinomura, M.; Kanai, Y.; Morohashi, K.-I.; Kawakami, K.; et al. Homeoproteins Six1 and Six4 Regulate Male Sex Determination and Mouse Gonadal Development. *Dev. Cell* **2013**, *26*, 416–430. [CrossRef]

109. Kobayashi, H.; Kawakami, K.; Asashima, M.; Nishinakamura, R. Six1 and Six4 are essential for Gdnf expression in the metanephric mesenchyme and ureteric bud formation, while Six1 deficiency alone causes mesonephric-tubule defects. *Mech. Dev.* **2007**, *124*, 290–303. [CrossRef]

110. Heikkilä, M.; Peltoketo, H.; Leppäluoto, J.; Ilves, M.; Vuolteenaho, O.; Vainio, S. Wnt-4 Deficiency Alters Mouse Adrenal Cortex Function, Reducing Aldosterone Production. *Endocrinology* **2002**, *143*, 4358–4365. [CrossRef]

111. Wilmouth, J.J.; Olabe, J.; Garcia-Garcia, D.; Lucas, C.; Guiton, R.; Roucher-Boulez, F.; Dufour, D.; Damon-Soubeyrand, C.; Sahut-Barnola, I.; Pointud, J.-C.; et al. Sexually dimorphic activation of innate antitumor immunity prevents adrenocortical carcinoma development. *Sci. Adv.* **2022**, *8*, eadd0422. [CrossRef] [PubMed]

112. Heikinheimo, M.; Scandrett, J.M.; Wilson, D. Localization of transcription factor GATA-4 to regions of the mouse embryo involved in cardiac development. *Dev. Biol.* **1994**, *164*, 361–373. [CrossRef] [PubMed]

113. Molkentin, J.D.; Lin, Q.; Duncan, S.A.; Olson, E.N. Requirement of the transcription factor GATA4 for heart tube formation and ventral morphogenesis. *Genes Dev.* **1997**, *11*, 1061–1072. [CrossRef] [PubMed]

114. Kuo, C.T.; Morrisey, E.E.; Anandappa, R.; Sigrist, K.; Lu, M.M.; Parmacek, M.S.; Soudais, C.; Leiden, J.M. GATA4 transcription factor is required for ventral morphogenesis and heart tube formation. *Genes Dev.* **1997**, *11*, 1048–1060. [CrossRef]

115. Moore, A.W.; Schedl, A.; McInnes, L.; Doyle, M.; Hecksher-Sorensen, J.; Hastie, N.D. YAC transgenic analysis reveals Wilms' Tumour 1 gene activity in the proliferating coelomic epithelium, developing diaphragm and limb. *Mech. Dev.* **1998**, *79*, 169–184. [CrossRef]

116. Furuhata, A.; Murakami, M.; Ito, H.; Gao, S.; Yoshida, K.; Sobue, S.; Kikuchi, R.; Iwasaki, T.; Takagi, A.; Kojima, T.; et al. GATA-1 and GATA-2 binding to 3′ enhancer of WT1 gene is essential for its transcription in acute leukemia and solid tumor cell lines. *Leukemia* **2009**, *23*, 1270–1277. [CrossRef]

117. Klattig, J.; Sierig, R.; Kruspe, D.; Makki, M.; Englert, C. WT1-Mediated Gene Regulation in Early Urogenital Ridge Development. *Sex. Dev.* **2007**, *1*, 238–254. [CrossRef]

118. Miyamoto, Y.; Taniguchi, H.; Hamel, F.; Silversides, D.W.; Viger, R.S. A GATA4/WT1 cooperation regulates transcription of genes required for mammalian sex determination and differentiation. *BMC Mol. Biol.* **2008**, *9*, 44. [CrossRef]

119. Wilhelm, D.; Englert, C. The Wilms tumor suppressor WT1 regulates early gonad development by activation of *Sf1*. *Genes Dev.* **2002**, *16*, 1839–1851. [CrossRef]

120. Tremblay, J.J.; Viger, R.S. GATA Factors Differentially Activate Multiple Gonadal Promoters through Conserved GATA Regulatory Elements. *Endocrinology* **2001**, *142*, 977–986. [CrossRef]

121. Dupont, J.; Holzenberger, M. Biology of insulin-like growth factors in development. *Birth Defects Res. C Embryo Today* **2003**, *69*, 257–271. [CrossRef] [PubMed]

122. Hammer, G.D.; Krylova, I.; Zhang, Y.; Darimont, B.D.; Simpson, K.; Weigel, N.L.; Ingraham, H.A. Phosphorylation of the nuclear receptor SF-1 modulates cofactor recruitment: Integration of hormone signaling in reproduction and stress. *Mol. Cell* **1999**, *3*, 521–526. [CrossRef]

123. França, M.M.; Ferraz-De-Souza, B.; Santos, M.G.; Lerario, A.M.; Fragoso, M.C.B.V.; Latronico, A.C.; Kuick, R.D.; Hammer, G.D.; Lotfi, C.F. POD-1 binding to the E-box sequence inhibits SF-1 and StAR expression in human adrenocortical tumor cells. *Mol. Cell. Endocrinol.* **2013**, *371*, 140–147. [CrossRef] [PubMed]

124. Tamura, M.; Kanno, Y.; Chuma, S.; Saito, T.; Nakatsuji, N. Pod-1/Capsulin shows a sex- and stage-dependent expression pattern in the mouse gonad development and represses expression of Ad4BP/SF-1. *Mech. Dev.* **2001**, *102*, 135–144. [CrossRef]

125. Garcia-Moreno, S.A.; Lin, Y.-T.; Futtner, C.R.; Salamone, I.M.; Capel, B.; Maatouk, D.M. CBX2 is required to stabilize the testis pathway by repressing Wnt signaling. *PLoS Genet.* **2019**, *15*, e1007895. [CrossRef]

126. Katoh-Fukui, Y.; Miyabayashi, K.; Komatsu, T.; Owaki, A.; Baba, T.; Shima, Y.; Kidokoro, T.; Kanai, Y.; Schedl, A.; Wilhelm, D.; et al. Cbx2, a Polycomb Group Gene, Is Required for Sry Gene Expression in Mice. *Endocrinology* **2012**, *153*, 913–924. [CrossRef]

127. Katoh-Fukui, Y.; Tsuchiya, R.; Shiroishi, T.; Nakahara, Y.; Hashimoto, N.; Noguchi, K.; Higashinakagawa, T. Male-to-female sex reversal in M33 mutant mice. *Nature* **1998**, *393*, 688–692. [CrossRef]

128. Kiiveri, S.; Liu, J.; Westerholm-Ormio, M.; Narita, N.; Wilson, D.B.; Voutilainen, R.; Heikinheimo, M. Transcription factors gata-4 and gata-6 during mouse and human adrenocortical development. *Endocr. Res.* **2002**, *28*, 647–650. [CrossRef]

129. Kiiveri, S.; Liu, J.; Westerholm-Ormio, M.; Narita, N.; Wilson, D.B.; Voutilainen, R.; Heikinheimo, M. Differential expression of GATA-4 and GATA-6 in fetal and adult mouse and human adrenal tissue. *Endocrinology* **2002**, *143*, 3136–3143. [CrossRef]

130. Chia, C.Y.; Madrigal, P.; Denil, S.; Martinez, I.; Garcia-Bernardo, J.; El-Khairi, R.; Chhatriwala, M.; Shepherd, M.H.; Hattersley, A.T.; Dunn, N.R.; et al. GATA6 Cooperates with EOMES/SMAD2/3 to Deploy the Gene Regulatory Network Governing Human Definitive Endoderm and Pancreas Formation. *Stem Cell Rep.* **2019**, *12*, 57–70. [CrossRef]

131. Liu, J.; Cheng, H.; Xiang, M.; Zhou, L.; Wu, B.; Moskowitz, I.; Zhang, K.; Xie, L. Gata4 regulates hedgehog signaling and Gata6 expression for outflow tract development. *PLoS Genet.* **2019**, *15*, e1007711. [CrossRef] [PubMed]

132. Ishibashi, T.; Yokura, Y.; Ohashi, K.; Yamamoto, H.; Maeda, M. Conserved GC-boxes, E-box and GATA motif are essential for GATA-4 gene expression in P19CL6 cells. *Biochem. Biophys. Res. Commun.* **2011**, *413*, 171–175. [CrossRef] [PubMed]

133. Zubair, M.; Ishihara, S.; Oka, S.; Okumura, K.; Morohashi, K.-I. Two-Step Regulation of *Ad4BP/SF-1* Gene Transcription during Fetal Adrenal Development: Initiation by a Hox-Pbx1-Prep1 Complex and Maintenance via Autoregulation by Ad4BP/SF-1. *Mol. Cell. Biol.* **2006**, *26*, 4111–4121. [CrossRef] [PubMed]

134. Bielohuby, M.; Herbach, N.; Wanke, R.; Maser-Gluth, C.; Beuschlein, F.; Wolf, E.; Hoeflich, A. Growth analysis of the mouse adrenal gland from weaning to adulthood: Time- and gender-dependent alterations of cell size and number in the cortical compartment. *Am. J. Physiol. Metab.* **2007**, *293*, E139–E146. [CrossRef] [PubMed]

135. Howard-Miller, E. A transitory zone in the adrenal cortex which shows age and sex relationships. *Am. J. Anat.* **1927**, *40*, 251–293. [CrossRef]

136. Furlan, A.; Dyachuk, V.; Kastriti, M.E.; Calvo-Enrique, L.; Abdo, H.; Hadjab, S.; Chontorotzea, T.; Akkuratova, N.; Usoskin, D.; Kamenev, D.; et al. Multipotent peripheral glial cells generate neuroendocrine cells of the adrenal medulla. *Science* **2017**, *357*, eaal3753. [CrossRef]

137. Hanemaaijer, E.S.; Margaritis, T.; Sanders, K.; Bos, F.L.; Candelli, T.; Al-Saati, H.; van Noesel, M.M.; Meyer-Wentrup, F.A.G.; van de Wetering, M.; Holstege, F.C.P.; et al. Single-cell atlas of developing murine adrenal gland reveals relation of Schwann cell precursor signature to neuroblastoma phenotype. *Proc. Natl. Acad. Sci. USA* **2021**, *118*, e2022350118. [CrossRef]

138. Kameneva, P.; Artemov, A.V.; Kastriti, M.E.; Faure, L.; Olsen, T.K.; Otte, J.; Erickson, A.; Semsch, B.; Andersson, E.R.; Ratz, M.; et al. Single-cell transcriptomics of human embryos identifies multiple sympathoblast lineages with potential implications for neuroblastoma origin. *Nat. Genet.* **2021**, *53*, 694–706. [CrossRef]

139. Britsch, S.; Li, L.; Kirchhoff, S.; Theuring, F.; Brinkmann, V.; Birchmeier, C.; Riethmacher, D. The ErbB2 and ErbB3 receptors and their ligand, neuregulin-1, are essential for development of the sympathetic nervous system. *Genes Dev.* **1998**, *12*, 1825–1836. [CrossRef]

140. Tomooka, Y.; Yasui, T. Electron microscopic study of the response of the adrenocortical X-zone in mice treated with sex steroids. *Cell Tissue Res.* **1978**, *194*, 269–277. [CrossRef]

141. Holmes, P.V.; Dickson, A.D. X-zone degeneration in the adrenal glands of adult and immature female mice. *J. Anat.* **1971**, *108*, 159–168. [PubMed]

142. Janat, M.F.; Shire, J.G. The adrenal X-zone of mice: Genetic analysis of its development with recombinant-inbred strains. *Exp. Biol.* **1987**, *46*, 217–221. [PubMed]

143. Ungar, F.; Stabler, T.A. 20α-Hydroxysteroid dehydrogenase activity and the x-zone of the female mouse adrenal. *J. Steroid Biochem.* **1980**, *13*, 23–28. [CrossRef]

144. Shibata, H.; Kurihara, I.; Kobayashi, S.; Yokota, K.; Suda, N.; Saito, I.; Saruta, T. Regulation of differential COUP-TF-coregulator interactions in adrenal cortical steroidogenesis. *J. Steroid Biochem. Mol. Biol.* **2003**, *85*, 449–456. [CrossRef]

145. Hall, J.G.; Pallister, P.D.; Clarren, S.K.; Beckwith, J.B.; Wiglesworth, F.W.; Fraser, F.C.; Cho, S.; Benke, P.J.; Reed, S.D. Congenital hypothalamic hamartoblastoma, hypopituitarism, imperforate anus and postaxial polydactyly—A new syndrome? Part I: Clinical, causal, and pathogenetic considerations. *Am. J. Med. Genet.* **1980**, *7*, 47–74. [CrossRef] [PubMed]

146. De Lau, W.B.; Snel, B.; Clevers, H.C. The R-spondin protein family. *Genome Biol.* **2012**, *13*, 242. [CrossRef]

147. Lin, Y.; Liu, A.; Zhang, S.; Ruusunen, T.; Kreidberg, J.A.; Peltoketo, H.; Drummond, I.; Vainio, S. Induction of ureter branching as a response to Wnt-2b signaling during early kidney organogenesis. *Dev. Dyn.* **2001**, *222*, 26–39. [CrossRef]
148. Lopez, J.P.; Brivio, E.; Santambrogio, A.; De Donno, C.; Kos, A.; Peters, M.; Rost, N.; Czamara, D.; Brückl, T.M.; Roeh, S.; et al. Single-cell molecular profiling of all three components of the HPA axis reveals adrenal ABCB1 as a regulator of stress adaptation. *Sci. Adv.* **2021**, *7*, eabe4497. [CrossRef]
149. Lyu, Q.; Wang, H.; Kang, Y.; Wu, X.; Zheng, H.S.; Laprocina, K.; Junghans, K.; Ding, X.; Huang, C.-C.J. RNA-Seq Reveals Sub-Zones in Mouse Adrenal Zona Fasciculata and the Sexually Dimorphic Responses to Thyroid Hormone. *Endocrinology* **2020**, *161*, bqaa126. [CrossRef]
150. Tee, M.K.; Speek, M.; Legeza, B.; Modi, B.; Teves, M.E.; McAllister, J.M.; Strauss, J.F., 3rd; Miller, W.L. Alternative splicing of DENND1A, a PCOS candidate gene, generates variant 2. *Mol. Cell. Endocrinol.* **2016**, *434*, 25–35. [CrossRef]
151. McAllister, J.M.; Modi, B.; Miller, B.A.; Biegler, J.; Bruggeman, R.; Legro, R.S.; Strauss, J.F. Overexpression of a DENND1A isoform produces a polycystic ovary syndrome theca phenotype. *Proc. Natl. Acad. Sci. USA* **2014**, *111*, E1519–E1527. [CrossRef] [PubMed]
152. Kulkarni, R.; Teves, M.E.; Han, A.X.; McAllister, J.M.; Strauss, J.F. Colocalization of Polycystic Ovary Syndrome Candidate Gene Products in Theca Cells Suggests Novel Signaling Pathways. *J. Endocr. Soc.* **2019**, *3*, 2204–2223. [CrossRef] [PubMed]
153. McAllister, J.M.; Legro, R.S.; Modi, B.; Strauss, J.F. Functional genomics of PCOS: From GWAS to molecular mechanisms. *Trends Endocrinol. Metab.* **2015**, *26*, 118–124. [CrossRef] [PubMed]
154. Quinn, T.A.; Ratnayake, U.; Dickinson, H.; Nguyen, T.-H.; McIntosh, M.; Castillo-Melendez, M.; Conley, A.J.; Walker, D.W. Ontogeny of the Adrenal Gland in the Spiny Mouse, With Particular Reference to Production of the Steroids Cortisol and Dehydroepiandrosterone. *Endocrinology* **2013**, *154*, 1190–1201. [CrossRef] [PubMed]
155. Mostaghel, E.A.; Zhang, A.; Hernandez, S.; Marck, B.T.; Zhang, X.; Tamae, D.; Biehl, H.E.; Tretiakova, M.S.; Bartlett, J.; Burns, J.F.; et al. Contribution of Adrenal Glands to Intratumor Androgens and Growth of Castration-Resistant Prostate Cancer. *Clin. Cancer Res.* **2019**, *25*, 426–439. [CrossRef]
156. Yaglova, N.V.; Obernikhin, S.S.; Nazimova, S.V.; Yaglov, V.V. Role of Transcription Factor Oct4 in Postnatal Development and Function of the Adrenal Cortex. *Bull. Exp. Biol. Med.* **2019**, *167*, 568–573. [CrossRef] [PubMed]
157. Moog, F.; Bennett, C.J.; Dean, C.M., Jr. Growth and cytochemistry of the adrenal gland of the mouse from birth to maturity. *Anat. Rec.* **1954**, *120*, 873–891. [CrossRef]
158. Callow, R.K.; Deanesly, R. Effect of androsterone and of male hormone concentrates on the accessory reproductive organs of castrated rats, mice and guinea-pigs. *Biochem. J.* **1935**, *29*, 1424–1445. [CrossRef]
159. Starkey, W.F.; Schmidt, E.C.H. THE EFFECT OF TESTOSTERONE-PROPIONATE ON THE X-ZONE OF THE MOUSE ADRENAL. *Endocrinology* **1938**, *23*, 339–344. [CrossRef]
160. Mukai, T.; Kusaka, M.; Kawabe, K.; Goto, K.; Nawata, H.; Fujieda, K.; Morohashi, K.-I. Sexually dimorphic expression of Dax-1 in the adrenal cortex. *Genes Cells* **2002**, *7*, 717–729. [CrossRef]
161. Kuser-Abali, G.; Alptekin, A.; Lewis, M.J.; Garraway, I.P.; Cinar, B. YAP1 and AR interactions contribute to the switch from androgen-dependent to castration-resistant growth in prostate cancer. *Nat. Commun.* **2015**, *6*, 8126. [CrossRef] [PubMed]
162. Audenet, F.; Méjean, A.; Chartier-Kastler, E.; Rouprêt, M. Adrenal tumours are more predominant in females regardless of their histological subtype: A review. *World J. Urol.* **2013**, *31*, 1037–1043. [CrossRef] [PubMed]
163. Ayala-Ramirez, M.; Jasim, S.; Feng, L.; Ejaz, S.; Deniz, F.; Busaidy, N.; Waguespack, S.G.; Naing, A.; Sircar, K.; Wood, C.G.; et al. Adrenocortical carcinoma: Clinical outcomes and prognosis of 330 patients at a tertiary care center. *Eur. J. Endocrinol.* **2013**, *169*, 891–899. [CrossRef] [PubMed]
164. Scollo, C.; Russo, M.; Trovato, M.A.; Sambataro, D.; Giuffrida, D.; Manusia, M.; Sapuppo, G.; Malandrino, P.; Vigneri, R.; Pellegriti, G. Prognostic Factors for Adrenocortical Carcinoma Outcomes. *Front. Endocrinol.* **2016**, *7*, 99. [CrossRef] [PubMed]
165. Dolfi, B.; Gallerand, A.; Firulyova, M.M.; Xu, Y.; Merlin, J.; Dumont, A.; Castiglione, A.; Vaillant, N.; Quemener, S.; Gerke, H.; et al. Unravelling the sex-specific diversity and functions of adrenal gland macrophages. *Cell Rep.* **2022**, *39*, 110949. [CrossRef]
166. Boyle, M.H.; Paranjpe, M.G.; Creasy, D.M. High Background Incidence of Spontaneous Subcapsular Adrenal Gland Hyperplasia of Tg.rasH2 Mice Used in 26-week Carcinogenicity Studies. *Toxicol. Pathol.* **2018**, *46*, 444–448. [CrossRef]
167. Petterino, C.; Naylor, S.; Mukaratirwa, S.; Bradley, A. Adrenal Gland Background Findings in CD-1 (Crl:CD-1(ICR)BR) Mice from 104-week Carcinogenicity Studies. *Toxicol. Pathol.* **2015**, *43*, 816–824. [CrossRef]
168. Yoshida, A.; Maita, K.; Shirasu, Y. Subcapsular cell hyperplasia in the mouse adrenal glands. *Jpn. J. Vet. Sci.* **1986**, *48*, 719–728. [CrossRef]
169. Bielinska, M.; Kiiveri, S.; Parviainen, H.; Mannisto, S.; Heikinheimo, M.; Wilson, D.B. Gonadectomy-induced Adrenocortical Neoplasia in the Domestic Ferret (*Mustela putorius furo*) and Laboratory Mouse. *Vet. Pathol.* **2006**, *43*, 97–117. [CrossRef]
170. Bielinska, M.; Parviainen, H.; Porter-Tinge, S.B.; Kiiveri, S.; Genova, E.; Rahman, N.; Huhtaniemi, I.T.; Muglia, L.J.; Heikinheimo, M.; Wilson, D. Mouse Strain Susceptibility to Gonadectomy-Induced Adrenocortical Tumor Formation Correlates with the Expression of GATA-4 and Luteinizing Hormone Receptor. *Endocrinology* **2003**, *144*, 4123–4133. [CrossRef]
171. Bielinska, M.; Genova, E.; Boime, I.; Parviainen, H.; Kiiveri, S.; Leppäluoto, J.; Rahman, N.; Heikinheimo, M.; Wilson, D.B. Gonadotropin-Induced Adrenocortical Neoplasia in NU/J Nude Mice. *Endocrinology* **2005**, *146*, 3975–3984. [CrossRef] [PubMed]
172. Looyenga, B.; Hammer, G.D. Origin and Identity of Adrenocortical Tumors in Inhibin Knockout Mice: Implications for Cellular Plasticity in the Adrenal Cortex. *Mol. Endocrinol.* **2006**, *20*, 2848–2863. [CrossRef] [PubMed]

International Journal of
Molecular Sciences

Article

Intronic Enhancer Is Essential for *Nr5a1* Expression in the Pituitary Gonadotrope and for Postnatal Development of Male Reproductive Organs in a Mouse Model

Yuichi Shima [1,2,*], Kanako Miyabayashi [1], Takami Mori [3], Koji Ono [2,4], Mizuki Kajimoto [5], Hae Lim Cho [6], Hitomi Tsuchida [7], Yoshihisa Uenoyama [7], Hiroko Tsukamura [7], Kentaro Suzuki [8], Man Ho Choi [6] and Kazunori Toida [2]

[1] Division of Microscopic and Developmental Anatomy, Department of Anatomy, Kurume University School of Medicine, 67 Asahi-machi, Kurume 830-0011, Fukuoka, Japan
[2] Department of Anatomy, Kawasaki Medical School, 577 Matsushima, Kurashiki 701-0192, Okayama, Japan
[3] Department of Medical Technology, Faculty of Health Science and Technology, Kawasaki University of Medical Welfare, 288 Matsushima, Kurashiki 701-0193, Okayama, Japan
[4] School of Medical Technology, Faculty of Health and Medical Care, Saitama Medical University, 1397-1 Yamane, Hidaka 350-1241, Saitama, Japan
[5] Pharmacokinetics and Bioanalysis Center, Shin Nippon Biomedical Laboratories, Ltd. (SNBL), 16-1 Minamiakasaka, Kainan-shi 642-0017, Wakayama, Japan
[6] Center for Advanced Biomolecular Recognition, Korea Institute of Science and Technology, 5 Hwarang-ro 14-gil, Seoul 02792, Republic of Korea
[7] Laboratory of Animal Reproduction, Graduate School of Bioagricultural Sciences, Nagoya University, Nagoya 464-8601, Aichi, Japan
[8] Faculty of Life and Environmental Sciences, University of Yamanashi, 4-4-37, Takeda, Kofu 400-8510, Yamanashi, Japan
* Correspondence: yshima@med.kurume-u.ac.jp; Tel.: +81-942-35-3311 (ext. 3150)

Citation: Shima, Y.; Miyabayashi, K.; Mori, T.; Ono, K.; Kajimoto, M.; Cho, H.L.; Tsuchida, H.; Uenoyama, Y.; Tsukamura, H.; Suzuki, K.; et al. Intronic Enhancer Is Essential for *Nr5a1* Expression in the Pituitary Gonadotrope and for Postnatal Development of Male Reproductive Organs in a Mouse Model. *Int. J. Mol. Sci.* **2023**, *24*, 192. https://doi.org/10.3390/ijms24010192

Academic Editor: Jacques J. Tremblay

Received: 18 November 2022
Revised: 15 December 2022
Accepted: 19 December 2022
Published: 22 December 2022

Abstract: Nuclear receptor subfamily 5 group A member 1 (NR5A1) is expressed in the pituitary gonadotrope and regulates their differentiation. Although several regulatory regions were implicated in *Nr5a1* gene expression in the pituitary gland, none of these regions have been verified using mouse models. Furthermore, the molecular functions of NR5A1 in the pituitary gonadotrope have not been fully elucidated. In the present study, we generated mice lacking the pituitary enhancer located in the 6th intron of the *Nr5a1* gene. These mice showed pituitary gland-specific disappearance of NR5A1, confirming the functional importance of the enhancer. Enhancer-deleted male mice demonstrated no defects at fetal stages. Meanwhile, androgen production decreased markedly in adult, and postnatal development of reproductive organs, such as the seminal vesicle, prostate, and penis was severely impaired. We further performed transcriptomic analyses of the whole pituitary gland of the enhancer-deleted mice and controls, as well as gonadotropes isolated from Ad4BP-BAC-EGFP mice. These analyses identified several genes showing gonadotrope-specific, NR5A1-dependent expressions, such as *Spp1*, *Tgfbr3l*, *Grem1*, and *Nr0b2*. These factors are thought to function downstream of NR5A1 and play important roles in reproductive organ development through regulation of pituitary gonadotrope functions.

Keywords: NR5A1; enhancer; pituitary gonadotrope; luteinizing hormone; follicle-stimulating hormone

1. Introduction

Sex hormone secretion from the gonads is controlled by pituitary gonadotropins, the production and secretion of which is controlled by the hypothalamic gonadotropin-releasing hormone (GnRH). This hierarchical sex hormone production control mechanism is called the hypothalamic-pituitary-gonadal (HPG) axis. Pituitary gonadotropins include luteinizing hormone (LH) and follicle-stimulating hormone (FSH), and they are composed of common α subunit and unique β subunits, LHβ and FSHβ, respectively. LH stimulates testicular Leydig cells to produce androgens, whereas FSH stimulates Sertoli cells to support spermatogenesis. Male LHβ and LH receptor knockout mice show normal masculinization at fetal periods but severely impaired postnatal reproductive organ development, indicating the physiological importance of LH in male reproductive function [1,2]. FSHβ knockout mice show milder phenotypes than LHβ/LH receptor knockout mice, indicating a minor role of FSH in male reproductive function [3].

One of the most important factors for HPG axis formation is the nuclear receptor subfamily 5 group A member 1 (NR5A1, also known as Ad4-binding protein (Ad4BP) or Steroidogenic Factor-1 (SF-1)). Although this factor is not expressed in the hypothalamic GnRH neurons, it is expressed not only in pituitary gonadotropes but also in the ventro-medial hypothalamic nucleus, adrenal cortex, Sertoli and Leydig cells of the testis, and granulosa and theca cells of the ovaries [4,5]. In mice, systemic *Nr5a1* gene disruption resulted in structural and functional abnormalities in all these tissues, indicating the pivotal roles of NR5A1 in each tissue [6]. However, because systemic *Nr5a1* knockout mice die in the neonatal period due to adrenal insufficiency, tissue-specific functions of NR5A1 are not well understood. In order to clarify the NR5A1 function in the pituitary, pituitary-specific conditional *Nr5a1* knockout mice have been generated using the αGSU-Cre lineage [7,8]. In these mice, the expression of *Lhb* and *Fshb* is abrogated, indicating that NR5A1 plays an important role in the functional differentiation of the pituitary gonadotropes.

As *Nr5a1* is expressed in multiple tissues, several research groups including our group have performed transgenic mouse assays to identify tissue-specific regulatory regions. These analyses have identified several enhancers such as the fetal adrenal enhancer (FAdE) [9], ventromedial hypothalamus enhancer (VE) [10], pituitary enhancer (PE) [11], and fetal Leydig enhancer (FLE) [12] in the *Nr5a1* gene locus. Although these enhancers have been identified by generating transgenic mice, functional importance of these enhancers has not been directly verified by genome deletion. We recently used genome editing to generate mice with FLE deletion, which showed fetal Leydig cell (FLC)-specific NR5A1 deficiency and severe defects in male reproductive organs from fetal stages, clearly demonstrating the indispensable role of FLE in FLC-specific *Nr5a1* expression [13]. Based on this result, in this study, we generated a mouse line lacking the PE of *Nr5a1* and confirmed that PE plays an essential role in pituitary-specific *Nr5a1* expression.

NR5A1 begins to express in the anterior pituitary at E13.5–E14.5, and analysis of *Nr5a1*-disrupted mice suggested that NR5A1 regulates gonadotropin production in the pituitary gonadotrope [14]. In addition, the results of in vitro analysis suggested that *Lhb* and *Cga* expression was directly controlled by NR5A1 [15,16]. On the other hand, the expression of LHβ and FSHβ was reduced but not completely lost in the pituitary gland–specific *Nr5a1* knockout mice [7,8]. Furthermore, the expression of LHβ and FSHβ is induced by GnRH stimulation in *Nr5a1* gene knockout mice [14], suggesting that NR5A1 is not essential for LHβ and FSHβ expression. Considering these results together, it is conceivable that there are other downstream genes that are directly regulated by NR5A1 in pituitary gonadotropes. In this study, we identified several candidate NR5A1 downstream genes. Some of these genes has not been previously linked to the pituitary gonadotrope functions and might be worth for further analyses in future studies.

2. Results

2.1. Deletion of Gonadotrope-Specific PE of Nr5a1

To confirm the functional significance of the PE, we adopted CRISPR/Cas9 genome editing to delete the PE region (Figure 1A). Genotyping PCR and direct sequencing confirmed that the PE region was successfully deleted from the mouse genome (Figures 1B and S1). Both male and female PE$^{+/-}$ (heterozygous deletion of the enhancer) mice were fertile, but PE$^{-/-}$ (homozygous deletion of the enhancer) mice were infertile in both males and females.

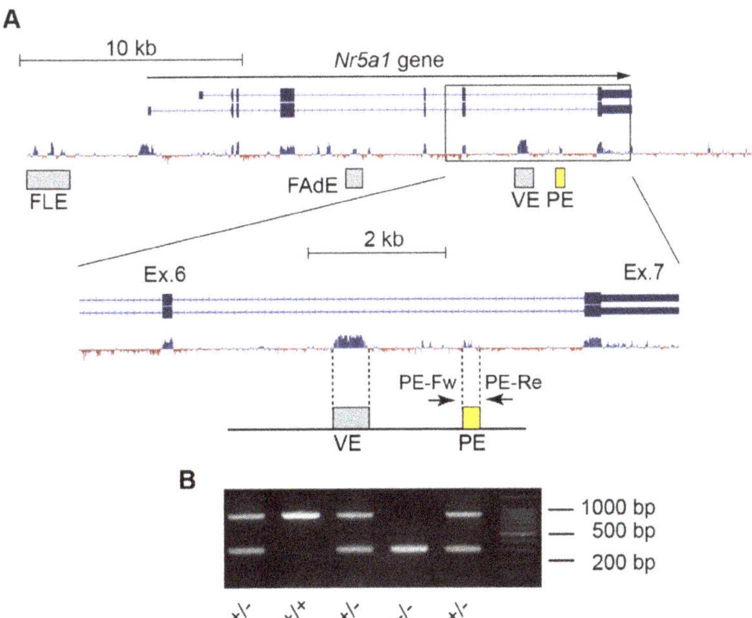

Figure 1. Deletion of the pituitary enhancer (PE) of *Nr5a1* by genome editing. (**A**) *Nr5a1* comprises seven exons. Several tissue-specific enhancers exist in the intronic and upstream regions of *Nr5a1*, such as the fetal Leydig enhancer (FLE), fetal adrenal enhancer (FAdE), ventromedial hypothalamic nucleus enhancer (VE), and PE. (**B**) The PE (yellow filled box) in the sixth intron of the gene was deleted by genome editing, and genotyping PCR with the primers PE-Fw and PE-Re confirmed that the region was successfully removed from the genome.

2.2. Normal Masculinization in Fetal Stages

2.2.1. Testis and Accessory Reproductive Organs

Because PE deficiency was expected to abolish pituitary Nr5a1 expression and reduce the function of the pituitary gonadotropes, we examined the phenotype of $PE^{-/-}$ male mice in comparison with that of control ($PE^{+/-}$) mice to analyze the effects of enhancer deficiency. Fetal $PE^{-/-}$ male mice (embryonic day 18.5; E18.5) showed normal descendance and size of the testis relative to those in the control mice (arrows in Figure 2A,B). In addition, vas deferens development and adrenal gland size (arrowheads in Figure 2A,B) were consistent between groups, confirming that the effect of enhancer deficiency was limited to the pituitary gonadotrope. Immunostaining revealed that NR5A1 was strongly expressed in Leydig cells in the interstitium of the testis and weakly expressed in Sertoli cells in the seminiferous tubules of the testis in $PE^{-/-}$ mice, showing no clear differences from expression patterns in control mice. Accordingly, no abnormality was observed in HSD3B expression in Leydig cells or SOX9 expression in Sertoli cells (Figure 2C–F).

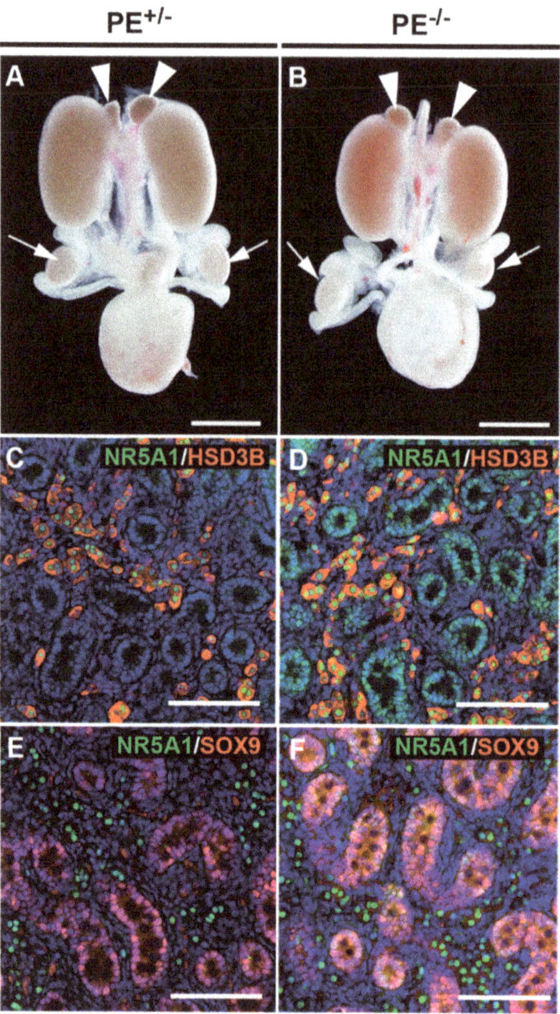

Figure 2. Fetal masculinization was not affected in PE$^{-/-}$ mice. (**A,B**) Macroscopic observation of the urogenital systems in the control mice (**A**) and PE$^{-/-}$ mice (**B**). Arrowheads indicate adrenal glands, whereas arrows indicate testes. (**C,D**) Double staining of the testis with antibodies for NR5A1 (green) and HSD3B (red). (**E,F**) Double staining of the testis with NR5A1 (green) and SOX9 (red) antibodies. Scale bar: 2 mm in (**A,B**), 100 μm in (**C–F**).

2.2.2. Steroids in Fetal Testes

The intratesticular concentration of steroid hormones was evaluated by GC-MS. Testosterone and androstenedione levels were slightly higher in PE$^{-/-}$ mice than in the control mice, but the difference was not significant (Table 1). Levels of other steroids showed no significant differences in concentration between control and PE$^{-/-}$ testis. There was also no significant difference between control and PE$^{-/-}$ testes in the metabolic ratio of enzymatic reactions required for androgen synthesis (Supplemental Table S1).

Table 1. Intratesticular steroids in PE$^{+/-}$ mice and PE$^{-/-}$ mice at E18.5.

	Amount (pg/Whole Tissue)				
	PE$^{+/-}$		PE$^{-/-}$		
Steroid Compound	Mean	SD	Mean	SD	p-Value [1]
Dehydroepiandrosterone	28.4	3.1	24.5	8.5	0.5476
Androstenediol	36.7	1.2	39.7	8.7	0.5000
Epitestosterone	12.2	2.1	15.8	9.5	0.6905
5α-androstane-3β, 17β-diol	18.7	11.2	14.9	1.8	0.6905
Androstenedione	324.0	133.2	381.4	156.0	0.5476
Testosterone	627.1	248.0	722.9	274.6	0.6905
Pregnenolone	18.8	4.6	15.5	0.6	0.2222
Progesterone	28.97	0.24	28.68	11.33	0.8413
17α-hydroxyprogesterone	100.3	35.3	98.4	33.6	>0.9999
11-deoxycortisol	49.9	7.9	62.8	16.5	0.1508

[1] Statical significance was determined using the Mann–Whitney U test.

2.3. Impaired Development of Reproductive Organs at Adult Stages

2.3.1. Testis

The testes of adult male PE$^{-/-}$ mice were significantly smaller than those of control mice (Figure 3A), whereas the size of the adrenal gland was unaffected (Figure S2). Hematoxylin and eosin (HE) staining of testis sections revealed that the diameter of the seminiferous tubules was clearly reduced in PE$^{-/-}$ mice relative to that in controls, and few mature spermatozoa were found within the seminiferous tubules (Figure 3B,C). The area of the testicular interstitium was also narrower in PE$^{-/-}$ than in control mice, and lipid droplets within the interstitial Leydig cells were reduced (Figure 3B′,C′).

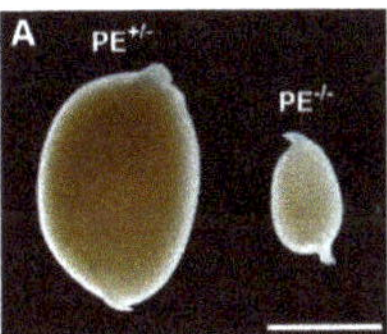

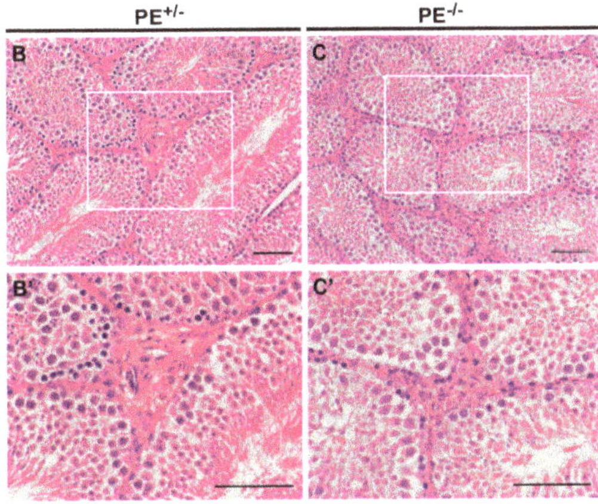

Figure 3. Severely impaired testicular architecture in PE$^{-/-}$ adult mice. (**A**) Macroscopic view of the testes collected from control and PE$^{-/-}$ mice. (**B,C**) Low magnification view of the HE stained section of testes collected from control (**B**) and PE$^{-/-}$ mice (**C**). (**B′,C′**) Magnified view of the areas enclosed by open rectangles in (**B,C**). Scale bar: 2 mm in (**A**), 100 μm in (**B–C′**).

Female PE$^{-/-}$ mice exhibited smaller ovaries than control mice (Figure S3A). When tissue sections were prepared and analyzed, the PE$^{-/-}$ mice did not present a large number of corpora lutea compared to those found in the ovaries of control mice (Figure S3B,B',C). This was thought to be because ovulation did not occur due to decreased LH secretion from the pituitary. In addition, many traces of closed follicles were observed in the ovaries (arrowheads in Figure S3C'), and it was speculated that the follicles could not be maintained due to the decrease in estrogen.

2.3.2. Seminal Vesicles, Prostate Gland, and Penis

Macroscopic observation of accessory reproductive organs showed that seminal vesicles were not apparent in PE$^{-/-}$ male mice (Figure 4A,B). HE staining revealed that most of the seminal vesicles and prostate had been replaced with adipose tissue, and only a few traces of the prostate glands were identified (Figure 4C–D'). The external genitalia were also clearly smaller in appearance in the PE$^{-/-}$ mice than in the controls (Figure 5A,B), and the size of the penis was also reduced (Figure 5C,D). Masson-trichrome staining clearly showed poor development of the penis in PE$^{-/-}$ males (Figure 5E,F).

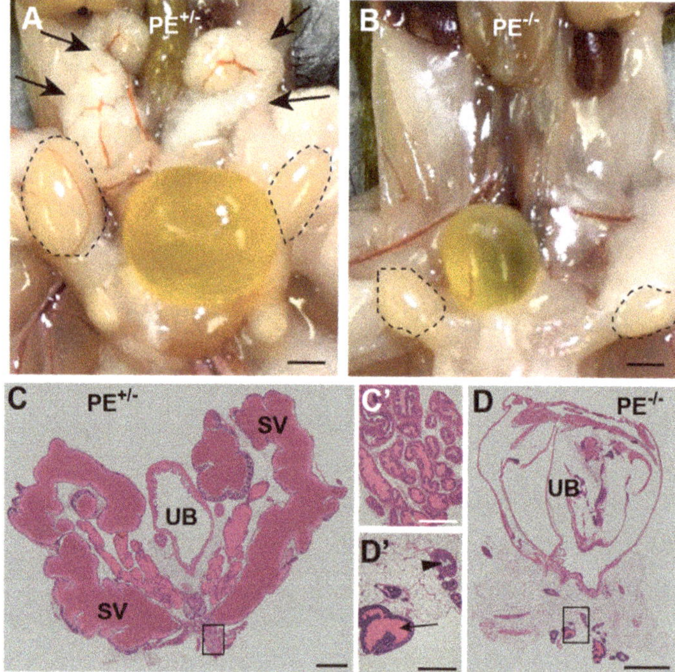

Figure 4. Hypoplastic development of reproductive organs in PE$^{-/-}$ male mice. (**A,B**) Macroscopic views of the lower abdominal organs in control (**A**) and PE$^{-/-}$ (**B**) mice. Testes are encircled by broken lines. Arrows in (**A**) indicate seminal vesicles, which were not observed in PE$^{-/-}$ mice (**B**). (**C,D**) HE-stained sections of the urinary bladder (UB) and seminal vesicles (SV) in control (**C**) and PE$^{-/-}$ (**D**) mice. (**C',D'**) Magnified view of the areas enclosed by open rectangles in (**C,D**). An arrow in (**D'**) indicates a rudimentary tissue of the prostate gland. An arrowhead in (**D'**) indicates the urethra. Scale bar: 2 mm in (**A–D**); 100 µm in (**C',D'**).

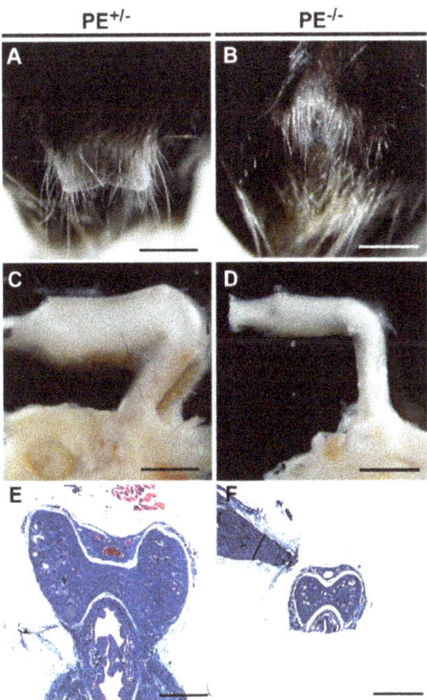

Figure 5. Hypoplastic penis in PE$^{-/-}$ male mice. (**A**,**B**) Macroscopic view of the penis of control (**A**) and PE$^{-/-}$ (**B**) mice. (**C**,**D**) Penile foreskin was removed, and penes were compared between control (**C**) and PE$^{-/-}$ (**D**) mice. (**E**,**F**) Masson-trichrome staining of the penile sections of control (**E**) and PE$^{-/-}$ (**F**) mice. Scale bar: 2 mm in (**A–D**); 500 μm in (**E**,**F**).

2.3.3. Steroid Levels in Adult Testes

GC-MS revealed the presence of several steroids in adult testes that were not detected in the fetal testes, such as 3α-androstane-3α,17β-diol, 7α-hydroxyandrostenedione, dihydrotestosterone, 6β-hydroxyandrostenedione, 6β-hydroxytestosterone, 16α-hydroxytestosterone, 16α-hydroxyandrostenedione, 17α-hydroxypregnenolone, tetrahydrodeoxycorticosterone, allo-tetrahydrodeoxycorticosterone, and corticosteron. Among detected steroids, dehydroepiandrosterone, androstenediol, 7α-hydroxyandrostenedione, androstenedione, and testosterone were significantly lower in the testes of PE$^{-/-}$ mice than in those of control mice (Table 2). From these results, it was speculated that the activities of 17α-hydroxylase/17,20-lyase, 3β-HSD, and 17β-HSD, enzymes involved in the synthesis of testosterone, were globally reduced. Indeed, comparison of metabolic ratios between control testes and PE$^{-/-}$ testes suggested that activities of enzymes, such as 21-hydroxylase, 17,20-lyase, 3β-HSD, 17β-HSD, 17α-HSD, 5α-reductase, and 3α-HSD were significantly decreased in PE$^{-/-}$ testes (Supplemental Table S2).

Table 2. Intratesticular steroids in adult PE$^{+/-}$ and PE$^{-/-}$ mice.

| | Concentration (pg/mg Tissue) | | | | |
| | PE$^{+/-}$ | | PE$^{-/-}$ | | |
Steroid Compound	Mean	SD	Mean	SD	*p*-Value [1]
3α-androstane-3α, 17β-diol	2.1	0.3	2.3	1.2	0.6905
Dehydroepiandrosterone	0.6	0.1	1.8	0.4	0.0079 *
Androstenediol	3.7	0.8	2.4	0.6	0.0317 *
Epitestosterone	0.9	0.9	1.2	0.8	>0.9999

Table 2. *Cont.*

| | Concentration (pg/mg Tissue) | | | | |
| | PE[+/−] | | PE[−/−] | | |
Steroid Compound	Mean	SD	Mean	SD	*p*-Value [1]
3α-androstane-3β, 17β-diol	1.1	0.5	1.6	0.5	0.2222
7α-hydroxyandrostenedione	3.1	0.9	12.3	2.7	0.0079 *
Dihydrotestosterone	1.7	0.8	2.3	0.5	0.2222
Androstenedione	54.0	21.7	10.9	5.5	0.0079 *
Testosterone	58.3	57.4	9.7	8.1	0.0079 *
6β-hydroxyandrostenedione	1.3	0.3	ND	ND	NA
6β-hydroxytestosterone	1.5	1.5	ND	ND	NA
Pregnenolon	4.7	1.1	2.8	2.0	0.0952
Progesterone	8.61	3.11	9.95	3.73	0.6905
16α-hydroxytestosterone	5.9	6.8	ND	ND	NA
16α-hydroxyandrostenedione	1.0	0.7	ND	ND	NA
17α-hydroxypregnenolone	1.4	0.3	ND	ND	NA
17α-hydroxyprogesterone	9.6	2.9	7.1	2.9	0.3095
Tetrahydrodeoxycorticosterone	ND	ND	3.9	2.1	NA
Allo-tetrahydrodeoxycorticosterone	0.9	0.4	1.0	0.3	0.5476
11-deoxycortisol	2.9	3.0	3.8	2.5	0.1508
Corticosterone	13.3	12.4	7.7	5.7	0.3095

[1] Significance was determined using the Mann–Whitney U test. * Statistically significant difference ($p < 0.05$). ND, not detected under the limit of quantification; NA, not applicable because analyte was not detected in either experimental group.

2.3.4. Quantitative Reverse Transcription (qRT)-PCR and Gonadotropin Immunostaining

The pituitary glands of PE[−/−] mice showed no apparent size difference compared to those of the control mice (Figure 6A). Plasma LH levels tended to be lower in PE[−/−] mice than in control mice, but no significant difference was detected because LH concentrations were generally low in control mice. FSH concentration was significantly lower in PE[−/−] mice than in the control group (Figure 6B).

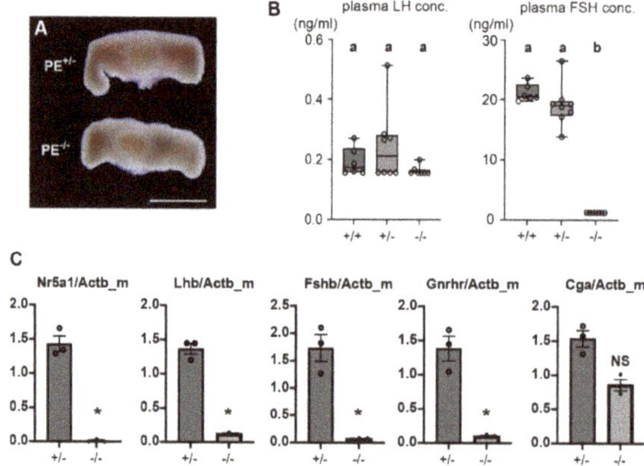

Figure 6. Changes in pituitary gonadotrope marker genes in PE[−/−] mice. (**A**) Macroscopic view of the pituitary glands of control and PE[−/−] mice. Scale bar: 1 mm (**B**) Plasma concentrations of LH and FSH in control (PE[+/+] and PE[+/−]) and PE[−/−] mice. Differences were evaluated by one-way ANOVA followed by Tukey's post hoc test at a significance level of $p < 0.05$. a and b: significant difference between different characters. (**C**) Relative expression of pituitary gonadotrope marker genes in control and PE[−/−] mice, as evaluated by qRT-PCR. Y-axis represents gene expression relative to that of *Actb*. Statistical significance between two experimental groups was evaluated by unpaired *t*-test. * significant difference ($p < 0.05$), NS: not significant.

RNA was extracted from the pituitary gland, and the expression of marker genes of gonadotropin-producing cells was analyzed by qRT-PCR. In male and female $PE^{-/-}$ mice, *Nr5a1* expression was almost completely absent, while that of *Lhb*, *Fshb*, and *Gnrhr* was detectable but significantly reduced relative to control values. *Cga* expression in male $PE^{-/-}$ mice was reduced to about half that of control mice; expression in females was reduced to about 70% of the control value (Figures 6C and S4).

Expression of LHβ, FSHβ, and thyroid stimulating hormone β (TSHβ) was examined by immunostaining. A considerable number of LHβ-expressing cells were present in the pituitary gland of $PE^{-/-}$ mice (Figure 7A,B). These LHβ-expressing cells did not show nuclear NR5A1 expression, suggesting that NR5A1 was not essential for LHβ expression (Figure 7A′,B′). The number of FSHβ-expressing cells was dramatically reduced in the $PE^{-/-}$ group relative to that in the controls (Figure 7C,D). However, cells weakly expressing FSHβ were still observed in the pituitary of $PE^{-/-}$ mice, suggesting that NR5A1 influenced FSHβ expression (Figure 7C′,D′). NR5A1 was not expressed in TSHβ-expressing cells, and no obvious abnormalities in TSHβ expression were observed in $PE^{-/-}$ mice relative to control expression (Figure 7E–F′).

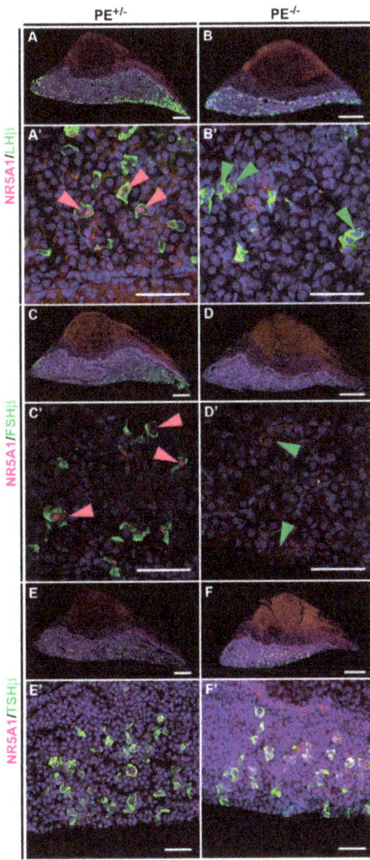

Figure 7. Levels of LHβ and FSHβ decreased in the $PE^{-/-}$ mouse pituitary; TSHβ levels did not change. (**A–B′**) Immunostaining of pituitary sections using antibodies for NR5A1 (red) and LHβ (green). (**C–D′**) Immunostaining of pituitary sections using antibodies for NR5A1 (red) and FSHβ (green). (**E–F′**) Immunostaining of pituitary sections using antibodies for NR5A1 (red) and TSHβ (green). Scale bars: 200 μm in (**A–F**), 50 μm in (**A′–F′**).

2.4. Transcriptome Analyses of Pituitaries and Isolated Gonadotropes

Because the expression of *Lhb* and *Fshb* was not completely lost in $PE^{-/-}$ mice, we searched for other downstream genes directly regulated by NR5A1. We first analyzed the transcriptome of the entire pituitary gland and extracted 43 genes with reduced pituitary expression in $PE^{-/-}$ mice compared to that in controls (pit_m_homo_down; Figure 8A). Thereafter, we analyzed the transcriptome of the isolated gonadotropes and compared it with that of the whole pituitary, identifying 189 highly expressed genes in the gonadotropes relative to whole pituitary expression (gonadotrope_m_up; Figure 8A). In this process, we noticed that one of the isolated gonadotrope samples (gonadotrope_m1) showed a distinct gene expression pattern from the other three (Figure S5) and excluded this sample from the analysis. By comparing the pit_m_homo_down and gonadotrope_m_up gene sets, we identified 16 genes with gonadotrope-specific, NR5A1-dependent expression (Figure 8B). Gene ontology (GO) analysis of these genes highlighted "regulation of bone remodeling," "GnRH signaling pathway," and "regulation of hormone levels" as highly enriched GO terms (Figure 8C). We performed the same analyses in female samples and identified nine genes enriched in "gonad development" and "neuroactive ligand-receptor interaction" (Figures S6 and S7).

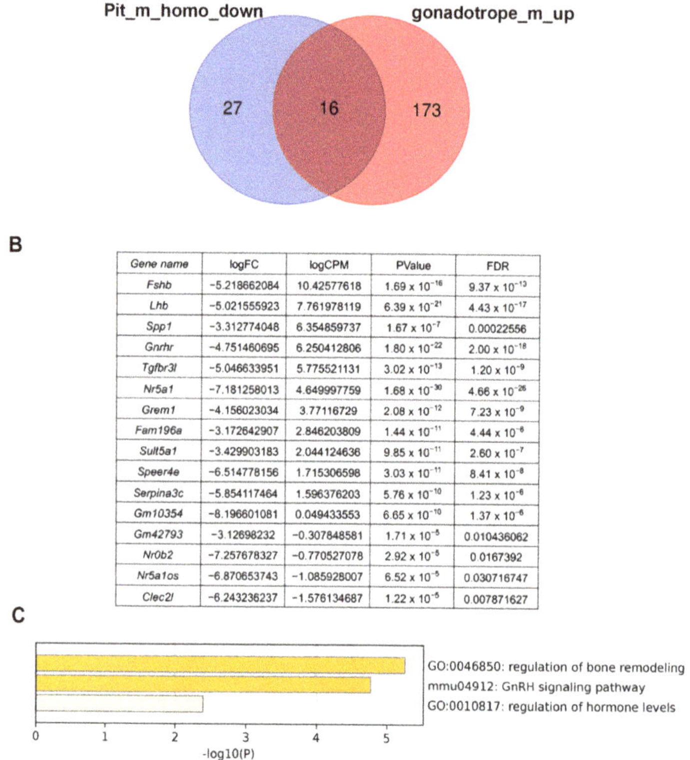

Gene name	logFC	logCPM	PValue	FDR
Fshb	−5.218662084	10.42577618	1.69×10^{-16}	9.37×10^{-13}
Lhb	−5.021555923	7.761978119	6.39×10^{-21}	4.43×10^{-17}
Spp1	−3.312774048	6.354859737	1.67×10^{-7}	0.00022556
Gnrhr	−4.751460695	6.250412806	1.80×10^{-22}	2.00×10^{-18}
Tgfbr3l	−5.046633951	5.775521131	3.02×10^{-13}	1.20×10^{-9}
Nr5a1	−7.181258013	4.649997759	1.68×10^{-30}	4.66×10^{-26}
Grem1	−4.156023034	3.77116729	2.08×10^{-12}	7.23×10^{-9}
Fam196a	−3.172642907	2.846203809	1.44×10^{-11}	4.44×10^{-6}
Sult5a1	−3.429903183	2.044124636	9.85×10^{-11}	2.60×10^{-7}
Speer4e	−6.514778156	1.715306598	3.03×10^{-11}	8.41×10^{-8}
Serpina3c	−5.854117464	1.596376203	5.76×10^{-10}	1.23×10^{-6}
Gm10354	−8.196601081	0.049433553	6.65×10^{-10}	1.37×10^{-6}
Gm42793	−3.12698232	−0.307848581	1.71×10^{-5}	0.010436062
Nr0b2	−7.257678327	−0.770527078	2.92×10^{-5}	0.0167392
Nr5a1os	−6.870653743	−1.085928007	6.52×10^{-5}	0.030716747
Clec2l	−6.243236237	−1.576134687	1.22×10^{-5}	0.007871627

Figure 8. Transcriptomic analyses of the whole pituitary gland and isolated gonadotropes in males. (**A**) Venn diagram showing an overlap between genes downregulated in $PE^{-/-}$ pituitary (Pit_m_homo_down) relative to the control and those showing higher expression in the gonadotropes than in the whole pituitary (gonadotrope_m_up). (**B**) List of the 16 overlapping genes of the two gene sets shown in (**A**); differential expression of 16 genes between control and $PE^{-/-}$ pituitary glands. (**C**) GO terms enriched in the 16 genes.

3. Discussion

3.1. Functional Importance of the PE in Nr5a1 Gene Regulation

Nr5a1 contains multiple internal and upstream regulatory regions (enhancers). Although these enhancers have been identified by generating transgenic mice [9–12], their functional importance has not been strictly defined. In our previous study, the PE of *Nr5a1* was identified in the sixth intron [11]. In this study, we demonstrated that deletion of this PE leads to cell-specific and complete NR5A1 deficiency. In a recent study by another group, the ATAC-sequence of pituitary gonadotrope–derived cell lines suggested that regions other than the PE (the FLE and a small region in the fourth intron) are also implicated in *Nr5a1* expression in the pituitary gonadotropes [17]. The involvement of these regions (especially functionally undefined region in the fourth intron) in pituitary-specific NR5A1 expression should be carefully investigated in future studies.

3.2. Dependence of Fetal and Adult Leydig Cells on Pituitary Gonadotropins

The phenotype of PE-deficient mice was essentially the same as that previously reported in mice with pituitary-specific *Nr5a1* gene disruption [7,8]. That is, adult male mice were infertile due to insufficient formation of reproductive organs and reduced production of androgens. In females, ovulation did not occur, and the corpus luteum did not form, causing infertility. Furthermore, no defects were observed in the masculinization of fetal $PE^{-/-}$ male mice. These data suggest that fetal masculinization proceeds in a pituitary-independent manner. Previous studies have shown that even when LHβ or LH receptors are deleted, fetal masculinization proceeds normally, but the production of androgens after birth declines and puberty does not occur, leading to defective spermatogenesis and hypoplastic male reproductive organs [1,2]. Another example is the *Kiss1* knockout mouse. In these mice, kisspeptin-induced GnRH production is absent and blood LH levels are decreased, but fetal androgen production is unaffected, whereas postnatal androgen production is markedly reduced [18]. These results were explained by the pituitary gland-independent development of Leydig cells in fetal testes, and LH-dependent Leydig cell development in postnatal testes [19]. To support this notion, our previous study showed that fetal Leydig cell–specific LH receptor knockout mice exhibited normal reproductive organs at the fetal stage [20].

The production of male hormones has been reported to be triphasic, comprising fetal, neonatal, and adolescent periods [21]. Fetal Leydig cells are responsible for the production of male hormones during the fetal period, and adult Leydig cells after puberty. In addition, transient HPG axis activation during the neonatal period is known to produce male sex hormones through a process called mini-puberty. Recent studies have focused on the influence of mini-puberty on spermatogenesis and male reproductive function at adult stages [22]. PE-deficient mice may represent a useful tool to clarify the physiological significance of mini-puberty.

3.3. Role of NR5A1 in the Pituitary Gonadotrope

Analysis of *Nr5a1*-disrupted mice suggested that NR5A1 is important for the functional differentiation of pituitary gonadotropes. Moreover, from the results of in vitro analysis, *Lhb* and *Cga* expression been reported to be directly controlled by NR5A1 [15,16]. However, in both this study and the previous works [7,8], the expression of LHβ and FSHβ was reduced but not completely lost in the pituitary gland–specific *Nr5a1* knockout mice. Furthermore, the expression of LHβ and FSHβ is also induced by GnRH stimulation in *Nr5a1* gene knockout mice [14], suggesting that NR5A1 is not essential for LHβ and FSHβ expression. These results suggested that there may be other downstream genes that are directly regulated by NR5A1 in pituitary gonadotropes. These genes might be related to GnRH responsiveness, signal transduction downstream of the GnRH receptor, or gonadotropin secretion, and several studies have been performed to identify such genes.

3.4. Candidate NR5A1 Downstream Genes

From the results of transcriptome analyses, 16 NR5A1-dependent genes with high expression in isolated gonadotropes were identified. GO analyses of these 16 genes identified *Fshb*, *Spp1*, and *Grem1* as related to "regulation of bone remodeling." The *Spp1* gene encodes secreted phosphoprotein 1, or osteopontin, which shows gonadotrope-specific pituitary expression and regulates the interaction between gonadotropes and extracellular matrices [23]. Interestingly, osteopontin shows higher expression in male than in female gonadotropes. In agreement with this, our results showed that *Spp1* was highly expressed in the male gonadotrope but not in the female gonadotrope. *Grem1* encodes Gremlin1, an antagonist of bone morphogenetic protein. In a previous study, *Grem2*-null mice showed irregular estrous cycles and subfertility [24]. Although *Grem2* is not expressed in the pituitary gland, these previous data suggested that Grem2 plays an important role in HPG axis regulation and reproductive function in females. Our study expands on this to suggest that Grem1 is a novel regulator of gonadotrope function in males. *Tgfbr3l*, which encodes transforming growth factor β receptor III-like protein and plays essential roles in the transduction of inhibin B signaling to the pituitary gonadotrope, was also included in the gene set. Recently generated *Tgfbr3l* gene-disrupted female mice showed increased FSH production and follicle development relative to controls, and double knockout of *Tgfbr3l* and betaglycan resulted in female infertility [25], indicating an important role of this factor in female reproductive function. Another recent study identified an NR5A1 binding sequence in the proximal promoter of the human and murine *Tgfbr3l* homologs, and in vitro analyses suggested that NR5A1 directly induces gonadotrope-specific *Tgfbr3l* gene expression [26]. Our study supported this finding and strongly suggested that NR5A1 directly regulates *Tgfbr3l* gene expression in vivo. The *Tgfbr3l* gene also shows gonadotrope-specific and NR5A1-dependent expression in males. However, its role in male reproductive function has not been clarified. *Nr0b2* encodes a small heterodimer partner (SHP), a factor known to regulate bile acid homeostasis [27]. Recent studies have focused on its function in the testes [27], but the physiological function of SHP in the pituitary gonadotrope has not been investigated so far. Another *Nr0b* family gene, *Nr0b1*, showed gonadotrope-specific and NR5A1-dependent expression in females. This gene encodes dosage-sensitive sex reversal, adrenal hypoplasia congenita critical region, on chromosome X, gene 1 (DAX-1) [27]. Previous studies have shown that DAX-1 expression overlaps with that of NR5A1 in various tissues, including pituitary gonadotropes [28]. Although several previous in vitro studies have suggested that DAX-1 is directly regulated by NR5A1 [29–31], ours is the first report to suggest that NR5A1 regulates DAX-1 expression in the pituitary gonadotropes in vivo. Overall, we identified several candidate NR5A1 downstream genes in the pituitary gonadotrope. Among these, several genes have not yet been linked to pituitary gonadotrope function and should be evaluated in future studies.

4. Materials and Methods

4.1. Mice

We previously identified a gonadotrope-specific PE of *Nr5a1* [11]. In this study, we deleted the PE region from the mouse genome following a published procedure [32]. Guide RNAs targeting the upstream and downstream regions of the PE were designed using CRISPR direct (http://crispr.dbcls.jp/, accessed on 17 March 2017). crRNA, tracrRNA, and Cas9 protein (Integrated DNA Technologies) were mixed to form an RNP complex and then introduced into the fertilized eggs by electroporation (Genome Editor, BEX). The eggs were then transferred into the oviducts of recipient mothers, and the genotypes of the resulting pups were determined by PCR. The sequences of genotyping primers are shown in Supplemental Table S3. Homozygous PE deletion mice were designated as PE$^{-/-}$ mice, whereas heterozygous PE deletion mice (PE$^{+/-}$ mice) were used as controls unless otherwise noted. Ad4BP-BAC-EGFP mice [33] were used to collect NR5A1-expressing gonadotropes from the pituitary gland via fluorescence-activated cell sorting (FACS).

4.2. Tissue Preparation, Histological Analyses, and Immunostaining

Mice were anesthetized with 0.3 mg/kg medetomidine hydrochloride (Nippon Zenyaku Kogyo, Fukushima, Japan), 4 mg/kg midazolam (Astellas Pharma, Tokyo, Japan), and 5 mg/kg butorphanol tartrate (Meiji Seika Pharma, Tokyo, Japan), and then perfused with PBS followed by 4% paraformaldehyde in 0.1 M phosphate buffer (pH 7.4) from the left ventricle. For histological analyses, tissues were embedded in paraffin wax, sectioned to 5 μm in thickness, and subjected to HE or Masson trichrome staining. Stained sections were observed using a BZ-X700 fluorescence microscope (Keyence, Osaka, Japan). For immunostaining, 50-μm thick sections were cut using a cryotome (Leica CM3050 S, Leica Camera AG, Wetzlar, Germany) and stained using the free-floating staining method [13]. The primary and secondary antibodies used in this study are listed in Supplemental Table S4. For nuclear staining, 4′6′-diamidino-2-phenylindole (DAPI) (Sigma-Aldrich, St. Louis, MO, USA) was used. Tissue sections were encapsulated in VECTASHIELD Mounting Medium (Vector Laboratories, Newark, CA, USA) and photographed with a LSM 700 laser scanning microscope (Carl Zeiss AG, Oberkochen, Germany).

4.3. RNA Preparation and Quantitative RT-PCR

Total RNA was prepared from the anterior pituitary of $PE^{+/-}$ (n = 3) and $PE^{-/-}$ (n = 3) male mice and subjected to reverse transcription with random hexamers (Superscript VILO master mix, Invitrogen, Carlsbad, CA, USA). Synthesized cDNA was used for quantitative PCR using the AriaMx Real Time PCR system (Agilent, Santa Clara, CA, USA) with gene-specific primers (Supplemental Table S5) and SYBR green qPCR master mix (Agilent, Santa Clara, CA, USA). Expression of the genes of interest was adjusted relative to that of *Actb*, the gene encoding β-actin.

4.4. Measurement of Plasma Gonadotropin

Blood samples were collected from the right ventricle of anesthetized $PE^{+/+}$ (n = 6), $PE^{+/-}$ (n = 8), and $PE^{-/-}$ (n = 7) male mice. Plasma LH and FSH concentrations were measured by a double-antibody radioimmunoassay (RIA) with mouse LH- and FSH-RIA kits provided by the National Hormone and Peptide Program (Torrance, CA, USA), as previously described [34,35]. LH and FSH concentrations were expressed in terms of mouse LH-RP (AFP-5306A) and FSH-RP (AFP-5308D), respectively. The lowest detectable level of LH in 25 μL plasma samples was 0.156 ng/mL, and the intra- and inter-assay coefficients of variation were 6.5 and 7.5%, respectively, at 2.8 ng/mL. The lowest detectable level of FSH in 25 μL plasma samples was 1.252 ng/mL, and the intra- and inter-assay coefficients of variation were 8.7 and 8.7%, respectively, at 17.1 ng/mL.

4.5. Measurement of Testicular Steroids

Testes were collected from control and $PE^{-/-}$ mice at the fetal stage (E18.5) and adult stage (8–10 weeks after birth), respectively (n = 5 in each experimental condition). Levels of testicular steroids were determined by gas chromatography-mass spectrometry (GC-MS) as previously described [36]. The concentration of steroid hormone was given in units of ng/tissue in the fetal testis, and ng/mg tissue in the adult testis. The metabolic ratio for each enzymatic reaction was calculated by dividing the metabolite concentration by the precursor concentration.

4.6. mRNA Sequencing, Data Processing, and Differentially Expressed Gene Analyses

mRNA sequencing analyses were performed as previously described [13]. Briefly, total RNAs were prepared from the whole pituitary gland (control mice and $PE^{-/-}$ mice) or from EGFP-positive cells sorted from Ad4BP-BAC-EGFP mouse pituitary glands, and were then subjected to library construction using NEBNext Single Cell/Low Input RNA Library Prep Kit for Illumina (New England Biolabs, Ipswich, MA, USA). Libraries were subjected to paired-end 150-bp sequencing on an Illumina series sequencer. After removing adapter sequences and low-quality reads using "cutadapt" (version 4.2) with default parameters,

FASTQ files were mapped to the mouse genome (mm10) by "STAR" (version 2.5.4a) with default parameters. Reads for each gene were counted using "featureCounts" (version 1.6.1) with default parameters, and gene expression matrix files were subjected to differentially expressed gene analyses using "EdgeR". Genes with reduced expression in $PE^{-/-}$ mice relative to controls ($\log_2 FC < -2$, *p*-value < 0.05, FDR < 0.05) were extracted. Genes with higher expression in the isolated gonadotropes than in the control whole pituitary were also extracted. We compared the two gene sets, and overlapping genes were then subjected to annotation analyses by "Metascape" [37].

4.7. Statistical Analyses

Quantitative RT-PCR data were presented as mean $\pm$ SEM, and statistical differences between experimental groups were examined by the two-tailed unpaired Student's *t*-test. Plasma gonadotropin levels were presented as mean $\pm$ SEM, and differences were evaluated by one-way ANOVA followed by Tukey's post hoc test. Intratesticular steroid levels and metabolic ratios were presented as mean and SD, and comparative levels of testicular steroids and metabolic ratios between control and $PE^{-/-}$ groups were evaluated by a non-parametric Mann–Whitney U test.

5. Conclusions

Intronic enhancer plays an essential role in pituitary gonadotrope-specific *Nr5a1* gene expression. NR5A1 regulates functional differentiation of pituitary gonadotropes, and thereby induces development of reproductive organs. This study identified candidate downstream genes of NR5A1 in the pituitary gonadotrope. Some of them have been already shown to be important for the pituitary gonadotrope function. However, we also identified several genes of which function in the pituitary gland is unclear. These genes may be the target of future studies to clarify the pathogenesis of human hypogonadotropic hypogonadism patients.

Supplementary Materials: The supporting information can be downloaded at: https://www.mdpi.com/article/10.3390/ijms24010192/s1.

Author Contributions: Conceptualization, Y.S.; Methodology, Y.S., Y.U., H.T. (Hiroko Tsukamura), K.S., M.H.C.; Investigation, Y.S., K.M., T.M., K.O., M.K., H.L.C., H.T. (Hitomi Tsuchida); Format analysis and Writing—Original Draft Preparation, Y.S.; Writing—Review and Editing, K.M., T.M., K.O., K.M., H.L.C., H.T. (Hitomi Tsuchida), Y.U., H.T. (Hiroko Tsukamura), K.S., M.H.C., K.T.; Supervision, Y.S., K.T.; Funding Acquisition, Y.S. All authors have read and agreed to the published version of the manuscript.

Funding: This work was supported by JSPS (Japan Society for the Promotion of Science) KAKENHI Grant numbers 21H00235 (Y.S.); Research Project Grant from Kawasaki Medical School (R02B-025 to Y.S.).

Institutional Review Board Statement: All animal protocols were approved by the Animal Care and Use Committee of Kawasaki Medical School and Kurume University School of Medicine.

Informed Consent Statement: Not applicable.

Data Availability Statement: Sequence data were deposited to the GEO repository with accession numbers GSE216466 and GSE2164685.

Acknowledgments: We deeply thank Gen Yamada (Wakayama Medical University, Japan) for his valuable advice regarding this work. We appreciate the technical support provided by the Central Research Center of Kawasaki Medical School. We also thank A. F. Parlow of the National Hormone and Peptide Program for providing antibodies against pituitary hormones.

Conflicts of Interest: The authors declare no conflict of interest.

References

1. Ma, X.; Dong, Y.; Matzuk, M.M.; Kumar, T.R. Targeted disruption of luteinizing hormone beta-subunit leads to hypogonadism, defects in gonadal steroidogenesis, and infertility. *Proc. Natl. Acad. Sci. USA* **2004**, *101*, 17294–17299. [CrossRef] [PubMed]
2. Lei, Z.M.; Mishra, S.; Zou, W.; Xu, B.; Foltz, M.; Li, X.; Rao, C.V. Targeted disruption of luteinizing hormone/human chorionic gonadotropin receptor gene. *Mol. Endocrinol.* **2001**, *15*, 184–200. [CrossRef] [PubMed]
3. Kumar, T.R.; Wang, Y.; Lu, N.; Matzuk, M.M. Follicle stimulating hormone is required for ovarian follicle maturation but not male fertility. *Nat. Genet.* **1997**, *15*, 201–204. [CrossRef] [PubMed]
4. Morohashi, K. Gonadal and extragonadal functions of Ad4BP/SF-1: Developmental aspects. *Trends Endocrinol. Metab.* **1999**, *10*, 169–173. [CrossRef] [PubMed]
5. Parker, K.L.; Schimmer, B.P. Steroidogenic factor 1: A key determinant of endocrine development and function. *Endocr. Rev.* **1997**, *18*, 361–377. [CrossRef]
6. Luo, X.; Ikeda, Y.; Lala, D.S.; Baity, L.A.; Meade, J.C.; Parker, K.L. A cell-specific nuclear receptor plays essential roles in adrenal and gonadal development. *Endocr. Res.* **1995**, *21*, 517–524. [CrossRef]
7. Zhao, L.; Bakke, M.; Krimkevich, Y.; Cushman, L.J.; Parlow, A.F.; Camper, S.A.; Parker, K.L. Steroidogenic factor 1 (SF1) is essential for pituitary gonadotrope function. *Development* **2001**, *128*, 147–154. [CrossRef]
8. Zhao, L.; Bakke, M.; Parker, K.L. Pituitary-specific knockout of steroidogenic factor 1. *Mol. Cell Endocrinol.* **2001**, *185*, 27–32. [CrossRef]
9. Zubair, M.; Ishihara, S.; Oka, S.; Okumura, K.; Morohashi, K. Two-step regulation of Ad4BP/SF-1 gene transcription during fetal adrenal development: Initiation by a Hox-Pbx1-Prep1 complex and maintenance via autoregulation by Ad4BP/SF-1. *Mol. Cell Biol.* **2006**, *26*, 4111–4121. [CrossRef]
10. Shima, Y.; Zubair, M.; Ishihara, S.; Shinohara, Y.; Oka, S.; Kimura, S.; Okamoto, S.; Minokoshi, Y.; Suita, S.; Morohashi, K. Ventromedial hypothalamic nucleus-specific enhancer of Ad4BP/SF-1 gene. *Mol. Endocrinol.* **2005**, *19*, 2812–2823. [CrossRef]
11. Shima, Y.; Zubair, M.; Komatsu, T.; Oka, S.; Yokoyama, C.; Tachibana, T.; Hjalt, T.A.; Drouin, J.; Morohashi, K. Pituitary homeobox 2 regulates adrenal 4 binding protein/steroidogenic factor-1 gene transcription in the pituitary gonadotrope through interaction with the intronic enhancer. *Mol. Endocrinol.* **2008**, *22*, 1633–1646. [CrossRef] [PubMed]
12. Shima, Y.; Miyabayashi, K.; Baba, T.; Otake, H.; Katsura, Y.; Oka, S.; Zubair, M.; Morohashi, K. Identification of an enhancer in the Ad4BP/SF-1 gene specific for fetal Leydig cells. *Endocrinology* **2012**, *153*, 417–425. [CrossRef] [PubMed]
13. Shima, Y.; Miyabayashi, K.; Sato, T.; Suyama, M.; Ohkawa, Y.; Doi, M.; Okamura, H.; Suzuki, K. Fetal Leydig cells dedifferentiate and serve as adult Leydig stem cells. *Development* **2018**, *145*. [CrossRef] [PubMed]
14. Ikeda, Y.; Luo, X.; Abbud, R.; Nilson, J.H.; Parker, K.L. The Nuclear Receptor Steroidogenic Factor 1 Is Essential for the Formation of the Ventromedial Hypothalamic Nucleus. *Mol. Endocrinol.* **1995**, *9*, 478–486. [CrossRef] [PubMed]
15. Fowkes, R.C.; Desclozeaux, M.; Patel, M.V.; Aylwin, S.J.B.; King, P.; Ingraham, H.A.; Burrin, J.M. Steroidogenic Factor-1 and the Gonadotrope-Specific Element Enhance Basal and Pituitary Adenylate Cyclase-Activating Polypeptide-Stimulated Transcription of the Human Glycoprotein Hormone Alpha-Subunit Gene in Gonadotropes. *Mol. Endocrinol.* **2003**, *17*, 2177–2188. [CrossRef]
16. Kaiser, U.B.; Halvorson, L.M.; Chen, M.T. Sp1, Steroidogenic Factor 1 (SF-1), and Early Growth Response Protein 1 (Egr-1) Binding Sites Form a Tripartite Gonadotropin-Releasing Hormone Response Element in the Rat Luteinizing Hormone-Beta Gene Promoter: An Integral Role for SF-1. *Mol. Endocrinol.* **2000**, *14*, 1235–1245. [CrossRef]
17. Pacini, V.; Petit, F.; Querat, B.; Laverriere, J.N.; Cohen-Tannoudji, J.; L'Hôte, D. Identification of a pituitary ERα-activated enhancer triggering the expression of Nr5a1, the earliest gonadotrope lineage-specific transcription factor. *Epigenet Chromatin* **2019**, *12*, 48. [CrossRef]
18. Chen, J.; Minabe, S.; Munetomo, A.; Magata, F.; Sato, M.; Nakamura, S.; Hirabayashi, M.; Ishihara, Y.; Yamazaki, T.; Uenoyama, Y.; et al. Kiss1-Dependent and Independent Release of Luteinizing Hormone and Testosterone in Perinatal Male Rats. *Endocr. J.* **2022**, *69*, 797–807. [CrossRef]
19. Shima, Y. Development of Fetal and Adult Leydig Cells. *Reprod. Med. Biol.* **2019**, *18*, 323–330. [CrossRef]
20. Shima, Y.; Matsuzaki, S.; Miyabayashi, K.; Otake, H.; Baba, T.; Kato, S.; Huhtaniemi, I.; Morohashi, K. Fetal Leydig cells persist as an androgen-independent subpopulation in the postnatal testis. *Mol. Endocrinol.* **2015**, *29*, 1581–1593. [CrossRef]
21. Prince, F.P. The triphasic nature of Leydig cell development in humans, and comments on nomenclature. *J. Endocrinol.* **2001**, *168*, 213–216. [CrossRef] [PubMed]
22. Shima, Y. Functional importance of mini-puberty in spermatogenic stem cell formation. *Front. Cell Dev. Biol.* **2022**, *10*, 907989. [CrossRef] [PubMed]
23. Bjelobaba, I.; Janjic, M.M.; Prévide, R.M.; Abebe, D.; Kucka, M.; Stojilkovic, S.S. Distinct expression patterns of osteopontin and dentin matrix protein 1 genes in pituitary gonadotrophs. *Front. Endocrinol.* **2019**, *10*, 248. [CrossRef] [PubMed]
24. Rydze, R.T.; Patton, B.K.; Briley, S.M.; Salazar Torralba, H.; Gipson, G.; James, R.; Rajkovic, A.; Thompson, T.; Pangas, S.A. deletion of gremlin-2 alters estrous cyclicity and disrupts female fertility in mice. *Biol. Reprod.* **2021**, *105*, 1205–1220. [CrossRef] [PubMed]
25. Brûlé, E.; Wang, Y.; Li, Y.; Lin, Y.-F.; Zhou, X.; Ongaro, L.; Alonso, C.A.I.; Buddle, E.R.S.; Schneyer, A.L.; Byeon, C.-H.; et al. TGFBR3L is an inhibin B co-receptor that regulates female fertility. *Sci. Adv.* **2021**, *7*, eabl4391. [CrossRef]
26. Lin, Y.-F.; Schang, G.; Buddle, E.R.S.; Schultz, H.; Willis, T.L.; Ruf-Zamojski, F.; Zamojski, M.; Mendelev, N.; Boehm, U.; Sealfon, S.C.; et al. Steroidogenic factor 1 regulates transcription of the inhibin B coreceptor in pituitary gonadotrope cells. *Endocrinology* **2022**, *163*, bqac131. [CrossRef] [PubMed]

27. Shin, D.-J.; Wang, L. Bile acid-activated receptors: A review on FXR and other nuclear receptors. *Handb. Exp. Pharmacol.* **2019**, *256*, 51–72. [CrossRef]
28. Ikeda, Y.; Takeda, Y.; Shikayama, T.; Mukai, T.; Hisano, S.; Morohashi, K.I. Comparative localization of Dax-1 and Ad4BP/SF-1 during development of the hypothalamic-pituitary-gonadal axis suggests their closely related and distinct functions. *Dev. Dyn.* **2001**, *220*, 363–376. [CrossRef]
29. Kawabe, K.; Shikayama, T.; Tsuboi, H.; Oka, S.; Oba, K.; Yanase, T.; Nawata, H.; Morohashi, K. Dax-1 as one of the target genes of Ad4BP/SF-1. *Mol. Endocrinol.* **1999**, *13*, 1267–1284. [CrossRef]
30. Yu, R.N.; Ito, M.; Jameson, J.L. The murine Dax-1 promoter is stimulated by SF-1 (steroidogenic factor-1) and inhibited by COUP-TF (chicken ovalbumin upstream promoter-transcription factor) via a composite nuclear receptor-regulatory element. *Mol. Endocrinol.* **1998**, *12*, 1010–1022. [CrossRef]
31. Burris, T.P.; Guo, W.; Le, T.; McCabe, E.R. Identification of a putative steroidogenic factor-1 response element in the DAX-1 promoter. *Biochem. Biophys. Res. Commun.* **1995**, *214*, 576–581. [CrossRef] [PubMed]
32. Jinek, M.; Chylinski, K.; Fonfara, I.; Hauer, M.; Doudna, J.A.; Charpentier, E. A programmable dual-RNA-guided DNA endonuclease in adaptive bacterial immunity. *Science* **2012**, *337*, 816–821. [CrossRef] [PubMed]
33. Miyabayashi, K.; Tokunaga, K.; Otake, H.; Baba, T.; Shima, Y.; Morohashi, K. Heterogeneity of Ovarian Theca and Interstitial Gland Cells in Mice. *PLoS ONE* **2015**, *10*, e0128352. [CrossRef] [PubMed]
34. Ikegami, K.; Goto, T.; Nakamura, S.; Watanabe, Y.; Sugimoto, A.; Majarune, S.; Horihata, K.; Nagae, M.; Tomikawa, J.; Imamura, T.; et al. Conditional Kisspeptin Neuron-Specific Kiss1 Knockout with Newly Generated Kiss1-Floxed and Kiss1-Cre Mice Replicates a Hypogonadal Phenotype of Global Kiss1 Knockout Mice. *J. Reprod. Dev.* **2020**, *66*, 359–367. [CrossRef] [PubMed]
35. Goto, T.; Hirabayashi, M.; Watanabe, Y.; Sanbo, M.; Tomita, K.; Inoue, N.; Tsukamura, H.; Uenoyama, Y. Testosterone supplementation rescues spermatogenesis and in vitro fertilizing ability of sperm in Kiss1 knockout mice. *Endocrinology* **2020**, *161*. [CrossRef]
36. Han, S.; Baba, T.; Yanai, S.; Byun, D.J.; Morohashi, K.-I.; Kim, J.-H.; Choi, M.H. GC-MS-based metabolic signatures reveal comparative steroidogenic pathways between fetal and adult mouse testes. *Andrology* **2021**, *9*, 400–406. [CrossRef]
37. Zhou, Y.; Zhou, B.; Pache, L.; Chang, M.; Khodabakhshi, A.H.; Tanaseichuk, O.; Benner, C.; Chanda, S.K. Metascape provides a biologist-oriented resource for the analysis of systems-level datasets. *Nat. Commun.* **2019**, *10*, 1523. [CrossRef]

International Journal of
Molecular Sciences

MDPI

Article

FSH Regulates YAP-TEAD Transcriptional Activity in Bovine Granulosa Cells to Allow the Future Dominant Follicle to Exert Its Augmented Estrogenic Capacity

Leonardo Guedes de Andrade [1,2], Valério Marques Portela [2], Esdras Corrêa Dos Santos [1], Karine de Vargas Aires [2], Rogério Ferreira [3], Daniele Missio [2], Zigomar da Silva [2], Júlia Koch [2], Alfredo Quites Antoniazzi [2], Paulo Bayard Dias Gonçalves [2,4] and Gustavo Zamberlam [1,*]

[1] Centre de Recherche en Reproduction et Fertilité, Faculté de Médecine Vétérinaire, Université de Montréal, Saint-Hyacinthe, QC J2S 7C6, Canada
[2] Laboratory of Biotechnology and Animal Reproduction (BioRep), Veterinary Hospital, Federal University of Santa Maria (UFSM), Santa Maria 97105-900, Brazil
[3] Department of Animal Science, Santa Catarina State University (UDESC), Chapecó 88035-901, Brazil
[4] Molecular and Integrative Physiology of Reproduction Laboratory (MINT), Federal University of Pampa (Unipampa), Uruguaiana 97501-970, Brazil
* Correspondence: gustavo.zamberlam@umontreal.ca; Tel.: +1-450-773-8521 (ext. 0196)

Citation: de Andrade, L.G.; Portela, V.M.; Dos Santos, E.C.; Aires, K.d.V.; Ferreira, R.; Missio, D.; da Silva, Z.; Koch, J.; Antoniazzi, A.Q.; Gonçalves, P.B.D.; et al. FSH Regulates YAP-TEAD Transcriptional Activity in Bovine Granulosa Cells to Allow the Future Dominant Follicle to Exert Its Augmented Estrogenic Capacity. *Int. J. Mol. Sci.* **2022**, 23, 14160. https://doi.org/10.3390/ijms232214160

Academic Editor: Jacques J. Tremblay

Received: 5 October 2022
Accepted: 14 November 2022
Published: 16 November 2022

Publisher's Note: MDPI stays neutral with regard to jurisdictional claims in published maps and institutional affiliations.

Abstract: The molecular mechanisms that drive the granulosa cells' (GC) differentiation into a more estrogenic phenotype during follicular divergence and establishment of follicle dominance have not been completely elucidated. The main Hippo signaling effector, YAP, has, however, emerged as a potential key player to explain such complex processes. Studies using rat and bovine GC demonstrate that, in conditions where the expression of the classic YAP-TEAD target gene tissue growth factor (*CTGF*) is augmented, *CYP19A1* expression and activity and, consequently, estradiol (E2) secretion are reduced. These findings led us to hypothesize that, during ovarian follicular divergence in cattle, FSH downregulates YAP-TEAD-dependent transcriptional activity in GC to allow the future dominant follicle to exert its augmented estrogenic capacity. To address this, we performed a series of experiments employing distinct bovine models. Our in vitro and ex vivo experiments indicated that indeed FSH downregulates, in a concentration-dependent manner, mRNA levels not only for *CTGF* but also for the other classic YAP-TEAD transcriptional target genes *ANKRD1* and *CYR61* by a mechanism that involves increased YAP phosphorylation. To better elucidate the functional importance of such FSH-induced YAP activity regulation, we then cultured GC in the presence of verteporfin (VP) or peptide 17 (P17), two pharmacological inhibitors known to interfere with YAP binding to TEADs. The results showed that both VP and P17 increased *CYP19A1* basal mRNA levels in a concentration-dependent manner. Most interestingly, by using GC samples obtained in vivo from dominant vs. subordinate follicles, we found that mRNA levels for *CTGF*, *CYR61*, and *ANKRD1* are higher in subordinate follicles following the follicular divergence. Taken together, our novel results demonstrate that YAP transcriptional activity is regulated in bovine granulosa cells to allow the increased estrogenic capacity of the selected dominant follicle.

Keywords: ovary; cow; Hippo; steroidogenesis; follicle deviation or divergence; yes-associated protein 1; CTGF; CYR61; ANKRD1

1. Introduction

Ovarian follicular development and growth in mammals is a complex and dynamic physiological process that requires the interaction of different hormones and cell signaling pathways [1]. In ruminants, as follicle development progresses, follicles gradually become more and more reliant on gonadotropins, first as gonadotropin-responsive follicles and then as gonadotropin-dependent follicles [2]. During the antral growth stage, the most advanced follicles in the pool emerge concomitantly with the increase in circulating follicle-stimulating hormone (FSH) levels to form what is commonly referred to as the cohort of gonadotropin-dependent follicles. During a certain period of this phase, granulosa cells (GC) present similar elevated proliferative capacity, which allows these follicles to grow at an approximately similar rate, until one follicle is selected for further growth [3,4]. In cattle, as in other monovulatory species such as human and equine, this process is known as follicle selection [5,6]. The moment when the selected follicle continues its growth, while the remaining follicles cease growing, is known as follicle deviation or follicle divergence [7].

Probably the most important characteristic of the dominant follicle is its greater capacity for estradiol (E2) production by its GC. After the wave emergence, E2 content in the follicular fluid of the growing dominant follicle increases at least 20-fold by the day of selection [4–8]. Such augmented steroidogenic capacity is mainly due to the fact that GC differentiate to produce more E2. A key steroidogenic enzyme to this process is cytochrome P450 aromatase family 19 subfamily A member 1 (CYP19A1). At the ruminant GC level, this enzyme can metabolize the theca-derived androgen testosterone into E2 and/or the theca-derived androstenedione into estrone (E1), which will then be metabolized into E2 by another steroidogenic enzyme known as 17β-hydroxysteroid dehydrogenase (HSD17B1) [9,10]. In cattle, as well as in other mammalian species, FSH can be considered one of the primary stimulators of GC CYP19A1 expression [11,12]. Despite years of research, the molecular mechanisms that drive the GC differentiation into a more estrogenic profile during follicular divergence have not been completely elucidated. The Hippo signaling has, however, emerged as a potential key player to explain such complex processes.

The core Hippo pathway consists of a kinase cascade that ultimately regulates the activity of the transcriptional activators yes-associated protein 1 (YAP) and transcriptional co-activator with PDZ-binding motif (TAZ). In a conserved manner, it is known that when Hippo signaling is inactive, YAP/TAZ accumulate in the nucleus and form complexes with numerous transcription factors, notably those of the TEA domain transcription factor (TEAD) family of transcription factors, resulting in the modulation of the transcriptional activity of several target genes, such as the classic tissue growth factor (*CTGF*, also known as *CCN2*), ankyrin repeat domain 1 (*ANKRD1*), and cysteine-rich protein 61 (*CYR61*, also known as *CCN1*). Conversely, when Hippo signaling is activated, YAP/TAZ are phosphorylated (at serine residues S127 and S397 for YAP, and at S89 for TAZ), resulting in their nuclear export to the cytoplasm where they will be retained and/or degraded, therefore compromising YAP/TAZ-dependent transcriptional activity [13,14]. A study employing mouse in vivo and in vitro models reported that timely expression and activation of the Hippo effector YAP in GC is critical for ovarian follicle development [15]. Briefly, it was demonstrated that while an increase of YAP-dependent transcriptional activity promotes mouse GC proliferation, it suppresses GC differentiation and steroidogenesis. Although a physiological correlation between FSH and ANKRD1 or CYR61 has not been reported, a study employing rat GC cultures demonstrated that *Ctgf* mRNA downregulation coincides with FSH-induced GC differentiation [16]. Interestingly, a recent study employing a well-defined bovine in vitro GC model clearly showed that when *CTGF* mRNA levels are increased in this cell type, there is a decrease in *CYP19A1* expression and, consequently, a significant reduction in E2 secretion levels [17]. Taken together, these findings led us to hypothesize that, during ovarian follicle divergence in cattle, FSH downregulates YAP-TEAD-dependent transcriptional activity in granulosa cells to allow the future dominant follicle to launch its augmented estrogenic capacity. To address this, we performed a series of experiments employing bovine in vitro, ex vivo, and in vivo models.

2. Results

2.1. FSH Downregulates, in a Concentration-Dependent Manner, mRNA Levels for CTGF and Other Classic YAP-TEAD Transcriptional Target Genes

To determine whether FSH regulates the expression of CTGF in bovine granulosa cells, we employed a non-luteinizing GC culture model in which cells were cultured in the presence of graded doses of FSH for the last four days of culture. This is a completely serum-free, long-term GC culture system that allows the induction/maintenance of CYP19A1 expression and E2 secretion in response to physiological doses of FSH [18,19]. As expected, FSH upregulated *CYP19A1* mRNA levels in a concentration-dependent manner ($p < 0.05$, Figure 1A) and, consequently, stimulated E2 secretion in a concentration-dependent manner, too ($p < 0.05$, Figure 1B).

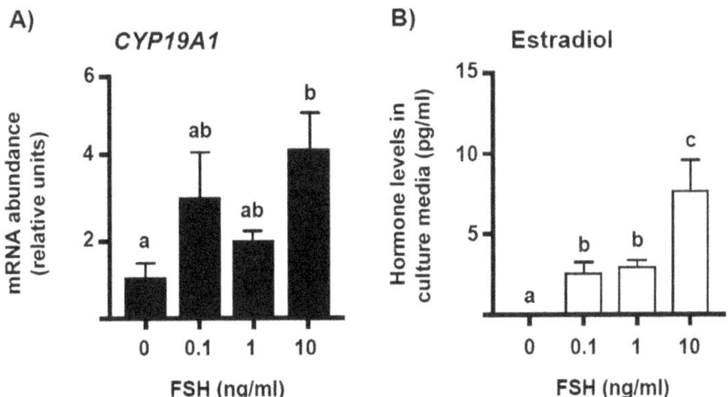

Figure 1. Effect of FSH on *CYP19A1* mRNA abundance and estradiol secretion in vitro. Granulosa cells (GC) were cultured for 6 days under non-luteinizing conditions and treated with graded doses of FSH for the last 4 days of culture (see Section 4 for details). (**A**) Messenger RNA (mRNA) abundance for *CYP19A1* was measured by real-time PCR and normalized to the housekeeping gene *H2AFZ*. (**B**) Estradiol (E2) secretion in culture media was measured by chemiluminescence. Data represent the mean ± SEM for three independent replicate cultures. Bars with different letters are significantly different ($p < 0.05$).

Conversely, FSH downregulated, in a concentration-dependent manner, the mRNA levels not only for *CTGF* ($p < 0.05$, Figure 2A) but also for *ANKRD1* ($p < 0.05$, Figure 2B) and *CYR61* (also known as *CCN1*; $p < 0.05$, Figure 2B), both considered classic YAP-TEAD target genes along with *CTGF*.

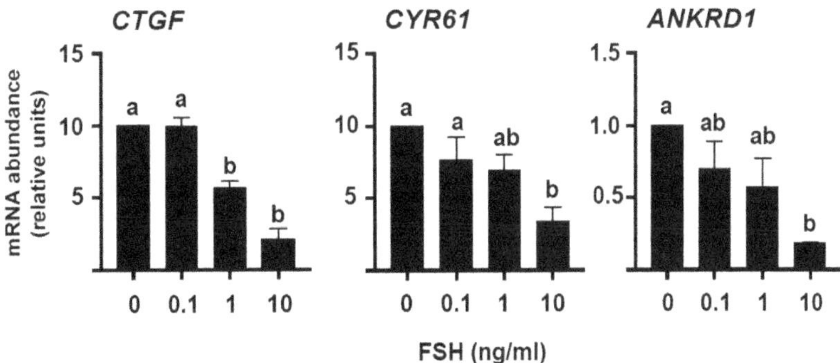

Figure 2. Effect of FSH on mRNA levels for *CTGF* and other classic YAP-TEAD transcriptional target genes in vitro. Granulosa cells (GC) were cultured for 6 days under non-luteinizing conditions and treated with graded doses of FSH for the last 4 days of culture (see Section 4 for details). Messenger RNA (mRNA) abundance for connective tissue growth factor (*CTGF*), ankyrin repeat domain 1 (*ANKRD1*), and cysteine-rich protein 61 (*CYR61*) was measured by real-time PCR and normalized to the housekeeping gene *H2AFZ*. Data represent the mean ± SEM for three independent replicate cultures. Bars with different letters are significantly different ($p < 0.05$).

2.2. YAP Phosphorylation in Granulosa Cells Increases following FSH Challenge and along the Bovine Follicle Growth

To determine if the mechanism by which FSH downregulates *CTGF* and the other YAP-TEAD target genes in bovine GC involves regulation of the YAP phosphorylation status, we then cultured cells in the presence of two doses of FSH (1 and 10 ng/mL) for 4 days, and samples were then collected at the end of the culture for Western blot (WB) analyses. The results indicated that FSH treatments did not alter total YAP protein levels ($p > 0.05$, Figure 3) but significantly promoted YAP phosphorylation on serine 127 (Ser127) in a concentration-dependent manner ($p < 0.05$, Figure 3). The phosphorylation of this serine is known to promote a binding site for 14-3-3 protein, leading to YAP-14-3-3 complex formation in the cytoplasm, and therefore affects the binding of this Hippo effector to its target transcription factors at the nucleus [20].

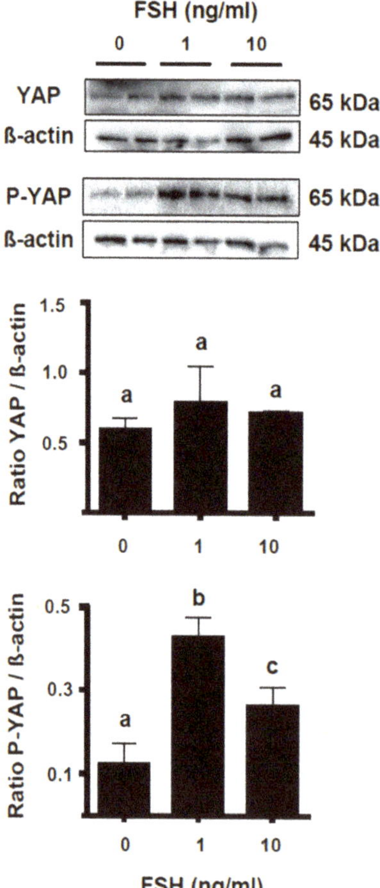

Figure 3. YAP phosphorylation is regulated by FSH in vitro. Granulosa cells (GC) were cultured for 6 days under non-luteinizing conditions and treated with 2 distinct doses of FSH for the last 4 days of culture (see Section 4 for details). Total (YAP) and phosphorylated YAP on serine 127 (P-YAP) protein levels were measured by Western blot and normalized to β-actin, as shown in representative blots (n = 2 replicates). Data are means ± SEM of four independent cultures. Bars with different letters are significantly different ($p < 0.05$).

To determine if the above-observed in vitro FSH-induced YAP phosphorylation pattern could be somehow observed in GC along the follicle growth, we then collected bovine ovarian follicles of different sizes (compatible with those found in emergence to dominance) for evaluation. The intensity of staining observed suggested that the total YAP expression pattern is similar in GC collected from small (<5 mm) and medium (5–10 mm) follicles (Figure 4A,B). However, while the positive signal for phospho-YAP (Ser127) was absent or barely detected in GC from small follicles (Figure 4A), the staining was easily observed in the cytoplasm of medium follicles (Figure 4B).

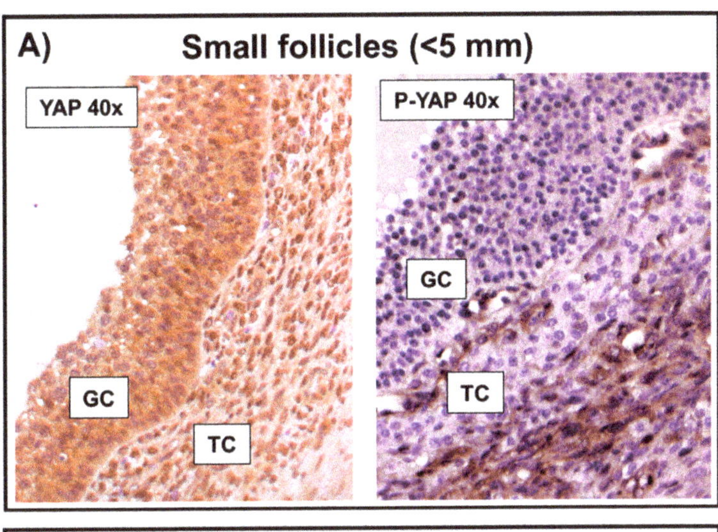

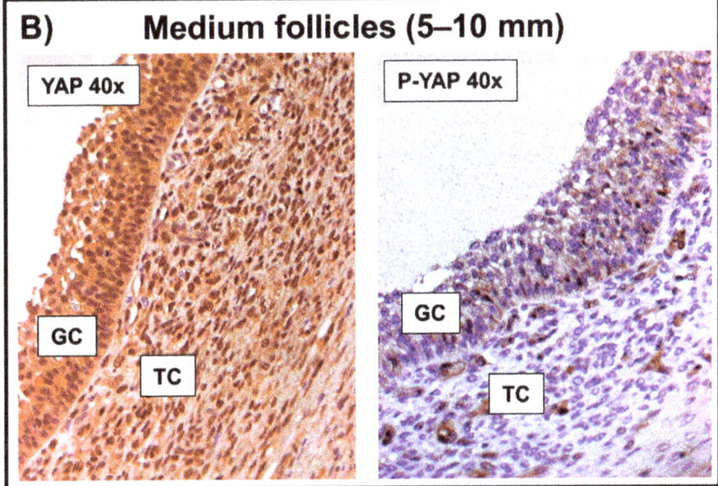

Figure 4. Localization and expression pattern of total (YAP) and phospho-YAP (Ser127) in granulosa cells from follicles of increasing size. Representative immunohistochemistry (IHC) micrographs (objective 40x) show the expression pattern of total (YAP) and phosphorylated YAP on serine 127 (P-YAP) in granulosa (GC) and theca cells (TC) from follicles of (**A**) small (<5 mm) and (**B**) medium (5–10 mm) sizes. While brown color represents a positive immunostaining signal for total and/or P-YAP (detected in nucleus and/or cytoplasm of GC), the counterstain is hematoxylin, which stains the cell nuclei in blue, contrasting with the brown.

2.3. The mRNA Abundance of CTGF, ANKRD1, and CYR61 Is Higher in Subordinate Follicles following the Follicular Divergence

The previously described in vitro and ex vivo findings suggested that FSH may promote YAP phosphorylation to decrease the expression of CTGF and other classic YAP-TEAD target genes to allow GC from growing follicles to better express their estrogenic capacity and, consequently, establish their dominance over subordinate follicles in the same cohort. To confirm that, an in vivo experiment was then performed to obtain the largest (F1: herein also referred to as dominant follicle) vs. the second largest (F2: herein also referred to as subordinate follicle) GC samples from ovaries collected at days 2 (D2), 3 (D3), and 4 (D4) of the first follicular wave (collection time points correspond to the days before, during, and after ovarian follicular follicle deviation in bovine, respectively). Although mRNA levels for *CTGF*, *CYR61*, and *ANKRD1* were not significantly higher in GC of subordinate follicles collected at D2 and D3 ($p > 0.05$, Figure 5), mRNA levels for those 3 genes were significantly higher in GC from subordinate follicles collected at D4 ($p < 0.05$, Figure 5). The latter day corresponds to the first day after divergence when the dominant follicle has established its dominance and presents much higher mRNA levels for *CYP19A1* in GC, and consequently, higher intra-follicular E2 levels, as it can be confirmed in a previous publication [8].

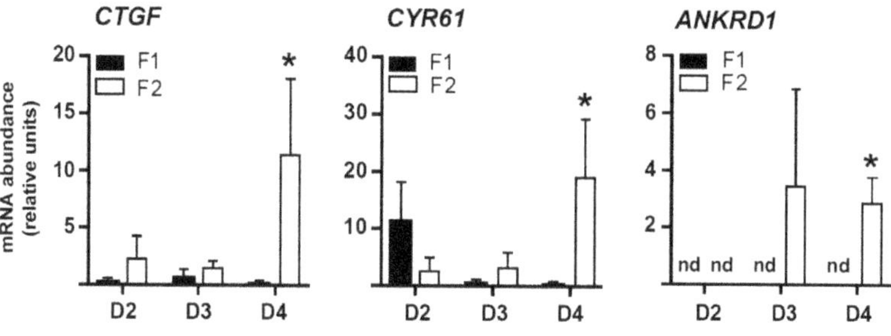

Figure 5. Expression of classic YAP-TEAD transcriptional target genes in granulosa cells during establishment of the dominant follicle in vivo. Granulosa cells (GC) were recovered from the largest (F1—dominant follicles, black bar) and the second largest (F2—subordinate follicles, white bar) follicles collected at days 2 (D2), 3 (D3), and 4 (D4) of the first follicular wave of the synchronized estrous cycle. D2 corresponds to the day before the divergence, D3 corresponds to the day of follicular divergence, and D4 corresponds to the first day after divergence. Messenger RNA abundance for connective tissue growth factor (*CTGF*), ankyrin repeat domain 1 (*ANKRD1*), and cysteine-rich protein 61 (*CYR61*) was measured by real-time PCR and normalized to the housekeeping gene *GAPDH*. Data represent the mean ± SEM of independent follicle samples (n = 4) per group in each time point. An asterisk (*) indicates significant difference between F2 (subordinate) and F1 (dominant) follicle groups over day-matched comparison ($p < 0.05$) and "nd" denotes non-detectable amplification in the real-time PCR analysis.

2.4. Pharmacological Inhibition of YAP-TEAD Interaction In Vitro Increases Basal Levels for mRNA Encoding CYP19A1

Some of the experiments previously performed in the present study clearly confirmed that, in bovine GC, there is a clear inverse relationship not only between the expression levels of *CYP19A1* and *CTGF*, but also an inverse relationship between *CYP19A1* and other YAP-TEAD classic target genes. To better elucidate the nature of such relationship (cause vs. consequence), we decided to perform a series of in vitro experiments using pharmacological inhibitors known to interfere with YAP binding to TEAD family transcription factors. In the first series of cultures, GC were cultured without or in the presence of different concentrations of verteporfin (VP), a well-known and commonly used YAP-TEAD inhibitor molecule that decreases basal levels of *CTGF*, *CYR61*, and *ANKRD1* in a concentration-dependent manner [21,22]. The results shown herein indicated that VP increased *CYP19A1* basal mRNA levels in a concentration-dependent manner ($p < 0.05$, Figure 6A).

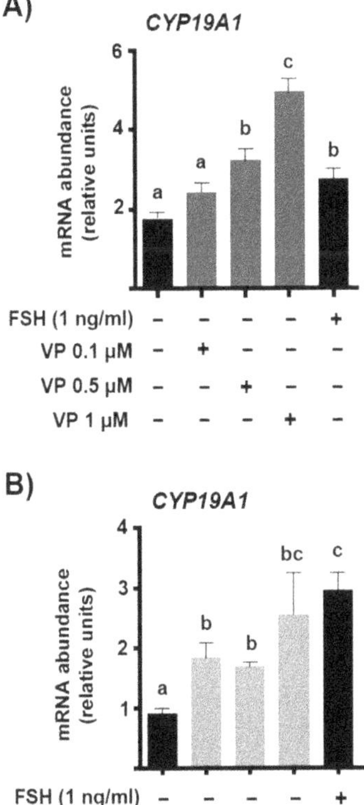

Figure 6. Pharmacological inhibition of YAP-TEAD interaction increases basal *CYP19A1* mRNA abundance in vitro. Granulosa cells (GC) were cultured for 6 days under non-luteinizing conditions (see Section 4 for details) and treated (for the last 4 days of culture) with different concentrations of (**A**) Verteporfin (VP: 0, 0.1, 0.5, and 1 µM) or (**B**) Peptide 17 (P17: 0, 1, 5, and 10 µM), molecules known to interfere with YAP binding to the TEAD family of transcription factors. A FSH treatment group (1 ng/mL) was included from day 2 to 6 in both experiments as a positive control for *CYP19A1* stimulation. Messenger RNA abundance for *CYP19A1* was measured by real-time PCR and normalized to the housekeeping gene *H2AFZ*. Data are means ± SEM of three independent cultures. Bars with different letters are significantly different ($p < 0.05$).

To further investigate whether VP action was specific or not, we then decided to use an alternative inhibitor, peptide 17 (P17), which is an engineered peptide that also disrupts the YAP-TEAD interaction [23,24]. Similar to what was observed for VP, *CYP19A1* basal mRNA levels were also increased in a concentration-dependent manner following P17 treatment ($p < 0.05$, Figure 6B). Together, these findings clearly indicate that CTGF and/or other YAP-TEAD target genes exert a direct or indirect inhibitory effect of *CYP19A1* transcriptional regulation in bovine GC.

3. Discussion

The establishment of ovarian follicle dominance in monovulatory species involves complex and dynamic processes. Despite years of research, some key molecular mechanisms that drive the GC differentiation into a more estrogenic phenotype during follicular divergence remain unclear. In the present study, we used bovine in vitro, ex vivo, and in vivo models to generate novel and exciting data, showing an important role of the Hippo effector YAP during the establishment of follicular dominance. Together, our data indicate that YAP transcriptional activity is downregulated in bovine granulosa cells by FSH to allow or facilitate the increased estrogenic capacity of the selected dominant follicle.

In terms of the expression pattern for the Hippo effector YAP in bovine ovarian follicles, a recent study in this species showed via WB analysis that total YAP expression levels were similar in GC isolated from all stages of follicle development (2–5, 5–10, >10 mm) [25]. These authors, however, did not include phospho-YAP (Ser127) in such WB comparison nor did they evaluate the effect of FSH or any other growth factor on total and phospho-YAP protein levels. Although their WB results for total YAP in GC isolated from different follicle sizes corroborate the total YAP expression stability pattern observed in our IHC analysis and following our FSH treatment in vitro, our study clearly contributes with novel data for this and other monovulatory species, particularly related to FSH-induced YAP phosphorylation aspects and its consequence to YAP-TEAD-related transcriptional activity during the establishment of follicle dominance.

FSH action in GC occurs mainly through the adenylyl cyclase (cAMP) pathway [11,26]. Forskolin is a pharmacological agonist of adenylyl cyclase which is widely used to mimic FSH in activating differentiation signaling in GC [27]. Treatment of mouse GC with forskolin induced phosphorylation of YAP protein at serine 127 faster than forskolin increased *Cyp19a1* mRNA abundance in this cell type [15]. In addition, these same authors employed a human ovarian granulosa cell-like tumor cell line (KGN) to show that constitutively transcriptional active YAP (YAPS127A) significantly suppresses E2 production by these cells. These authors, nevertheless, did not assess (in neither their mouse models nor in their human models) any classic YAP-TEAD target genes, nor did they discuss in the latter experiment whether the expression of CTGF, CYR61, or ANKRD1 could be directly or indirectly related to *CYP19A1* transcriptional regulation in GC.

Although a previous study in rat GC has reported that FSH downregulates *Ctgf* in the same conditions that it stimulates E2 secretion [16], such study only showed a negative correlation and never confirmed if CTGF by itself can indeed alter *Cyp19a1* transcription in this cell type. Similarly, a study in bovine GC [17] used the same cell culture system that we employed in the present study to also observe such inverted correlation between CTGF expression and *CYP19A1*/E2 levels in bovine species. Nevertheless, no functional experiment was performed by these authors. To the best of our knowledge, our study shows the first evidence that disrupting the YAP–TEAD interaction, and consequently affecting CTGF expression, leads to basal *CYP19A1* mRNA levels' augmentation in GC. Such pharmacological inhibition, however, does not alter only *CTGF*, but also affects the basal levels of the other classic YAP-TEAD target genes *ANKRD1* and *CYR61*. Studies showing the physiological roles exerted by these two proteins in bovine ovary are, nevertheless, scarce. While the expression of *ANKRD1* in bovine GC has been associated with a period of decreasing oocyte competence and ANKRD1 was pointed out as a gene that can lead to apoptosis and atresia [28], CYR61 is known to be expressed in bovine granulosa-derived luteal cells and it has been identified as a potential molecular mediator of angiogenesis in the CL [29]. Interestingly, our in vivo experiment showed that indeed not only *CTGF* but also *CYR61* and *ANKRD1* are significantly higher in the largest subordinate follicles at the day after ovarian follicular deviation in bovine was established. Together, our in vivo findings and our in vitro experiments, using pharmacological YAP-TEAD inhibitors and challenging GC with FSH doses, suggest that not only CTGF but other YAP-TEAD-related genes might be involved not only with the transcriptional machinery responsible for *CYP19A1* regulation in GC, but also with the GC differentiation process required for it. The precise effects and respective mechanisms of action of CTGF, CYR61, and ANKRD1 in bovine GC, nevertheless, remain to be further investigated.

Even though our results employing distinct bovine models complement each other's findings, a puzzling question also remains to be better addressed: does the expression of these YAP-TEAD target genes decrease in GC from the selected dominant follicle, or is the expression of those genes augmented in GC from subordinate follicles during the divergence process? Based on the expression pattern for these genes in dominant vs. subordinate follicles collected at each of the days tested herein (collection time points correspondent to the days before, during, and after ovarian follicular follicle deviation in bovine, respectively), it is most likely that the mRNA levels for these YAP-TEAD targets started increasing in subordinate follicles during the divergence to then be significantly augmented after ovarian follicular deviation was established. This possibility is supported by the fact that fibroblast growth factor 2 (FGF2), a FGF related to bovine follicle atresia, augments *CTGF* mRNA abundance in a dose- and time-dependent manner in bovine GC cultured in vitro [17]. Curiously, FGF2 is known for inhibiting steroidogenesis in bovine GC by suppressing *CYP19A1* expression [30,31]. On the other hand, taking into consideration the facts that FSH increases YAP phosphorylation in bovine GC (data shown herein) and that the future dominant follicle is known for being more responsive to FSH as its circulating levels increase [32,33], it is plausible to suggest that YAP-dependent transcriptional activity is inhibited or, at least, transiently controlled in the selected dominant follicle until it establishes its estrogenic dominance over other follicles from the same cohort. To better understand such puzzle, nevertheless, it is important to take into consideration which main functions are normally attributed to Hippo effectors along the follicle development dynamics in mammals, particularly related to early stages of follicle development.

In murine models, it was demonstrated that induced ovarian fragmentation promotes follicle growth, which is related to decreased phospho-YAP levels, increased nuclear localization of YAP, and consequently, enhanced expression of CTGF [34]. Briefly, ovaries from juvenile mice (containing secondary and smaller follicles) were cut in 3–4 fragments and then allo-transplanted under kidney capsules of adult hosts. Histological analyses and follicle counting of grafts indicated an augmentation in the percentage of late secondary and antral/preovulatory follicles accompanied by decreases in primordial follicles. In addition, these authors also demonstrated that such fragmentation-induced follicle growth was partially blocked by CTGF antibodies or by verteporfin, and that CTGF and CYR61 recombinant proteins promoted the development of primary follicles to the late secondary/antral stage in ovarian explants. In both circumstances, however, the authors attributed a key role to the fact that CCN growth factors (CTGF and CYR61) can promote GC proliferation. Indeed, another study in mice demonstrated that stimulation of YAP-dependent transcriptional activity promotes mouse GC proliferation, however such induction consequently suppresses GC differentiation and steroidogenesis [15]. Curiously, a recent study in mice showed that YAP-induced transcriptional activity in large antral follicles is essential for LH-induced ovulatory cascade [35]. Taken together, these findings in murine models strongly indicate that the expression and activation of the Hippo effector YAP in murine GC may vary along the follicle development/growth to exert timely, distinct, required physiological functions. Based on that, we then hypothesized that in rodents, YAP target genes contribute to the initial follicle growth (involving high GC proliferation rates) until the follicles become gonadotropin-dependent, and therefore, require transitory YAP nuclear export for the final maturation/differentiation of the follicle until the return of this Hippo effector to the nucleus, where its transcriptional activity is critical for ovulation. Interestingly, it seems that the same hypothesis proposed for rodents can also be proposed for monovulatory species, particularly for the model employed herein, bovine.

In a recent study by our research group, we demonstrated by in vitro and in vivo approaches that YAP transcriptional activity in preovulatory bovine GC is critical for the LH-induced ovulation in this species [36]. These findings indicate that, in large bovine dominant preovulatory follicles ($\geq$12 mm), YAP must remain unphosphorylated and transcriptionally active in the nucleus to allow LH-induced pre-ovulatory signaling. The present study, nevertheless, shows strong evidence that during the follicle divergence and establishment of dominance, FSH increases YAP phosphorylation to allow the future dominant follicle to increase or accelerate its GC estrogenic capacity. Once the dominance is established, it is most likely that the YAP phosphorylation status returns to basal levels in the periovulatory period. To confirm this latter hypothesis, an ongoing investigation of our group is assessing YAP-TEAD target genes' expression patterns at later time points along the follicle wave in vivo and, most importantly, we are also evaluating the effects of insulin-like growth factor 1 (IGF1) on YAP activity in bovine GC in vitro. One of the reasons by which the selected dominant follicle continues its growth is directly related to the IGF system. IGF1 increases the sensitivity of small follicles (around 5 mm in cattle) to gonadotropins and simulates their transition from the gonadotropin-responsive to the gonadotropin-dependent stages [37]. This growth factor not only induces E2 secretion in GC, but also synergizes with FSH to promote final differentiation of GC until the luteinizing hormone (LH) surge, which is required for ovulation of the mature dominant follicle [38].

In summary, we provided novel evidence that YAP-TEAD-related transcriptional activity plays an important role in the molecular mechanisms that drive the GC differentiation into a more estrogenic profile during follicular divergence and the establishment of follicular dominance. By regulating YAP activity in bovine granulosa cells, FSH alters the expression of *CTGF* and other classic YAP-TEAD target genes and contributes to the augmented estrogenic capacity of the selected dominant follicle.

4. Material and Methods

4.1. In Vitro Studies

The reagents used for in vitro cultures were obtained from Thermo Fisher Scientific, except where otherwise stated. The granulosa cell (GC) culture employed herein is a completely serum-free, long-term GC culture system, also described as a GC differentiation culture system [18,31]. In such conditions, GC are responsive to FSH and maintain an estrogenic phenotype with a minimum of luteinization along the culture [39,40]. Briefly, bovine ovaries were collected in local abattoirs from random adult cows and were transported to the laboratory in PBS containing penicillin (100 IU/mL) and streptomycin (100 µg/mL). Follicles between 2 and 5 mm in diameter were dissected from the ovarian stroma and sectioned in Dulbecco's Modified Eagle Medium Nutrient Mixture F-12 (DMEM/F12). GC were then collected by rinsing the follicle walls with DMEM/F12, washed twice by centrifugation at $980 \times g$ for 20 min each, and filtered through a Cell Dissociation Sieve—Tissue Grinder Kit/150 Mesh (Sigma-Aldrich, Oakville, ON, Canada). Finally, GC were suspended in basal culture media composed by DMEM/F12 supplemented with sodium bicarbonate (10 mM), sodium selenite (4 ng/mL), BSA (1 mg/mL), penicillin (100 IU/mL), streptomycin (100 µg/mL), human transferrin (5 ng/mL), non-essential amino acid mix (10 mM), androstenedione (A4; 10^{-7} M at start of culture, and 10^{-6} Mat each medium change), and insulin (10 ng/mL). The number of cells was counted with a hemocytometer and the viable cells were assessed by the dye exclusion method using 0.4% Trypan Blue. For cultures, GC were seeded into 24-well tissue culture plates (Sarstedt Inc., St-Leonard, QC, Canada) at a density of 1×10^6 viable cells per well in 1 mL of medium. Cultures were maintained at 37 °C in 5% CO_2 in air for 6 days with 70% (700 µL) medium being replaced every 2 days and treatments added from day 2 on (on days 2 and 4 of culture). Although insulin (10 ng/mL) was added since day 0 and at each medium change (day 2 and day 4), depending on the experiment, cells were also treated for the last 4 days of culture with human FSH (1 or 10 ng/mL) or with distinct concentrations of the pharmacological inhibitors Verteporfin (VP; Sigma-Aldrich) or Peptide 17 (P17; Selleck Chemicals, Houston, TX, USA). Medium samples were collected on day 6 and stored at −20 °C until the steroid assay, and cells were collected on day 6 in Trizol or M-PER® mammalian protein extraction reagent and stored at −80 °C until RNA or protein extraction, respectively. All series of cultures were performed on at least three different pools of cells collected on different occasions.

4.1.1. Steroid Assay

Estradiol (E2) was measured from culture media samples collected on day 6 of culture. The concentration was determined by a chemiluminescence kit (ADVIA Centaur, Siemens, Munich, Germany) in a specialized clinical analysis laboratory following the manufacturer's recommendations.

4.1.2. Western Blotting

Total protein from GC was extracted using M-PER® mammalian protein extraction reagent according to the manufacturer's instructions and protein levels were quantified using the Pierce™ BCA Protein Assay Kit. Halt™ Protease and Phosphatase Inhibitor Cocktails were added to the samples' final solutions to avoid protein degradation. Samples (20–40 µg) were resolved on 12% sodium dodecyl sulfate-polyacrylamide gels and transferred to Hybond-P PVDF membrane (GE Amersham, Mississauga, ON, Canada). Membranes were then probed at 4 °C overnight in 5% BSA in TTBS with different primary antibodies (details and dilutions for each antibody are indicated in Table 1). After washing three times with TTBS, membranes were incubated for 1 h at room temperature with anti-rabbit HRP-conjugated IgG diluted in 5% non-fat dry milk in TTBS. Protein bands were visualized by chemiluminescence (ECL; Millipore, Billerica, MA, USA) and quantified using a ChemiDoc MP detection system (Bio-Rad, Hercules, CA, USA) and Image Lab™ software.

Table 1. List of antibodies used for IHC and WB.

Name of Antibody	Manufacturer (Cat. No.)	Type	Dilution WB	Dilution IHC
ß-actin (C4)	Santa Cruz (sc-47778 HRP)	CkM	1:10,000	
YAP (D8H1X)	Cell signaling (14074)	RbM	1:1000	1:250
Phospho-YAP (Ser127) (D9W2I)	Cell signaling (13008)	RbM	1:1000	1:250
Anti-Rabbit IgG–HRP Conjugate	Promega (W401B)	Rb	1:1000	

CkM: Chicken monoclonal; RbM: rabbit monoclonal; Rb: rabbit.

4.2. Ex Vivo Study

The ex vivo study used reagents obtained from Thermo Fisher Scientific (Saint-Laurent, QC, Canada), except where otherwise stated.

4.2.1. Tissue Sampling

Bovine ovaries were collected on different days from random adult cows at a local abattoir and were transported to the laboratory in PBS at 35 °C containing penicillin (100 IU/mL), streptomycin (100 µg/mL), and fungizone (1 µg/mL). At least five ovaries from different animals that each contained small (<5 mm) and medium (5–10 mm) follicles concomitantly (compatible with those found in emergence to follicle dominance) were selected for further analysis.

4.2.2. Immunohistochemistry

For immunohistochemistry (IHC) evaluation, bovine ovaries were selected as described above. Entire ovaries were then fixed in 10% formaldehyde solution for 24 h, rinsed, and dehydrated in alcohol until they were embedded in paraffin. Serial sections were prepared (at a thickness of 3 µm), followed by deparaffinization, rehydration, sodium citrate heat-mediated antigen retrieval, peroxidase block, and protein blocking (10% goat for 30 min), and then slides were probed with primary antibody against total and phosphorylated forms of YAP (Table 1) overnight at 4 °C. Protein detection was then performed with the Vectastain Elite ABC HRP Kit (VECTPK6101, Vector Laboratories, Burlingame, CA, USA) and stained with the DAB substrate kit (VECTSK4100, Vector Laboratories). Slides were then counterstained with hematoxylin and dehydrated with graded alcohols prior to mounting. Negative controls were included in the IHC analysis and consisted of slides for which the primary antibodies (for both total and phosphorylated YAP) were omitted.

The results confirmed the specificity of our second antibody (not shown). Photomicrographs were taken using a Carl Zeiss Axio Imager M1 microscope (Carl Zeiss, Toronto, ON, Canada) at ×1000 magnification and using the Zen 2012 blue edition software (Carl Zeiss).

4.3. In Vivo Study

The reagents used for the in vivo experiment were obtained from Sigma-Aldrich Co., except where otherwise stated. All experimental procedures using cattle were reviewed and approved by the Federal University of Santa Maria Animal Care and Use Committee (ACUC No. 23081.009594/2007-41). To obtain GC from the largest (F1) and the second largest (F2) follicles (also referred to herein as dominant and subordinate follicles, respectively), ovaries were collected from the first follicular growth wave of the estrous cycle. For this, thirty-six weaned beef cows (predominantly Hereford and Aberdeen Angus) were injected with two doses of PGF2α analogue (Cloprostenol, 125 µg; Schering-Plough Animal Health, Kenilworth, NJ, USA) intramuscularly (i.m.), 12 h apart. They were then observed in estrus within 3–5 days after PGF2α. Ovaries were then examined once a day by transrectal ultrasonography, using an 8 MHz linear-array transducer (Aquila Vet scanner, Pie Medicals, Maastricht, The Netherlands), and all follicles larger than 5 mm were drafted using three to five virtual slices of the ovary, allowing a three-dimensional localization of follicles and monitoring individual ovarian follicles' location during the follicular wave [3]. The day of the follicular emergence was designated as day 0 (D0) of the wave and it was retrospectively identified as the last day on which the dominant follicle was 4 or 5 mm in diameter [8]. The cows were then randomly assigned to be ovariectomized by colpotomy at days 2 (D2), 3 (D3), or 4 (D4) of the follicular wave (four cows per group for each day) to recover the largest (F1: herein also referred to as dominant follicle) and the second largest (F2: herein also referred to as subordinate follicle) follicles from each cow. After ovariectomy, GC were recovered from F1 and F2 follicles and stored at –80 °C until RNA extraction for RT-qPCR analysis.

4.4. RNA Extraction, Reverse Transcription, and Quantitative PCR (qPCR) for In Vitro and In Vivo Studies

Total RNA from in vitro culture samples was extracted using the PureLink™ RNA Mini Kit according to the manufacturer's instructions. Total RNA from the in vivo samples was extracted using the silica column-based protocol (Qiagen, Mississauga, ON, Canada) according to the manufacturer's instructions. For reverse transcription reaction (RT), total RNA (0.2 µg from both in vitro and in vivo samples) was first treated with 1U DNase (Promega, Madison, WI, USA) at 37 °C for 5 min to digest any contaminating DNA. The RNA was then reverse-transcribed in the presence of 1 mM of oligo (dT) primer and 4U Omniscript Rtase (Qiagen), 0.25 mM of dideoxy-nucleotide triphosphate (dNTP) mix, and 19.33U RNase Inhibitor (GE Healthcare, Chicago, IL, USA) in a volume of 20 µL at 37 °C for 1 h. The reaction was terminated by incubation at 93 °C for 3 min. Real-time PCR was conducted in an ABI Prism 7300 instrument in a 25 µL reaction volume containing 12.5 µL of 2 × Power SYBR Green PCR Master Mix (Applied Biosystems, Waltham, MA, USA), 9.5 µL of water, and 1 µL of each sample cDNA and bovine-specific primers (Table 2). Cycling conditions were 3 min at 95 °C, followed by 40 cycles of 15 s at 95 °C, 30 s at 60 °C, and 30 s at 72 °C. In each run, melting curve analysis was used to verify that a single product was amplified. Each reaction was performed in duplicate, and the average threshold cycle (Ct) value was used to calculate relative mRNA abundance of target genes relative to the housekeeping genes *H2AFZ* (for in vitro samples) and *GAPDH* (for in vivo samples) and with the 2−ΔΔCt method and correction for amplification efficiency [41]. Primers not published previously were designed based on sequences from GenBank, using the Primer-BLAST platform, and their respective amplicons were sequenced to confirm their specificity.

Table 2. Sequences of primers used in the expression analysis of target genes.

Gene	Sequence 5′→3′	Accession Number
ANKRD1	F: ATCAGTGCGCGGGATAAGTT R: GGGAGTATCTCCTTCCCGGT	NM_001034378.2
CTGF	F: AGCTGAGCGAGTTGTGTACC R: TCCGAAAATGTAGGGGGCAC	[42]
CYP19A1	F: CTGAAGCAACAGGAGTCCTAAATGTACA R: AATGAGGGGCCCAATTCCCAGA	[43]
CYR61	F: GGCTCCCCGTTTTGGAATG R: TCATTGGTAACGCGTGTGGA	NM_001034340.2
GAPDH	F: GATTGTCAGCAATGCCTCCT R: CGTTCTCTGCCTTGACTGTG	[36]
H2AFZ	F: GAGGAGCTGAACAAGCTGTTG R: TTGTGGTGGCTCTCAGTCTTC	[43]

Forward (F) and reverse (R) primers used in RT-qPCR.

4.5. Statistical Analysis

The statistical analyses for all experiments were performed using JMP Software (SAS Institute Inc., Cary, NC, USA). Data that were not normally distributed (Shapiro–Wilk test) were transformed to natural logarithms. For mRNA abundance or target protein levels, ANOVA was used to test for the main effect (treatment) and culture replicate was included as a random effect. Multiple comparisons were tested using the Tukey–Kramer honestly significant difference (HSD) test to compare all treatment groups within the same experiment. All data were presented as means ± SEM and variables were considered statistically significant at $p < 0.05$, represented with different letters. For the in vivo experiment, the day-match differences in continuous data between the dominant (F1) and the subordinate (F2) were assessed by a paired Student's t test using the cow as the subject. The in vivo data were presented as means ± SEM and variables were considered statistically significant at $p < 0.05$, represented with an asterisk symbol (*).

Author Contributions: L.G.d.A., V.M.P., P.B.D.G. and G.Z. were involved in conceptualization and design of the study; L.G.d.A., V.M.P., E.C.D.S., K.d.V.A., R.F., D.M., Z.d.S., J.K. and A.Q.A. performed experiments and/or were involved in the acquisition, analyses, and interpretation of data; A.Q.A., V.M.P., P.B.D.G. and G.Z. contributed with resources and/or for funding acquisition; L.G.d.A. and G.Z. wrote the original draft; L.G.d.A., V.M.P., P.B.D.G. and G.Z. reviewed and edited the manuscript for publication. All authors have read and agreed to the published version of the manuscript.

Funding: This work was supported by the Natural Sciences and Engineering Research Council of Canada (NSERC), Discovery Grant RGPIN-2018-06470 (to Zamberlam), and by the Grant 19/2551-0002275-1 from Rio Grande do Sul State Research Support Foundation (FAPERGS), Brazil (to Dr. Gonçalves). Mr. Andrade's graduate research program was supported by a scholarship from Coordination for the Improvement of Higher Education Personnel (CAPES; Brazil).

Institutional Review Board Statement: All in vivo experimental procedures using cattle were reviewed and approved by the Federal University of Santa Maria Animal Care and Use Committee (ACUC No. 23081.009594/2007-41). Ex vivo and in vitro experiments used abattoir-derived bovine ovaries.

Informed Consent Statement: Not applicable.

Data Availability Statement: Not applicable.

Acknowledgments: The authors would like to thank Meggie Girard (UdeM) and Fernando Mesquita (UNIPAMPA) for technical support and Frigorífico Silva (RS-Brazil) for the ovaries donation.

Conflicts of Interest: The authors declare no conflict of interest.

References

1. Gupta, P.S.P.; Folger, J.K.; Rajput, S.K.; Lv, L.; Yao, J.; Ireland, J.J.; Smith, G.W. Regulation and Regulatory Role of WNT Signaling in Potentiating FSH Action during Bovine Dominant Follicle Selection. *PLoS ONE* **2014**, *9*, e0100201. [CrossRef] [PubMed]
2. Scaramuzzi, R.J.; Baird, D.T.; Campbell, B.K.; Driancourt, M.-A.; Dupont, J.; Fortune, J.E.; Gilchrist, R.B.; Martin, G.B.; McNatty, K.P.; McNeilly, A.S.; et al. Regulation of Folliculogenesis and the Determination of Ovulation Rate in Ruminants. *Reprod. Fertil. Dev.* **2011**, *23*, 444. [CrossRef] [PubMed]
3. Jaiswal, R.S.; Singh, J.; Adams, G.P. Developmental Pattern of Small Antral Follicles in the Bovine Ovary. *Biol. Reprod.* **2004**, *71*, 1244–1251. [CrossRef] [PubMed]
4. Adams, G.P.; Jaiswal, R.; Singh, J.; Malhi, P. Progress in Understanding Ovarian Follicular Dynamics in Cattle. *Theriogenology* **2008**, *69*, 72–80. [CrossRef]
5. Evans, A.C.O.; Fortune, J.E. Selection of the Dominant Follicle in Cattle Occurs in the Absence of Differences in the Expression of Messenger Ribonucleic Acid for Gonadotropin Receptors. *Endocrinology* **1997**, *138*, 2963–2971. [CrossRef]
6. Beg, M.A.; Bergfelt, D.R.; Kot, K.; Ginther, O.J. Follicle Selection in Cattle: Dynamics of Follicular Fluid Factors during Development of Follicle Dominance 1. *Biol. Reprod.* **2002**, *126*, 120–126. [CrossRef]
7. Ginther, O.J.; Kot, K.; Kulick, L.J.; Wiltbank, M.C. Emergence and Deviation of Follicles during the Development of Follicular Waves in Cattle. *Theriogenology* **1997**, *48*, 75–87. [CrossRef]
8. Ferreira, R.; Gasperin, B.; Santos, J.; Rovani, M.; Santos, R.A.; Gutierrez, K.; Oliveira, J.F.; Reis, A.M.; Gonçalves, P.B. Angiotensin II Profile and MRNA Encoding RAS Proteins during Bovine Follicular Wave. *JRAAS—J. Renin-Angiotensin-Aldosterone Syst.* **2011**, *12*, 475–482. [CrossRef]
9. Conley, A.J.; Bird, I.M. The Role of Cytochrome P450 17α-Hydroxylase and 3β-Hydroxysteroid Dehydrogenase in the Integration of Gonadal and Adrenal Steroidogenesis via the Δ5 and Δ4 Pathways of Steroidogenesis in Mammals. *Biol. Reprod.* **1997**, *56*, 789–799. [CrossRef]
10. Lapointe, E.; Boerboom, D. WNT Signaling and the Regulation of Ovarian Steroidogenesis. *Front. Biosci.—Sch.* **2011**, *3*, 276–285. [CrossRef]
11. Silva, J.M.; Price, C.A. Effect of Follicle-Stimulating Hormone on Steroid Secretion and Messenger Ribonucleic Acids Encoding Cytochromes P450 Aromatase and Cholesterol Side-Chain Cleavage in Bovine Granulosa Cells In Vitro. *Biol. Reprod.* **2000**, *62*, 186–191. [CrossRef] [PubMed]
12. Silva, J.M.; Hamel, M.; Sahmi, M.; Price, C.A. Control of Oestradiol Secretion and of Cytochrome P450 Aromatase Messenger Ribonucleic Acid Accumulation by FSH Involves Different Intracellular Pathways in Oestrogenic Bovine Granulosa Cells in Vitro. *Reproduction* **2006**, *132*, 909–917. [CrossRef] [PubMed]
13. Halder, G.; Johnson, R.L. Hippo Signaling: Growth Control and Beyond. *Development* **2011**, *138*, 9–22. [CrossRef] [PubMed]
14. Meng, Z.; Moroishi, T.; Guan, K.L. Mechanisms of Hippo Pathway Regulation. *Genes Dev.* **2016**, *30*, 1–17. [CrossRef]
15. Lv, X.; He, C.; Huang, C.; Wang, H.; Hua, G.; Wang, Z.; Zhou, J.; Chen, X.; Ma, B.; Timm, B.K.; et al. Timely Expression and Activation of YAP1 in Granulosa Cells Is Essential for Ovarian Follicle Development. *FASEB J.* **2019**, *39*, 10049–10064. [CrossRef]
16. Harlow, C.R.; Davidson, L.; Burns, K.H.; Changning, Y.; Matzuk, M.M.; Hillier, S.G. FSH and TGF-β Superfamily Members Regulate Granulosa Cell Connective Tissue Growth Factor Gene Expression in Vitro and in Vivo. *Endocrinology* **2002**, *143*, 3316–3325. [CrossRef]
17. Han, P.; Relav, L.; Price, C.A. Regulation of the Early Growth Response-1 Binding Protein NAB2 in Bovine Granulosa Cells and Effect on Connective Tissue Growth Factor Expression. *Mol. Cell. Endocrinol.* **2020**, *518*, 111041. [CrossRef]
18. Zamberlam, G.; Portela, V.; de Oliveira, J.F.C.; Gonçalves, P.B.D.; Price, C.A. Regulation of Inducible Nitric Oxide Synthase Expression in Bovine Ovarian Granulosa Cells. *Mol. Cell. Endocrinol.* **2011**, *335*, 189–194. [CrossRef]
19. Cao, M.; Nicola, E.; Portela, V.M.; Price, C.A. Regulation of Serine Protease Inhibitor-E2 and Plasminogen Activator Expression and Secretion by Follicle Stimulating Hormone and Growth Factors in Non-Luteinizing Bovine Granulosa Cells in Vitro. *Matrix Biol.* **2006**, *25*, 342–354. [CrossRef]
20. Wang, C.; Zhu, X.; Feng, W.; Yu, Y.; Jeong, K.; Guo, W.; Lu, Y.; Mills, G.B. Verteporfin Inhibits YAP Function through Up-Regulating 14-3-3σ Sequestering YAP in the Cytoplasm. *Am. J. Cancer Res.* **2016**, *6*, 27.
21. Liu-Chittenden, Y.; Huang, B.; Shim, J.S.; Chen, Q.; Lee, S.J.; Anders, R.A.; Liu, J.O.; Pan, D. Genetic and Pharmacological Disruption of the TEAD–YAP Complex Suppresses the Oncogenic Activity of YAP. *Genes Dev.* **2012**, *26*, 1300–1305. [CrossRef] [PubMed]
22. Feng, J.; Gou, J.; Jia, J.; Yi, T.; Cui, T.; Li, Z. Verteporfin, a Suppressor of YAP–TEAD Complex, Presents Promising Antitumor Properties on Ovarian Cancer. *Onco. Targets. Ther.* **2016**, *9*, 5371–5381. [CrossRef] [PubMed]
23. Zhang, Z.; Lin, Z.; Zhou, Z.; Shen, H.C.; Yan, S.F.; Mayweg, A.V.; Xu, Z.; Qin, N.; Wong, J.C.; Zhang, Z.; et al. Structure-Based Design and Synthesis of Potent Cyclic Peptides Inhibiting the YAP-TEAD Protein-Protein Interaction. *ACS Med. Chem. Lett.* **2014**, *5*, 993–998. [CrossRef]
24. Zhou, Z.; Hu, T.; Xu, Z.; Lin, Z.; Zhang, Z.; Feng, T.; Zhu, L.; Rong, Y.; Shen, H.; Luk, J.M.; et al. Targeting Hippo Pathway by Specific Interruption of YAP-TEAD Interaction Using Cyclic YAP-like Peptides. *FASEB J.* **2015**, *29*, 724–732. [CrossRef] [PubMed]
25. Plewes, M.R.; Hou, X.; Zhang, P.; Wang, C.; Davis, J.S. Yes-Associated Protein (YAP1) Is Required for Proliferation and Function of Bovine Granulosa Cells in Vitro. *Biol. Reprod.* **2019**, *101*, 1001–1017. [CrossRef]

26. Hunzicker-Dunn, M.; Maizels, E.T. FSH signaling pathways in immature granulosa cells that regulate target gene expression: Branching out from protein kinase A. *Cell Signal.* **2006**, *18*, 1351–1359. [CrossRef] [PubMed]
27. Ranta, T.; Knecht, M.; Darbon, J.-M.; Baukal, A.J.; Catt, K.J. Induction of Granulosa Cell Differentiation by Forskolin: Stimulation of Adenosine 3′,5′-Monophosphate Production, Progesterone Synthesis, and Luteinizing Hormone Receptor Expression. *Endocrinology* **1984**, *114*, 845–850. [CrossRef] [PubMed]
28. Nivet, A.L.; Vigneault, C.; Blondin, P.; Sirard, M.A. Changes in Granulosa Cells' Gene Expression Associated with Increased Oocyte Competence in Bovine. *Reproduction* **2013**, *145*, 555–565. [CrossRef]
29. Zhang, B.; Tsang, P.C.W.; Pate, J.L.; Moses, M.A. A Role for Cysteine-Rich 61 in the Angiogenic Switch during the Estrous Cycle in Cows: Regulation by Prostaglandin F2alpha. *Biol. Reprod.* **2011**, *85*, 261. [CrossRef]
30. Vernon, R.K.; Spicer, L.J. Effects of Basic Fibroblast Growth Factor and Heparin on Follicle-Stimulating Hormone-Induced Steroidogenesis by Bovine Granulosa Cells. *J. Anim. Sci.* **1994**, *72*, 2696–2702. [CrossRef]
31. Portela, V.M.; Dirandeh, E.; Guerrero-Netro, H.M.; Zamberlam, G.; Barreta, M.H.; Goetten, A.F.; Price, C.A. The Role of Fibroblast Growth Factor-18 in Follicular Atresia in Cattle. *Biol. Reprod.* **2015**, *92*, 1–8. [CrossRef] [PubMed]
32. Adams, G.P.; Matteri, R.L.; Kastelic, J.P.; Ko, J.C.; Ginther, O.J. Association between surges of follicle-stimulating hormone and the emergence of follicular waves in heifers. *J Reprod Fertil.* **1992**, *94*, 177–188. [CrossRef] [PubMed]
33. Rodgers, R.J.; Irving-Rodgers, H.F. Morphological Classification of Bovine Ovarian Follicles. *Reproduction* **2010**, *139*, 309–318. [CrossRef]
34. Kawamura, K.; Cheng, Y.; Suzuki, N.; Deguchi, M.; Sato, Y.; Takae, S.; Ho, C. HipKAWAMURA, Kazuhiro e Colab. Hippo Signaling Disruption and Akt Stimulation of Ovarian Follicles for Infertility Treatment. *Proc. Natl. Acad. Sci. USA* **2013**, *110*, 17474–17479. [CrossRef]
35. Godin, P.; Tsoi, M.F.; Morin, M.; Gévry, N.; Boerboom, D. The Granulosa Cell Response to Luteinizing Hormone Is Partly Mediated by YAP1-Dependent Induction of Amphiregulin. *Cell Commun. Signal.* **2022**, *20*, 72. [CrossRef] [PubMed]
36. Dos Santos, E.C.; Lalonde-Larue, A.; Antoniazzi, A.Q.; Barreta, M.H.; Price, C.A.; Dias Gonçalves, P.B.; Portela, V.M.; Zamberlam, G. YAP Signaling in Preovulatory Granulosa Cells Is Critical for the Functioning of the EGF Network during Ovulation. *Mol. Cell. Endocrinol.* **2022**, *541*, 111524. [CrossRef]
37. Mazerbourg, S.; Bondy, C.A.; Zhou, J.; Monget, P. The Insulin-like Growth Factor System: A Key Determinant Role in the Growth and Selection of Ovarian Follicles? A Comparative Species Study. *Reprod. Domest. Anim.* **2003**, *38*, 247–258. [CrossRef]
38. Spicer, L.J.; Aad, P.Y. Insulin-Like Growth Factor (IGF) 2 Stimulates Steroidogenesis and Mitosis of Bovine Granulosa Cells through the IGF1 Receptor: Role of Follicle-Stimulating Hormone and IGF2 Receptor. *Biol. Reprod.* **2007**, *77*, 18–27. [CrossRef]
39. Portela, M.; Ph, D.; Zamberlam, G.; Sc, M.; Price, C.A. Cell Plating Density Alters the Ratio of Estrogenic to Progestagenic Enzyme Gene Expression in Cultured Granulosa Cells. *Fertil. Steril.* **2010**, *93*, 2050–2055. [CrossRef]
40. Sahmi, M.; Nicola, E.S.; Silva, J.M.; Price, C.A. Expression of 17β- and 3β-Hydroxysteroid Dehydrogenases and Steroidogenic Acute Regulatory Protein in Non-Luteinizing Bovine Granulosa Cells in Vitro. *Mol. Cell. Endocrinol.* **2004**, *223*, 43–54. [CrossRef]
41. Pfaffl, M.W. A New Mathematical Model for Relative Quantification in Real-Time RT-PCR. *Nucleic Acids Res.* **2001**, *29*, e45. [CrossRef] [PubMed]
42. Koch, J.; Portela, V.M.; Dos Santos, E.C.; Missio, D.; de Andrade, L.G.; da Silva, Z.; Gasperin, B.G.; Antoniazzi, A.Q.; Gonçalves, P.B.D.; Zamberlam, G. The Hippo Pathway Effectors YAP and TAZ Interact with EGF-like Signaling to Regulate Expansion-Related Events in Bovine Cumulus Cells in Vitro. *J. Assist. Reprod. Genet.* **2022**, *39*, 481–492. [CrossRef] [PubMed]
43. Portela, V.M.; Machado, M.; Jr, J.B.; Zamberlam, G.; Amorim, R.L.; Goncalves, P.; Price, C.A.; De Fisiologia, D.; Biocie, I. De Expression and Function of Fibroblast Growth Factor 18 in the Ovarian Follicle in Cattle 1. *Biol. Reprod.* **2010**, *346*, 339–346. [CrossRef] [PubMed]

International Journal of
Molecular Sciences

Article

MnHR4 Functions during Molting of *Macrobrachium nipponense* by Regulating 20E Synthesis and Mediating 20E Signaling

Huwei Yuan [1], Wenyi Zhang [2], Hui Qiao [2], Shubo Jin [2], Sufei Jiang [2], Yiwei Xiong [2], Yongsheng Gong [2] and Hongtuo Fu [1,2,*]

[1] Wuxi Fisheries College, Nanjing Agricultural University, Wuxi 214081, China
[2] Key Laboratory of Freshwater Fisheries and Germplasm Resources Utilization, Ministry of Agriculture and Rural Affairs, Freshwater Fisheries Research Center, Chinese Academy of Fishery Sciences, Wuxi 214081,China
* Correspondence: fuht@ffrc.cn; Tel.: +86-510-8555-8835

Abstract: *HR4*, a member of the nuclear receptor family, has been extensively studied in insect molting and development, but reports on crustaceans are still lacking. In the current study, the *MnHR4* gene was identified in *Macrobrachium nipponense*. To further improve the molting molecular mechanism of *M. nipponense*, this study investigated whether *MnHR4* functions during the molting process of *M. nipponense*. The domain, phylogenetic relationship and 3D structure of *MnHR4* were analyzed by bioinformatics. Quantitative real-time PCR (qRT-PCR) analysis showed that *MnHR4* was highly expressed in the ovary. In different embryo stages, the highest mRNA expression was observed in the cleavage stage (CS). At different individual stages, the mRNA expression of *MnHR4* reached its peak on the fifteenth day after hatching (L15). The in vivo injection of 20-hydroxyecdysone (20E) can effectively promote the expression of the *MnHR4* gene, and the silencing of the *MnHR4* gene increased the content of 20E in *M. nipponense*. The regulatory role of *MnHR4* in 20E synthesis and 20E signaling was further investigated by RNAi. Finally, the function of the *MnHR4* gene in the molting process of *M. nipponense* was studied by counting the molting frequency. After knocking down *MnHR4*, the molting frequency of *M. nipponense* decreased significantly. It was proved that *MnHR4* plays a pivotal role in the molting process of *M. nipponense*.

Keywords: *MnHR4*; 20-hydroxyecdysone (20E); *Macrobrachium nipponense*; molt; RNA interference

Citation: Yuan, H.; Zhang, W.; Qiao, H.; Jin, S.; Jiang, S.; Xiong, Y.; Gong, Y.; Fu, H. *MnHR4* Functions during Molting of *Macrobrachium nipponense* by Regulating 20E Synthesis and Mediating 20E Signaling. *Int. J. Mol. Sci.* **2022**, 23, 12528. https://doi.org/10.3390/ijms232012528

Academic Editor: Jacques J. Tremblay

Received: 30 September 2022
Accepted: 17 October 2022
Published: 19 October 2022

Publisher's Note: MDPI stays neutral with regard to jurisdictional claims in published maps and institutional affiliations.

1. Introduction

Endocrine signals play a central role in animal growth and maturation. Although vertebrate growth and development are primarily controlled by thyroid hormones and sex steroids, insect development is controlled by several key hormones and neuropeptides, of which the steroid hormone ecdysteroid is the main regulator [1]. 20-Hydroxyecdysone (20E) is a steroid hormone that was originally discovered in plants and arthropods [2]. The accumulation of research over the past two decades has shown that cholesterol is catalyzed into 20E by the Halloween family of genes (i.e., *Spook, Phantom, Disembodied, Shadow* and *Shade*) [3–5]. 20E converts hormonal signals into transcriptional responses through members of the nuclear receptor family, and when 20E binds to the EcR/USP complex, the transcriptional cascade leads to the onset of ecdysis and metamorphosis [6,7]. *Drosophila*, as a classic model organism, has become a model for elucidating the regulatory role of 20E. The 20E pulse directs each molt and metamorphosis of the *Drosophila melanogaster*'s life cycle [8]. Many downstream nuclear receptor transcription factors activated by 20E have been identified in insects [9].

Nuclear receptors contain two common structural elements—a DNA-binding domain (DBD) and a ligand-binding domain (LBD)—which are involved in the regulation of myriad biological processes [10]. In the holometabolous insect *D. melanogaster*, some genes, such as nuclear receptor E75 (*E75*), nuclear receptor E78 (*E78*), hormone receptor 3 (*HR3*), hormone receptor 4 (*HR4*), ecdysis-triggering hormone receptor (*ETHR*) and Fushi tarazu factor-1 (*Ftz-f1*), are transcriptionally regulated by 20E and play a central role in transducing the molting signals [11,12]. The expressions of *HR3* and *E75B* increased with the increase in the 20E titer during *D. melanogaster* pupation, and *βFtz-f1* was activated after the decrease in 20E [13]. In the hemimetabolous insect *Blattella germanica*, the regulatory relationship between nuclear receptor genes *E75*, *HR3*, *HR4* and *Ftz-f1* changed with the fluctuation in the 20E titer [14]. In insects, the function of nuclear receptors activated by 20E has been reported in detail. For example, an *E75A* mutation causes developmental arrest and molting defects in *Drosophila* and the *E75C*-mutant *Drosophila* die in adulthood, whereas *E75B*-mutant individuals survive and reproduce normally [15]. The knockdown of *BgE75* results in *B. germanica* prothoracic gland degeneration and ecdysteroid deficiency [14]. *HR3* is involved in the regulation of chitin synthesis and degradation during *Locusta migratoria* molting [16]. Silencing of *BgFtz-f1* prevents normal molting and development of *B. germanica* [17], and *Ftz-f1* is involved in the regulation of *Leptinotarsa decemlineata*'s pupation by regulating 20E and JH titers [18]. *HR4* plays an important role in *Tribolium castaneum* molting and ovulation [19] and plays a central role in coordinating growth and the maturation of *D. melanogaster* [20]. In addition, the *HR4* gene has been identified in some crustacean transcriptomes. In *Litopenaeus vannamei* and *Daphnia*, *HR4* was identified as an ecdysone signaling response gene [21,22]. In *Callinectes sapidus*, increased ecdysteroid concentration induced *HR4* expression [23]. Altogether, nuclear receptor genes have been extensively studied in insect molting and development, but reports on crustaceans are still lacking.

Macrobrachium nipponense (Crustacea, Decapoda) is an important freshwater economic prawn in China [24]. Molting is a pivotal event in the growth of crustaceans [25]. *M. nipponense* grows by molting, but the mechanism of molting is still poorly understood. Therefore, it is of great scientific significance to study the molting mechanism of *M. nipponense* for breeding and increasing production. We previously demonstrated the functions of *Spook* and *Ftz-f1* genes in the molting and ovarian development of *M. nipponense* [5,26]. Transcriptome analysis of the different molting stages of *M. nipponense* revealed that *MnHR4* is an important differential gene. To further improve the molting molecular mechanism of *M. nipponense*, this study investigated whether *MnHR4* functions during the molting process of *M. nipponense*. In the current study, the *MnHR4* gene was identified in *M. nipponense*. The domain, phylogenetic relationship and 3D structure of *MnHR4* were analyzed by bioinformatics. The expression patterns of the *MnHR4* gene in different tissues and developmental stages of *M. nipponense* were detected by qRT-PCR. The expression of the *MnHR4* gene was detected by qRT-PCR after a 20E injection in vivo. After knockdown of the *MnHR4* gene by the RNA interference, the content of 20E in *M. nipponense* was detected by an ELISA. The regulatory role of *MnHR4* in 20E synthesis and 20E signaling was further investigated by RNAi. Finally, the function of the *MnHR4* gene in the molting process of *M. nipponense* was studied by counting the molting frequency.

2. Results

2.1. Sequence Analysis and Phylogeny of MnHR4

The rapid amplification of the cDNA ends (RACE) (TaKaRa, Kyoto, Japan) of the *HR4* fragment yielded a cDNA sequence with a putative open reading frame of 2901 bp, encoding a total of 966 amino acids, named *MnHR4* (Figure 1). A phylogenetic tree of the *HR4* amino acids of different species was constructed, indicating that insecta and crustacea form two independent clades, which is consistent with the traditional taxonomy of species. *M. nipponense* is the most closely related to *Penaeus chinensis*, followed by *Armadillidium vulgare* (Figure 2). A comparison of the *HR4* amino acid sequences between *M. nipponense* and other crustaceans using DNAMAN 6.0 showed that *MnHR4* contained a conserved C4 zinc finger (ZnF_C4) domain. Zinc finger domains are relatively small protein motifs containing different binding-specific finger-like protrusions, usually in clusters. The *HR4* amino acid identity of *M. nipponense* and *P. chinensis* was 49.64%, followed by *A. vulgare* at 30.18% (Figure 3). The amino acid sequence of *MnHR4* was analyzed using the iterative threading assembly refinement (I-TASSER) server. Figure 4A shows a three-dimensional (3D) atom model of *M. nipponense*'s *MnHR4* generated by I-TASSER. The predicted binding ligands are the pale-green spheres, and the binding residues are the blue ball and stick (Figure 4B).

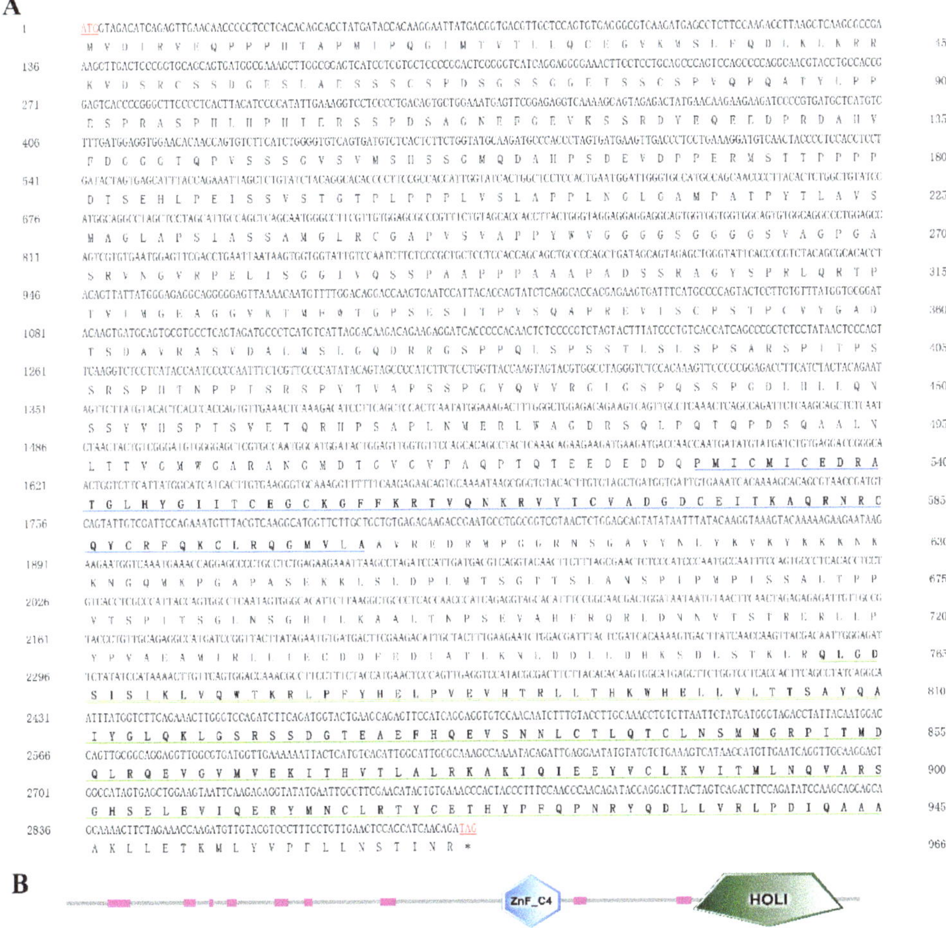

Figure 1. (**A**) Nucleotide and deduced amino acid sequences of *MnHR4* in *M. nipponense*. The numbers on the left and right refer to the coordinates of the nucleotide and amino acid sequences, respectively. Blue and green underscores represent the ZnF_C4 and HOLI domains, respectively. The termination signals are indicated with an asterisk (*). (**B**) Domain architecture organization of *MnHR4* as predicted by SMART (http://smart.embl-heidelberg.de/ (accessed on 20 April 2022)). The conserved domains (i.e., ZnF_C4 and HOLI) of MnHR4 are shown by the different shapes and colors.

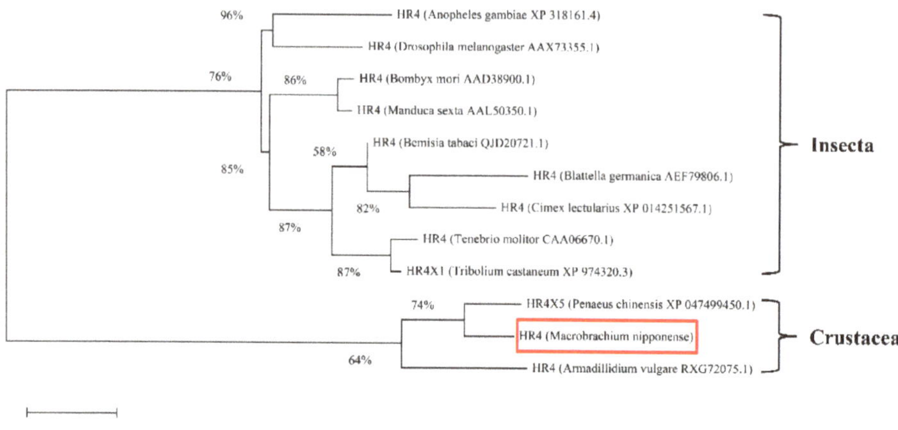

Figure 2. Phylogenetic tree of the amino acid sequences of *HR4* from various species. The numbers shown at the branches indicate the bootstrap values (%). The red rectangles indicate the position of *M. nipponense* in the phylogenetic tree. The species and GenBank accession numbers for constructing the phylogenetic tree are listed below: *Anopheles gambiae* (XP_318161.4), *D. melanogaster* (AAX73355.1), *Bombyx mori* (AAD38900.1), *Manduca sexta* (AAL50350.1), *Bemisia tabaci* (QJD20721.1), *B. germanica* (AEF79806.1), *Cimex lectularius* (XP_014251567.1), Tenebrio molitor (CAA06670.1), *Tribolium castaneum* (XP_974320.3), *P. chinensis* (XP_047499450.1), *M. nipponense* and *A. vulgare* (RXG72075.1).

Figure 3. Sequence alignment of the *HR4* amino acids between *M. nipponense* and other crustaceans. The blue and green underscores represent the ZnF_C4 and HOLI domains, respectively.

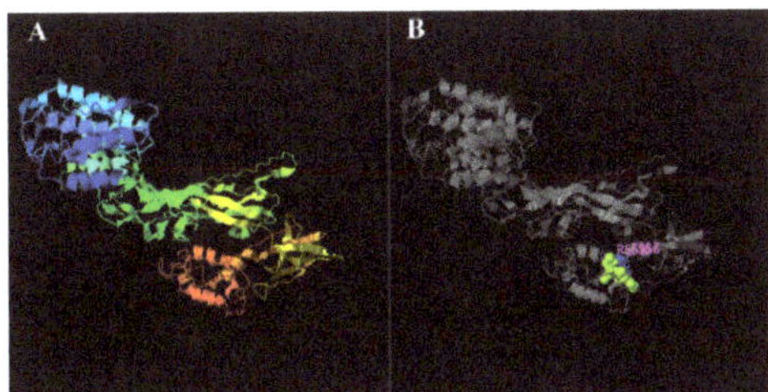

Figure 4. The 3D structures of *MnHR4* predicted by I-TASSER: (**A**) predicted function of *MnHR4* 3D structures using COFACTOR and COACH, where the pale-green sphere represents the predicted binding ligand; (**B**) molecules and ions that bind to anchoring proteins are called ligands.

2.2. Expression of the MnHR4 Gene in Different Tissues and Different Developmental Stages

The expression levels of *MnHR4* mRNA in different tissues and different developmental stages of *M. nipponense* were investigated by qRT-PCR. The highest mRNA expression of *MnHR4* was observed in the ovary, followed by the eyestalk, with the lowest expression in the hepatopancreas. The expression levels of *MnHR4* mRNA in the ovary and eyestalk were significantly higher than those in other tissues, and the expression level in the ovary was 4.47-fold higher than that in the eyestalk ($p < 0.05$) (Figure 5A). The expression of *MnHR4* mRNA showed no significant difference in different ovary stages ($p > 0.05$) (Figure 5B). In different embryo stages, the highest mRNA expression was observed in the cleavage stage (CS), followed by the blastula stage (BS), and the lowest was observed in the zoea stage (ZS). The expression level in the CS was 4.53-fold that in the BS and 23.22-fold that in the ZS ($p < 0.05$) (Figure 5C). In different individual stages, the mRNA expression of *MnHR4* gradually decreased from the first day after hatching (L1) to L10 and reached a peak at L15. *MnHR4* mRNA expression was lowest on the first day post-larvae (PL1) and showed a significant difference ($p < 0.05$) (Figure 5D).

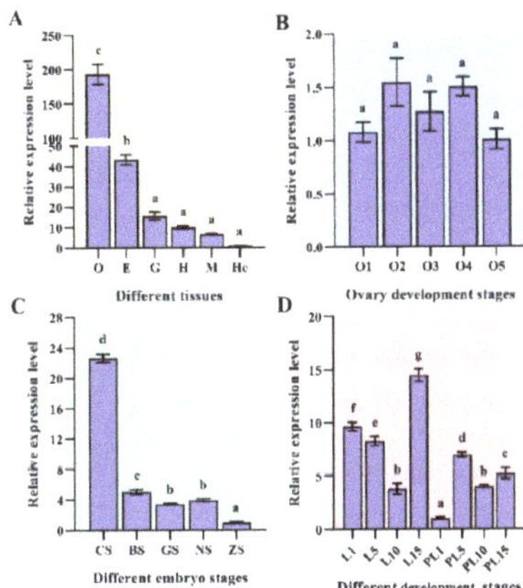

Figure 5. Expression of *MnHR4* mRNA in different tissues and different developmental stages of *M. nipponense*. Samples (at different developmental stages) were collected at the experimental site in Dapu, according to previous criteria [27,28]. The expression of *MnHR4* mRNA was normalized to the *EIF* transcript level. (**A**) Different tissues: O, ovary; E, eyestalk; G, gill; H, heart; M, muscle; He, hepatopancreas. (**B**) Different ovary stages: O1, undeveloped stage; O2, developing stage; O3, nearly ripe stage; O4, ripe stage; O5, spent stage. (**C**) Different embryo stages: CS, cleavage stage; BS, blastula stage; GS, gastrula stage; NS, nauplius stage; ZS, zoea stage. (**D**) Different development stages: L1, the first day after hatching; PL1, the first day post-larvae, etc. Statistical analyses were performed by one-way ANOVA. Data are shown as the mean ± SEM ($n = 6$). Bars with different letters indicate significant differences ($p < 0.05$).

2.3. Interaction between 20E and MnHR4

Referring to previous studies, 20E (5 μg/g) was injected into *M. nipponense* [26]. The effect of 20E on the expression of *MnHR4* was detected using qRT-PCR (Figure 6A). The results show that there was no significant difference in the expression level of the *MnHR4* gene between the experimental group and the control group at 0 h (no injection) and 3 h after injection ($p > 0.05$). The expression level of the *MnHR4* gene in the experimental group was significantly higher than that in the control group at 6 and 12 h after injection, and it reached a peak at 12 h in the experimental group ($p < 0.05$). There was no significant difference in the expression of *MnHR4* between the two groups at the 24th hour after injection ($p > 0.05$). After knockdown of *MnHR4*, the content of 20E in *M. nipponense* was measured by an ELISA (Figure 6B). The results showed that there was no significant difference in the content of 20E in *M. nipponense* on the first day ($p > 0.05$). The content of 20E in the experimental group of *M. nipponense* was significantly higher than that in the control group on the 5th day. Compared with the control group, the content of 20E in the experimental group increased by 37.79% ($p < 0.05$).

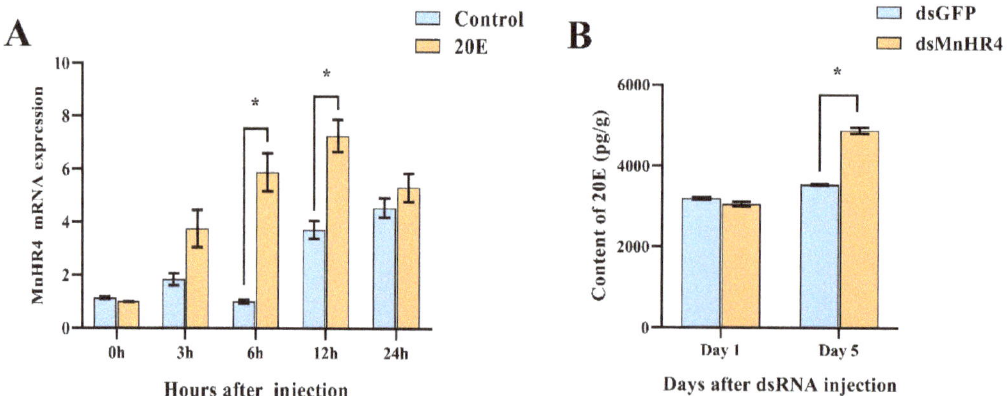

Figure 6. (**A**) Expression of *MnHR4* mRNA in the ovary under the influence of 20E. The expression of *MnHR4* mRNA was normalized to the *EIF* transcript level. (**B**) The content of 20E in *M. nipponense* after knockdown of *MnHR4*. Control, injection of solvent; 20E, injection of 20E. Data are expressed as the mean ± SEM ($n = 6$). Significant differences between the experimental group and the control group were determined using Student's *t*-test (* $p < 0.05$).

2.4. Effects of RNAi MnHR4 on the Expression of Mn-Spook, Phantom, HR3, E75b, ETHR and Mnftz-f1

The regulatory role of *MnHR4* in 20E synthesis and 20E signaling was further investigated by RNAi. After knockdown of *MnHR4*, the expression levels of genes catalyzing 20E synthesis (i.e., *Mn-Spook* and *Phantom*) and downstream genes conducting 20E signaling (i.e., *HR3*, *E75b*, *ETHR* and *MnFtz-f1*) were detected by qRT-PCR (Figure 7). Compared to the control, the expression of *MnHR4* mRNA in the experimental group decreased by 17.45%, 69.25%, 73.88% and 69.97% at 24, 48, 96 and 120 h after ds*MnHR4* administration, respectively ($p < 0.05$) (Figure 7A). After the knockdown of *MnHR4*, the expression levels of *Mn-Spook* and *Phantom* in the experimental group significantly increased. At the 120th hour after the knockdown of *MnHR4*, the expression levels of *Mn-Spook* and *Phantom* in the experimental group were 24.4-fold and 2.1-fold higher than those in the control group, respectively (Figure 7B,C). The results show that the expression levels of *HR3* and *E75b* in the experimental group also significantly increased at the 24th and 48th hours after *MnHR4* gene silencing. At the 120th hour after silencing, the expression of *HR3* in the experimental group was 20.55-fold that of the control group, and the expression of *E75b* in the experimental group was 2.49-fold that of the control group (Figure 7D,E). On the contrary, the expressions of *ETHR* and *MnFtz-f1* in the experimental group decreased to different degrees compared with the control group after *MnHR4* gene silencing. The expression of *ETHR* in the experimental group decreased by 31.13% and 38.91% at the 96th and 120th hour of *MnHR4* gene silencing, respectively (Figure 7F). Compared with the control group, the expression of *MnFtz-f1* in the experimental group decreased by 79.36% at the 120th hour (Figure 7G).

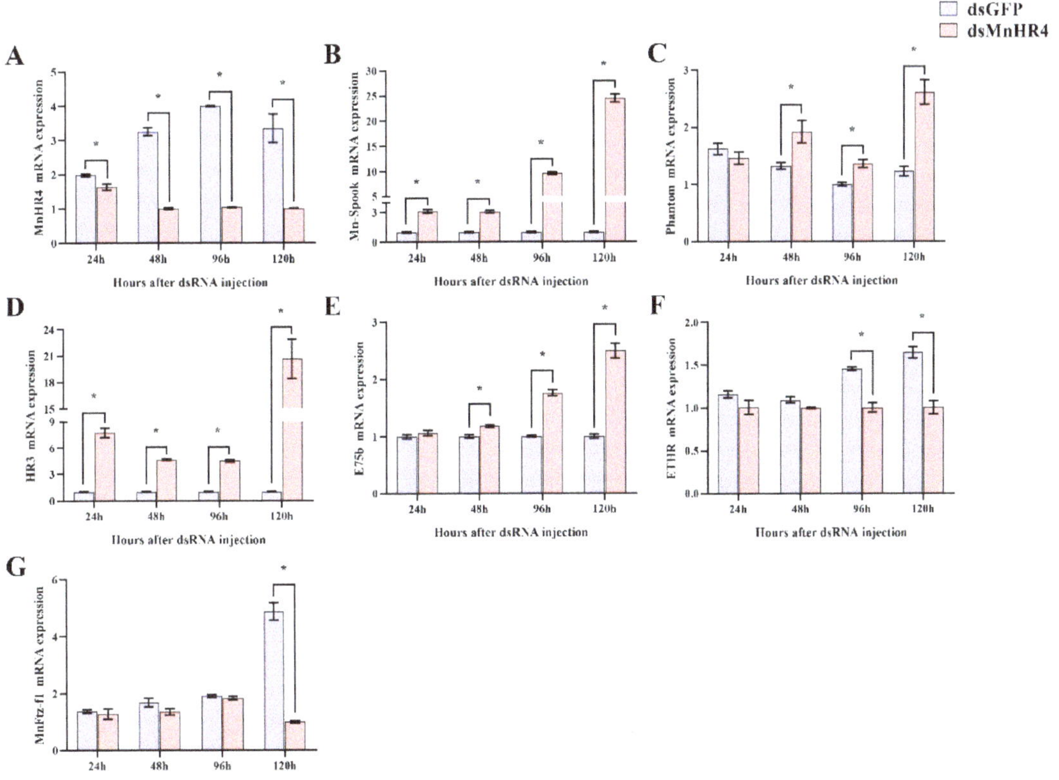

Figure 7. Knockdown of *MnHR4* on the expression of other 20E-related genes in the ovary of *M. nipponense*. The expression of *MnHR4* mRNA was normalized to the *EIF* transcript level: (**A**) *MnHR4*; (**B**) *Mn-Spook*; (**C**) *Phantom*; (**D**) *HR3*; (**E**) *E75b*; (**F**) *ETHR*; (**G**) *Mn-Ftz-f1*. Data are expressed as the mean ± SEM (*n* = 6). Significant differences between the experimental group and the control group were determined using Student's *t*-test (* *p* < 0.05).

2.5. Effect of MnHR4 Knockdown on the Molting Frequency of M. nipponense

Figure 8 shows the molting frequency of *M. nipponense* in the control and experimental groups after *MnHR4* knockdown. *M. nipponense* starts molting on the second day and completes one round of molting on the 12th day. The results show that there was no significant difference in the frequency of molting between the experimental and control groups during the first round of molting (*p* > 0.05). From the 21st day, the control group of *M. nipponense* began the second round of concentrated molting, which was significantly higher than the experimental group of *M. nipponense*'s molting frequency (*p* < 0.05).

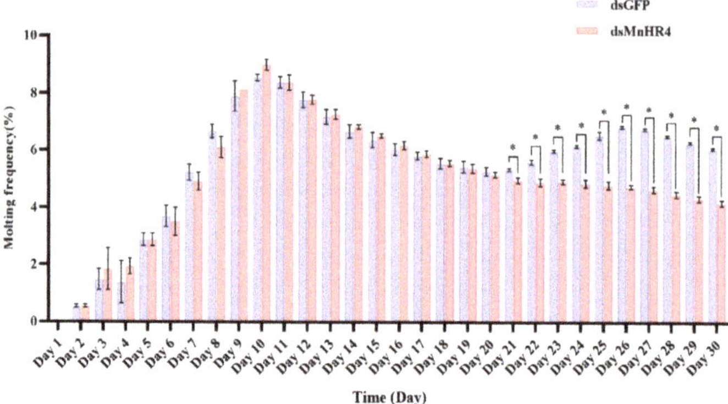

Figure 8. Molting frequency of *M. nipponense* in the experimental and control groups after knocking down *MnHR4*. Molting frequency = (Nm/Ns)/D, where Nm is the total number of molts; Ns is the number of prawns in the aquarium; D is the number of experimental days. Data are expressed as the mean ± SEM. Significant differences between the experimental group and the control group were determined using Student's *t*-test (* $p < 0.05$).

3. Discussion

Nuclear receptor genes function in many biological processes such as embryonic development, sex determination, insect metamorphosis and molting [29–31]. Using RNAi, we previously demonstrated that *MnFtz-f1* played a pivotal role in the molting and ovulation process of *M. nipponense* [26]. Previous studies have demonstrated that *HR4* plays an important role in both holometabolous and hemimetabolous insects [11,14], but its function in *M. nipponense* is still unclear.

In the present study, we identified the nuclear receptor gene *MnHR4* from the transcriptome of *M. nipponense* at different molting stages. The predicted *MnHR4* coding region encodes a total of 966 amino acids, including two conserved domains: the c4 zinc finger in nuclear hormone receptors (ZnF_C4) and ligand-binding domain of hormone receptors (HOLI). Zinc finger domains are relatively small protein motifs with a stable structure and are involved in a wide range of physiological functions including controlling embryonic development and cell differentiation [32,33]. HOLI is located in the LBD region of the nuclear receptor gene, which acts as a molecular switch to turn on transcriptional activity by binding to ligands [34]. In the phylogenetic tree, there is a clear boundary between crustaceans and insects, indicating that *HR4* is more conserved among its class.

MnHR4 was expressed in multiple tissues of *M. nipponense*, with higher expression levels in the ovary, indicating that *MnHR4* has different functions in *M. nipponense*. In insects, *HR4* expression was also detected in multiple tissues [35]. The extremely high expression of *HR4* in the ovaries is similar to other studies on *Drosophila*. Recent studies on *Drosophila* have shown that *HR4* is strongly expressed in the ovaries and is required for *Drosophila* to oogenesis [36]. There was no significant difference in the *MnHR4* in different ovarian development stages of *M. nipponense*. Therefore, we speculate that *MnHR4* may be required to function in the whole ovarian development cycle. The expression of *MnHR4* in the cleavage stage was significantly higher than that in other stages of embryonic development, suggesting it has a function during cell division. This result is similar to the expression trend of nuclear receptor gene *MnFtz-f1* in *M. nipponense* [26]. Additionally, *MnHR4* expression levels were the highest at L15 during the larval developmental stages, suggesting that *MnHR4* may function during the metamorphosis of *M. nipponense* [37].

To explore the relationship between 20E and *MnHR4*, the expression level of *MnHR4* was detected after 20E injection in *M. nipponense*, and the content of 20E in *M. nipponense* was measured after *MnHR4* knockdown. The expression of *MnHR4* significantly increased after 20E injection in vivo, which proves that 20E can induce the expression of *MnHR4* in *M. nipponense*. King-jones et al. demonstrated that feeding 20E for 3–4 h induced the peak expression of *HR4* in *Drosophila* [20]. Similarly, the injection of 20E upregulated *HR4*'s expression in *B. germanica* and *L. decemlineata* [38,39]. We further investigated the effect of the knockdown of *MnHR4* on the 20E content. The results show that the knockdown of *MnHR4* increased the 20E titer in *M. nipponense*. In *L. decemlineata*, 20E titers also increased after *HR4* knockdown, which is consistent with our conclusions [39]. Conversely, 20E titers decreased after silencing *HR4* in *L. migratoria* [35], whereas 20E titers were not affected by silencing *HR4* in *B. germanica* [38]. In conclusion, *HR4* may have different regulatory effects on 20E in different species, and its molecular mechanism needs to be further investigated.

RNAi is an effective method to explore the regulatory relationship between genes. We further investigated the molecular mechanism of *MnHR4* in 20E synthesis and signaling by RNAi (Figure 9). *Spook* and *Phantom* are Halloween family member genes that function in the 20E biosynthetic pathway [5,40]. To investigate the molecular mechanism of *MnHR4* on 20E synthesis in *M. nipponense*, the expressions of *Spook* and *Phantom* were detected after knockdown of *MnHR4*. The expressions of *Spook* and *Phantom* significantly increased after knockdown of *MnHR4*, indicating that *MnHR4* has an inhibitory effect on *Spook* and *Phantom*. Similar studies have shown that *HR4* inhibits ecdysone synthesis by regulating cytochrome P450 genes [41]. *HR4* is a repressor of early ecdysone-induced regulatory genes [20]. There are many nuclear receptor genes in the 20E signaling pathway, whose main function is to transmit upstream signals [42]. To further investigate the role of *MnHR4* in 20E signaling, the expression of other genes (i.e., *HR3*, *E75b*, *ETHR* and *MnFtz-f1*) was examined after knockdown of *MnHR4*. The results show that the expressions of *HR3* and *E75b* significantly increased after silencing *MnHR4*, indicating that *MnHR4* had an inhibitory effect on *HR3* and *E75b*. In *L. decemlineata*, the expressions of *HR3* and *E75* were also affected by *HR4*, and the knockdown of *HR4* significantly upregulated the expressions of *HR3* and *E75* [39]. In addition, the expressions of *ETHR* and *MnFtz-f1* were decreased after knockdown of *MnHR4*, indicating that *MnHR4* had a promoting effect on them. Previous studies have demonstrated that *ETHR* functions during the molting process of *M. nipponense* [43], but the regulatory relationship between *HR4* and *ETHR* is rarely reported. Our results demonstrate that *MnHR4* positively regulates *ETHR*, which provides a reference for future research. *Ftz-f1* plays a central role in coordinating different molting processes [44]. Consistent with our conclusion, the abundant results demonstrate that *HR4* can induce the expression of *Ftz-f1* [20,38,41]. In conclusion, *MnHR4* plays a regulatory role in 20E synthesis and signaling. We further investigated whether the molting of the *M. nipponense* occurred after the knockdown of *MnHR4*. It was proved that *MnHR4* plays a pivotal role in the molting process of *M. nipponense*. The molting function of *HR4* has been demonstrated in some insects. In *B. Germanica*, interference with *HR4* resulted in molting failure and eventual death [38]. *HR4* is required for pupal molting and adult oogenesis of *Tribolium castaneum* [19]. The knockdown of *MnHR4* increased 20E titers, which significantly inhibited the molting of *M. nipponense*. Our previous study showed that knockdown of *MnFtz-f1* reduced 20E titers, similarly, leading to the failure of *M. nipponense*'s molting [26]. In *B. mori*, increasing or decreasing 20E titers could affect the normal physiological phenomena of the larvae or even lead to death [45]. Precise regulation of 20E titers is also important for the molting process in *D. melanogaster* [1]. This suggests that an appropriate 20E titer is important for regulating successful molting in insects or crustaceans.

In summary, the *MnHR4* gene was identified in *M. nipponense*. The *MnHR4* gene was comprehensively analyzed using bioinformatics, qRT-PCR, RNAi, ELISA, etc. Our results strongly demonstrate that *MnHR4* functions in the molting process of *M. nipponense* by regulating 20E synthesis and 20E signaling. This study further enriched the molecular regulatory mechanism of molting in *M. nipponense* and could be useful for future gene-editing breeding.

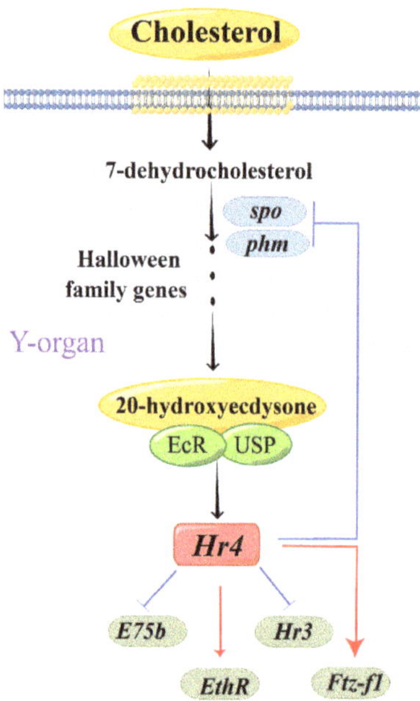

Figure 9. Model for *MnHR4* function, developed using Figdraw (www.figdraw.com). The Y-organ is a pair of secretory glands of crustaceans, the site of the synthesis of cholesterol into ecdysone.

4. Materials and Methods

4.1. Experimental Prawns and Conditions

Experimental prawns (2.15 ± 0.63 g) were obtained from the Dapu experimental base, Freshwater Fisheries Research Center, Chinese Academy of Fishery Sciences. Briefly, the experimental prawns were transferred from Dapu to the laboratory's constant-temperature water-circulation system and allowed to acclimatize for a week. Tissues (i.e., ovaries, muscles, gills, hepatopancreas, eyestalks and hearts) were harvested, frozen in liquid nitrogen and stored at −80 °C for RNA extraction. Samples were also collected at different stages of embryo, individual and ovarian development according to previous criteria [27,28]. All sampling was performed in triplicate (*n* = 6). The prawns in this study were handled according to the guidelines of the Institutional Animal Care and Use Ethics Committee of the Freshwater Fisheries Research Center, Chinese Academy of Fishery Sciences (Wuxi, China).

4.2. Nucleotide Sequence and Bioinformatics Analysis of MnHR4

Total RNA was extracted using the RNAiso Plus kit (TaKaRa, Shiga, Japan), as described previously [26]. DNase I (Sangon, Shanghai, China) was used to eliminate possible DNA contamination, and 1.2% agarose gel and a NanoDrop ND2000 (NanoDrop Technologies, Wilmington, DE, USA) were used to detect RNA quality and concentration, respectively, with an A260/A280 ratio of 1.9–2.0.

The *MnHR4* cDNA fragments were screened from the transcriptome of *M. nipponense* at different molting stages. The *MnHR4* cDNA fragments were analyzed using the GenBank BLASTX and BLASTN programs (https://blast.ncbi.nlm.nih.gov/Blast.cgi (accessed on 20 April 2022)). The open reading frame (ORF) of *MnHR4* was predicted by using ORF Finder (https://www.ncbi.nlm.nih.gov/orffinder/ (accessed on 20 April 2022)). Molecular Evolutionary Genetics Analysis (MEGA-X) software was used to construct a phylogenetic tree, and the bootstrapping replications were 1000 [46,47]. DNAMAN 6.0 was used for translating and aligning amino acid sequences. The spatial structure and function of *MnHR4* amino acids were predicted by I-TASSER (https://zhanglab.ccmb.med.umich.

edu/I-TASSER/ (accessed on 22 April 2022)) [48]. The *HR4* amino acid sequences of other species investigated in this study were downloaded from the GenBank database (http://www.ncbi.nlm.nih.gov/ (accessed on 20 April 2022)).

4.3. Quantitative Real-Time PCR (qRT-PCR) Analysis

Gene expression patterns were evaluated using qRT-PCR on a Bio-Rad iCycler iQ5 Real-Time PCR System (Bio-Rad, Carlsbad, CA, USA). The primers were designed using NCBI's Primer-Blast tool (http://www.ncbi.nlm.nih.gov/tools/primer-blast/ (accessed on 25 April 2022)) and synthesized by exsyn-bio Technology Co., Ltd. (Shanghai, China). The reaction system and procedure of qRT-PCR have been described in previous studies [5]. The internal reference gene, eukaryotic translation initiation factor 5A (*EIF*), was used as a control for data normalization [49], and the relative expression levels of the genes were calculated with the $2^{-\Delta\Delta CT}$ method [50].

4.4. RNA Interference (RNAi)

The design and synthesis of *MnHR4*-interfering primers followed a method used in previous studies [26]. The green fluorescent protein gene (GFP) was used as a control [51]. A total of 180 healthy female prawns in the pre-molting stage were randomly divided into two groups (i.e., experimental group and control group) in triplicate ($n = 30$). The experimental group and control group were injected with *MnHR4* dsRNA and *dsGFP*, respectively (8 μg/g of body weight). Prawn tissues were collected in triplicate at 24, 48, 96 and 120 h after injection for RNA extraction and an interference efficiency assessment ($n = 6$). On the basis of the significant interference efficiency of the experimental group and control group, 300 prawns in the pre-molting stage were divided into two groups using the same method and injected with *MnHR4* dsRNA and *dsGFP*, respectively ($n = 50$). The number of molting prawns per day was counted, and the molting frequency was calculated. The number of molts was counted for 30 days, and dsRNA was injected every five days. Molting frequency = (Nm/Ns)/D, where Nm is the total number of molts; Ns is the number of prawns in the aquarium; and D is the number of experimental days [5,26].

4.5. ELISA

After knocking down the *MnHR4* gene, the Shrimp EH ELISA Kit (lot number: m1963525-J; Meibo, Shanghai, China) was used to detect the content of 20E in *M. nipponense*, according to the manufacturer's protocols. In brief, the tissue was first washed with pre-cooled PBS to remove the residual blood. Next, the tissue was weighed and cut into pieces, and it was added to the PBS at a ratio of 1:9. Then, it was put into a glass homogenizer and fully ground on ice to lyse the tissue's cells. The tissue homogenate was centrifuged at 5000 rpm for 5–10 min, and the supernatant was removed until use. All reagents and samples were prepared before the experiment: (1) The standard wells and sample wells were set on the microtiter plate. Fifty microliters of different concentrations of standards (i.e., 2000, 1000, 500, 250, 125 and 62.5 pg/mL) were added to the standard wells, and 50 μL of the samples to be tested were added to the sample wells; blank wells were not added. (2) Then, 100 μL of enzyme ligands was added to the standard and sample wells, and the reaction wells were sealed with a closure plate membrane and incubated for 60 min at 37 °C in a water bath or incubator. (3) The microtiter plate was rinsed 4–5 times, and then, 50 μL of substrate A and B was added to each well. They were mixed gently and incubated at 37 °C for 15 min. (4) After adding 50 μL of stop solution to each well and tapping the plate to ensure adequate mixing, the OD values of each well were measured at a wavelength of 450 nm within 15 min. (5) Taking the OD value of the standard as the abscissa and the concentration value of the standard as the ordinate, the standard curve was drawn using Excel software, and the linear regression equation was obtained. The OD value of the sample was substituted into the equation to calculate the concentration of the sample.

4.6. 20E Treatments

A total of 120 pre-molting *M. nipponense* were divided into the experimental group and the control group, in triplicate (*n* = 20). *M. nipponense* in the experimental group was injected with 20E (Sigma-Aldrich, St. Louis, MO, USA; 5 μg/g), and the control group was injected with an equal volume of solvent (ethanol) [26]. Prawn tissues were collected at 3, 6, 12, 24 and 48 h after injection of 20E and stored at −80 °C after liquid nitrogen quick-freezing for mRNA extraction. The relative expression levels of the *MnHR4* gene in the experimental and control groups were evaluated by qRT-PCR.

4.7. Data Analysis

Statistical analyses were performed using SPSS 20.0 software (IBM, New York, NY, USA). One-way ANOVA was used for comparisons between multiple sample means. The differences between the two groups were compared using the independent sample *t*-test. All data are presented as the mean ± SEM. The significance level for the data was set at $p < 0.05$.

Author Contributions: Investigation, methodology, conceptualization, and writing—original draft, H.Y.; formal analysis and data curation, W.Z., H.Q. and S.J. (Shubo Jin); resources and investigation, S.J. (Sufei Jiang), Y.X. and Y.G.; supervision, funding acquisition, and writing—review and editing, H.F. All authors have read and agreed to the published version of the manuscript.

Funding: This research was supported by grants from the Central Public-interest Scientific Institution Basal Research Fund CAFS (2020TD36); the seed industry revitalization project of Jiangsu province (JBGS [2021] 118); Jiangsu Agricultural Industry Technology System; the earmarked fund for CARS-48; and the New cultivar breeding Major Project of Jiangsu province (PZCZ201745).

Acknowledgments: Our thanks goes to the Germplasm Bank of Jiangsu Provincial Science and Technology Resources (Agricultural Germplasm Resources) Coordination Service Platform for providing the *Macrobrachium nipponense*.

Conflicts of Interest: The authors declare no conflict of interest.

References

1. Yamanaka, N.; Rewitz, K.F.; O'Connor, M.B. Ecdysone control of developmental transitions: Lessons from Drosophila research. *Annu. Rev. Entomol.* **2013**, *58*, 497–516. [CrossRef] [PubMed]
2. Festucci-Buselli, R.A.; Contim, L.A.; Barbosa, L.C.A.; Stuart, J.; Otoni, W.C. Biosynthesis and potential functions of the ecdysteroid 20-hydroxyecdysone—A review. *Botany* **2008**, *86*, 978–987. [CrossRef]
3. Petryk, A.; Warren, J.T.; Marqués, G.; Jarcho, M.P.; Gilbert, L.I.; Kahler, J.; Parvy, J.P.; Li, Y.; Dauphin-Villemant, C.; O'Connor, M.B. Shade is the Drosophila P450 enzyme that mediates the hydroxylation of ecdysone to the steroid insect molting hormone 20-hydroxyecdysone. *Proc. Natl. Acad. Sci. USA* **2003**, *100*, 13773–13778. [CrossRef] [PubMed]
4. Rewitz, K.; Rybczynski, R.; Warren, J.T.; Gilbert, L.I. The Halloween genes code for cytochrome P450 enzymes mediating synthesis of the insect moulting hormone. *Biochem. Soc. Trans.* **2006**, *34*, 1256–1260. [CrossRef] [PubMed]
5. Yuan, H.; Qiao, H.; Fu, Y.; Fu, H.; Zhang, W.; Jin, S.; Gong, Y.; Jiang, S.; Xiong, Y.; Hu, Y. RNA interference shows that Spook, the precursor gene of 20-hydroxyecdysone (20E), regulates the molting of *Macrobrachium nipponense*. *J. Steroid Biochem. Mol. Biol.* **2021**, *213*, 105976. [CrossRef] [PubMed]
6. Yao, T.P.; Forman, B.M.; Jiang, Z.; Cherbas, L.; Chen, J.; McKeown, M.; Cherbas, P.; Evans, R.M. Functional ecdysone receptor is the product of EcR and Ultraspiracle genes. *Nature* **1993**, *366*, 476–479. [CrossRef]
7. Koelle, M.R.; Talbot, W.S.; Segraves, W.A.; Bender, M.T.; Cherbas, P.; Hogness, D.S. The Drosophila EcR gene encodes an ecdysone receptor, a new member of the steroid receptor superfamily. *Cell* **1991**, *67*, 59–77. [CrossRef]
8. Riddiford, L.M. Hormones and Drosophila development. *Dev. Drosoph. Melanogaster* **1993**, *2*, 899–939.
9. Ou, Q.; King-Jones, K. What goes up must come down: Transcription factors have their say in making ecdysone pulses. *Curr. Top. Dev. Biol.* **2013**, *103*, 35–71.
10. Bain, D.L.; Heneghan, A.F.; Connaghan-Jones, K.D.; Miura, M.T. Nuclear receptor structure: Implications for function. *Annu. Rev. Physiol.* **2007**, *69*, 201–220. [CrossRef]
11. King-Jones, K.; Thummel, C.S. Nuclear receptors—A perspective from Drosophila. *Nat. Rev. Genet.* **2005**, *6*, 311–323. [CrossRef] [PubMed]
12. Broadus, J.; McCabe, J.R.; Endrizzi, B.; Thummel, C.S.; Woodard, C.T. The Drosophila βFTZ-F1 orphan nuclear receptor provides competence for stage-specific responses to the steroid hormone ecdysone. *Mol. Cell* **1999**, *3*, 143–149. [CrossRef]

13. White, K.P.; Hurban, P.; Watanabe, T.; Hogness, D.S. Coordination of Drosophila metamorphosis by two ecdysone-induced nuclear receptors. *Science* **1997**, *276*, 114–117. [CrossRef] [PubMed]

14. Mané-Padrós, D.; Cruz, J.; Vilaplana, L.; Pascual, N.; Bellés, X.; Martín, D. The nuclear hormone receptor BgE75 links molting and developmental progression in the direct-developing insect *Blattella germanica*. *Dev. Biol.* **2008**, *315*, 147–160. [CrossRef] [PubMed]

15. Bialecki, M.; Shilton, A.; Fichtenberg, C.; Segraves, W.A.; Thummel, C.S. Loss of the ecdysteroid-inducible E75A orphan nuclear receptor uncouples molting from metamorphosis in Drosophila. *Dev. Cell* **2002**, *3*, 209–220. [CrossRef]

16. Zhao, X.; Qin, Z.; Liu, W.; Liu, X.; Moussian, B.; Ma, E.; Li, S.; Zhang, J. Nuclear receptor HR3 controls locust molt by regulating chitin synthesis and degradation genes of *Locusta migratoria*. *Insect Biochem. Mol. Biol.* **2018**, *92*, 1–11. [CrossRef] [PubMed]

17. Cruz, J.; Nieva, C.; Mané-Padrós, D.; Martín, D.; Bellés, X. Nuclear receptor BgFTZ-F1 regulates molting and the timing of ecdysteroid production during nymphal development in the hemimetabolous insect *Blattella germanica*. *Dev. Dyn.* **2008**, *237*, 3179–3191. [CrossRef]

18. Vickers, N.J. Animal communication: When I'm calling you, will you answer too? *Curr. Biol.* **2017**, *27*, R713–R715. [CrossRef]

19. Tan, A.; Palli, S.R. Identification and characterization of nuclear receptors from the red flour beetle, *Tribolium castaneum*. *Insect Biochem. Mol. Biol.* **2008**, *38*, 430–439. [CrossRef]

20. King-Jones, K.; Charles, J.P.; Lam, G.; Thummel, C.S. The ecdysone-induced DHR4 orphan nuclear receptor coordinates growth and maturation in Drosophila. *Cell* **2005**, *121*, 773–784. [CrossRef]

21. Liu, J.; Zhou, T.; Wang, C.; Chan, S.; Wang, W. Deciphering the molecular regulatory mechanism orchestrating ovary development of the Pacific whiteleg shrimp *Litopenaeus vannamei* through integrated transcriptomic analysis of reproduction-related organs. *Aquaculture* **2021**, *533*, 736160. [CrossRef]

22. Miyakawa, H.; Sato, T.; Song, Y.; Tollefsen, K.E.; Iguchi, T. Ecdysteroid and juvenile hormone biosynthesis, receptors and their signaling in the freshwater microcrustacean Daphnia. *J. Steroid Biochem. Mol. Biol.* **2018**, *184*, 62–68. [CrossRef] [PubMed]

23. Legrand, E.; Bachvaroff, T.; Schock, T.B.; Chung, J.S. Understanding molt control switches: Transcriptomic and expression analysis of the genes involved in ecdysteroidogenesis and cholesterol uptake pathways in the Y-organ of the blue crab, Callinectes sapidus. *PLoS ONE* **2021**, *16*, e0256735. [CrossRef] [PubMed]

24. Hongtuo, F.; Sufei, J.; Yiwei, X. Current status and prospects of farming the giant river prawn (*Macrobrachium rosenbergii*) and the oriental river prawn (*Macrobrachium nipponense*) in China. *Aquac. Res.* **2012**, *43*, 993–998. [CrossRef]

25. Chang, E.S.; Mykles, D.L. Regulation of crustacean molting: A review and our perspectives. *Gen. Comp. Endocrinol.* **2011**, *172*, 323–330. [CrossRef]

26. Yuan, H.; Zhang, W.; Fu, Y.; Jiang, S.; Xiong, Y.; Zhai, S.; Gong, Y.; Qiao, H.; Fu, H.; Wu, Y. *MnFtz-f1* Is Required for Molting and Ovulation of the Oriental River Prawn *Macrobrachium nipponense*. *Front. Endocrinol.* **2021**, *12*, 798577. [CrossRef] [PubMed]

27. Chen, Y.Z.; Chen, H.; Zhu, X.; Cui, Z.; Qiu, G. The morphological and histological observation of embryonic development in the oriental river prawn *Macrobrachium nipponense*. *J. Shanghai Ocean Univ.* **2012**, *21*, 33–40.

28. Qiao, H.; Fu, H.; Xiong, Y.; Jiang, S.; Sun, S.; Jin, S.; Gong, Y.; Wang, Y.; Shan, D.; Li, F. Molecular insights into reproduction regulation of female Oriental River prawns *Macrobrachium nipponense* through comparative transcriptomic analysis. *Sci. Rep.* **2017**, *7*, 12161. [CrossRef]

29. Kastner, P.; Mark, M.; Chambon, P. Nonsteroid nuclear receptors: What are genetic studies telling us about their role in real life? *Cell* **1995**, *83*, 859–869. [CrossRef]

30. Thummel, C.S. From embryogenesis to metamorphosis: The regulation and function of Drosophila nuclear receptor superfamily members. *Cell* **1995**, *83*, 871–877. [CrossRef]

31. Iyer, A.K.; McCabe, E.R. Molecular mechanisms of DAX1 action. *Mol. Genet. Metab.* **2004**, *83*, 60–73. [CrossRef] [PubMed]

32. Gronemeyer, H.; Laudet, V. Transcription factors 3: Nuclear receptors. *Protein Profile* **1995**, *2*, 1173–1308. [PubMed]

33. Schwabe, J.W.; Rhodes, D. Beyond zinc fingers: Steroid hormone receptors have a novel structural motif for DNA recognition. *Trends Biochem. Sci.* **1991**, *16*, 291–296. [CrossRef]

34. Edwards, D.P. The role of coactivators and corepressors in the biology and mechanism of action of steroid hormone receptors. *J. Mammary Gland Biol. Neoplasia* **2000**, *5*, 307–324. [CrossRef] [PubMed]

35. Liu, X.; Li, J.; Sun, Y.; Liang, X.; Zhang, R.; Zhao, X.; Zhang, M.; Zhang, J. A nuclear receptor HR4 is essential for the formation of epidermal cuticle in the migratory locust, *Locusta migratoria*. *Insect Biochem. Mol. Biol.* **2022**, *143*, 103740. [CrossRef] [PubMed]

36. Weaver, L.N.; Drummond-Barbosa, D. Hormone receptor 4 is required in muscles and distinct ovarian cell types to regulate specific steps of *Drosophila oogenesis*. *Development* **2021**, *148*, dev198663. [CrossRef] [PubMed]

37. Zhang, Y.; Qiao, H.; Zhang, W.; Sun, S.; Jiang, S.; Gong, Y.; Xiong, Y.; Jin, S.; Fu, H. Molecular cloning and expression analysis of two sex-lethal homolog genes during development in the oriental river prawn, *Macrobrachium nipponense*. *Genet. Mol. Res.* **2013**, *12*, 4698–4711. [CrossRef]

38. Mané-Padrós, D.; Borràs-Castells, F.; Belles, X.; Martín, D. Nuclear receptor HR4 plays an essential role in the ecdysteroid-triggered gene cascade in the development of the hemimetabolous insect *Blattella germanica*. *Mol. Cell. Endocrinol.* **2012**, *348*, 322–330. [CrossRef]

39. Xu, Q.Y.; Meng, Q.W.; Deng, P.; Guo, W.C.; Li, G.Q. Leptinotarsa hormone receptor 4 (HR4) tunes ecdysteroidogenesis and mediates 20-hydroxyecdysone signaling during larval-pupal metamorphosis. *Insect Biochem. Mol. Biol.* **2018**, *94*, 50–60. [CrossRef]

40. Niwa, R.; Matsuda, T.; Yoshiyama, T.; Namiki, T.; Mita, K.; Fujimoto, Y.; Kataoka, H. CYP306A1, a cytochrome P450 enzyme, is essential for ecdysteroid biosynthesis in the prothoracic glands of Bombyx and Drosophila. *J. Biol. Chem.* **2004**, *279*, 35942–35949. [CrossRef]

41. Ou, Q.; Magico, A.; King-Jones, K. Nuclear receptor DHR4 controls the timing of steroid hormone pulses during Drosophila development. *PLoS Biol.* **2011**, *9*, e1001160. [CrossRef] [PubMed]

42. Fahrbach, S.E.; Smagghe, G.; Velarde, R.A. Insect nuclear receptors. *Annu. Rev. Entomol.* **2012**, *57*, 83–106. [CrossRef] [PubMed]

43. Liang, G.X.; Fu, H.T.; Qiao, H.; Sun, S.M.; Zhang, W.Y.; Jin, S.B.; Gong, Y.S.; Jiang, S.F.; Xiong, Y.W.; Wu, Y. Molecular Cloning and Characterization of a Putative Ecdysis-triggering Hormone Receptor (ETHR) Gene from *Macrobrachium nipponense*. *J. World Aquacult. Soc.* **2018**, *49*, 1081–1094. [CrossRef]

44. Pick, L.; Anderson, W.R.; Shultz, J.; Woodard, C.T. The Ftz-F1 family: Orphan nuclear receptors regulated by novel protein–protein interactions. *Adv. Dev. Biol.* **2006**, *16*, 255–296.

45. You, L.; Li, Z.; Zhang, Z.; Hu, B.; Yu, Y.; Yang, F.; Tan, A. Two dehydroecdysone reductases act as fat body-specific 20E catalyzers in Bombyx mori. *Insect Sci.* **2022**, *29*, 100–110. [CrossRef] [PubMed]

46. Bairoch, A.; Bucher, P.; Hofmann, K. The PROSITE database, its status in 1997. *Nucleic Acids Res.* **1997**, *25*, 217–221. [CrossRef]

47. Combet, C.; Blanchet, C.; Geourjon, C.; Deleage, G. NPS@: Network protein sequence analysis. *Trends Biochem. Sci.* **2000**, *25*, 147–150. [CrossRef]

48. Yang, J.; Zhang, Y. I-TASSER server: New development for protein structure and function predictions. *Nucleic Acids Res.* **2015**, *43*, W174–W181. [CrossRef]

49. Hu, Y.; Fu, H.; Qiao, H.; Sun, S.; Zhang, W.; Jin, S.; Jiang, S.; Gong, Y.; Xiong, Y.; Wu, Y. Validation and evaluation of reference genes for quantitative real-time PCR in *Macrobrachium nipponense*. *Int. J. Mol. Sci.* **2018**, *19*, 2258. [CrossRef]

50. Livak, K.J.; Schmittgen, T.D. Analysis of relative gene expression data using real-time quantitative PCR and the $2^{-\Delta\Delta CT}$ method. *Methods* **2001**, *25*, 402–408. [CrossRef]

51. Qian, Z.; He, S.; Liu, T.; Liu, Y.; Hou, F.; Liu, Q.; Wang, X.; Mi, X.; Wang, P.; Liu, X. Identification of ecdysteroid signaling late-response genes from different tissues of the Pacific white shrimp, *Litopenaeus vannamei*. *Comp. Biochem. Physiol. A-Mol. Integr. Physiol.* **2014**, *172*, 10–30. [CrossRef] [PubMed]

MDPI
St. Alban-Anlage 66
4052 Basel
Switzerland
Tel. +41 61 683 77 34
Fax +41 61 302 89 18
www.mdpi.com

International Journal of Molecular Sciences Editorial Office
E-mail: ijms@mdpi.com
www.mdpi.com/journal/ijms